AF761674

CORONARY ARTERY DISEASE

CORONARY ARTERY DISEASE

From Biology to Clinical Practice

DIMITRIS TOUSOULIS

ACADEMIC PRESS
An imprint of Elsevier

Academic Press is an imprint of Elsevier
125 London Wall, London EC2Y 5AS, United Kingdom
525 B Street, Suite 1800, San Diego, CA 92101-4495, United States
50 Hampshire Street, 5th Floor, Cambridge, MA 02139, United States
The Boulevard, Langford Lane, Kidlington, Oxford OX5 1GB, United Kingdom

Library of Congress Cataloging-in-Publication Data
A catalog record for this book is available from the Library of Congress

British Library Cataloguing-in-Publication Data
A catalogue record for this book is available from the British Library

ISBN: 978-0-12-811908-2

For information on all Academic press publications visit our website at https://www.elsevier.com/books-and-journals

Publisher: Mica Haley
Acquisition Editor: Stacy Masucci
Editorial Project Manager: Kathy Padilla
Production Project Manager: Priya Kumaraguruparan
Designer: Matthew Limbert

Typeset by TNQ Books and Journals

CONTENTS

CHAPTER 2 CORONARY ARTERY CIRCULATION: BASIC PRINCIPLES 119

CONTRIBUTORS

Constantina Aggeli
1st Cardiology Clinic, 'Hippokration' General Hospital, National and Kapodistrian University of Athens, School of Medicine, Athens, Greece

Charalambos Antoniades
University of Oxford, Oxford, United Kingdom

Alexios S. Antonopoulos
1st Cardiology Clinic, 'Hippokration' General Hospital, National and Kapodistrian University of Athens, School of Medicine, Athens, Greece

Konstantinos Aznaouridis
1st Cardiology Clinic, 'Hippokration' General Hospital, National and Kapodistrian University of Athens, School of Medicine, Athens, Greece

Alexandros Briasoulis
Mayo Clinic, Rochester, MN, United States

Evangelos Diamantis
1st Cardiology Clinic, 'Hippokration' General Hospital, National and Kapodistrian University of Athens, School of Medicine, Athens, Greece

Maria Drakopoulou
1st Cardiology Clinic, 'Hippokration' General Hospital, National and Kapodistrian University of Athens, School of Medicine, Athens, Greece

Debbie Falconer
University College London Hospital, London, United Kingdom

Petros Fountoulakis
1st Cardiology Clinic, 'Hippokration' General Hospital, National and Kapodistrian University of Athens, School of Medicine, Athens, Greece

Christos Georgakopoulos
1st Cardiology Clinic, 'Hippokration' General Hospital, National and Kapodistrian University of Athens, School of Medicine, Athens, Greece

Athina Goliopoulou
1st Cardiology Clinic, 'Hippokration' General Hospital, National and Kapodistrian University of Athens, School of Medicine, Athens, Greece

Odysseas Kaitozis
1st Cardiology Clinic, 'Hippokration' General Hospital, National and Kapodistrian University of Athens, School of Medicine, Athens, Greece

Konstantinos Kalogeras
1st Cardiology Clinic, 'Hippokration' General Hospital, National and Kapodistrian University of Athens, School of Medicine, Athens, Greece

Theodore Kalos
1st Cardiology Clinic, 'Hippokration' General Hospital, National and Kapodistrian University of Athens, School of Medicine, Athens, Greece

Dimitrios Konstantinidis
1st Cardiology Clinic, 'Hippokration' General Hospital, National and Kapodistrian University of Athens, School of Medicine, Athens, Greece

Georgios Latsios
1st Cardiology Clinic, 'Hippokration' General Hospital, National and Kapodistrian University of Athens, School of Medicine, Athens, Greece

Georgios Lazaros
1st Cardiology Clinic, 'Hippokration' General Hospital, National and Kapodistrian University of Athens, School of Medicine, Athens, Greece

Emmanouil Mantzouranis
1st Cardiology Clinic, 'Hippokration' General Hospital, National and Kapodistrian University of Athens, School of Medicine, Athens, Greece

Konstantina Masoura
1st Cardiology Clinic, 'Hippokration' General Hospital, National and Kapodistrian University of Athens, School of Medicine, Athens, Greece

Sofia Mavrogeni
Onassis Cardiac Surgery Center, Athens, Greece

Konstantinos Mourouzis
1st Cardiology Clinic, 'Hippokration' General Hospital, National and Kapodistrian University of Athens, School of Medicine, Athens, Greece

Evangelos Oikonomou
1st Cardiology Clinic, 'Hippokration' General Hospital, National and Kapodistrian University of Athens, School of Medicine, Athens, Greece

Nikolaos Papageorgiou
1st Cardiology Clinic, 'Hippokration' General Hospital, National and Kapodistrian University of Athens, School of Medicine, Athens, Greece
St Bartholomew's Hospital, London, United Kingdom

Aggelos Papamikroulis
1st Cardiology Clinic, 'Hippokration' General Hospital, National and Kapodistrian University of Athens, School of Medicine, Athens, Greece

Dimitris Pollalis
1st Cardiology Clinic, 'Hippokration' General Hospital, National and Kapodistrian University of Athens, School of Medicine, Athens, Greece

Gerasimos Siasos
1st Cardiology Clinic, 'Hippokration' General Hospital, National and Kapodistrian University of Athens, School of Medicine, Athens, Greece

Andreas Synetos
1st Cardiology Clinic, 'Hippokration' General Hospital, National and Kapodistrian University of Athens, School of Medicine, Athens, Greece

Dimitris Tousoulis
1st Cardiology Clinic, 'Hippokration' General Hospital, National and Kapodistrian University of Athens, School of Medicine, Athens, Greece

Konstantinos Toutouzas
1st Cardiology Clinic, 'Hippokration' General Hospital, National and Kapodistrian University of Athens, School of Medicine, Athens, Greece

Sotirios Tsalamandris
1st Cardiology Clinic, 'Hippokration' General Hospital, National and Kapodistrian University of Athens, School of Medicine, Athens, Greece

Vasiliki Tsigkou
1st Cardiology Clinic, 'Hippokration' General Hospital, National and Kapodistrian University of Athens, School of Medicine, Athens, Greece

Costas Tsioufis
1st Cardiology Clinic, 'Hippokration' General Hospital, National and Kapodistrian University of Athens, School of Medicine, Athens, Greece

Sofia Vaina
1st Cardiology Clinic, 'Hippokration' General Hospital, National and Kapodistrian University of Athens, School of Medicine, Athens, Greece

Emmanouil Vavouranakis
1st Cardiology Clinic, 'Hippokration' General Hospital, National and Kapodistrian University of Athens, School of Medicine, Athens, Greece

Charalambos Vlachopoulos
1st Cardiology Clinic, 'Hippokration' General Hospital, National and Kapodistrian University of Athens, School of Medicine, Athens, Greece

Georgia Vogiatzi
1st Cardiology Clinic, 'Hippokration' General Hospital, National and Kapodistrian University of Athens, School of Medicine, Athens, Greece

Effimia Zacharia
1st Cardiology Clinic, 'Hippokration' General Hospital, National and Kapodistrian University of Athens, School of Medicine, Athens, Greece

Marina Zaromitidou
1st Cardiology Clinic, 'Hippokration' General Hospital, National and Kapodistrian University of Athens, School of Medicine, Athens, Greece

PREFACE BY FILIPPO CREA

I am delighted to write the preface of this excellent book authored by Professor Tousoulis. Despite the considerable advances in the diagnosis, medical treatment, and interventional strategies, ischemic heart disease remains the main cause of mortality and morbidity worldwide, whereas risk factors explain only part of the disease spectrum. A better knowledge of ischemic heart disease pathophysiology might have a considerable impact on clinical management. The purpose of this book is to link basic science (genetic and molecular mechanisms) with clinical practice for the benefit of both patients and physicians. To achieve these goals, this book is organized into three chapters.

The first chapter is focused on the mechanisms leading to atheroma formation and progression with specific regard to the complex interaction among genetics, risk factors, inflammation, and oxidative stress.

The second chapter is focused on the physiology and pathophysiology of coronary circulation with specific regard to the mechanisms responsible for the mismatch between perfusion and function, which is responsible for myocardial ischemia. The mechanisms leading to acute coronary thrombosis are also elucidated.

The third chapter is focused on the therapeutic challenges and unmet needs in the treatment of stable ischemic heart disease and of acute coronary syndromes. The initial sections of this chapter are dedicated to potential new treatments targeting inflammation and oxidative stress not yet tested in large randomized trials but supported by strong pathophysiological observations. The later sections in the chapter address the potential of gene and stem cell therapy for patients not amenable to conventional anti-ischemic treatments. In addition, this chapter addresses the utilization of invasive and noninvasive techniques for the assessment of anatomic and functional alterations of coronary circulation. Finally, this chapter critically discusses traditional and novel coronary revascularization strategies.

I strongly recommend this book to students, internists, and cardiologists with an interest in ischemic heart disease from prevention to the management of acute coronary syndromes. The translational approach, the emphasis on pathophysiology, and the focus on unmet needs and potential innovative treatments make this book very appealing for both the present and future of ischemic heart disease.

Filippo Crea
Professor of Cardiology
Head of the Department of Cardiovascular and Thoracic Sciences
Catholic University
Rome

PREFACE BY JUAN CARLOS KASKI

I am delighted to write a preface for this excellent monographic work by Professor D. Tousoulis, whose work in the field of pathogenesis of cardiovascular disease has helped in improving our understanding of this problematic condition. Coronary artery disease is a worldwide disease causing significant mortality, morbidity, and disabilities and affects millions of people. Despite the considerable advance in the diagnosis, management, and interventional strategies, cardiovascular disease still features prominently among the most devastating conditions affecting men and women worldwide. A better understanding of the pathophysiology and pathogenesis of cardiovascular disease and the true role of conventional and novel risk factors is essential to identify effective preventative and therapeutic measures. Cardiologists and general practitioners who come across the burden of atherosclerotic cardiovascular entities in their daily practice will most likely treasure this volume edited by Professor Tousoulis.

This book aims to provide insight into the challenges raised by the complex and multifaceted presentation of atherosclerotic cardiovascular diseases and a comprehensive view of the pathophysiology and etiology of coronary artery disease. In addition, this volume deals with novel therapeutic developments, including invasive strategies to manage effectively the most difficult clinical cases.

The first part of the book is devoted to the basic pathophysiologic mechanisms of atherosclerosis providing data on the role of both established and novel risk factors and useful diagrams on the mechanisms responsible for the development and progression of atherosclerosis. Endothelial function, oxidative stress, and genetic aspects of atherosclerosis—often representing obscure mechanisms for the clinical practitioner are presented in a clear fashion in the book, which is specifically directed to assist those physicians who manage patients with advanced atherosclerosis in real life clinical practice. In particular, the biology of the vessel wall and the central role of the endothelium in the process of atherosclerosis are clearly presented in the initial sections of the book. Moreover, the detrimental role of dyslipidemia, smoking, hypertension, and diabetes mellitus is very appropriately highlighted in this monographic work, giving useful insight regarding the management of these frequent and problematic entities.

The second part of the book focuses on the basic principles of the coronary artery circulation and on the basic concepts of stable and acute coronary syndromes, which require optimal medical treatment and/or invasive revascularization. Herein, the reader will come across excellent explanations regarding the mechanism implicated in the genesis of typical and atypical symptoms and useful tips regarding which is the best management strategy in different patients. This section of the book deals with the principles of functional anatomy, the factors driving the balance between oxygen delivery and consumption, and the mechanisms leading to stable angina or acute coronary syndromes such as plaque rupture, erosion, the development of intramural hematoma, and intima dissection.

The final section of the book provides an overview of novel and experimental treatments for atherosclerosis and coronary artery disease. The role of gene and stem cell therapy is presented very clearly by the author, highlighting both current and possible future application. This section also discusses novel developments in both the diagnosis of coronary artery disease and the evaluation of transient myocardial ischemia. The roles of stress echocardiography and nuclear imaging are discussed in a scholarly but practical fashion together with the indications and usefulness of intravascular ultrasound and optical coherence tomography. Very importantly, the invasive, functional assessment of myocardial ischemia, and the clinical usefulness of calculating fractional flow reserve for the assessment of the hemodynamic effects of coronary stenosis are presented very clearly in this section of the book. The different types of stents available in clinical practice are presented toward the end of this section where the author discusses the appropriate criteria for the use of self-expandable stents, bio-absorbable stents, and polymer-free drug-coated stents for the management of advanced coronary artery disease.

I have no doubt whatsoever that this book will be treasured by its readers and will represent the ideal companion to the physician who wishes to understand the complexities of the modern management of coronary artery disease. The book will also be an invaluable tool for trainees in the field of interventional cardiology given the practical and comprehensive content of this publication.

Juan Carlos Kaski, DSc, DM(Hons), MD, FRCP, FESC, FACC, FAHA, FRSM
Professor of Cardiovascular Science
Molecular and Clinical Sciences Research Institute
St. George's, University of London
London, United Kingdom

PREFACE BY EMMANOUIL S. BRILAKIS

Coronary artery disease is the most common cardiovascular disease and remains the leading cause of death in the world. It has involved all of us by affecting a family member or a friend. Although steady and significant improvements in diagnosis and treatment have improved the outcomes of patients with coronary artery disease, much remains to be accomplished.

Understanding the underlying mechanisms of coronary artery disease is necessary for developing the next generation of treatments. This is a Herculean task, given its broad scope and rapid evolution of the field. Dr. Tousoulis and his colleagues successfully accomplished this task by creating *Coronary Artery Disease: From Biology to Practice*, a comprehensive, up-to-date, and state-of-the-art book.

Chapter 1 provides an in-depth overview of the pathophysiology of coronary atherosclerosis, covering both traditional risk factors and novel mechanisms and processes, such as inflammation, oxidative stress, and genetics. Chapter 2 describes the basic mechanisms of coronary artery circulation, as well as the pathophysiology of stable angina and acute coronary syndromes. Chapter 3 reviews in detail both established and novel/experimental atherosclerosis treatments.

This book is an invaluable resource for everyone engaged in the fight against coronary artery disease, including physicians, nurses, technicians, students, scientists, and researchers. It provides answers to basic and advanced questions and builds a broad, yet detailed understanding of the field. By creating a solid foundation, *Coronary Artery Disease: From Biology to Practice* will spearhead the "beginning of the end" of coronary artery disease, at least as we know it today.

Emmanouil S. Brilakis
Director
Center for Advanced Coronary Interventions
Minneapolis Heart Institute
Adjunct Professor of Medicine
University of Texas Southwestern Medical School at Dallas

CHAPTER 1

Pathophysiology of Atherosclerosis

CHAPTER 1.1

Biology of the Vessel Wall

Evangelos Oikonomou, Sotirios Tsalamandris, Konstantinos Mourouzis, Dimitris Tousoulis
1st Cardiology Clinic, 'Hippokration' General Hospital, National and Kapodistrian University of Athens, School of Medicine, Athens, Greece

Contents

In a simplistic view, the role of arteries is to transfer blood from one site to another merely functioning as tubes. However, this concept has long been overcome and since the mid-19th century it has been recognized that the arterial wall serves multiple functions and can actively regulate blood flow and pressure, tissue perfusion or ischemia, and coagulation, while also participating in atherogenesis formation and progression. Therefore the biology of vessel wall and arterial anatomy has gained research interest.

ARTERIAL ANATOMY

Three layers can be identified in the arterial wall from inside to outside: the intima, media, and adventitia, and atherosclerosis process involves primarily pathologic changes in the intima with reactive changes in the media and adventitia.

Coronary Artery Disease
ISBN 978-0-12-811908-2, http://dx.doi.org/10.1016/B978-0-12-811908-2.00001-5

Intima Layer

The intima layer is made up of the endothelial surface, the basement membrane, and the internal elastic lamina. In the absence of atherosclerosis intima is extremely thin and serves several unique roles orchestrating the functional characteristics and properties of the arterial wall.

Endothelium

Normal endothelium is a thin monolayer of endothelial cells. Its surface in the human body is estimated to be 350 m^2 with a mass of only 110 g. Many vasoactive substances (i.e., nitric oxide, endothelin 1) are produced by endothelial cells with vasodilating and antithrombogenic actions regulating vascular homeostasis. Importantly, endothelial cells have the unique ability to keep blood liquid and to inhibit activation of the coagulation cascade by the expression of the anticoagulants thrombomodulin, heparin sulfate, and prostacyclin on their surface. On the contrary, in the presence of endothelial damage and dysfunction the balance shifts toward a prethrombotic state and proatherosclerotic changes ensue (Fig. 1.1.1). Interestingly, bone marrow-derived endothelial progenitor cells circulating in the blood stream contribute to neovascularization and re-endothelialization of the injured vessel maintaining vascular function and homeostasis.

Therefore the central role of the endothelium in vascular homeostasis and atherosclerosis progression is discussed in detail in a separate section of this book.

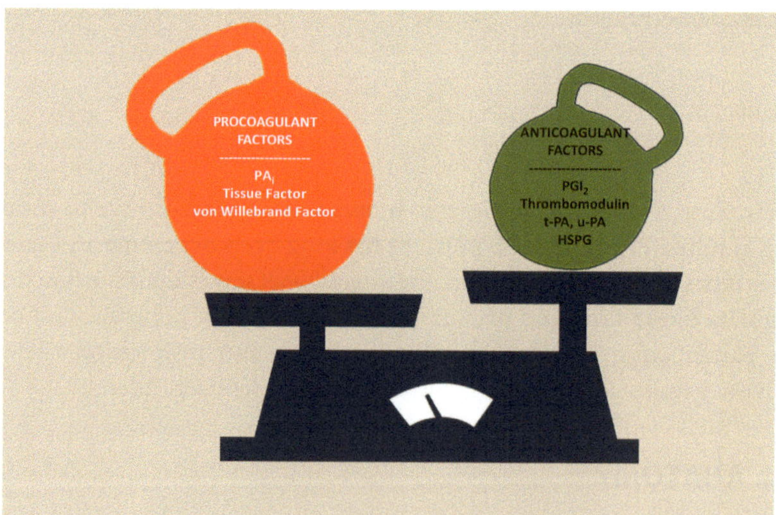

Figure 1.1.1 Endothelial-derived procoagulant and anticoagulant factors. *HSPG*, heparan sulfate proteoglycan; *PA_i*, plasminogen activator inhibitor; *PGI_2*, prostacyclin; *t-PA*, tissue plasminogen activator; *u-PA*, urokinase-type plasminogen activator.

Basement Membrane

Basement membrane supports the endothelial monolayer and contains collagen, laminin, fibronectin, and other extracellular matrix molecules. Aging process changes the composition of basement membrane with presentation of smooth muscle cells and fibrillar forms of collagen (collagen types I and III). The extracellular contents of the basement membrane are produced by smooth muscle cells. Intimal thickening characterizes most adult human arteries and does not necessarily accompany atherosclerosis.

Elastic Membrane

Elastic membrane serves as a connection of the intima with the media layer.

Media Layer

This layer is composed of a series of concentric layers of smooth muscle cells interchangeable with layers of extracellular matrix rich in elastin especially in the great elastic arteries such as aorta. Of note, this structure changes as we progress from great arteries to small muscular arteries where the layers of muscular and extracellular contents cannot be readily recognized. In great arteries, the described structure of media tunica serves to store kinetic energy powered by left ventricular systole in a dynamic form, which is attributed latter during diastole.

Adventitia

The adventitia mainly consists of collagen fibrils and a sparse cellular population consisting of fibroblasts and mast cells. In this layer nerve endings and vasa vasorum are observed.

Coronary Perivascular Adipose Tissue

Adipose tissue plays a specific, well-known role in cardiovascular system physiology through the systemic effects of active adipokines that are released into blood circulation or through paracrine signaling of the fat secretome that surrounds the vascular wall. Our recent studies have underlined the significance of a bidirectional interaction between the arterial wall and perivascular adipose tissue and between the epicardial adipose tissue and the myocardium.

A layer of epicardial fat surrounds the major epicardial arteries, which is critical for their function and coronary atherosclerosis progression. Human epicardial adipose tissue biopsies have shown that its gene expression profile changes to a proinflammatory phenotype in the presence of coronary atherosclerosis. Recently, we have proved that perivascular adipose tissue has defense mechanisms whose activation is induced by "rescue signals" sent by the arterial wall in the presence of cardiovascular disease; for example, the expression of adiponectin, an antiinflammatory adipokine, is upregulated in the human perivascular adipose tissue during high oxidative stress and increased lipid peroxidation, and counteracts high oxidative stress in the vessels.

PERFUSION OF ARTERIAL WALL AND THE ROLE OF VASA VASORUM

The normal arterial wall is perfused and nourished by the diffusion of oxygen through endothelial cells. Vessels that are thicker than 250 μm are additionally supplied by vasa vasorum course into the adventitia. In coronary arteries, vasa vasorum mostly come from the side branches and less directly from the main lumen.

Reacting to chemical and paracrine signals the network of vasa vasorum may grow to increase the delivery of oxygen and nutritional substances in augmenting atherosclerotic lesions. From another viewpoint, vasa vasorum may initiate the progression of atherosclerotic plaques. Apart from their role as a cause or result of atherosclerosis development, the network of vasa vasorum seems to be involved in the atherosclerotic process and a rich network may be linked with plaque vulnerability, rupture, and consequently, acute coronary syndromes (ACSs).

BIOLOGY OF CORONARY ATHEROMATOUS PLAQUES

The severity of coronary artery stenosis is linked with stable angina and symptoms of ischemia, whereas its prognosis is mostly influenced by acute coronary events. It is very important to note that the incidence of ACSs is not connected to the extent of stenosis and that an important percentage of unstable plaques lead to stenosis in less than 50% cases. Hence the critical question is why plaques modified after a long inactive period and expressed vulnerable characteristics and caused thrombus formation, acute lumen occlusion, and ischemic symptoms (Fig. 1.1.2).

Historically, the plaque characteristics that predispose to vulnerability and rupture have been identified based on several pathological studies and are described by the general term "thin-capped fibroatheroma." Among these characteristics of vulnerable plaques are (1) a large atheromatous core high lipid content, (2) a thin fibrous cap, (3) outward remodeling, (4) infiltration of the plaque with macrophages and lymphocytes, and (5) thinning of the media and decrease in the number of smooth muscle cells (Fig. 1.1.3).

Especially the role of inflammation in the induction of atherosclerosis and the development of stable atherosclerotic plaques to unstable ones has been studied in detail over the recent years. Indeed, the connection between common community-acquired infectious diseases and ACS has been well studied in the past, and it might be linked with the mobilization of vascular inflammatory cytokines like the soluble intercellular adhesion molecule-1, soluble vascular cell adhesion molecule-1, and interleukin-6. In addition, experimental models have demonstrated that exposure to lipopolysaccharide may destabilize atherosclerotic plaques, whereas decrease of neutrophils may induce plaque rupture. Of note, the role of matrix metalloproteinases and their inhibitors (tissue inhibitors of matrix metalloproteinases) and the relative balance between their activation

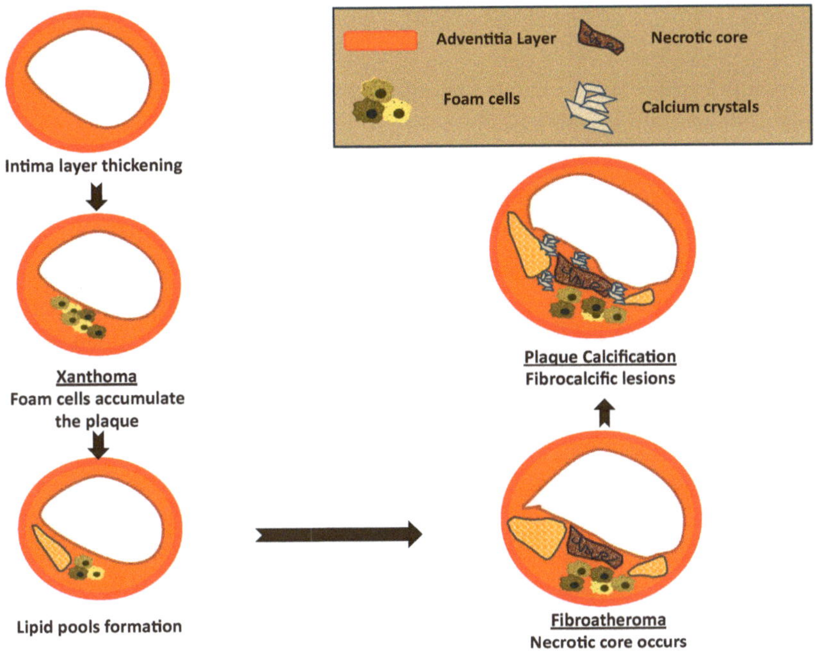

Figure 1.1.2 Stages of atheromatic plaque progression.

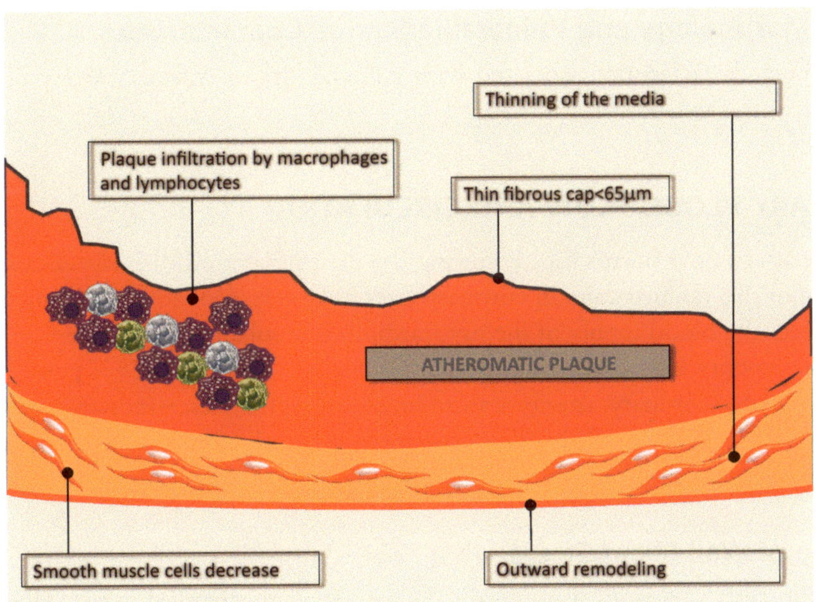

Figure 1.1.3 Vulnerable plaque characteristics.

orchestrate in a favorable or unfavorable way extracellular matrix remodeling and modification of the vessel wall.

Another important feature characterizing plaque stability is the calcium deposition of atheromatic plaques. In general, heavily calcified plaques are more stable, although calcified coronary arteries may have lost their ability to autoregulate the coronary flow depending on the heart requirements. Interestingly, several biomarkers associated with calcium metabolism may be associated with increase in cardiovascular events.

As far as morphologic features and identification of the local coronary inflammation are concerned, new modalities such as positron emission tomography—computed tomography with tracers like ^{18}F-NaF (18F sodium fluoride) have been tested, and it was shown that they have the ability to noninvasively identify ruptured and high-risk coronary artery atherosclerotic lesions.

Plaque rupture is the most common mechanism that leads to thrombus formation resulting in ST elevation myocardial infarction, whereas plaque erosion results in non-ST elevation myocardial infarction. The less common mechanism of thrombus development is a calcified nodule that progresses in the coronary lumen. These calcified nodules, which are characterized by surface large calcium deposits and negative remodeling, are the underlying pathophysiology of ACSs, especially in older patients with arterial hypertension and renal insufficiency. Finally, in approximately 20% of patients, coronary instability occurs in the absence of a coronary thrombus implying a functional modification of large epicardial vessels or of the coronary microcirculation.

Plaque Morphology and Vulnerable Plaque Characteristics

Morphologic characteristics associated with vulnerability of coronary artery plaques are summarized in Table 1.1.1.

CORONARY BLOOD FLOW AND CIRCULATION

Coronary artery flow occurs mostly during the diastolic period. It is significant that the regulation of the vascular tone in coronary arteries largely depends on endothelial function. Indeed, the blood supply of the myocardium is regulated by the vasodilation of the epicardial coronary arteries reacting to multiple stimuli. Nitric oxide is a possible vasodilator released by coronary arteries controlling guanylyl-cyclase—induced relaxation of vascular smooth muscle cells in the coronary wall. Radical modifications in the tone of coronary arteries may result in multiple changes in coronary blood flow.

The Role of Wall Shear Stress

Cellular function of various tissues greatly depends on mechanical forces. Significantly, normal endothelial function is associated with pulsatile blood flow and shear stress.

Table 1.1.1 Plaque morphologic characteristics associated with vulnerability and ACSs

Studies	Prognostic implications
Varnava et al. [1]	Outward positive remodeling was associated with higher lipid content, macrophage count, and plaque vulnerability
Burke et al. [2]	Thin fibrous cap (<100 or 65 μm) is associated with rupture of vulnerable plaques
Moreno et al. [3]	Macrophage infiltration is associated with unstable angina and acute coronary events
Kolodgie et al. [4]	Abundant intraplaque vasa vasorum and/or hemorrhage are associated with vulnerable coronary plaques
Tian et al. [5]	Combination of thin fibrous cap, increased plaque burden, and luminal narrowing determines the coronary plaque rupture
Stefanadis et al. [6]	Larger plaque thermal heterogeneity is associated with ACSs
Otsuka et al. [7]	Superficial erosion is associated with ACSs
Mauriello et al. [8]	Plaque infiltration with inflammatory cells in a diffuse pattern is associated with myocardial infraction mortality
Joshi et al. [9]	(18)F-NaF uptake is detected in culprit lesions in patients with ACS and in the lesions of patients with stable CAD with vulnerable plaque characteristics as assessed with IVUS

(18)F-NaF, 18F-sodium fluoride; *ACSs*, acute coronary syndromes; *CAD*, coronary artery disease; *IVUS*, intravascular ultrasound.

However, in situations such as low flow, flow reversal, or oscillating flow, the endothelium is damaged predisposing to uptake of low-density lipoprotein cholesterol; activation of redox-sensitive intracellular pathways inducing vascular inflammation and mobilization of inflammatory cytokines, oxidative stress, vascular smooth muscle cell proliferation, and neovascularization; as well as gradually the creation of atherosclerotic plaques.

Endothelial shear stress participates in endothelial health and a favorable transcriptomic profile of the vascular wall. Clinical studies have shown that low endothelial shear stress in the coronary artery network is linked with atherosclerosis progression and high-risk plaque features.

Collaterals and Microcirculation

Another exceptional feature of coronary circulation is that even though collateral channels preexist, they are not developed in the normal heart. Collaterals evolve when stenosis of important epicardial coronary arteries occurs. Therefore coronary lesion severity, apart from genetic factors, is the only independent pathogenetic variable related to collateral flow. However, as in most ACSs, the responsible lesions are not hemodynamically important before thrombus formation and the network of collaterals is insufficient, whereas a history of stable angina pectoris can provide angiogenetic stimuli that leads to collateral development.

CONCLUSIONS

Coronary artery circulation and coronary syndromes significantly depend on the biology of the vessel wall. Integrity of the vessel wall not only ensures appropriate function of arteries under physiologic stresses and adaptation to heart muscle requirements but also ensures prevention from thrombus formation and unobscured flow in the entire coronary arterial network. On the contrary, pathologic changes in the vessel wall not only limit flow in the myocardium but also may enhance atherosclerosis progression, plaque rapture or erosion, coagulation cataract, and clinical manifestations of ACSs. Especially adverse remodeling of the coronary artery wall should be taken into consideration when a firm estimation of cardiovascular risk is the requisite, and evolving research and imaging modalities in the last years have focused on this issue to elucidate the unidentified mechanism provoking coronary syndromes.

REFERENCES

[1] Varnava AM, Mills PG, Davies MJ. Relationship between coronary artery remodeling and plaque vulnerability. Circulation 2002;105:939–43.

[2] Burke AP, Farb A, Malcom GT, Liang YH, Smialek J, Virmani R. Coronary risk factors and plaque morphology in men with coronary disease who died suddenly. N Engl J Med 1997;336:1276–82.

[3] Moreno PR, Falk E, Palacios IF, Newell JB, Fuster V, Fallon JT. Macrophage infiltration in acute coronary syndromes. Implications for plaque rupture. Circulation 1994;90:775–8.

[4] Kolodgie FD, Gold HK, Burke AP, Fowler DR, Kruth HS, Weber DK, et al. Intraplaque hemorrhage and progression of coronary atheroma. N Engl J Med 2003;349:2316–25.

[5] Tian J, Ren X, Vergallo R, Xing L, Yu H, Jia H, et al. Distinct morphological features of ruptured culprit plaque for acute coronary events compared to those with silent rupture and thin-cap fibroatheroma: a combined optical coherence tomography and intravascular ultrasound study. J Am Coll Cardiol 2014;63:2209–16.

[6] Stefanadis C, Diamantopoulos L, Vlachopoulos C, Tsiamis E, Dernellis J, Toutouzas K, et al. Thermal heterogeneity within human atherosclerotic coronary arteries detected in vivo: a new method of detection by application of a special thermography catheter. Circulation 1999;99:1965–71.

[7] Otsuka F, Joner M, Prati F, Virmani R, Narula J. Clinical classification of plaque morphology in coronary disease. Nat Rev Cardiol 2014;11:379–89.

[8] Mauriello A, Sangiorgi G, Fratoni S, Palmieri G, Bonanno E, Anemona L, et al. Diffuse and active inflammation occurs in both vulnerable and stable plaques of the entire coronary tree: a histopathologic study of patients dying of acute myocardial infarction. J Am Coll Cardiol 2005;45:1585–93.

[9] Joshi NV, Vesey AT, Williams MC, Shah AS, Calvert PA, Craighead FH, et al. 18F-fluoride positron emission tomography for identification of ruptured and high-risk coronary atherosclerotic plaques: a prospective clinical trial. Lancet 2014;383:705–13.

FURTHER READING

[1] Pasterkamp G, Galis ZS, de Kleijn DP. Expansive arterial remodeling: location, location, location. Arterioscler Thromb Vasc Biol 2004;24:650–7.

[2] Braunwald E, Bonow RO. Braunwald's heart disease: a textbook of cardiovascular medicine. 9th ed. Philadelphia: Saunders; 2012. xxiv, 1961 pp.

[3] Pries AR, Kuebler WM. Normal endothelium. Handb Exp Pharmacol 2006:1–40.

[4] Ghule AE, Kandhare AD, Jadhav SS, Zanwar AA, Bodhankar SL. Omega-3-fatty acid adds to the protective effect of flax lignan concentrate in pressure overload-induced myocardial hypertrophy in rats via modulation of oxidative stress and apoptosis. Int Immunopharmacol 2015;28:751–63.
[5] Siasos G, Tousoulis D, Kioufis S, Oikonomou E, Siasou Z, Limperi M, et al. Inflammatory mechanisms in atherosclerosis: the impact of matrix metalloproteinases. Curr Top Med Chem 2012;12:1132–48.
[6] Tousoulis D, Siasos G, Maniatis K, Oikonomou E, Vlasis K, Papavassiliou AG, et al. Novel biomarkers assessing the calcium deposition in coronary artery disease. Curr Med Chem 2012;19:901–20.
[7] Tousoulis D, Siasos G, Maniatis K, Oikonomou E, Kioufis S, Zaromitidou M, et al. Serum osteoprotegerin and osteopontin levels are associated with arterial stiffness and the presence and severity of coronary artery disease. Int J Cardiol 2013;167:1924–8.
[8] Antonopoulos AS, Margaritis M, Coutinho P, Shirodaria C, Psarros C, Herdman L, et al. Adiponectin as a link between type 2 diabetes and vascular NADPH oxidase activity in the human arterial wall: the regulatory role of perivascular adipose tissue. Diabetes 2015;64:2207–19.
[9] Antonopoulos AS, Margaritis M, Verheule S, Recalde A, Sanna F, Herdman L, et al. Mutual regulation of epicardial adipose tissue and myocardial redox state by PPAR-gamma/adiponectin signalling. Circ Res 2016;118:842–55.
[10] Shimabukuro M, Hirata Y, Tabata M, Dagvasumberel M, Sato H, Kurobe H, et al. Epicardial adipose tissue volume and adipocytokine imbalance are strongly linked to human coronary atherosclerosis. Arterioscler Thromb Vasc Biol 2013;33:1077–84.
[11] Margaritis M, Antonopoulos AS, Digby J, Lee R, Reilly S, Coutinho P, et al. Interactions between vascular wall and perivascular adipose tissue reveal novel roles for adiponectin in the regulation of endothelial nitric oxide synthase function in human vessels. Circulation 2013;127:2209–21.
[12] Douglas PS, Pontone G, Hlatky MA, Patel MR, Norgaard BL, Byrne RA, et al. Clinical outcomes of fractional flow reserve by computed tomographic angiography-guided diagnostic strategies vs. usual care in patients with suspected coronary artery disease: the prospective longitudinal trial of FFRCT: outcome and resource impacts study. Eur Heart J 2015;36:3359–67.
[13] Motoyama S, Ito H, Sarai M, Kondo T, Kawai H, Nagahara Y, et al. Plaque characterization by coronary computed tomography angiography and the likelihood of acute coronary events in mid-term follow-up. J Am Coll Cardiol 2015;66:337–46.
[14] Thygesen K, Alpert JS, Jaffe AS, Simoons ML, Chaitman BR, White HD, et al. Third universal definition of myocardial infarction. Eur Heart J 2012;33:2551–67.
[15] Schaar JA, Muller JE, Falk E, Virmani R, Fuster V, Serruys PW, et al. Terminology for high-risk and vulnerable coronary artery plaques. Report of a meeting on the vulnerable plaque, June 17 and 18, 2003, Santorini, Greece. Eur Heart J 2004;25:1077–82.
[16] Nikolopoulou A, Tousoulis D, Antoniades C, Petroheilou K, Vasiliadou C, Papageorgiou N, et al. Common community infections and the risk for coronary artery disease and acute myocardial infarction: evidence for chronic over-expression of tumor necrosis factor alpha and vascular cells adhesion molecule-1. Int J Cardiol 2008;130:246–50.
[17] Tousoulis D, Antoniades C, Nikolopoulou A, Koniari K, Vasiliadou C, Marinou K, et al. Interaction between cytokines and sCD40L in patients with stable and unstable coronary syndromes. Eur J Clin Invest 2007;37:623–8.
[18] Jaw JE, Tsuruta M, Oh Y, Schipilow J, Hirano Y, Ngan DA, et al. Lung exposure to lipopolysaccharide causes atherosclerotic plaque destabilisation. Eur Respir J 2016;48(1).
[19] Niccoli G, Montone RA, Di Vito L, Gramegna M, Refaat H, Scalone G, et al. Plaque rupture and intact fibrous cap assessed by optical coherence tomography portend different outcomes in patients with acute coronary syndrome. Eur Heart J 2015;36:1377–84.
[20] Arbustini E, Dal Bello B, Morbini P, Burke AP, Bocciarelli M, Specchia G, et al. Plaque erosion is a major substrate for coronary thrombosis in acute myocardial infarction. Heart 1999;82:269–72.
[21] Higuma T, Soeda T, Abe N, Yamada M, Yokoyama H, Shibutani S, et al. A combined optical coherence tomography and intravascular ultrasound study on plaque rupture, plaque erosion, and calcified nodule in patients with ST-segment elevation myocardial infarction: incidence, morphologic characteristics, and outcomes after percutaneous coronary intervention. JACC Cardiovasc Interv 2015;8:1166–76.

[22] Jia H, Abtahian F, Aguirre AD, Lee S, Chia S, Lowe H, et al. In vivo diagnosis of plaque erosion and calcified nodule in patients with acute coronary syndrome by intravascular optical coherence tomography. J Am Coll Cardiol 2013;62:1748–58.
[23] Niccoli G, Liuzzo G, Montone RA, Crea F. Advances in mechanisms, imaging and management of the unstable plaque. Atherosclerosis 2014;233:467–77.
[24] Carter HH, Atkinson CL, Heinonen IH, Haynes A, Robey E, Smith KJ, et al. Evidence for shear stress-mediated dilation of the internal carotid artery in humans. Hypertension 2016;68:1217–24.
[25] Davies PF. Flow-mediated endothelial mechanotransduction. Physiol Rev 1995;75:519–60.
[26] Chatzizisis YS, Coskun AU, Jonas M, Edelman ER, Feldman CL, Stone PH. Role of endothelial shear stress in the natural history of coronary atherosclerosis and vascular remodeling: molecular, cellular, and vascular behavior. J Am Coll Cardiol 2007;49:2379–93.
[27] Chatzizisis YS, Toutouzas K, Giannopoulos AA, Riga M, Antoniadis AP, Fujinom Y, et al. Association of global and local low endothelial shear stress with high-risk plaque using intracoronary 3D optical coherence tomography: introduction of 'shear stress score'. Eur Heart J Cardiovasc Imaging 2016. http://dx.doi.org/10.1093/ehjci/jew134.
[28] Pohl T, Seiler C, Billinger M, Herren E, Wustmann K, Mehta H, et al. Frequency distribution of collateral flow and factors influencing collateral channel development. Functional collateral channel measurement in 450 patients with coronary artery disease. J Am Coll Cardiol 2001;38:1872–8.
[29] Fujita M, Nakae I, Kihara Y, Hasegawa K, Nohara R, Ueda K, et al. Determinants of collateral development in patients with acute myocardial infarction. Clin Cardiol 1999;22:595–9.

CHAPTER 1.2

Endothelial Function

Petros Fountoulakis, Evangelos Oikonomou, Georgios Lazaros, Dimitris Tousoulis
1st Cardiology Clinic, 'Hippokration' General Hospital, National and Kapodistrian University of Athens, School of Medicine, Athens, Greece

Contents

ENDOTHELIAL FUNCTION AND VASCULAR HOMEOSTASIS

For many years, the endothelium was supposed to be a simple semipermeable membrane separating the vascular lumen from the vascular wall. Today it is recognized as an important endocrine organ and the primary regulator of the vascular homeostasis. The endothelium responds to the incoming natural and chemical stimuli via the production of a multitude of intermediary substances regulating the vascular tone, the diameter of the lumen, the protection from the thrombogenesis and the restriction of the cellular proliferation and the inflammatory response [1,2]. The balance among the specific impacts of the intermediary molecules is of major importance for the normal endothelial function.

One of the most important and scrutinized intermediary endothelial molecules is the nitric oxide (NO) [3]. It is produced from L-arginine with the presence of the enzyme endothelial NO synthase (NOS) and cofactors such as tetrahydrobiopterine (BH4) whose availability is influenced by the genetic variation in GTP-cyclohydrolase I (GCH1) haplotype which encodes GCH1 [4]. The NO is diffused to the smooth muscle cells and activates the guanylyl cyclase through the cyclic guanosine monophosphate

Coronary Artery Disease
ISBN 978-0-12-811908-2, http://dx.doi.org/10.1016/B978-0-12-811908-2.00002-7

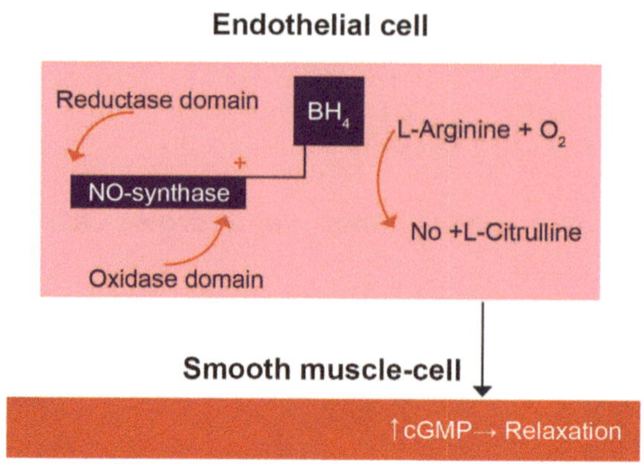

Figure 1.2.1 Production of nitric oxide and smooth muscle cell effect.

(cGMP) pathway leading to vasodilatation (Fig. 1.2.1). The endothelial NOS is triggered by the shear stress adjusting the tissue blood flow to the changes of the cardiac output. Furthermore, the enzyme is activated by a number of intermediary molecules such as bradycinine, adenocine, vascular endothelial growth factor (VEGF) in response to the hypoxia, and serotonin which is released during the platelet aggregation [5]. Reduced NO activity is related to hypercholesterolemia [6].

Besides NO vasodilation, the endothelium has the ability to evoke hyper polarization of the smooth muscle cells through factors not totally clarified such as molecules induced by the cytochrome and possibly the natriuretic peptide type C. In this way, the conductivity of potassium is increased and the polarization of the smooth muscle cell is expanded maintaining the tone of vasodilation especially to the microcirculation [5].

The endothelium also affects vasoconstriction with the production of endothelin, vasoconstrictor prostanoids and the conversion of angiotensin I to angiotensin II on the endothelial surface. The action of these vasoconstrictor agents is mainly local and there may be a role on the arterial remodeling [7].

Under normal circumstances, the endothelium offers protection from the inflammation and thrombosis. In this direction, the NO plays an important role in maintaining the balance of the vascular wall through the nitrosylation of a cysteine of a multitude of proteins leading to the reduction of their biological activity [8]. The main factor of this vascular balance phenotype is the shear stress [5].

ENDOTHELIAL DYSFUNCTION AND ATHEROMATOSIS

Risk Factors and Atheromatosis

Endothelial dysfunction is defined as the alteration of the endothelial phenotype from "calm" to a more activated condition in response to the invasion of an antigen in the organism and the influence of cardiovascular risk factors. Nonmodifiable risk factors include older age and male sex. Modifiable risk factors include hypercholesterolemia,

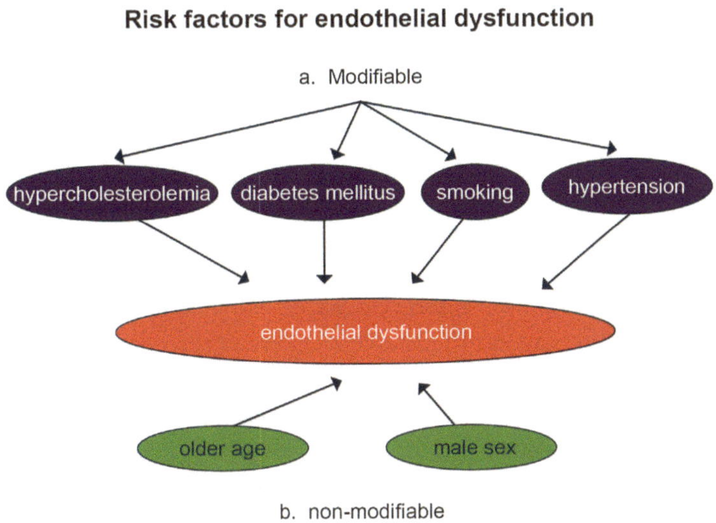

Figure 1.2.2 Risk factors alternating endothelial function.

which appears as one of the most important contributors to atherosclerotic disease progression, and others such as smoking, hypertension, and diabetes, are shown as major risk factors for coronary and cerebral events irrespective of age and sex.

Recent risk factors associated with the western way of life have also been described. Among them obesity, atherogenic diet, and physical inactivity are important problems in this process. In addition, infection and chronic inflammation may trigger endothelial dysfunction. Furthermore, serological evidence of exposure to multiple intracellular pathogens, particularly in the context of a low grade inflammatory response, was associated with increased coronary atherosclerosis. More recently mild nonspecific viral infections have been found to have a detrimental effect on the vascular endothelium [9] (Fig. 1.2.2).

Pathophysiology of Atheromatosis

The endothelium holds a main role on inflammation and in all phases of the atherosclerotic process. Endothelial mechanisms are activated leading to the expression of cytokines, chemokines and adhesion molecules through interactions with leucocytes and platelets, thus encouraging the inflammatory process of the vascular wall [10]. Atherosclerosis is promoted by decreased shear stress, as it is associated with suppression of functions taking place on the vascular wall, such as endothelial NOS (eNOS) production and endothelial cell (EC) repair. In the presence of systemic risk factors, there is an increased tendency for atherosclerotic plaque formation, which, once formed, further disrupts flow and facilitates growth of the fibro-inflammatory lipid plaque [11].

The atherosclerotic events start with the activation of protein kinase C and transcriptional messenger nuclear factor κB. This leads to upregulation of genes encoding and inducing angiotensin converting enzyme activity, local production of angiotensin II, and expression of EC surface adhesion molecules. In this way, cellular and subcellular responses are initiated and amplified in conduit coronary arteries that lead to endothelial dysfunction. This may lead to intimal thickening, plaque formation and disruption.

On cellular level, inflammatory cells are recruited from the circulation and migrate through the endothelium, which is mediated by cellular adhesion molecules. They are expressed on the vascular endothelium and on circulating leucocytes in response to several inflammatory stimuli. Adhesion is a process that starts with leucocyte rolling on the endothelial surface. This is due to selectin ligation, whereas the subsequent firm adhesion depends on interactions between immunoglobulin-like molecules (vascular cell adhesion molecule 1 [VCAM-1], intercellular cell adhesion molecule 1 [ICAM-1]) on the endothelium and integrins on the leucocyte surface.

Recruitment of monocytes into the arterial wall is one of the earliest events in atherosclerosis. Intimal monocytes develop into macrophages mediating to the inflammation and the innate immune response in atherosclerotic lesions. Macrophages contribute to the local inflammatory responses through production of cytokines, free oxygen radicals, proteases, and complement factors. Macrophages uptake modified lipoproteins leading to the accumulation of cholesterol esters and formation of macrophage derived foam cells, the hallmark of the fatty streak. Macrophages also contribute to lesion remodeling and to plaque rupture by secreting matrix metalloproteinases and thus contributing to the evolution of atherosclerotic disease.

A diverse lymphocyte population is found in atherosclerotic lesions with substantial number of T lymphocytes detected. They may enter the vessel wall before monocytes during the earliest stages of lesion formation and become activated The presence of activated lymphocytes at each stage of human lesion formation produces compelling evidence in the role of this cell type in the orchestration of the disease process (Fig. 1.2.3).

Major role on the endothelial activation maintains the eNOS, which under certain circumstances can evoke oxidative stress. The free oxygen radicals enable the production of hydrogen peroxide, which is quickly diffused to the cells affecting the cysteine of the proteins and furthermore their functionality [12]. The chronic production of free oxygen radicals may surpass the antacid mechanisms favoring the endothelial activation and the vascular malfunction. An important source of free radicals is the mitochondria, in which the production of free oxygen radicals and the influence of the mitochondrial dismutase maintain the balance during the oxidative phosphorylation. This balance can be disturbed during hypoxia or conditions enabling the increase of the enzyme reaction such as obesity or diabetes mellitus type 2 through the hyperglycemia and increase of the free fat acids [13]. Another important source of oxidative stress is the NAD(P)H (nicotinamide adenine dinucleotide phosphate) oxidase and the xanthine oxidase which present an increased reactivity in patients with coronary artery disease [5].

The prolonged or/and the repetitive exposure to the cardiovascular risk factors depletes the endogenous anti-inflammatory mechanisms of the endothelium resulting in endothelial dysfunction and loss of the continuity on the vascular wall [14]. The

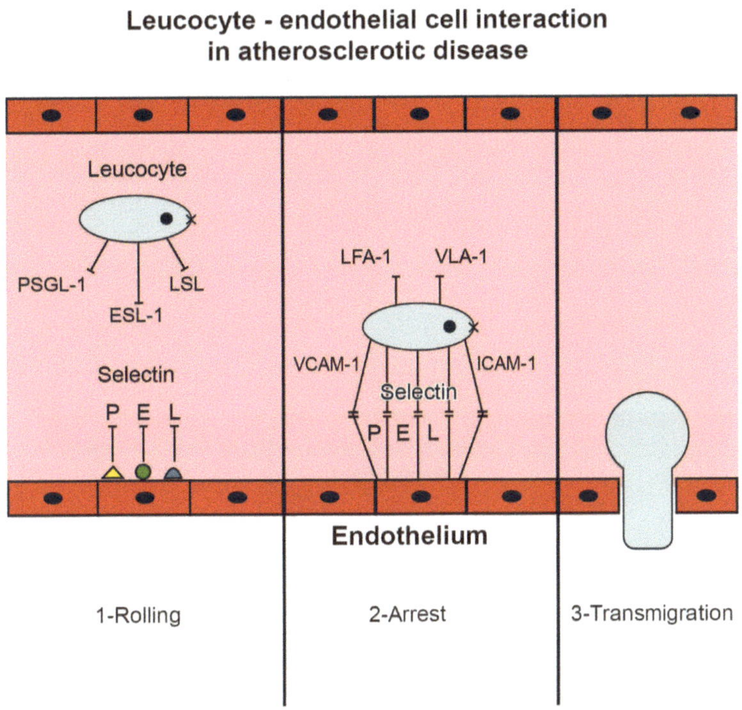

Figure 1.2.3 Leucocyte—endothelial cell interaction in atherosclerotic disease. The rolling of leucocytes on the endothelial surface is facilitated by selectins (E-selectin, L-selectin, P-selectin). Leucocytes express on the cell surface molecules, such as leucocyte function associated antigen (LFA) 1 and very late antigen (VLA) 4, vascular cell adhesion molecule (VCAM-1) and intercellular cell adhesion molecule (ICAM-1), thus promoting their adhesion to the vascular wall. Finally, leucocytes transmigrate to the inner intima. *ESL-1*, E-selectin ligand 1; *LSL*, L-selectin ligand; *PSGL-1*, P-selectin glycoprotein ligand 1.

mature ECs have the ability of local reproduction and restoration of the damaged cells. There is an alternative mechanism of preservation of the endothelial continuity described recently. Human endothelial progenitor cells (EPCs), a subset of bone marrow-derived cells have been of particular interest, as these cells were suggested to home to sites of neovascularization and neoendothelialization and differentiate into ECs in situ, a process referred to as postnatal vasculogenesis [15] (Fig. 1.2.4). The expression of the cells depending on the NO is reduced in patients with cardiovascular risk factors. On the other hand, factors which increase the NO bioavailability such as exercise and statin therapy have a positive impact on the expression of EPCs [5]. The significance of the balance between exposure to cardiovascular risk factors and restoration of the endothelial damage is outlined on the observation that people with increased number of EPCs present preserved endothelial function despite exposure to risk factors [16].

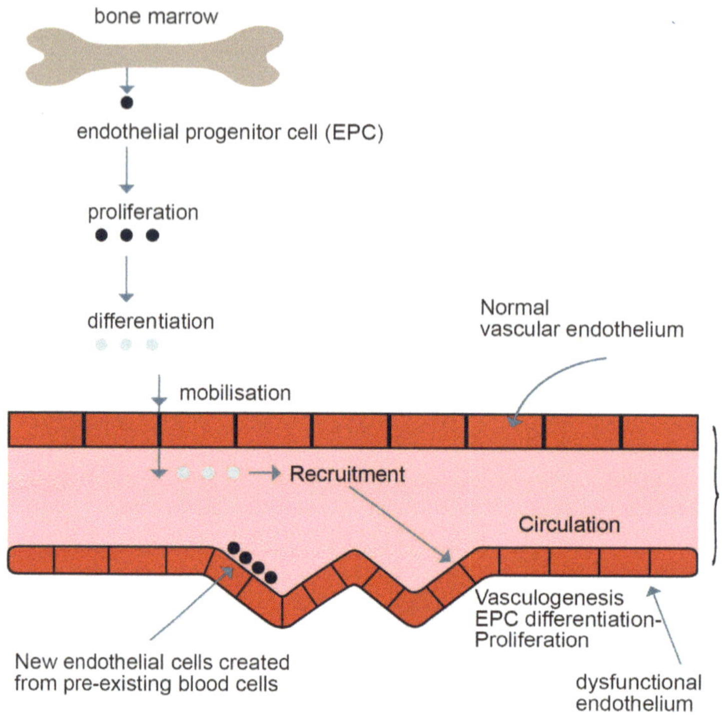

Figure 1.2.4 Functions of human endothelial progenitor cells on the vascular surface.

METHODS FOR EVALUATING ENDOTHELIAL FUNCTION

In recent years, the progress on understanding the vascular homeostasis allowed the development of methods for evaluating the function of the normal and the activated endothelium. However, none of these methods is a gold standard and there has to be a combination of specific information of different examinations for concluding on the complex endothelial biology and the prognostic significance of its disorders.

Biomarkers

The classic proinflammatory cytokines, interleukin 1, and tumor necrosis factor α (TNF-α), typically mediate proatherogenic processes, whereas interleukin 10 mediates antiatherogenic pathways [9].

Biomarkers relating to the classical cardiovascular disease risk factors are high-sensitivity C reactive protein (hs-CRP) and oxidized low-density lipoprotein (OX-LDL). The hs-CRP has the greatest prognostic value and is incorporated in models of risk stratification of cardiovascular disease. It seems that CRP is not only a cardiovascular risk factor and predictor of future or recurrent cardiovascular events, but also acts as a proatherogenic molecule, interferes in the composition of NO, inhibits angiogenesis and affects arterial remodeling [17,18]. In addition, CRP, a marker of underlying inflammation, may have a direct role in the pathophysiology of atherosclerosis. Thus, in the presence of CRP, uptake of low density lipoprotein cholesterol by macrophages is increased and contributes to foam cell formation. CRP can activate complement in atherosclerotic plaques leading potentially to plaque instability. It can induce adhesion molecule expression on human coronary ECs. Last, increased CRP is also associated with endothelial dysfunction and the progression of atherosclerosis.

Another way of evaluating the endothelial function is estimating the levels of molecules related to the NO biology (nitrosylated proteins, asymmetric dimethylarginine), inflammatory cytokines and adhesion molecules (E-selectin, P-selectin, VCAM-1, ICAM-1), thrombotic factors (tissue plasminogen activator (t-PA), plasminogen activator inhibitor-1 (PAI-1), Von Willebrand factor (vWF), and markers of endothelial damage and restoration. The clinical utility for most of these markers is restricted by technical difficulties and is not cost effective. The best studied and easily used markers are E-selectin, the most specific marker of endothelial activation, P-selectin, vascular cellular adhesion molecule-1, intercellular adhesion molecule-1, whose levels are increased in the presence of cardiovascular risk factors and is related to unfavorable prognosis and vWF [5]. Biomarkers such as interleukin 6 (IL-6), soluble VCAM-1 (sVCAM) and soluble ICAM-1 (sICAM-1) were related to inflammation and impaired endothelial function in transfusion dependent patients with beta-thalassemia major [19].

Another interesting and promising category of biomarkers on research participating to the development of atherosclerosis are microRNAs (miRNAs). They are a class of small, endogenous RNAs of 21—25 nucleotides (nts) in length and play an important regulatory role by targeting specific miRNAs to degradation or translation repression. Far from being simple intracellular regulators, miRNAs have recently been involved in intercellular communication and have been shown to circulate in the bloodstream in stable forms. They participate in cardiovascular disease pathogenesis including atherosclerosis and endothelial dysfunction and this may have important clinical implications. During the process of plaque development different sets of miRNAs have been found in different stages of plaque progression and miRNA dysregulation plays a crucial role in the destabilization and rupture of atherosclerotic plaques [20,21].

There is particular scientific interest in the endothelial microparticles (EMPs) and human EPCs, nevertheless further understanding of their pathophysiology are necessary before the passage in clinical practice. EMPs belong to a family of extracellular vesicles that are dynamic, mobile, biological effectors capable of mediating vascular physiology and function. The release of EMPs can impart autocrine and paracrine effects on target cells through surface interaction, cellular fusion, and, possibly, the delivery of intravesicular cargo. They are derived from activated or apoptotic ECs and indicate cellular damage. These markers are elevated in patients with coronary artery disease, peripheral arterial disease and inflammatory diseases with increased vascular risk such as systemic lupus erythematosus and rheumatoid arthritis [5,22].

Human EPCs have been generally defined as circulating cells produced from the bone marrow that express a variety of cell surface markers similar to those expressed by vascular ECs, adhere to endothelium at sites of hypoxia/ischemia, and participate in new vessel formation [23]. EPCs are recognized by specific surface markers and are detected with flow cytometry and represent the ability of endothelial endogenous restoration [24]. Recent evidence from clinical studies suggests that inflammatory and oxidative changes influence EPC mobilization. Drugs with anti-inflammatory and antioxidant properties, administered to patients with cardiovascular diseases, such as statins, exerted beneficial effects on EPC biology [25,26].

Invasive Methods

In clinical practice, the most useful method of evaluating endothelial function is the endothelium dependent alteration of the vascular tone in response to natural or pharmacological stimuli. The first clinical trials involved the coronary circulation with infusion of acetylcholine locally and calculation of the vascular diameter with coronography (Intracoronary infusions of vasoactive agents). The acetylcholine releases NO from the vessels with normal endothelium resulting in vasodilation as a result of its direct action to the muscarinic receptors of the smooth muscle cells. However, this molecule results in vasoconstriction at vessels with endothelial impairment. Several endothelial agonists, such as adenosine, bradykinin, and Substance P (SP) were used for this purpose. However, the use of NO antagonists (NG-monomethyl L-arginine, L-NMMA) highlighted the alteration of the vascular tone [5,27,28]. They have the disadvantage of being invasive and expensive, with the risks of coronary catheterization, and cannot be used as a screening test in the general population [29].

The same principles were also applied to the peripheral circulation with catheterization of the brachial artery and evaluation of the changes of the vascular tone with venous occlusion plethysmography [30]. This method is highly reproducible, reliable and easily applied. However, it still remains an invasive method with potential side effects such as injuries of the median nerve or the brachial artery [29]. In a clinical trial conducted with this method, was discovered that antioxidant treatment with vitamins C and E at high doses, improved endothelial function and decreased plasma levels of PAI-1, vWF, and PAI-1/tPA ratio in chronic smokers [31].

Nevertheless, these invasive methods were restricted as the long-term follow up of the patients was complicated.

Noninvasive Methods

Gauge-Strain Plethysmography

Another noninvasive technique for evaluating the endothelial function in the brachial artery is the evaluation of the changes in forearm blood flow during reactive hyperemia, by the use of gauge-strain plethysmography.

With this technique, endothelial function is evaluated after the determination of the maximum blood flow during reactive hyperemia, the percentage change of flow from rest to the maximum hyperemic flow, as well as by the analysis of the time-flow curves during reactive hyperemia [29].

This technique is noninvasive, simple, and reproducible. Even though this method leads to less specific results than flow-mediated dilation (FMD), it is preferred because it is easily applied and does not require highly trained personnel, and its results are less observer dependent.

This technique was used in numerous clinical trials. In transfusion dependent patients, beta-thalassemia major was associated with impaired endothelial function [32]. Same results were demonstrated to another population of patients after successfully repaired coarctation of the aorta [33].

On genetic level, the polymorphism G894T on eNOS affects endothelial function in patients with a history of premature myocardial infarction [19].

Thanks to another clinical trial based on gauge-strain plethysmography, it was found that consumption of maize oil leads to impaired endothelial function, whereas soya oil and cod liver oil ameliorate endothelial function [34]. Acute consumption of alcoholic beverages such as red wine and beer was proven to have a positive impact on endothelial function and the balance of thrombosis and fibrinolysis system [35].

Clinical trials also proved the benefits of medication on vascular function. Treatment with low dose atorvastatin in patients with ischemic heart failure improved endothelial function and reduced inflammatory process which was partially depressed by vitamin E [36,37]. Combined treatment with metformin and atorvastatin for 6 weeks partly prevented the glucose-induced endothelial dysfunction for the diabetic population [38].

Flow-Mediated Dilation

The dominant method of evaluating endothelial function in clinical practice is FMD, which is an ultrasound method.

Peripheral arteries, especially the brachial artery, respond to physical and chemical stimuli by adjusting vascular tone and regulating blood flow. Increased blood flow in peripheral arteries leads to increased shear stress stimuli, increased NO production, and vasodilation. The vasodilatory response of the brachial artery to increased shear stress is called FMD, and reflects the ability of vascular endothelium to produce NO [29].

It evaluates the alterations of the diameter of the brachial artery in response to the endothelium dependent hyperemia and compares them with the alteration of the vascular diameter after the ingestion of endothelium independent agonists, such as glycerin trinitrate [17]. FMD technique is used widely on clinical research and allows the studying of the influence of the hygienic-dietary and medical regimens on endothelial function, when atherosclerotic disease is reversible.

The vasoactive drugs should be discontinued at least 12 h before the study, and the patient should refrain from smoking for at least 6 h. After 12 h fasting, the patient is placed supine in a quiet room at a constant temperature, with his hand in a comfortable position, which should remain stable for at least 1 min. The brachial artery is shown above the antercubital fossa on the longitudinal plane, using a linear array transducer (frequency 07.12 MHz) connected to a high quality ultrasound system. The diameter of the brachial artery is determined initially at rest, and blood flow is calculated with the pulse Doppler as an average volume of the artery specimen. After determination of the base diameter of the brachial artery, ischemia is produced by inflating a cuff balloon, which is placed in the distal forearm, at a pressure of 50 mmHg higher than the systolic blood pressure. The release of the tourniquet after 5 min leads to an increase in blood flow to the forearm, resulting in vasodilation brachial artery. The maximum blood flow rate is detected by analyzing the pulse Doppler signal within the artery immediately following or up to 15 s after cuff release, the maximum diameter of the brachial determined about 60 s after release, or 45—60 s after the peak hyperemia flow (Fig. 1.2.5). FMD has been defined as the percent change in the brachial artery diameter at rest in diameter 60 s after release of the tourniquet, as the brachial artery continuously monitored 30 s before and 120 s after the release of slinging.

The method of FMD is reproducible, with a coefficient of variation for repeated measurements of brachial artery diameter about 3%—4% in short term (2 h interval) as well as in longer term (3-week interval). Even though, FMD is noninvasive, safer, and faster than invasive methods, it is highly operator dependent, requires excellent patient cooperation, and has relatively poor resolution relative to arterial size [29].

The method of FMD is reproducible, with a coefficient of variation for repeated measurements of brachial artery diameter about 3%—4% in short term (2 h interval) as well as in longer term (3-week interval). Even though, FMD is noninvasive, safer, and faster than invasive methods, it is highly operator dependent, requires excellent patient cooperation, and has relatively poor resolution relative to arterial size [29].

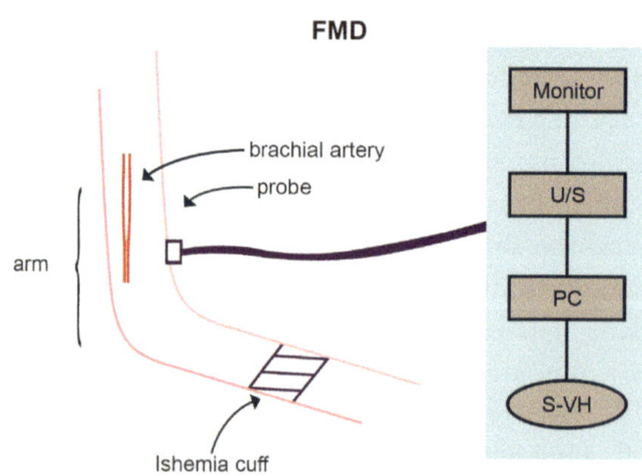

Figure 1.2.5 Method of flow-mediated dilation.

Prognostic Value of the Endothelial Function

Endothelial dysfunction is an early and potentially reversible disorder of the atherosclerotic process. Atherosclerosis begins from childhood in response to genetic and environmental risk factors, evolves silently during a long-term preclinical stage, and finally is manifested usually on the middle-aged population. Additionally, endothelium dependent vascular biology is related to alterations of the composition and biology of the atherosclerotic plaque leading to local or total hypoperfusion of the target organs and finally increased incidence of ischemic episodes [39].

Several clinical trials demonstrated that the evaluation of the endothelial function to the coronary and peripheral circulation has a prognostic value to a wide range of clinical cases from patients with atypical symptoms to those with documented coronary disease or peripheral arterial disease [40—43]. Second, endothelial dysfunction is a prognostic factor of restenosis after angioplasty and complications of vascular surgeries [40]. However, the evaluation of the endothelial function is more valuable on early stages of the vascular disease. A recent meta-analysis proved the correlation between FMD results and cardiovascular risk on low risk patients [44]. FMD represents a prognostic factor of cardiovascular risk to the elder population regardless of age, sex, smoking, diabetes mellitus or arterial hypertension, although its contribution to the accuracy of the classical models of cardiovascular risk prediction is up to 1% [45].

FMD has been shown to be impaired in a range of classical and novel risk factors from the first decade of life according to numerous clinical trials as follows:

1. Acute systemic inflammation induced by vaccination against the H1N1 influenza virus resulted in degradation of the endothelial function in seropositive (HIV+) patients for more than 48 h [46].
2. Nonalcoholic fatty liver disease is associated with arterial stiffness and endothelial dysfunction leading to increased cardiovascular risk [47].
3. The presence of multiple cardiovascular risk factors (hypertension, hyperlipidemia, smoking, and diabetes) is associated with accelerated progression of vascular aging [48].
4. Diabetic retinopathy and vascular dysfunction coexist in diabetic patients. The progress of the disease and the development of microvascular complications were strongly associated with the further deterioration of endothelial function and arterial stiffness. Furthermore, there was a negative predictive value of endothelial function in the presence of ocular lesions in diabetic patients [49].
5. The diabetic foot in type 2 diabetes is associated with endothelial dysfunction and increased stiffness of the large arteries. Endothelial dysfunction is installed at an early stage of the course of diabetes and impaired vascular elasticity follows a gradual process [50].
6. Loss of polymorphism of CYP2C19 * 2 gene had no effect on endothelial dysfunction and inflammatory profile coronary patients treated with clopidogrel [51].

7. The ocular disease of sarcoidosis is associated with impaired endothelial dysfunction and increased arterial stiffness [52].
8. Methionine-induced homocysteinemia alternates endothelial function in hypertensive patients via endothelin-1 activation [53].
9. Rs1501299 polymorphism of adiponectin gene affects endothelial function in healthy Presence of the T variant at rs1501299 on adiponectin gene is independently associated with increased myocardial infarction risk [54]. Circulating adiponectin was demonstrated as an independent predictor of endothelial function in well phenotyped patients with coronary artery disease, which was aggravated by genetically lower circulating levels of adiponectin. Furthermore, adiponectin increased vascular Tetrahydrobiopterin (BH4), thus increasing the enzymatic coupling of eNOS [55].

Impaired endothelial dysfunction is associated with a more rapid progression of structural arterial disease and medial thickness of the carotid artery in asymptomatic middle-aged population.

A large number of clinical trials are expected to clarify the prognostic value of endothelial dysfunction to the general population. A therapeutic approach in which the estimation of genetic predisposition and exposure to risk factors, the evaluation of the endothelial function and imaging of the arterial disease in early stages would benefit the public health.

Restoration of the Endothelial Function

The restoration of the endothelial function is a target of therapeutic approaches to a wide range of patients suffering from vascular disease.

FMD can rapidly be changed and restored with lifestyle changes. Continuous aerobics and high-intensity aerobics can have favorable effect on endothelial function, suggesting a cardioprotective function of acute exercise [56]. Exercise training, physical activity along with chronic coffee consumption in the elderly population improved endothelial function [57,58].

Most clinical trials demonstrated that the improvement of endothelial function is related to therapies reducing the cardiovascular events, such as inhibitors of the renin—angiotensin system, statins and Ω-3 fatty acids, leading to the stabilization of atherosclerotic disease [59—62]. Treatment with low-dose atorvastatin modulates inflammatory response and ameliorates endothelial function in patients with recent acute ST-elevated myocardial infarction and on indices of left ventricular remodeling in the heart failure population [63,64]. Patients with congestive heart failure improved vascular function under rosuvastatin treatment [65].

Furthermore, specific medical treatments have shown a direct effect on cardiovascular risk factors. The endothelial function of nondiabetic patients with coronary artery disease was ameliorated with thiazolidinediones. Treatment with calcium channel blockers also revealed a positive effect on endothelial function to patients with normal arterial pressure and hypercholesterolemia [62,66,67].

A recent study on population with rheumatoid arthritis proved that the inhibition of interleukin-1 by the human recombined antagonist of the interleukin-1 receptor anakinra reduces the markers of oxidative stress and endothelin and improves endothelial function based on FMD results [68]. At the same time, small doses of the oral endothelin receptor antagonist bosentan improved vascular function in patients with systemic sclerosis (SSc; Scleroderma) [69]. Patients with chronic psoriases demonstrated significant improvement on the endothelial function under adalimumab treatment, an anti-TNF-α [70].

Eventually, novel therapies depending on the understanding of endothelial biology, such as dietary supplements with L-arginine offered favorable results [5]. Additionally, short term administration of L-arginine ameliorates endothelial function and arterial elasticity in healthy smokers [71]. In some elderly populations with overt endothelial dysfunction and reduced L-arginine stores, L-arginine improved vascular function via NO augmentation, its ancillary effects on systemic hormonal modulation (i.e., growth hormone augmentation) and anti-inflammatory/antioxidant phenotypic changes in ECs [72]. Treatment with BH4 and asymmetric dimethylarginine has been implicated to improve endothelial function on diabetic population [73].

CONCLUSION

To sum up, endothelium the inner layer of the arterial wall, represent an active organ with multiple functions orchestrating vasodilation and vasoconstriction of the arterial wall and the activation of the thrombosis/fibrinolysis system. Endothelial dysfunction caused by the chronic impact of cardiovascular risk factors represents the first step in the process of atheromatosis and atherosclerosis. Evaluation of endothelial function can be used to identify patients with increased cardiovascular risk who may be benefit from intensive risk management modification.

REFERENCES

[1] Deanfield J, Donald A, Ferri C, Giannattasio C, Halcox J, Halligan S, Lerman A, Mancia G, Oliver JJ, Pessina AC, Rizzoni D, Rossi GP, Salvetti A, Schiffrin EL, Taddei S, Webb DJ, Working Group on Endothelin, Endothelial Factors of the European Society of Hypertension. Endothelial function and dysfunction. Part I: Methodological issues for assessment in the different vascular beds: a statement by the Working Group on Endothelin and Endothelial Factors of the European Society of Hypertension. J Hypertens January 2005;23(1):7–17.

[2] Brunner H, Cockcroft JR, Deanfield J, Donald A, Ferrannini E, Halcox J, Kiowski W, Lüscher TF, Mancia G, Natali A, Oliver JJ, Pessina AC, Rizzoni D, Rossi GP, Salvetti A, Spieker LE, Taddei S, Webb DJ, Working Group on Endothelins, Endothelial Factors of the European Society of Hypertension. Endothelial function and dysfunction. Part II: Association with cardiovascular risk factors and diseases. A statement by the Working Group on Endothelins and Endothelial Factors of the European Society of Hypertension. J Hypertens Feburary 2005;23(2):233–46.
[3] Palmer RM, Ashton DS, Moncada S. Vascular endothelial cells synthesize nitric oxide from L-arginine. Nature June 16, 1988;333(6174):664–6.
[4] Antoniades C, Shirodaria C, Assche TV, Cunnington C, Tegeder I, Lötsch J, Guzik TJ, Leeson P, Diesch J, Tousoulis D, Stefanadis C, Costigan M, Woolf CJ, Alp NJ, Channon KM. GCH1 haplotype determines vascular and plasma biopterin availability in coronary artery disease. Effects on vascular superoxide production and endothelial function. J Am Coll Cardiol July 8, 2008;52(2):158–65.
[5] Deanfield JE, Halcox JP, Rabelink TJ. Endothelial function and dysfunction: testing and clinical relevance. Circulation March 13, 2007;115(10):1285–95.
[6] Tentolouris C, Tousoulis D, Davies G, Stefanadis C, Trikas A, Goumas G, Toutouzas P. The impact of risk factors for atherosclerosis on the vasomotor effects of inhibition of nitric oxide synthesis in patients with normal angiograms. Cardiology 2000;94(1):26–30.
[7] Kinlay S, Behrendt D, Wainstein M, Beltrame J, Fang JC, Creager MA, Selwyn AP, Ganz P. Role of endothelin-1 in the active constriction of human atherosclerotic coronary arteries. Circulation September 4, 2001;104(10):1114–8.
[8] Stamler JS, Lamas S, Fang FC. Nitrosylation. The prototypic redox-based signaling mechanism. Cell September 21, 2001;106(6):675–83.
[9] Tousoulis D, Charakida M, Stefanadis C. Endothelial function and inflammation in coronary artery disease. Heart April 2006;92(4):441–4.
[10] Hansson GK. Inflammation, atherosclerosis, and coronary artery disease. N Engl J Med April 21, 2005;352(16):1685–95.
[11] Siasos G, Tousoulis D, Siasou Z, Stefanadis C, Papavassiliou AG. Shear stress, protein kinases and atherosclerosis. Curr Med Chem 2007;14(14):1567–72.
[12] Rhee SG. Cell signaling. H2O2, a necessary evil for cell signaling. Science June 30, 2006;312(5782):1882–3.
[13] Evans JL, Goldfine ID, Maddux BA, Grodsky GM. Oxidative stress and stress-activated signaling pathways: a unifying hypothesis of type 2 diabetes. Endocr Rev October 2002;23(5):599–622.
[14] Woywodt A, Bahlmann FH, De Groot K, Haller H, Haubitz M. Circulating endothelial cells: life, death, detachment and repair of the endothelial cell layer. Nephrol Dial Transpl October 2002;17(10):1728–30.
[15] Asahara T, Murohara T, Sullivan A, Silver M, van der Zee R, Li T, Witzenbichler B, Schatteman G, Isner JM. Isolation of putative progenitor endothelial cells for angiogenesis. Science Feburary 14, 1997;275(5302):964–7.
[16] Hill JM, Zalos G, Halcox JP, Schenke WH, Waclawiw MA, Quyyumi AA, Finkel T. Circulating endothelial progenitor cells, vascular function, and cardiovascular risk. N Engl J Med Feburary 13, 2003;348(7):593–600.
[17] Lane HA, Smith JC, Davies JS. Noninvasive assessment of preclinical atherosclerosis. Vasc Health Risk Manag 2006;2(1):19–30.
[18] Tousoulis D, Papageorgiou N, Latsios G, Siasos G, Antoniades C, Stefanadis C. C-reactive protein and endothelial dysfunction: gazing at the coronaries. Int J Cardiol October 6, 2011;152(1):1–3.
[19] Antoniades C, Tousoulis D, Vasiliadou C, Pitsavos C, Chrysochoou C, Panagiotakos D, Tentolouris C, Marinou K, Koumallos N, Stefanadis C. Genetic polymorphism on endothelial nitric oxide synthase affects endothelial activation and inflammatory response during the acute phase of myocardial infarction. J Am Coll Cardiol September 20, 2005;46(6):1101–9.
[20] Economou EK, Oikonomou E, Siasos G, Papageorgiou N, Tsalamandris S, Mourouzis K, Papaioanou S, Tousoulis D. The role of microRNAs in coronary artery disease: from pathophysiology to diagnosis and treatment. Atherosclerosis August 2015;241(2):624–33.

[21] Tousoulis D, Androulakis E, Papageorgiou N, Siasos G, Latsios G, Charakida M, Kampoli AM, Oikonomou E, Stefanadis C. Novel biomarkers assessing endothelial dysfunction: role of microRNAs. Curr Top Med Chem 2013;13(13):1518–26.
[22] Mallat Z, Benamer H, Hugel B, Benessiano J, Steg PG, Freyssinet JM, Tedgui A. Elevated levels of shed membrane microparticles with procoagulant potential in the peripheral circulating blood of patients with acute coronary syndromes. Circulation Febuary 29, 2000;101(8):841–3.
[23] Briasoulis A, Tousoulis D, Antoniades C, Papageorgiou N, Stefanadis C. The role of endothelial progenitor cells in vascular repair after arterial injury and atherosclerotic plaque development. Cardiovasc Ther April 2011;29(2):125–39.
[24] Goon PK, Boos CJ, Lip GY. Circulating endothelial cells: markers of vascular dysfunction. Clin Lab 2005;51(9–10):531–8.
[25] Tousoulis D, Andreou I, Antoniades C, Tentolouris C, Stefanadis C. Role of inflammation and oxidative stress in endothelial progenitor cell function and mobilization: therapeutic implications for cardiovascular diseases. Atherosclerosis December 2008;201(2):236–47.
[26] Oikonomou E, Siasos G, Zaromitidou M, Hatzis G, Mourouzis K, Chrysohoou C, Zisimos K, Mazaris S, Tourikis P, Athanasiou D, Stefanadis C, Papavassiliou AG, Tousoulis D. Atorvastatin treatment improves endothelial function through endothelial progenitor cells mobilization in ischemic heart failure patients. Atherosclerosis Feburary 2015;238(2):159–64.
[27] Tousoulis D, Tentolouris C, Crake T, Stefanadis C, Toutouzas P, Davies G. Evidence of partially preserved endothelial dilator function in diseased coronary arteries. Heart October 1999;82(4):471–6.
[28] Goodhart DM, Anderson TJ. Role of nitric oxide in coronary arterial vasomotion and the influence of coronary atherosclerosis and its risks. Am J Cardiol 1998;82(9):1034–9.
[29] Tousoulis D, Antoniades C, Stefanadis C. Evaluating endothelial function in humans: a guide to invasive and non-invasive techniques. Heart April 2005;91(4):553–8.
[30] Joannides R, Bellien J, Thuillez C. Clinical methods for the evaluation of endothelial function: a focus on resistance arteries. Fundam Clin Pharmacol June 2006;20(3):311–20.
[31] Antoniades C, Tousoulis D, Tentolouris C, Toutouza M, Marinou K, Goumas G, Tsioufis C, Toutouzas P, Stefanadis C. Effects of antioxidant vitamins C and E on endothelial function and thrombosis/fibrinolysis system in smokers. Thromb Haemost June 2003;89(6):990–5.
[32] Aggeli C, Antoniades C, Cosma C, Chrysohoou C, Tousoulis D, Ladis V, Karageorga M, Pitsavos C, Stefanadis C. Endothelial dysfunction and inflammatory process in transfusion-dependent patients with beta-thalassemia major. Int J Cardiol 2005;105(1):80–4.
[33] Brili S, Tousoulis D, Antoniades C, Aggeli C, Roubelakis A, Papathanasiu S, Stefanadis C. Evidence of vascular dysfunction in young patients with successfully repaired coarctation of aorta. Atherosclerosis September 2005;182(1):97–103. Epub Feburary 24, 2005.
[34] Tousoulis D, Papageorgiou N, Antoniades C, Giolis A, Bouras G, Gounari P, Stefanadi E, Miliou A, Psaltopoulou T, Stefanadis C. Acute effects of different types of oil consumption on endothelial function, oxidative stress status and vascular inflammation in healthy volunteers. Br J Nutr January 2010; 103(1):43–9. http://dx.doi.org/10.1017/S0007114509991346. Epub August 25, 2009.
[35] Tousoulis D, Ntarladimas I, Antoniades C, Vasiliadou C, Tentolouris C, Papageorgiou N, Latsios G, Stefanadis C. Acute effects of different alcoholic beverages on vascular endothelium, inflammatory markers and thrombosis fibrinolysis system. Clin Nutr August 2008;27(4):594–600.
[36] Antoniades C, Tousoulis D, Stefanadis C. Smoking in Asians: it doesn't stop at vascular endothelium. Int J Cardiol August 18, 2008;128(2):151–3.
[37] Tousoulis D, Antoniades C, Vassiliadou C, Toutouza M, Pitsavos C, Tentolouris C, Trikas A, Stefanadis C. Effects of combined administration of low dose atorvastatin and vitamin E on inflammatory markers and endothelial function in patients with heart failure. Eur J Heart Fail December 2005; 7(7):1126–32. Epub July 27, 2005.
[38] Tousoulis D, Koniari K, Antoniades C, Miliou A, Noutsou M, Nikolopoulou A, Papageorgiou N, Marinou K, Stefanadi E, Stefanadis C. Impact of 6 weeks of treatment with low-dose metformin and atorvastatin on glucose-induced changes of endothelial function in adults with newly diagnosed type 2 diabetes mellitus: a single-blind study. Clin Ther September 2010;32(10):1720–8.

[39] Perticone F, Ceravolo R, Pujia A, Ventura G, Iacopino S, Scozzafava A, Ferraro A, Chello M, Mastroroberto P, Verdecchia P, Schillaci G. Prognostic significance of endothelial dysfunction in hypertensive patients. Circulation July 10, 2001;104(2):191–6.
[40] Anderson TJ. Prognostic significance of brachial flow-mediated vasodilation. Circulation 2007;115: 2373–5.
[41] Fathi R, Haluska B, Isbel N, Short L, Marwick TH. The relative importance of vascular structure and function in predicting cardiovascular events. J Am Coll Cardiol February 2004;43(4):616–23. No. 4.
[42] Karatzis EN, Ikonomidis I, Vamvakou GD, Papaioannou TG, Protogerou AD, Andreadou I, Voidonikola PT, Karatzi KN, Papamichael CM, Lekakis JP. Long-term prognostic role of flow-mediated dilatation of the brachial artery after acute coronary syndromes without ST elevation. Am J Cardiol December 1, 2006;98(11):1424–8. Epub October 2, 2006.
[43] Shimbo D, Grahame-Clarke C, Miyake Y, Rodriguez C, Sciacca R, Di Tullio M, Boden-Albala B, Sacco R, Homma S. The association between endothelial dysfunction and cardiovascular outcomes in a population-based multi-ethnic cohort. Atherosclerosis May 2007;192(1):197–203. Epub June 8, 2006.
[44] Witte DR, Westerink J, de Koning EJ, van der Graaf Y, Grobbee DE, Bots ML. Is the association between flow-mediated dilation and cardiovascular risk limited to low-risk populations? J Am Coll Cardiol June 21, 2005;45(12):1987–93.
[45] Yeboah J, Crouse JR, Hsu FC, Burke GL, Herrington DM. Brachial flow-mediated dilation predicts incident cardiovascular events in older adults: the Cardiovascular Health Study. Circulation May 8, 2007;115(18):2390–7. Epub April 23, 2007.
[46] Vlachopoulos C, Xaplanteris P, Sambatakou H, Mariolis E, Bratsas A, Christoforatou E, Miliou A, Aznaouridis K, Stefanadis C. Acute systemic inflammation induced by influenza A (H1N1) : vaccination causes a deterioration in endothelial function in HIV-infected patients. HIV Med November 2011;12(10):594–601.
[47] Vlachopoulos C, Manesis E, Baou K, Papatheodoridis G, Koskinas J, Tiniakos D, Aznaouridis K, Archimandritis A, Stefanadis C. Increased arterial stiffness and impaired endothelial function in: nonalcoholic fatty liver disease: a pilot study. Am J Hypertens November 2010;23(11):1183–9. http://dx.doi.org/10.1038/ajh.2010.144. Epub July 15, 2010.
[48] Vlachopoulos C, Terentes-Printzios D, Xaplanteris P, Ioakeimidis N, Aznaouridis K, Pietri P, Stefanadis C, Tousoulis D. Athens medical School, 1st Cardiology Department, Hippokration Hospital, Athens, Greece. Oral Sess 4D. 4D.02. Copyright 2015. Wolters Kluwer Health, Inc.
[49] Siasos G, Gouliopoulos N, Moschos MM, Oikonomou E, Kollia C, Konsola T, Athanasiou D, Siasou G, Mourouzis K, Zisimos K, Papavassiliou AG, Stefanadis C, Tousoulis D. Role of endothelial dysfunction and arterial stiffness in the development of diabetic retinopathy. Diabetes Care 2015;38: e9–10. http://dx.doi.org/10.2337/dc14-1741.
[50] Antonopoulos AS, Siasos G, Konsola T, Oikonomou E, Tentolouris N, Kollia C, Gouliopoulos N, Zografos T, Papavassiliou AG, Tousoulis D. Arterial wall elastic properties and endothelial dysfunction in the diabetic foot syndrome in patients with type 2 diabetes. Diabetes Care 2015;38:e180–1. http://dx.doi.org/10.2337/dc15-1042.
[51] Siasos G, Zarotimidou M, Oikonomou E, Mourouzis K, Tsalamandris S, Kioufis S, Kokkou E, Vavuranakis M, Zografos T, Antonopoulos A, Dimitropoulos S, Stefanadis C, Papavassiliou AG, Tousoulis D. Effects of CYP2C19 polymorphism on endothelial function, arterial stiffness and inflammation in coronary artery disease patients under clopidogrel treatment. Curr Pharm Des 2015;21(34): 5041–6.
[52] Siasos G, Paraskevopoulos T, Gialafos E, Rapti A, Oikonomou E, Zaromitidou M, Mourouzis K, Siasou G, Gouliopoulos N, Tsalamandris S, Vlassis K, Stefanadis C, Papavassileiou AG, Tousoulis D. Vascular function and ocular : involvement in sarcoidosis. Microvasc Res July 2015; 100:54–8. http://dx.doi.org/10.1016/j.mvr.2015.03.008. Epub April 29, 2015.
[53] Tousoulis D, Bouras G, Antoniades C, Marinou K, Miliou A, Papageorgiou N, Chatzis G, Tentolouris C, Tsioufis C, Stefanadis C. The activation of endothelin-1 pathway during methionine-induced homocysteinemia mediates endothelial dysfunction in hypertensive individuals. J Hypertens May 2010;28(5):925–30.

[54] Antonopoulos AS, Tousoulis D, Antoniades C, Miliou A, Hatzis G, Papageorgiou N, Demosthenous M, Tentolouris C, Stefanadis C. Genetic variability on adiponectin gene affects myocardial infarction risk: the role of endothelial dysfunction. Int J Cardiol September 20, 2013;168(1):326–30.
[55] Margaritis M, Antonopoulos AS, Digby J, Lee R, Reilly S, Coutinho P, Shirodaria C, Sayeed R, Petrou M, De Silva R, Jalilzadeh S, Demosthenous M, Bakogiannis C, Tousoulis D, Stefanadis C, Choudhury RP, Casadei B, Channon KM, Antoniades C. Interactions between vascular wall and perivascular adipose tissue reveal novel roles for adiponectin in the regulation of endothelial nitric oxide synthase function in human vessels. Circulation June 4, 2013;127(22):2209–21.
[56] Siasos G, Athanasiou D, Terzis G, Stasinaki A, Oikonomou E, Tsitkanou S, Kolokytha T, Spengos K, Papavassiliou AG, Tousoulis D. Acute effects of different types of aerobic exercise on endothelial function and arterial stiffness. Eur J Prev Cardiol April 27, 2016;23. pii:2047487316647185.
[57] Siasos G, Chrysohoou C, Tousoulis D, Oikonomou E, Panagiotakos D, Zaromitidou M, Zisimos K, Marinos G, Mazaris S, Kampaksis M, Papavassiliou AG, Pitsavos C, Stefanadis C. The impact of physical activity on endothelial function in middle-aged and elderly subjects: the Ikaria study. Hell J Cardiol March–April 2013;54(2):94–101.
[58] Siasos G, Oikonomou E, Chrysohoou C, Tousoulis D, Panagiotakos D, Zaromitidou M, Zisimos K, Kokkou E, Marinos G, Papavassiliou AG, Pitsavos C, Stefanadis C. Consumption of a boiled Greek type of coffee is associated with improved endothelial function: the Ikaria study. Vasc Med April 2013;18(2):55–62.
[59] Mancini GB, Henry GC, Macaya C, O'Neill BJ, Pucillo AL, Carere RG, Wargovich TJ, Mudra H, Lüscher TF, Klibaner MI, Haber HE, Uprichard AC, Pepine CJ, Pitt B. Angiotensin-converting enzyme inhibition with quinapril improves endothelial vasomotor dysfunction in patients with coronary artery disease. The TREND (Trial on Reversing ENdothelial Dysfunction) Study. Circulation August 1, 1996;94(3):258–65.
[60] Anderson TJ, Meredith IT, Yeung AC, Frei B, Selwyn AP, Ganz P. The effect of cholesterol-lowering and antioxidant therapy on endothelium-dependent coronary vasomotion. N Engl J Med Feburary 23, 1995;332(8):488–93.
[61] Ruilope LM, Redón J, Schmieder R. Cardiovascular risk reduction by reversing endothelial dysfunction: ARBs, ACE inhibitors, or both? Expectations from the ONTARGET Trial Programme. Vasc Health Risk Manag March 15, 2007;3(1):1–9.
[62] Antoniades C, Bakogiannis C, Leeson P, Guzik TJ, Zhang MH, Tousoulis D, Antonopoulos AS, Demosthenous M, Marinou K, Hale A, Paschalis A, Psarros C, Triantafyllou C, Bendall J, Casadei B, Stefanadis C, Channon KM. Rapid, direct effects of statin treatment on arterial redox state and nitric oxide bioavailability in human atherosclerosis via tetrahydrobiopterin-mediated endothelial nitric oxide synthase coupling. Circulation July 19, 2011;124(3):335–45.
[63] Stefanadi E, Tousoulis D, Antoniades C, Katsi V, Bosinakou E, Vavuranakis E, Triantafyllou G, Marinou K, Tsioufis C, Papageorgiou N, Latsios G, Stefanadis C. Early initiation of low-dose atorvastatin treatment after an acute ST-elevated myocardial infarction, decreases inflammatory process and prevents endothelial injury and activation. Int J Cardiol April 3, 2009;133(2):266–8. http://dx.doi.org/10.1016/j.ijcard.2007.11.025. Epub January 9, 2008.
[64] Tousoulis D, Oikonomou E, Siasos G, Chrysohoou C, Zaromitidou M, Kioufis S, Maniatis K, Dilaveris P, Miliou A, Michalea S, Papavassiliou AG, Stefanadis C. Dose-dependent effects of short term atorvastatin treatment on arterial wall properties and on indices of left ventricular remodeling in ischemic heart failure. Atherosclerosis April 2013;227(2):367–72.
[65] Gounari P, Tousoulis D, Antoniades C, Kampoli AM, Stougiannos P, Papageorgiou N, Roulia G, Stefanadi E, Siasos G, Tsioufis C, Stefanadis C. Rosuvastatin but not ezetimibe improves endothelial function in patients with heart failure, by mechanisms independent of lipid lowering. Int J Cardiol June 25, 2010;142(1):87–91.
[66] Sidhu JS, Kaposzta Z, Markus HS, Kaski JC. Effect of rosiglitazone on common carotid intima-media thickness progression in coronary artery disease patients without diabetes mellitus. Arterioscler Thromb Vasc Biol May 2004;24(5):930–4. Epub March 4, 2004.

[67] Verhaar MC, Honing ML, van Dam T, Zwart M, Koomans HA, Kastelein JJ, Rabelink TJ. Nifedipine improves endothelial function in hypercholesterolemia, independently of an effect on blood pressure or plasma lipids. Cardiovasc Res June 1999;42(3):752–60.
[68] Ikonomidis I, Lekakis JP, Nikolaou M, Paraskevaidis I, Andreadou I, Kaplanoglou T, Katsimbri P, Skarantavos G, Soucacos PN, Kremastinos DT. Inhibition of interleukin-1 by anakinra improves vascular and left ventricular function in patients with rheumatoid arthritis. Circulation May 20, 2008;117(20):2662–9. http://dx.doi.org/10.1161/CIRCULATIONAHA.107.731877. Epub May 12, 2008.
[69] Sfikakis PP, Papamichael C, Stamatelopoulos KS, Tousoulis D, Fragiadaki KG, Katsichti P, Stefanadis C, Mavrikakis M. Improvement of vascular endothelial function using the oral endothelin receptor antagonist bosentan in patients with systemic sclerosis. Arthritis Rheum June 2007;56(6): 1985–93.
[70] Avgerinou G, Tousoulis D, Siasos G, Oikonomou E, Maniatis K, Papageorgiou N, Paraskevopoulos T, Miliou A, Koumaki D, Latsios G, Therianiou A, Trikas A, Kampoli AM, Stefanadis C. Anti-tumor necrosis factor á treatment with adalimumab improves significantly endothelial function and decreases inflammatory process in patients with chronic psoriasis. Int J Cardiol September 15, 2011;151(3):382–3. http://dx.doi.org/10.1016/j.ijcard.2011.06.112. Epub July 18, 2011.
[71] Siasos G, Tousoulis D, Vlachopoulos C, Antoniades C, Stefanadi E, Ioakeimidis N, Andreou I, Zisimos K, Papavassiliou AG, Stefanadis C. Short-term treatment with L-arginine prevents the smoking-induced impairment of endothelial function and vascular elastic properties in young individuals. Int J Cardiol June 6, 2008;126(3):394–9. Epub June 22, 2007.
[72] Heffernan KS, Fahs AS, Ranadive SM, Eshan A, Patvardhan EA. L-arginine as a nutritional prophylaxis against vascular endothelial dysfunction with aging. J Cardiovasc Pharmacol Ther March 2010; 15(1):17–23.
[73] Tousoulis D, Kampoli AM, Stefanadis C. Diabetes mellitus and vascular endothelial dysfunction: current perspectives. Curr Vasc Pharmacol January 2012;10(1):19–32.

CHAPTER 1.3

Atherosclerotic Plaque

Evangelos Oikonomou, Georgios Latsios, Georgia Vogiatzi, Dimitris Tousoulis
1st Cardiology Clinic, 'Hippokration' General Hospital, National and Kapodistrian University of Athens, School of Medicine, Athens, Greece

Contents

INTRODUCTION

Atherosclerotic plaque constitutes the underlying entity of atherosclerotic diseases including coronary artery disease (CAD). As already we have described in Chapter 1 of this textbook, arteries are consisted of three layers from inner to outer: intima, media and adventitia (Fig. 1.3.1).

Endothelial cells are organized in a monolayer with significant biologic actions. They orchestrate vasoconstriction and vasodilation and they secure inhibition of coagulation keeping blood in a circulating liquid form. Endothelial cells are in alignment with the direction of flow except in areas of flow disturbance such as in bifurcation or branches (Fig. 1.3.2).

Intima is normally extremely thin except at sites of branches and bifurcations and in the proximal segment of the left anterior descending coronary artery where it may appear thick as smooth muscle cells may cluster at one or two layers (Fig. 1.3.3).

Coronary Artery Disease
ISBN 978-0-12-811908-2, http://dx.doi.org/10.1016/B978-0-12-811908-2.00003-9

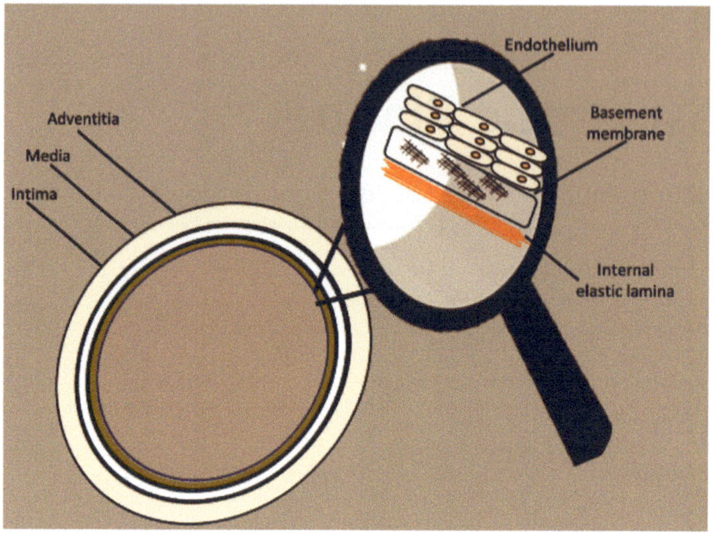

Figure 1.3.1 Schematic representation of the layers of the arterial wall, focus on the details of intima layer.

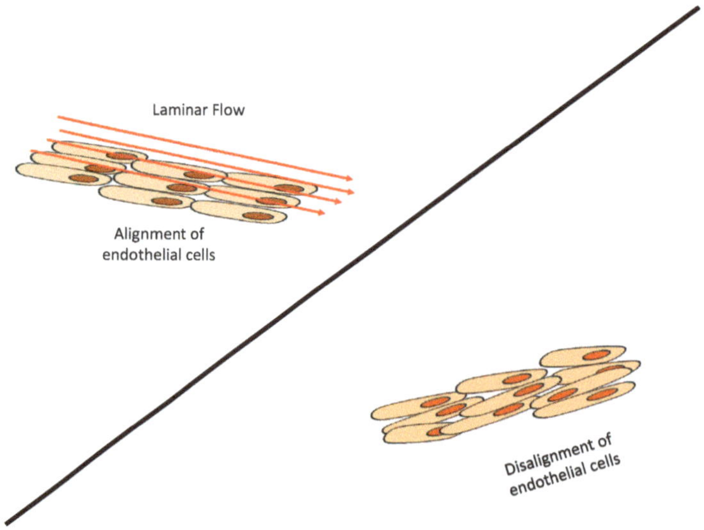

Figure 1.3.2 Flow disturbances affect orientation of endothelial cells.

Of note atherosclerosis is characterized by primary alterations in the intima layer with corresponding changes in the media and adventitia.

GENESIS AND ADVANCEMENT OF ATHEROMATIC PLAQUES

Atheromatic plaques are constitute from accumulation of degenerative materials inside arterial wall. These accumulations are mainly consisted from macrophages, lipid debris, connective tissue and fibrous materials which are calcified in various

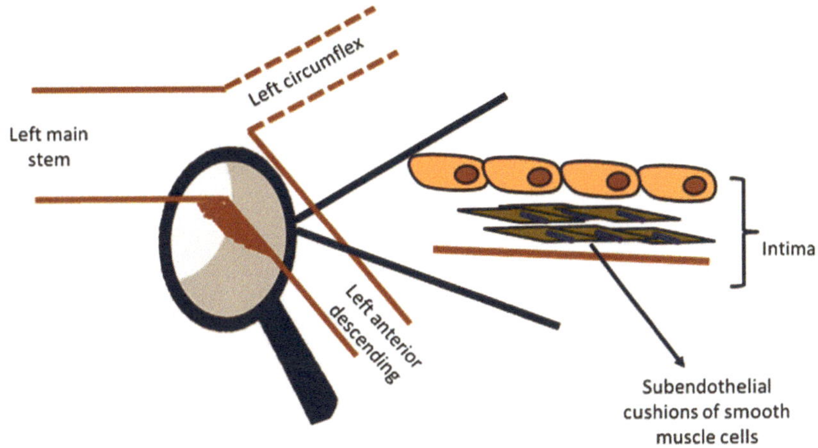

Figure 1.3.3 Subendothelial cushions of smooth muscle cells in regions of intimal thickening.

degree depending on regional wall and systematic properties as well as from the chronicity of the lesions.

Initiation of atheromatic plaque formation cannot definitively define but several post-mortem and animal data suggests that the formation of the early lesions happens in the first or second decade of life.

Lipid Accumulation

Under the appropriate circumstances (lipid rich diet, lipemic serum) lipoprotein particles of minimum diameter passes the endothelial layer and accumulate in the extracellular sub-endothelial space. Subendothelial lipoproteins tent to bind with proteoglycans trapping them and causing increase residency time.

Lipoproteins bind to proteoglycans are susceptible to oxidative modification which may promote following steps in atherosclerosis progression. Of note oxidative modification is enhance by nicotinamide adenine dinucleotide (reduced form)/nicotinamide adenine dinucleotide phosphate (NADH/NADPH) oxidases and myeloperoxidase, which may expressed by vascular cells or mainly by leukocytes infiltrate the subendothelial space [1,2] (Fig. 1.3.4).

Leucocyte Accumulation and Recruitment

Leucocyte recruitment and accumulation constitute another hallmark of atherosclerotic plaque progression and generation. Monocytes and T lymphocytes are the main immune cells implicated in the process of atherosclerosis. However, endothelial cells when intact does not allowed entering of immune cells in the subendothelial space. Interestingly, lipid environment promotes functional and structural changes in endothelial monolayer potentiating adhesion of leukocytes on endothelial surface or even penetration [3].

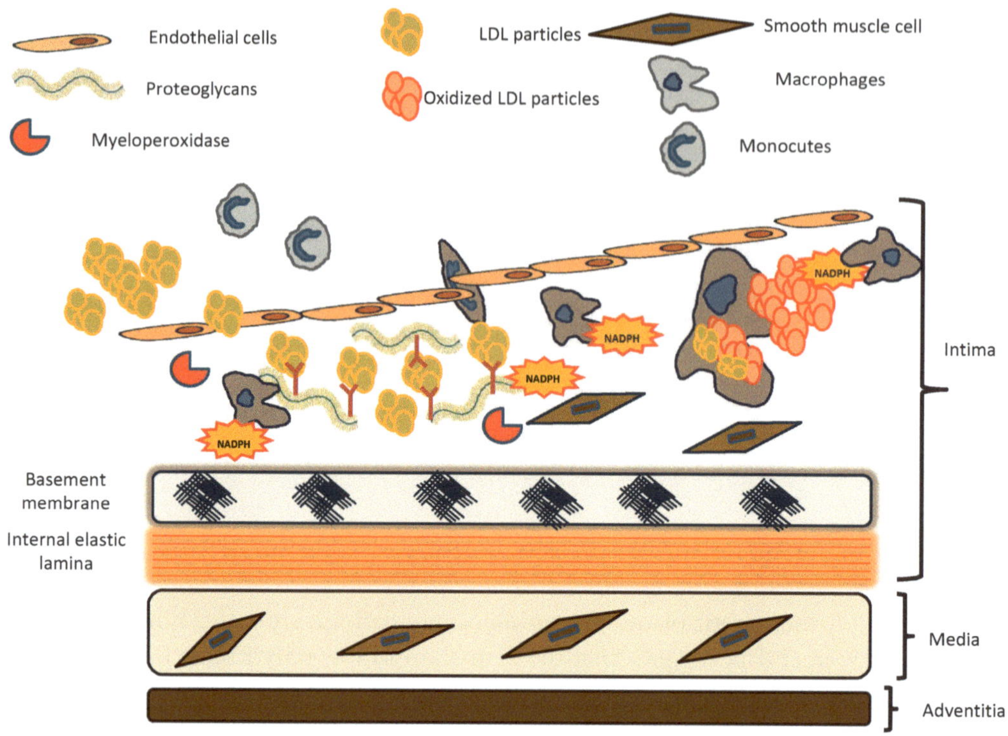

Figure 1.3.4 Mechanisms of subendothelial lipid accumulation and oxidative modification of low-density lipoprotein particles.

Endothelial cell expression of adhesion molecules orchestrate and regulate the recruitment and penetration of immune cells [4] (Fig. 1.3.5).

Adhesion molecules may be part of the immunoglobulin superfamily such as vascular cell adhesion molecule 1 (VCAM-1) and intracellular adhesion molecule 1. These molecules interact with integrins expressed in the surface of immune cells. Selectins constitute the other type of adhesion molecules. They are not present in early atheroma but may have significant role in later stages of atherogenesis as under inflammatory circumstances. Especially E-selectin may adhere polymorphonuclear leucocytes and p-selectin may have a role in platelet adhesion, accumulation and aggregation.

Of course, penetration of leucocytes in the subendothelial space is a distinct procedure from adhesion and it is regulated by proteins named chemokines and cytokines that promote migration. Especially the role of monocyte chemoattractant protein 1 (MCP) has been recognized. MCP-1 is produced by endothelium and by other arterial wall cells (i.e., smooth muscle cells) in response to oxidative stress and inflammation [5,6].

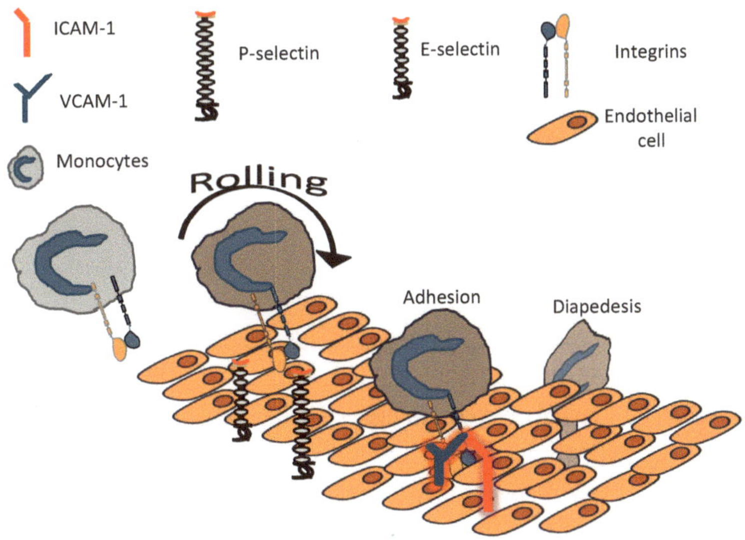

Figure 1.3.5 Mechanisms of adhesion and penetration of immune cells in the subendothelial space.

Foam Cell Formation

Once monocytes engaged in the subendothelial space of the intima layer can absorb lipids (lipoproteins, especially oxidized low-density lipoprotein [LDL] particles) and it is transformed in a lipid rich macrophage namely foam cell. Lipid uptake (endocytosis) by macrophages does not mediated by LDL receptors. Instead scavengers receptors determine excessive lipid accumulation in foam cells.

Foam cells in the intima can replicate, an action mediated by mitogens such as macrophage colony stimulating factor and to a lesser degree by granulocyte macrophage colony stimulating factor [7].

Arterial Networks/Regional Factors

Atherosclerosis is considered a systematic rather a regional disease. In addition, the recognized risk factors associated with atherosclerosis progression and initiation as well as the mechanisms discuss just previously involved in atheromatic plaque formation have a systematic rather a focal nature. However, we cannot ignore the fact that atheromatic lesions tent to cluster in some arterial networks. Moreover, atheromatic lesions mainly happens in the proximal portions of arteries, following sites of bifurcations, where flow is disturbed suggesting a hemodynamic component.

A few hypotheses have been generated regarding the hemodynamic components and mechanisms altering progression of atheromatic plaques. Shear stress seems to play a significant role in these regional differences observed in the rhythm of atheromatic plaque progression. In places of unobstructed laminar flow shear stress is increased interacting

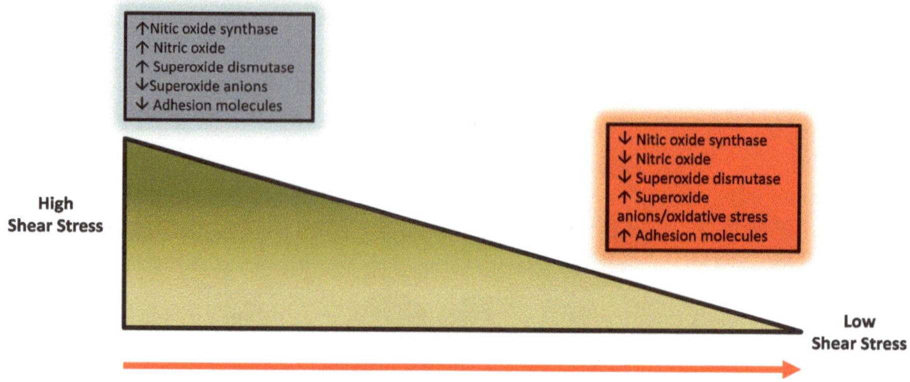

Figure 1.3.6 Impact of low and high shear stress in endothelial cells.

with the endothelial surface favoring favorable alteration and enhancing gene expression regulating superoxide dismutase and nitric oxide synthase production. Opposing changes are observed in cases of disturbed flow with decrease shear stress (Fig. 1.3.6).

EVOLUTION OF ATHEROMATIC PLAQUE

Inflammation and Atherogenesis

During the last two decades, the role of inflammation in the evolution of atheromatic plaque has been established and studied intentionally. As this issue is further discuss in dedicate chapter of this book at this point we briefly goes through the basic concepts of innate and adaptive immunity.

Innate immunity is an inflammatory response independent from inflammatory response. The basic component of this type of immunity consisted from the macrophage foam cells. They produce not only proinflammatory cytokines and chemokines but accumulate and oxidative species [8].

Adaptive immunity is antigen specific. At the atheromatic plaque environment, macrophages, dendritic cells and endothelial cells can present antigens to T cells. Therefore, T cells are activated and produce large amounts of cytokines that further impact atherogenesis [9]. The basic components of these concepts are presented in Fig. 1.3.7.

The Role of Smooth Muscle Cells

As we already have mentioned normally Smooth muscle cells are found in the media layer of arterial wall. Few smooth muscle cell can be observed in the intima early in life. However, as atheromatic plaque evolve beyond recruitment of macrophages and changes in endothelial cells function smooth muscle cells recruited from the media to the intima.

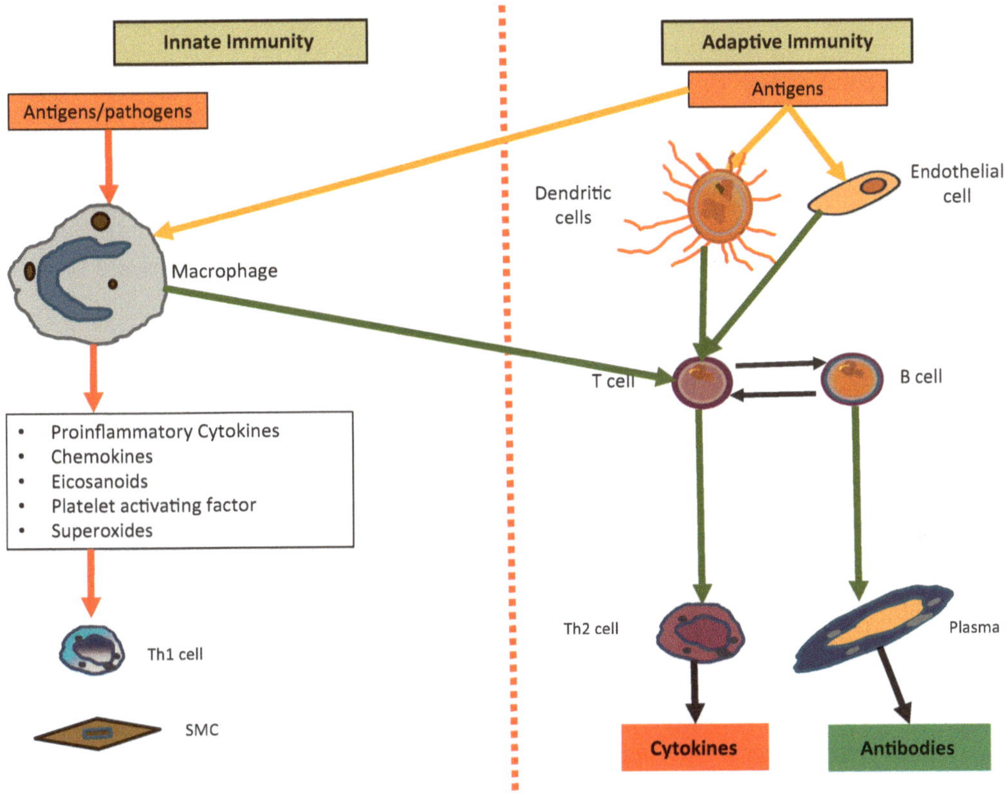

Figure 1.3.7 Simplified concepts of innate and adaptive immunity implicated in the mechanism of atherogenesis.

Smooth muscle cells in the intima are less differentiated exhibited an embryonic phenotype with embryonic myosin isoform and rough endoplasmic reticulum. Concerning replication of Smooth muscle cells although it appears infrequent in stable atherosclerotic plaques in vulnerable plaques may exacerbate contributing to aggravation of atherosclerotic plaques.

Interestingly inflammatory cytokines and immune cells found in the intima may potentiate programmed smooth muscle cells death. Indeed, a balance between smooth muscle cell migration, proliferation and apoptosis regulate the amount of these cells in the evolving atheromatic plaques diverting the functional futures of the plaques to either a vulnerable or stable phenotype [10].

Extracellular Matrix

The volume of an atherosclerotic plaque is much dependent on the extracellular matrix which mainly consist from collagen fiber (type I and III), elastin and proteoglycans. These molecules are produced by smooth muscle cells under the stimuli of chemokines like platelet derived growth factor and transforming growth factor-β.

Of note composition of this extracellular matrix components it depends on a balance between matrix metalloproteinases (MMPs), which catabolized components of this extracellular network, and their inhibitors-tissue inhibitors of MMPs (TIMPs).

MMPs and TIMPs play a major role not only in extracellular matrix degradation but also in mediating cell migration, proliferation, tissue remodeling. Smooth muscle cells migration through a dense layer of extracellular components it depends on this balance between MMPs and TIMPs.

Moreover regulation of extracellular synthesis orchestrate atherosclerotic lesion expands toward the adventitia until a critical point is reached, at which time the lesion can no longer grow outward at the expense of the normal media and then begins to encroach on the lumen [11].

Calcification of the Atherosclerotic Plaque

Calcium mineralization of the lumen in the atherosclerotic lesion of the artery enhances and solidifies plaque creation causing the stenosis of the vessel. The process of atherosclerosis is mostly driven by CD4+ T cells, CD40L, macrophages, foam cells with elevated transcription of many matrix metalloproteinases, osteoblasts, cytokines, selectins, myeloperoxidases, vascular adhesion molecules (VCAM), and smooth muscle cells. Arterial calcification begins from two different types, the medial-elastin dependent and the intimal, both of which are directly connected to atherosclerosis because of osteoblast differentiation of vascular smooth muscle cells. The deposition of minerals in the form of calcium ($Ca(2+)$) begins from the inorganic mineral octacalcium phosphate [$Ca_8H_2(PO_4)_6.5H_2O$] to the form of hydroxylapatite [$Ca_{10}(PO_4)_6(OH)_2$].

ATHEROMA COMPLICATIONS PLAQUE RAPTURE AND THROMBOSIS

The prevalence of acute coronary events is not associated with the degree of stenosis or plaque volume. Interestingly, a significant proportion of acute coronary events happens in atherosclerotic plaques with less than 50% stenosis [12,13]. The question remain why after a long, lethargic period plaques develops vulnerable characteristics leading to thrombus formation, acute lumen occlusion and ischemic events [14].

Traditionally, plaques characteristics predispose to vulnerability have been identified based on pathological studies and are described by the general term "thin-cap fibroatheroma." The main characteristics of such a vulnerable plaque consist of: (1) a large atheromatous core with lipid-rich content, (2) a thin fibrous cap, (3) outward remodeling, (4) infiltration of the plaque with macrophages and lymphocytes, and (5) thinning of the media and decrease in the number of smooth muscle cells [15–17].

Recently, there has been an extensive study for the role of inflammation in the process of atherosclerosis and the evolution of stable atherosclerotic plaques to unstable. Indeed, the association of common community acquired infections and acute coronary syndromes have been well documented in the past. This has been probably associated with the mobilization of vascular inflammatory cytokines, such as soluble vascular cell adhesion molecule-1 and soluble intercellular adhesion molecule-1, intereleukin-6, etc. [18,19]. Moreover, in experimental studies the exposure to lipopolysaccharide may destabilize plaque, whereas the reduction of neutrophils may attenuate plaque rupture [20].

Though, questions still remain for the underline pathology of a vulnerable plaque predispose to thrombus formations. It has long been suggested that acute thrombosis happens in three main occasions.

Plaque Rapture

Rupture of the plaque is the most common cause of coronary thrombosis, which is commonly happens at the shoulders of the plaque. The rapture expose the highly thombogenic lipid core of the plaque to blood circulation causing thrombus formation [21–23]. Notably, as a number of studies based on optical coherence tomography, has revealed that acute coronary syndromes caused by plaque rupture, has adverse prognosis compared to acute coronary syndromes caused by other etiologies [24].

Plaque Erosion

Plaque erosion is a less frequent situation associated with intracoronary thrombus formation and acute coronary events. In this pattern, there is no contact of lipid core with the blood circulation. The dysfunction and the injury of the endothelial layer is the main lesion in this case [23]. This situation is often encountered in women, younger patients, and smokers. Patients usually present with non-ST elevation myocardial infraction [17]. Lately it has been recognized the role of neutrophils and toll like receptor-2 to potentiate endothelial stress, apoptosis and detachment implicating in the erosion of atheromatic plaques [25]. Remarkably, as recent advances have revealed from the imaging of atherosclerotic lesions, in cases of acute coronary syndromes caused by eroded plaques, there is no total occlusion of the coronary lumen as opposed to rupture plaques [23]. Moreover, intravascular ultrasound has shown that in patients with acute coronary syndromes and intact fibrous cap (as in cases of plaque erosion) there is less remodeling and better long-term prognosis [26].

Calcified Nodule

An infrequent mechanism of thrombus formation is a calcified nodule protruding into the coronary lumen. Especially in older patients with hypertension and renal insufficiency, calcified nodule is the underline etiology of an acute coronary syndrome [27]. Lately, Higuma et al. concluded that calcified nodule lesions are characterized by superficial large calcium and negative remodeling [28].

Intraplaque Hemorrhage

Intraplaque hemorrhage is an established mechanism of coronary plaque extension which in some cases may account for acute coronary syndromes [29]. Notably, intraplaque hemorrhage in carotid arteries may explain most of the cases of symptomatic CAD that approaching, according to some reports, the 84%. Moreover, the reported plaque hemorrhage is also high in asymptomatic cases [30,31].

CONCLUSION

Atheromatic plaques constitute the underlying pathology of arterial atheromatic disease. Atherosclerotic plaques are caused on a dysfunctional endothelium by a sequence of multiple pathophysiological events, such as low sear stress, inflammation, oxidative stress leading to accumulation of cholesterol, macrophages, foam cells, smooth muscle cells, mineral crystals and finally to plaque growth. Progression of atheromatic plaques may even disturb blood flow or changes in plaque characteristics may transform a stable plaque to a vulnerable one which may rapture causing acute coronary events.

REFERENCES

[1] Braunwald E, Bonow RO. Braunwald's heart disease: a textbook of cardiovascular medicine. 9th ed. Philadelphia: Saunders; 2012xxiv. 1961.
[2] Kruth HS. Sequestration of aggregated low-density lipoproteins by macrophages. Curr Opin Lipidol 2002;13:483–8.
[3] Muller WA. Mechanisms of transendothelial migration of leukocytes. Circ Res 2009;105:223–30.
[4] Galkina E, Ley K. Vascular adhesion molecules in atherosclerosis. Arterioscler Thromb Vasc Biol 2007; 27:2292–301.
[5] Libby P. Inflammation in atherosclerosis. Nature 2002;420:868–74.
[6] Libby P, Nahrendorf M, Pittet MJ, Swirski FK. Diversity of denizens of the atherosclerotic plaque: not all monocytes are created equal. Circulation 2008;117:3168–70.
[7] Moore KJ, Freeman MW. Scavenger receptors in atherosclerosis: beyond lipid uptake. Arterioscler Thromb Vasc Biol 2006;26:1702–11.
[8] Libby P, Ridker PM, Hansson GK, Leducq Transatlantic Network on Atherothrombosis. Inflammation in atherosclerosis: from pathophysiology to practice. J Am Coll Cardiol 2009;54:2129–38.
[9] Tsimikas S, Brilakis ES, Miller ER, et al. Oxidized phospholipids, Lp(a) lipoprotein, and coronary artery disease. N Engl J Med 2005;353:46–57.
[10] Bentzon JF, Weile C, Sondergaard CS, Hindkjaer J, Kassem M, Falk E. Smooth muscle cells in atherosclerosis originate from the local vessel wall and not circulating progenitor cells in ApoE knockout mice. Arterioscler Thromb Vasc Biol 2006;26:2696–702.
[11] Dollery CM, Libby P. Atherosclerosis and proteinase activation. Cardiovasc Res 2006;69:625–35.
[12] Falk E, Shah PK, Fuster V. Coronary plaque disruption. Circulation 1995;92:657–71.

[13] Motoyama S, Ito H, Sarai M, et al. Plaque characterization by coronary computed tomography angiography and the likelihood of acute coronary events in mid-term follow-up. J Am Coll Cardiol 2015; 66:337−46.
[14] Thygesen K, Alpert JS, Jaffe AS, et al. Third universal definition of myocardial infarction. Eur Heart J 2012;33:2551−67.
[15] Schaar JA, Muller JE, Falk E, et al. Terminology for high-risk and vulnerable coronary artery plaques. Report of a meeting on the vulnerable plaque, June 17 and 18, 2003, Santorini, Greece. Eur Heart J 2004;25:1077−82.
[16] Varnava AM, Mills PG, Davies MJ. Relationship between coronary artery remodeling and plaque vulnerability. Circulation 2002;105:939−43.
[17] Arbustini E, Dal Bello B, Morbini P, et al. Plaque erosion is a major substrate for coronary thrombosis in acute myocardial infarction. Heart 1999;82:269−72.
[18] Nikolopoulou A, Tousoulis D, Antoniades C, et al. Common community infections and the risk for coronary artery disease and acute myocardial infarction: evidence for chronic over-expression of tumor necrosis factor alpha and vascular cells adhesion molecule-1. Int J Cardiol 2008;130:246−50.
[19] Tousoulis D, Antoniades C, Nikolopoulou A, et al. Interaction between cytokines and sCD40L in patients with stable and unstable coronary syndromes. Eur J Clin Invest 2007;37:623−8.
[20] Jaw JE, Tsuruta M, Oh Y, et al. Lung exposure to lipopolysaccharide causes atherosclerotic plaque destabilisation. Eur Respir J 2016;48.
[21] Shimamura K, Ino Y, Kubo T, et al. Difference of ruptured plaque morphology between asymptomatic coronary artery disease and non-ST elevation acute coronary syndrome patients: an optical coherence tomography study. Atherosclerosis 2014;235:532−7.
[22] Toutouzas K, Benetos G, Karanasos A, Chatzizisis YS, Giannopoulos AA, Tousoulis D. Vulnerable plaque imaging: updates on new pathobiological mechanisms. Eur Heart J 2015;36:3147−54.
[23] Saia F, Komukai K, Capodanno D, et al. Eroded versus ruptured plaques at the culprit site of STEMI: in vivo pathophysiological features and response to primary PCI. JACC Cardiovasc Imaging 2015;8: 566−75.
[24] Niccoli G, Montone RA, Di Vito L, et al. Plaque rupture and intact fibrous cap assessed by optical coherence tomography portend different outcomes in patients with acute coronary syndrome. Eur Heart J 2015;36:1377−84.
[25] Quillard T, Araujo HA, Franck G, Shvartz E, Sukhova G, Libby P. TLR2 and neutrophils potentiate endothelial stress, apoptosis and detachment: implications for superficial erosion. Eur Heart J 2015;36: 1394−404.
[26] Yonetsu T, Lee T, Murai T, et al. Plaque morphologies and the clinical prognosis of acute coronary syndrome caused by lesions with intact fibrous cap diagnosed by optical coherence tomography. Int J Cardiol 2016;203:766−74.
[27] Jia H, Abtahian F, Aguirre AD, et al. In vivo diagnosis of plaque erosion and calcified nodule in patients with acute coronary syndrome by intravascular optical coherence tomography. J Am Coll Cardiol 2013;62:1748−58.
[28] Higuma T, Soeda T, Abe N, et al. A combined optical coherence tomography and intravascular ultrasound study on plaque rupture, plaque erosion, and calcified nodule in patients with ST-segment elevation myocardial infarction: incidence, morphologic characteristics, and outcomes after percutaneous coronary intervention. JACC Cardiovasc Interv 2015;8:1166−76.
[29] Kolodgie FD, Gold HK, Burke AP, et al. Intraplaque hemorrhage and progression of coronary atheroma. N Engl J Med 2003;349:2316−25.
[30] Carr S, Farb A, Pearce WH, Virmani R, Yao JS. Atherosclerotic plaque rupture in symptomatic carotid artery stenosis. J Vasc Surg 1996;23:755−65. Discussion 765−756.
[31] Kolodgie FD, Nakazawa G, Sangiorgi G, Ladich E, Burke AP, Virmani R. Pathology of atherosclerosis and stenting. Neuroimaging Clin N Am 2007;17. 285−301, vii.

CHAPTER 1.4

Risk Factors of Atherosclerosis: Pathophysiological Mechanisms

Costas Tsioufis, Emmanouil Mantzouranis, Theodore Kalos, Dimitrios Konstantinidis, Dimitris Tousoulis

1st Cardiology Clinic, 'Hippokration' General Hospital, National and Kapodistrian University of Athens, School of Medicine, Athens, Greece

Contents

Coronary Artery Disease
ISBN 978-0-12-811908-2, http://dx.doi.org/10.1016/B978-0-12-811908-2.00004-0

TRADITIONAL RISK FACTORS

Age

Age is a strong nonmodifiable risk factor for cardiovascular disease (CVD) and atherosclerosis. It depicts the chronic cumulative exposure to all other risk factors [1]. On the other hand, arterial aging is synonymous to arteriosclerosis, an age-related degenerative process that increases arterial stiffness. Arteriosclerosis is a remodeling procedure of the medial layer induced as a response to repetitive hemodynamic stress. Although it is a distinct condition from a histopathological and pathophysiological perspective, arterial aging can predispose, aggravate and accelerate atherosclerotic disease by promoting proatherogenic blood flow conditions (*extended reference in Hypertension section*) [2–4].

Gender

Evidence has demonstrated that within the age spectrum of 40–60 years old incidence of CVD and death attributed to atherosclerosis is at least twice as high in men compared to women. Menopause has been found to be the equalizer. These observations have raised a hypothesis regarding the atheroprotective role of female sex hormones. Indeed, estrogens have been associated with increased high-density lipoprotein (HDL) and decreased low-density lipoprotein (LDL) levels, as well as diverse antioxidant, vasoprotective and antithrombotic effects. However, despite these correlations, research failed to demonstrate a beneficial cardiovascular effect of hormone replacement therapy after menopause [5–7].

Hypertension

Hypertension is a well-established risk factor for atherosclerosis [8–13]. Being an independent risk factor, elevated blood pressure is believed to have a complex precursor role in the development of atherosclerosis and a crucial impact on the rupture of the vulnerable thin capped plaques. Despite the vast amount of evidence linking independently hypertension to atherosclerotic disease, the underlying pathophysiological mechanisms are still not fully elucidated.

Arterial hypertension is a chronic systemic disease involving many different functional and structural macrovascular and microvascular alterations. Vasculature modifications depict an adaptive response to hemodynamic stress and compose a vicious cycle where hypertension is often seen as both a cause and a consequence [14]. On the other hand, emerging evidence links hypertension to oxidative stress and inflammation via diverse biological pathways. The role of hypertension as a risk factor for atherosclerosis is thus mainly determined by its impact on hemodynamic conditions and metabolic burden throughout the vascular network. Endothelial dysfunction induced by chronic inflammation and redox state, is believed to be the compelling link between the two pathological entities [15].

Hemodynamic Mechanisms

Normal pulsatile blood flow subjects arterial wall under two different kinds of hemodynamic strain named "shear stress" and "tensile stress." Both components stimulate endothelial and smooth muscle cells (SMCs) to act as mechanosensors and biochemical transducers sensing complex hemodynamic signals and transforming them into biological responses [16,17]. These responses consist of diverse transcription regulation pathways interfering with changes in cell phenotype, inflammation, production of vasoactive mediators, oxidative stress state, SMC, and leukocyte migration, extracellular matrix production [18]. This hemodynamic-mediated transcription regulation increases metabolic burden and promotes vascular remodeling.

Shear Stress Macrovascular Impact

"Shear stress" is a tangential frictional force implicated on the endothelial layer and created by linear blood flow [2]. Blood flow pattern, which is greatly affected by vessel anatomy, determines the type of shear stress. Normal blood flow, observed at atherosclerosis-resistant straight segments, causes uniform laminar shear stress, whereas turbulent flow observed at bifurcations, curves and branches is characterized by high oscillatory (bidirectional) shear stress [16]. These nonlinear arterial sites are in fact the ones showing a predisposition for the atherosclerotic lesions [1,5].

At the molecular level, research has unveiled two distinct pathways of flow-mediated transcriptional regulation associated with atheroprotective or atheroprone endothelial phenotypes (Fig. 1.4.1). The atheroprotective pathway involves the upregulation of

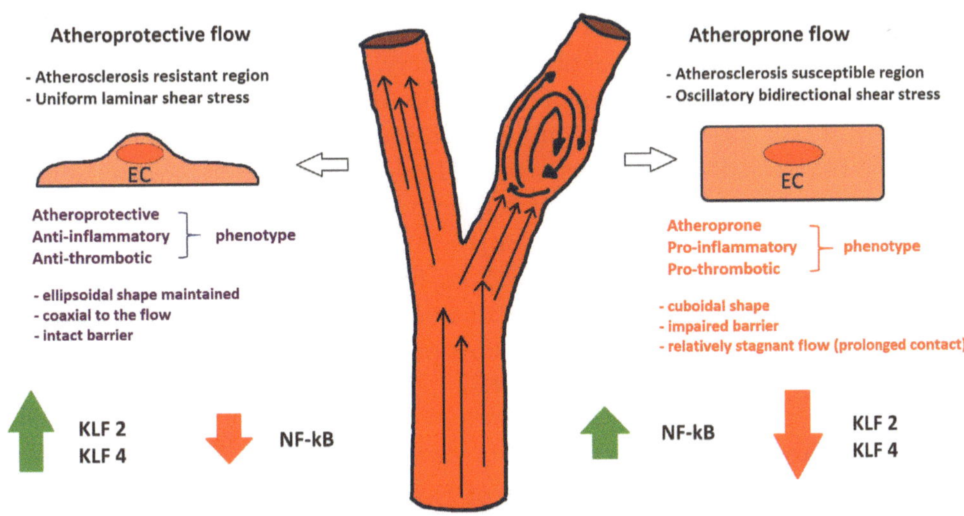

Figure 1.4.1 Main differences between flow-mediated endothelial regulation (bifurcation of carotid artery). *EC*, endothelial cell; *KLF*, Kruppel-like factor; *NF-kB*, nuclear factor kappa B.

transcriptional integrators KLF2 (Kruppel-like factor 2) and KLF4 whereas the athero-prone one involves the activation of the NF-κB (nuclear factor kappa B). Interestingly, these two pathways act antagonistically [16]. Main features of flow-mediated endothelial regulation are presented below:

Atheroprotective flow features:

- Atherosclerosis-resistant sites (straight vessels).
- Uniform laminar shear stress.
- Activation of endothelial nitric oxide synthase (eNOS) augmenting the production of NO, a known vasodilator and endogenous mediator with pleiotropic anti-inflammatory and atheroprotective properties [16].
- NO-mediated and flow-mediated upregulation of transcriptional integrators KLF2 and KLF4. The latter promote an atheroprotective, anti-inflammatory, antithrombotic endothelial phenotype. Endothelial cells maintain an ellipsoidal morphology, stay coaxial to the flow alignment and form a thick intact glycocalyx layer, features that enhance the barrier function against macromolecules, such as lipoproteins [16].
- Antagonism of NF-κB pathway inhibits the expression of adhesion molecules such as vascular cell adhesion molecule-1 (VCAM-1), reduces secreted and membrane-associated chemokines, such as monocyte chemoattractant protein (MCP-1) and prothrombotic mediators [e.g., tissue factor (TF), von Willebrand factor (vWF), and plasminogen activator inhibitor (PAI-1)] (Fig. 1.4.2) [18,19].

Atheroprone flow features:

- Atherosclerosis-susceptible sites (bifurcations, curves, and branches).
- Oscillatory (bidirectional) shear stress.
- Activation of vascular NADPH oxidase, increased production of reactive oxygen species (ROS) and decreased NO bioavailability [20].
- Flow-mediated upregulation of proinflammatory, prothrombotic, atheropromoting NF-κB pathway and antagonism of transcriptional integrators KLF2 and KLF4 [18].
- Endothelial cells transform into a proinflammatory phenotype driven by the activation of NF-κB and obtain a different cuboidal geometry related to apoptosis, impaired endothelial barrier function and increased permeability for lipid macromolecules and leukocytes into the subendothelial layer [16,18]. Barrier function is also impaired by the relatively stagnant flow, which accounts for delayed local blood clearance and thus prolonged contact of potentially atherogenic molecules with the intima [5]. Finally, there is evidence that turbulent flow per se could cause mechanical endothelial injury and dysfunction [18] (Fig. 1.4.2).

Tensile Stress Macrovascular Impact

"Tensile stress," the pulsatile component of blood pressure, describes the repetitive circumferential stretch experienced by the arterial wall as it expands and contracts during each cardiac cycle. It is a force perpendicular to the arterial wall, which is dependent on

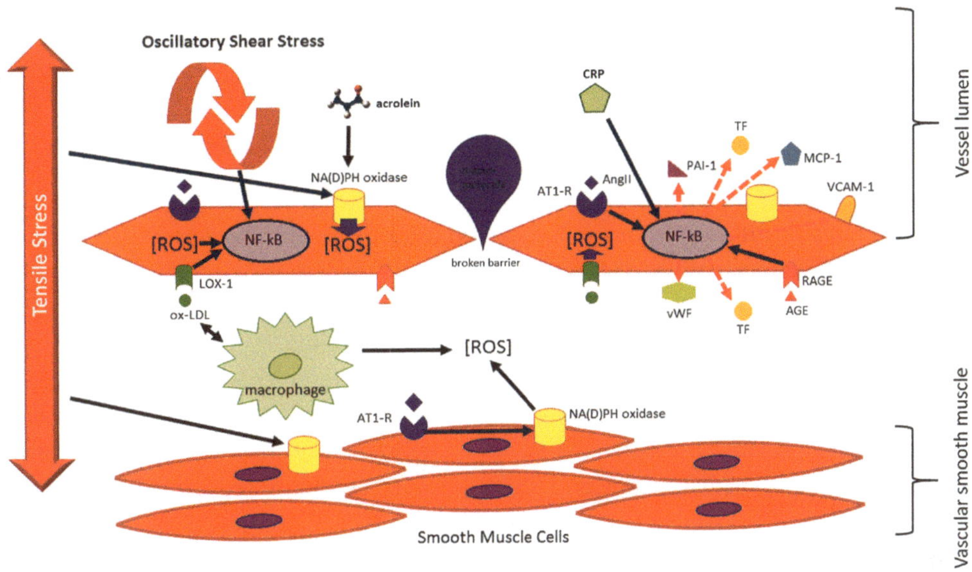

Figure 1.4.2 Schematic of nuclear factor kappa B (NF-κB) pathway activation. *Black arrows* depict triggering mechanisms and *red dashed arrows* show the final proinflammatory, proatherogenic and prothrombotic products of NF-κB cascade. *AGE*, advanced glycation end products; *AngII*, angiotensin II; *AT1-R*, angiotensin type-1 receptor; *CRP*, C reactive protein; *LOX-1*, lectin-like ox-LDL receptor-1; *MCP-1*, monocyte chemoattractant protein; *ox-LDL*, oxidized low density lipoprotein; *PAI-1*, plasminogen activator inhibitor; *RAGE*, transmembrane receptor for AGEs; *ROS*, reactive oxygen species; *TF*, tissue factor; *VCAM-1*, vascular cell adhesion molecule-1; *vWF*, vonWillebrand factor.

the arterial elasticity and assessed by pulse pressure (PP) [21,22]. Hypertension is strongly and independently associated with aortic stiffness, thus wide PP is a frequent finding [2].

Arterial wall structure has the ability to adapt to different tensile stress conditions. Although acute blood pressure changes lead to transient modifications of the vasomotor tone and the lumen diameter, the established disease induces alterations in the geometry and composition of the arterial wall. Medial hypertrophy is a necessary response of large conduit arteries to increase wall-to-lumen ratio in an attempt to normalize tensile stress compensating for the lack of myogenic constriction ability. On the other hand, muscular resistance small arteries (<100 μm) do not need to undergo hypertrophic remodeling because they can easily normalize tensile stress by myogenic constriction and lumen diameter reduction [23]. Arterioles in the middle zone (100−300 μm) that show some compliance features undergo a combination of the above alterations [2,14].

Medial hypertrophy features three major atheropromoting consequences:

1. Narrowing of the lumen can itself further aggravate hemodynamic conditions.
2. A thicker media is translated into longer distance for oxygen diffusion causing inevitably local hypoxia and promoting oxidative stress [24].

3. SMCs gradually transform from contractile muscle cells into a secretory and proliferating phenotype. The latter is responsible for an increase in collagen and elastin components of the extracellular matrix resulting in an increase in not only arterial wall thickness but also vascular stiffness [14].

The Role of Vascular Stiffness

Regarding vascular stiffness, evidence has shown its frequent coexistence with atherosclerotic disease. Their causative relationship or simple coexistence remains a matter of debate. On the contrary, the proposed theory of a vicious cycle is widely accepted. The part of aortic stiffness in the cycle is summarized below:

- Adverse effects of aortic stiffness further expand into the microcirculation, where the wide PP means more pulsatile pressure in vessels designed for a thoroughly regulated flow [25].
- Vascular stiffness leads to increased shear stress fluctuations reducing NO bioavailability [23].
- Wide PP is a marker of higher tensile stress. Interestingly higher values of PP have been associated with elevated markers of inflammation such as C reactive protein (CRP) [22].

Tensile Stress and Endothelial Dysfunction

Apart from medial remodeling, increased circumferential stress is believed to directly promote oxidative stress and inflammation. Vascular reduced nicotinamide dinucleotide hydrogen phosphate (NADPH) oxidase is an enzyme responsible for the production of ROS. It can be found in both endothelial and SMCs and its function is regulated by many biochemical, mechanical and genetic stimuli, one of which has proven to be tensile stress [26] (Fig. 1.4.2). Endothelial cells seem to exhibit a significant proinflammatory response stimulated by cyclic strain considering the recent data showing increased gene expression of proinflammatory chemokine MCP-1, intracellular adhesion molecule-1 (ICAM-1), VCAM-1, E-selectin, and matrix metalloproteinase-2 (MMP-2) in endothelial cells exposed to increased tensile stress [22].

Microvascular Alterations

The potential pathophysiological mechanism that links microvascular hemodynamic strain with the development of atherosclerotic disease is the role of microvasculature per se. Acting as resistance vessels and defenders against the deleterious effects of turbulent pulsatile blood flow on downstream capillaries, arterioles adapt to harmful hemodynamic conditions by myogenic vasoconstriction. The latter leaves the arterial wall properties intact, but leads to a significant narrowing of the lumen and a subsequent progressive reduction in capillary perfusion. Reduced perfusion initiates a capillary rarefaction process. At first, rarefaction is limited to a functional pattern (increased number of nonperfused capillaries) but overtime it progresses toward the permanent anatomical disappearance of nonperfused capillaries. This structural rarefaction is a consequence of the loss of beneficial shear stress into the nonperfused capillaries which inevitably reduces NO bioavailability. These conditions are in favor of a redox state environment leading to endothelial cell apoptosis. What is more, small resistance arteries chronic adaptation to hypertension promotes a remodeling process, which may be responsible for long-term systemic vascular resistance augmentation and enhanced aortic stiffness [14].

Contribution of Vasa Vasorum

There is a direct link between large vessels and microvasculature, which seems to play a critical role in atherosclerotic development. Vasa vasorum are the arterioles responsible to nourish large arterial wall as an alternative source to intraluminal oxygen diffusion. Considering the fact that vasa vasorum display the exact same architecture as arterioles of peripheral tissues, they adapt to hemodynamic stress similarly. Vasa vasorum rarefaction and concomitant local hypoxia initiate a defective neovascularization process which produces faulty vessels that allow the intrusion of proatherogenic plasma components into the large artery wall. Oxidative stress, chronic inflammation, enhanced arterial stiffness and endothelial dysfunction compose the spectrum of consequences but this time directly upon the target organ of atherosclerotic disease [27].

The Complex Role of Angiotensin II

Renin—angiotensin system (RAS) activation has a vital role in the pathophysiology of hypertension. Angiotensin II (Ang II), a major component of RAS, is a well-documented vasoconstrictor. Apart from this role, increasing evidence unveils its association with atherosclerosis development and acceleration as well, implying mechanisms independent of its hemodynamic contribution. Endothelium is the site of production for both Ang II and NO and it also happens to be the place of their antagonistic interactions. The latter are believed to define the course of atherosclerotic disease in the presence of hypertension.

First, Ang II seems to interfere with the expression and uncoupling of eNOS, whereas NO downregulates the expression of AT-1 receptor, via which Ang II initiates all of its atheropromoting actions [28].

A suggested pathway toward oxidative stress is the AT-1-induced activation of vascular NAD(P)H oxidase (Fig. 1.4.2) and xanthine oxidase, an endothelial enzyme which catalyzes the oxidation of hypoxanthine and xanthine to provide uric acid and ROS [29].

As far as eNOS is concerned, under specific conditions that induce enzyme "uncoupling," it provides ROS instead of NO. Uncoupling is usually a consequence of the lack of critical co-factors L-arginine and tetrahydrobiopterin (BH4). Ang II-induced hypertension can reduce BH4 levels due to increased oxidation to its inactive form

7,8-dihydrobiopterin, and also via downregulation of the dihydrofolate reductase (DHFR), which could reverse this procedure [30,31].

Finally, Ang II was found related to the activation of NF-κB pathway and therefore the total spectrum of its proinflammatory and prothrombotic effects (Fig. 1.4.2) [18,28].

Smoking

Smoking has long been established as a traditional modifiable risk factor for CVD [8,10,12,32]. Active smoking is recognized as the leading cause of preventable death [33] accounting for up to 10% of all cardiovascular deaths according to WHO [34] and for approximately 6,000,000 deaths per year [35] worldwide. More shocking is the fact that the 1.2 billion of smokers globally expose 40% of children and 34% of nonsmoker adults to the deadly effects of secondhand smoke [36] considering the fact that in the United States more than 50,000 cardiovascular deaths annually are attributed to passive smoking [33].

Tobacco usage is recognized as the most complex risk factor of atherosclerosis [35]. Its complexity is a combination of many facts. First, tobacco smoke contains more than 4000 identified and 100,000 unidentified components [37]. What is more, these substances undergo a modification process by human detoxification systems of the CYP family. We also know that negative effects of smoking are generally dose dependent [35]. Finally, there is evidence that genetic background can further influence the susceptibility to the adverse effects of smoking [35,38,39].

Smoking plays a crucial role in both the development and aggravation of the disease. Cigarette smoke is believed to exert its adverse effects by promoting endothelial dysfunction through oxidative stress and inflammation, modifying the lipid profile and inducing a highly prothrombotic environment [37,40]. Interestingly, current evidence implies a synergy with other risk factors both traditional (hypertension, dyslipidemia, diabetes mellitus [DM], menopause) and emerging (homocysteine, fibrinogen, infection) [1,41]. A fortunate observation is that all relevant adverse effects are rapidly reversed with cessation, equating the CV risk of the former smoker to the one of a longtime nonsmoker within 3 years of abstention [1].

Hemodynamics

Cigarette smoke exhibits significant hemodynamic effects, mediated mainly by nicotine and associated with the stimulation of sympathetic nervous system and the increased levels of blood catecholamines. Acute hemodynamic effects of this adrenergic response include: increased heart rate, blood pressure and myocardial contractility, as well as vasoconstriction [41].

Although the particular effects are quite similar to those following mild exercise, one should not forget that, in the setting of chronic smoking, they take place in the presence of atherosclerotic lesions, in which case they are likely to aggravate coronary heart disease and even provoke an acute coronary event [42].

Apart from the obvious hemodynamic role of nicotine in the aggravation of atherosclerosis, the exact same effects may be involved in the initiation of the disease as well. Acute heart rate accelerations and blood pressure elevations cause blood flow disturbances and trigger mechanisms of endothelial dysfunction related to tensile and shear stress similar to the ones discussed in hypertension [42].

The Role of Smoking in the Initiation of Atherosclerosis

Endothelial Dysfunction

The association between smoking and endothelial dysfunction was established after a vast amount of evidence demonstrated a reversible impairment of flow-mediated dilatation [35]. Two possible ways link smoking with endothelial injury:

- Acute and frequent hemodynamic fluctuations act as mechanical stimuli on vascular cells triggering biological responses that induce oxidative stress and inflammation (discussed in Hypertension section).
- Chemical components of smoke interfere with molecular signaling and gene expression pathways that reduce NO bioavailability and promote oxidative stress and inflammation.

Oxidative Stress and Inflammation

Sources of Free Radicals in Chronic Smoking

- ROS included in tobacco smoke.
- Circulation or in situ activated macrophages and neutrophils.
- Endogenous production of ROS.

Tobacco smoke has a high concentration of superoxide anion. Superoxide interacts with NO reducing its bioavailability. Furthermore, the same interaction provides peroxynitrite, a strong oxidative radical [38,39]. Smoking can elevate peripheral blood leukocyte count by up to 25% [38] usually in favor of the monocyte fraction [41]. Furthermore, tobacco chemical components interact with proinflammatory pathways which promote leukocyte recruitment, adhesion and activation at the site of the new-forming lesions. Both circulating monocytes and activated macrophages produce ROS and contribute to the redox state.

Endogenous production systems of ROS are eNOS, NADPH, xanthine oxidase. In the setting of chronic smoking, eNOS is paradoxically upregulated. However, relevant trials demonstrated decreased activity of the enzyme due to smoking-induced acetylation. Even worse, eNOS entering the vicious cycle of redox state is susceptible to enzyme uncoupling, due to oxidation of BH4 cofactor. Uncoupled eNOS displays an altered function providing superoxide instead of NO [30,35,43–45].

Unsaturated aldehydes of tobacco smoke, such as acrolein and crotonaldehyde, display remarkable stability and thus they can exert their effects across a large extent of vasculature. By reaction with thiol groups, acrolein is believed to activate NADPH and xanthine oxidases resulting in the increase of endogenous ROS production (Fig. 1.4.2) [35,39,46].

Apart from comprising a source of ROS, cigarette smoke can also impair antioxidant defenses of the organism enhancing its susceptibility to oxidative stress. Paraoxonase is an enzyme that protects LDL from oxidation and its activity is found diminished in the serum of smokers. Furthermore, smokers seem to lack the antioxidant effects of vitamins (C, E, b-carotene), due to decreased dietary intake and potentially smoke-induced deactivation [41].

Connections Between Redox State, Inflammation and Thrombosis Oxidative environment of chronic smoking offers compelling links with proinflammatory and prothrombotic mechanisms via activation of proinflammatory NF-κB pathway [35,39,43]. Indeed, chronic smoking has been associated with a state of chronic inflammation, as proven by the increased plasma levels of neutrophils and inflammation markers, such as C-reactive protein, interleukin 6 (IL-6), and tumor necrosis factor-α [35,38].

The Role of Smoking in the Aggravation of Atherosclerosis

Plaque Destabilization

Except for the already described nicotine-mediated hemodynamic mechanisms precipitating to plaque disruption and acute thrombosis, chronic smoking appears to also act as a mediator of plaque vulnerability. A smoker's plaque is characterized by a lipid-rich necrotic core. The highly oxidative setting results in the apoptosis of migrated SMCs, endothelial cells and leukocytes, releasing oxidized phagocytosed lipids and proteolytic enzymes. On the other hand, ROS, acrolein and nicotine co-operate to increase the expression of MMPs and attenuate the activity of TIMPs. The concomitant degradation of extracellular matrix proteins transforms rapidly an otherwise stable disease to an unstable rupture-prone type of atherosclerosis [35,39].

Another feature of atherosclerotic plaques after chronic smoking is the frequent intraplaque hemorrhage deriving from immature, leaky neovessels [39]. The latter depict a consequence of pathological vasa vasorum neovascularization induced by nicotine and enhanced by MMPs [43].

Promotion of a Pro-thrombotic Setting

Platelet Adhesion and Aggregation Smokers exhibit an increase in platelet activation and adhesiveness as the number of newly formed reticulated platelets and mean platelet volume are found significantly increased in their serum [43]. Evidence has also demonstrated an increase in plasma concentrations of thromboxane B2, platelet factor-4, and thromboglobulin after smoking [41]. Smoking-induced oxidative stress is also believed to play a critical role in platelet aggregation through the triggering of NF-κB pathway [43]. Initiation of hemostasis can be attributed to the local release of TF from damaged endothelial cells, as well as the elevated levels of circulating TF observed in smokers [39].

Finally, experimental data reveal a significant increase of cyclooxygenase-2 (COX-2) activity in smokers. Normally COX-2 is responsible for the production of proinflammatory prostaglandin E2 (PGE-2), whereas COX-1 provides TXA2, a mediator of platelet activation. However, there is a chance that the upregulated COX-2 can contribute to the TXA2 production [47]. COX-1 and COX-2 can also generate PGI-2, which antagonizes TXA2 and inhibits platelet activation. However, both PGI-2 synthesis and platelet response to it were found impaired in the setting of chronic smoking [39].

Coagulation cascade and Fibrinolysis Smoking seems to promote a hypercoagulable state by altering the balance between anticoagulant and procoagulant factors towards the latter. Procoagulant factors such as von Willebrand factor and thrombomodulin are overgenerated in parallel with a rise in the production of thrombin and fibrinogen. On the other hand, smoking dramatically impairs fibrinolysis through an increase in the levels and activity of plasminogen activator inhibitor 1 (PAI-1) and a parallel inhibition of tissue-type plasminogen activator. Regarding blood viscosity, the significant increase observed in smokers is believed to result from the elevated hematocrit and the high concentration of circulating fibrinogen. The raised white blood cell count and the aggregation phenomenon of red and white blood cells may further contribute to blood viscosity [39].

DM and Insulin Resistance

Another strong traditional risk factor for CVD is DM. Although DM develops often in coexistence with other well-established risk factors, such as dyslipidemia and hypertension, diabetic patients exhibit a higher level of cardiovascular risk unexplained by assessing these factors separately. Apart from that, diabetic coronary artery disease (CAD) tends to manifest in a more severe multivessel pattern. Diabetes appears to also significantly worsen the prognosis after percutaneous coronary interventions. In light of the foregoing, DM is considered as a "coronary risk equivalent" rather than a risk factor [1,48].

DM has a critical impact on both pathogenesis and progression of atherosclerosis, as the latter contains the primary pathophysiological process underlying CAD. Diabetic proatherogenic effects involve complex mechanisms based on the three main metabolic disorders of the disease: hyperglycemia, insulin resistance (IR), and diabetic dyslipidemia [48].

The Role of DM in the Development of Atherosclerosis

Kick off in the development of atherosclerosis is once again the tripod "endothelial dysfunction—oxidative stress—inflammation." DM begins to play a role as soon as the aforementioned metabolic abnormalities promote major proinflammatory pathways, oxidative stress and endothelial dysfunction.

Hyperglycemia

Hyperglycemia is considered as the signature metabolic disturbance of DM. Glucose excess exerts critical adverse effects on the homeostatic balance of endothelial function. Four main pathways are described and all of them are associated with oxidative stress, inflammation and acceleration of atherosclerotic plaque formation.

The Polyol-sorbitol Pathway In the polyol pathway, aldose reductase (AR) catalyzes the metabolism of glucose to sorbitol and sorbitol dehydrogenase catalyzes the metabolism of sorbitol to fructose. The first step is accompanied by the oxidation of NADPH to NADP+ and the second by the reduction of NAD+ to NADH. Excessive glucose results in the depletion of NADPH, which is an essential cofactor in the production of glutathione. The latter being a major antioxidant, it becomes evident how the polyol-sorbitol pathway promotes oxidative stress in the setting of hyperglycemia [48].

$$\text{Glucose} \xrightarrow{\textbf{NADPH}\rightarrow\textbf{NADP+}} \text{Sorbitol} \xrightarrow{\text{NAD+}\rightarrow\text{NADH}} \text{Fructose}$$

The Hexosamine Pathway In the hexosamine pathway, fructose-6-phosphate is converted by L-glutamine:D-fructose-6-phosphate aminotransferase to uridine diphosphate (UDP)-*N*-acetyl glucosamine. The latter promotes O-linked glycosylation of transcription factor Sp-1, which becomes hyperactive and causes increased production of transforming growth factor-b (TGF-b) and PAI-1 [48,49].

$$\text{Fructose} - 6 - \text{phosphate} \xrightarrow{GFAT} \text{Glucosamine} - 6 - \text{phosphate}$$
$$+ \textbf{UDP-}\boldsymbol{N}\textbf{-acetyl glucosamine}$$

Nonenzymatic Glycosylation and Advanced Glycation End Products Nonenzymatic glycosylation is a physiological reaction between glucose and proteins of the arterial wall giving birth to advanced glycation end products (AGEs). AGEs naturally accumulate on the vessel wall following the natural vascular aging. However, excessive glucose can radically accelerate this procedure. AGE formation can be accelerated either directly by glucose-mediated mechanisms or indirectly as a result of redox state-mediated mechanisms. Proatherogenic adverse effects are exerted via two pathways depending on the involvement or not of cellular receptors.

Non-receptor mediated mechanisms: AGEs have the ability to diffuse out of cells. Studies have demonstrated that diffused AGEs interact with extracellular matrix proteins and circulating proteins modifying their structure and altering their function.

Regarding the extracellular matrix, the most affected molecules seem to be type 4 collagen, laminin, and fibronectin, which are found overexpressed under the influence of hyperglycemia. AGE-promoted alterations are irreversible and render the aforementioned molecules resistant to proteolysis. The latter results in the fibrosis of the vessel wall and increased arterial stiffness.

As far as circulating proteins are concerned, apolipoprotein B is the most critically affected representative. AGE-mediated modification of ApoB increases the uptake of LDL particle by macrophages and excessive glycation of LDL renders the latter more susceptible to oxidation. Both effects significantly enhance the proatherogenic potential of LDL [48].

Receptor-mediated mechanisms: Transmembrane receptor for AGEs (RAGE) is present in macrophages, endothelial cells and SMC. Binding of AGEs activates the two major proinflammatory cascades of protein kinase C (PKC) and NF-κB (Fig. 1.4.2).

Protein Kinase C Pathway PKC is a member of protein kinase enzyme family, which regulates the function of other proteins via phosphorylation. PKC can be activated in endothelial cells either directly by hyperglycemia-induced triggering of PKC activator diacylglycerol or indirectly by RAGE and polyol-sorbitol pathways. It promotes a diverse spectrum of intracellular signaling responses associated with the development of atherosclerosis, such as:

- Enhanced growth factor transcription:
  - Increased expression of TGF-b that promotes extracellular matrix production.
  - Stimulation of vasa vasorum pathological angiogenesis.
- Activation of a proinflammatory setting via expression of cytokines and adhesion molecules.
- Promotion of oxidative stress by triggering NADPH oxidase.
- Impairment of fibrinolysis through increased expression of PAI-1.
- Triggering apoptosis of endothelial cells resulting in increased endothelial permeability.
- Proliferation and hypertrophy of vascular SMC (VSMC) accelerating the plaque formation.

More Hyperglycemia-Mediated Effects Apart from the above distinct pathways, hyperglycemia can promote redox state via other mechanisms. In the setting of DM, glucose is prone to auto-oxidation and can also induce the production of ROS by the mitochondrial chain. Furthermore, it is believed that excessive glucose may inhibit eNOS, which is already exposed to the "uncoupling" effects of the oxidative environment. Antioxidant systems seem the last resort of the organism, although they are also compromised in diabetics.

Glutathione is depleted due to polyol-sorbitol pathway and vitamins E and C are found either in reduced levels or inactive due to the effects of glycation [48].

IR and Diabetic Dyslipidemia

Insulin regulates carbohydrate, lipid and protein metabolism and interferes with DNA replication, protein synthesis, and enzyme activity. Under normal circumstances insulin plays a beneficial role featuring major antithrombotic, anti-inflammatory, and antiatherogenic effects, which are negated in a state of IR. IR is associated with the development of atherosclerotic disease even in the absence of established DM.

When insulin binds to its receptor, two pathways can be triggered. The phosphoinositol-3 kinase/Akt pathway (PI(3)K/Akt) which regulates the insulin-dependent glucose uptake and the mitogen-activated protein kinase (MAPK) pathway, which accounts for the mitogenic effects of insulin. PI(3)K/Akt pathway is also linked with eNOS activity and SMC differentiation. In the setting of IR (impaired glucose tolerance, type 2 DM, obesity) hyperglycemia results in hyperinsulinemia. Although MAPK pathway stays intact and overstimulated, PI3K/Akt is significantly impaired. Protective effects of insulin are replaced with proatherogenic ones, such as increased production of proinflammatory cytokines and growth factors, SMC proliferation and migration, reduced fibrinolysis and reduction in NO synthesis [48].

IR is also associated with a proatherogenic lipid profile resulting from the resistance of lipoprotein lipase (LPL) to the suppressive effects of insulin. Hence, IR conditions are characterized by deficient removal of very LDL (VLDL) from peripheral tissues accompanied by increased lipolysis and release of free fatty acids (FFAs) from the adipose tissue. FFAs trigger hepatic VLDL synthesis further increasing its concentration and worsening the lipid profile.

Diabetic dyslipidemia presents with a distinct pattern characterized by hypertriglyceridemia, low levels of HDL and increased small dense LDL particles, a combination known as the "lipid triad." Overt diabetic disease is not essential for the manifestation of dyslipidemia, because as already mentioned IR is the primary cause. Small dense LDL particles are considered as the most proatherogenic lipid molecules because of their reduced affinity for LDL receptor, the easier invasion into the arterial wall and their high susceptibility to oxidation [48].

The Role of DM in the Aggravation of Atherosclerosis

Destabilization of Atherosclerotic Plaque

Metabolic setting in DM has a significantly negative impact on plaque stability. Redox state and inflammatory burden result in overexpression of MMPs and inhibition of MMPIs. Apoptosis of endothelial cells, SMC and macrophages releases proteolytic enzymes and further increases oxidative burden. Severe degradation of extracellular matrix and fibrous cap is inevitable and combined with an ox-LDL-rich necrotic core render the atherosclerotic lesion extremely susceptible to abruption. Last but not least, vasa vasorum display an accelerated but faulty angiogenetic process resulting in intraplaque hemorrhages and thus enlargement of the core [48].

Prothrombotic State

Platelet Activation and Aggregation It has been demonstrated that hyperglycemia promotes platelet activation and aggregation via surface protein glycation and interference with intraplatelet calcium homeostasis. Platelets exposed to the impact of hyperglycemia adopt a stellate shape and overexpress surface glycoproteins Gp Ib, Gp IIb/IIIa. Both effects enhance platelet adhesion and aggregation. On the other hand, IR accounts for reduced production of the potent platelet antiaggregants NO and PGI2.

Coagulation and Fibrinolysis DM induces a procoagulant state by shifting the homeostatic balance of the coagulation cascade. Hyperglycemia/IR-mediated mechanisms upregulate the expression of factor VII, thrombin and vWF, whereas at the same time the levels of antithrombin III (ATIII) and protein C are significantly reduced. Additionally, oxidative stress and inflammation account for the increased release of TF at the lesion sites.

Prothrombotic environment is further maintained by the impairment of fibrinolytic mechanism due to overexpression of PAI-1 via MAPK, PKC, and hexosamine pathways [48].

Dyslipidemia

Dyslipidemia is one of the strongest risk factors for the development of atherosclerotic disease. All known components of an abnormal lipid profile, such as high levels of total cholesterol, LDL, and triglycerides (TG) and low levels of HDL, have displayed independent associations with CVD and atherosclerosis [1,8].

All lipoprotein subclasses consist of TGs, cholesterol and apolipoproteins and differ in size, composition, origin and function. Liver produces and secretes VLDL particles, which consist of apolipoprotein B assembled with a high concentration of TGs and a much lower of cholesterol. Circulating LPL hydrolyzes TGs and releases FFAs to supply tissues and renew adipose storage. As VLDL undergoes consecutive hydrolyses, the ratio of TG/cholesterol content decreases and the lipoprotein changes subclass to IDL and finally LDL [50]. HDL is known for its atheroprotective potential [51].

Role of LDL

In the setting of oxidative stress, lipid molecules containing polyunsaturated fatty acids in circulating LDL get oxidized. ox-LDL particles invade the endothelial layer and either bind on endothelial lectin-like oxidized low-density lipoprotein receptor-1 (LOX-1) or get engulfed via scavenger receptors by the macrophages. Incorporation of oxidized cholesterol transforms the latter into foam cells, the hallmarks in plaque formation. Activation of foam cells and LOX-1 binding are recognized as the key step in the initiation of atherosclerotic disease because they exert diverse proatherogenic adverse effects, such as:

- Activation of NF-κB proinflammatory pathway (Fig. 1.4.2).
- Promotion of migration, proliferation, and apoptosis of SMCs.
- Impairment of NO production [18,19,50,52].

Role of HDL

HDL is believed to exert its atheroprotective effects via three main mechanisms:

1. The Reverse Cholesterol Transport process

   Nascent HDL particles scavenge free cholesterol from peripheral tissues and possibly atherosclerotic lesions. Lecithin—cholesterol acyltransferase (LCAT) converts free cholesterol into cholesteryl ester and matures HDL particles. Cholesteryl esters are afterwards exchanged by cholesteryl ester transfer protein (CETP) to VLDL and LDL for TGs. The remaining cholesteryl esters are directly delivered to the liver by HDL [51].

2. Anti-inflammatory properties

   It has been demonstrated that HDL can interfere and inhibit NF-κB pathway. Furthermore, HDL can bind and regulate sphingosine-1-phosphate (S1P), a lipid mediator with anti-inflammatory properties. The result is suppression of the inflammation by an increased ratio of Treg/T1 helpers [51,53].

3. Antioxidant properties

   HDL exhibits a significant antioxidant potential, as it can transfer and metabolize lipid peroxides. In particular, HDL particles absorb ox-LDL and degrade it via their intrinsic enzymes paraoxonase, LCAT and lipoprotein-associated phospholipase A2. In addition, HDL is believed to induce the expression and the activation of eNOS via triggering the PI3K/Akt signaling pathway and increase the production of NO [51,54].

Role of TG

TGs have not been observed in atherosclerotic lesions nor have a direct proatherogenic effect. However, they have been associated with CVD and atherosclerosis because they represent the concentration of triglyceride rich lipoproteins (TRL), such as VLDL and its remnants [55,56].

TRL remnants are formed from successive hydrolyses of the TG-rich VLDL by LPL. Although the large size of VLDL prevents endothelial invasion, TRL remnants are much smaller and cross the barrier. Notably, unlike LDL, they do not need to undergo oxidation in order to get engulfed by macrophages and they carry more cholesterol. Thus they form foam cells with the same, if not more, ease [57].

What is more, TG-rich lipoproteins can affect other lipoprotein subclasses as well. Their elevated levels seem associated with a reduction in HDL levels and an increase in small dense LDL particles. This is the consequence of hypertriglyceridemia-induced triggering of CETP. The latter transfers TGs from TRL to HDL and LDL in exchange for cholesteryl esters. As a result, TRLs increase their cholesterol load, whereas HDL and LDL their TG content. Increased TG content means more LPL processing and catabolism, resulting in HDL with shorter half-lives and the extremely atherogenic small dense LDL particles [55].

ApoC-III consists a glycoprotein synthesized by the liver and intestines and a major component of VLDL, LDL, and TRL remnants. ApoC-III atherogenicity derives from the fact that it can activate NF-κB and induce the hepatic secretion of VLDL [56].

EMERGING RISK FACTORS

Homocysteine

The first association between homocysteine and CVD was attempted in 1969 by Kilmer McCully who noticed severe extensive atherosclerotic disease to infants with a rare genetic disorder of methionine metabolism [58]. Nowadays it is estimated that elevated plasma levels of homocysteine account for a 10% of the population's risk for atherosclerotic disease and that 40% of patients with CVD exhibit such high levels [59,60].

Homocysteine is a nonessential, sulfur-containing amino acid provided by demethylation of the essential amino acid methionine. Normal values of homocysteine are between 5 and 15 μmol/L in healthy adults. Hyperhomocysteinemia is classified as moderate (15–30 μmol/L), intermediate (31–100 μmol/L), and severe (>100 μmol/L) [61]. The latter, fortunately resulting only at rare genetic disorders, accounts for a 50% chance of developing a premature CVD event if left untreated. Mild hyperhomocysteinemia, on the other hand, is a frequent finding in the general population and shows a significant but modest association with CVD [62]. Hyperhomocysteinemia is caused by genetic defects, vitamin deficiencies, drugs, renal dysfunction, chronic hepatic disease and other systemic diseases [61].

Oxidative Stress and Endothelial Dysfunction

Elevated plasma levels of homocysteine are associated with a significant deterioration of redox state. A quite obvious explanation would be that the thiol group of homocysteine undergoes auto-oxidation and generates ROS. However, we know that cysteine, which is found in plasma at much higher concentrations and gets easily auto-oxidized is not correlated with endothelial dysfunction. Having rejected the auto-oxidation theory, other mechanisms are proposed. It is believed that hyperhomocysteinemia accounts for the downregulation of crucial antioxidant enzymes, such as glutathione peroxidase, as well as the reduction of NO bioavailability. The latter is a result of epigenetic upregulation of asymmetric dimethylarginine, an endogenous inhibitor of eNOS [61,63,64].

Another way that high homocysteine levels disrupt the normal endothelial function is through the impairment of endothelial cell growth cycle. Research has demonstrated that homocysteine interferes with the expression of cyclin A, thus intercepting normal growth and replacement of damaged endothelial cells [65].

Inflammation and Smooth Muscle Cell Proliferation

It has been demonstrated that hyperhomocysteinemia can activate the NF-κB proinflammatory pathway via epigenetic DNA hypomethylation [63].

Apart from the promotion of an inflammatory environment, elevated homocysteine levels are believed to induce VSMCs proliferation and migration further accelerating the formation of atheroma. The proposed mechanism involves the overexpression of c-myb and c-fos genes, which interfere with protein kinase C cascade [65].

Endoplasmic Reticulum Stress

Endoplasmic reticulum includes a specialized network of proteins, called chaperones, which functions as a quality control system. Accumulation of misfolded or dysfunctional proteins causes a condition called endoplasmic reticulum (ER) stress. The latter initiates the unfolded protein response (UPR) procedure, which reassures the cell survival by pausing protein translation and breaking down the faulty proteins. Excess in homocysteine results in increased ER stress, by impairing proper protein folding and disulfide bond formation. UPR fails to cope with this situation and endothelial cells are inevitably led to apoptosis [63,64].

Interactions With Plasma Lipids

It has recently been shown that higher levels of homocysteine (>100 μmol/L) are associated with an otherwise unexpected increase in the plasma levels of total cholesterol, free cholesterol and cholesteryl ester accompanied by a decrease in ApoE expression. These effects are believed to occur as a result of epigenetic DNA methylation [61].

Furthermore, homocysteine thiolactone, a by-product of homocysteine metabolism, is known to react with apoB providing LDL aggregates. Lipoprotein aggregates invade the endothelial barrier and get easily incorporated by macrophages. Foam cells that include these aggregates are more atherogenic due to increased oxidative potential. The aforementioned homocysteinylated lipoproteins can also aggregate with microorganisms. The resulting complex particles are believed to be responsible for obstructions of vasa vasorum. The latter disrupt the integrity of arterial wall and destabilize the atherosclerotic plaque by causing microischemias, microhemorrhages, and microabscesses [59,66].

Prothrombogenic Effects

Elevated plasma levels of homocysteine have recently been associated with a wide spectrum of prothrombotic adverse effects. Homocysteine interferes with arachidonic acid in platelets and enhances thromboxane-mediated platelet aggregation. Regarding the coagulation cascade, homocysteine is believed to inhibit the expression of protein C and thrombomodulin while increasing the plasma concentrations of factors V and VIII [1,59].

C Reactive Protein

CRP represents an acute phase protein of the pentraxin family and it is considered a major part of the innate immune system, as it facilitates bacterial opsonization and complement activation. CRP is mainly produced by the liver in response to IL-6. However, it is also secreted locally in the vascular intima by SMCs, macrophages and atheroma foam cells [67].

Recently high sensitivity-CRP (hs-CRP) has emerged as an independent predictor of CVD. Compared to the other inflammatory markers hs-CRP has exhibited the strongest association with atherosclerotic disease. The latter fact combined with the recent observation of CRP localizing the atherosclerotic lesions raised the hypothesis of a causative relationship between CRP and atherosclerosis [1,68].

Role of CRP in Endothelial Dysfunction and Inflammation

As a mediator of inflammation CRP exerts potent effects that initiate and accelerate the atherosclerotic process. It has been demonstrated that high levels of CRP account for a diminished expression and activity of eNOS and therefore a reduction of NO bioavailability. The progression of atherosclerotic lesions requires active inflammation, which is maintained by CRP-mediated activation of complement, promotion of NF-κB pathway (Fig. 1.4.2) and augmented expression of several other adhesion and chemotactic molecules such as endothelial E-selectin, ICAM-1 and macrophage colony stimulating factor (M-CSF) [67,69].

CRP and Increased Uptake of Ox-LDL

CRP can increase the uptake of ox-LDL by both macrophages and endothelial cells and accelerate the formation of foam cells. Increased uptake by macrophages is induced by the interaction of CRP with the scavenger receptor CD36, whereas ox-LDL uptake by endothelial cells is promoted by CRP-mediated upregulation of lectin-like LOX-1 [67,69].

Interaction With Angiotensin II Type 1 Receptor

Another direct effect of CRP on the development of atherosclerosis involves the Angiotensin II type 1 receptor (AT1-R). Evidence demonstrated that CRP is associated with the up-regulation of AT1-R in VSMCs. Increased expression of AT1-R and thus enhanced activity of Angiotensin II accounts for diverse proatherogenic effects, described in Hypertension section [28,67,69].

Lipoprotein(a)

Lipoprotein (a) or Lp(a) is a circulating lipoprotein similar to LDL regarding lipid composition and surface apolipoprotein B-100. Their difference is the binding of glycoprotein apo(a) to B-100 by a single disulfide bridge. Normal function of Lp(a) is not yet fully elucidated; however, it is speculated that it contributes to tissue healing by binding fibrin and delivering cholesterol. Lp(a) is rapidly gaining ground as an independent risk factor of CVD [68,70].

Proatherogenic Effects of Lp(a)

Similarly to other risk factors, Lp(a) has been associated with endothelial dysfunction and impaired endothelial barrier function which is induced by the inhibition of eNOS. Lp(a) particles entering the atheroma show a tendency to either self-aggregate or bind to LDL and accumulate into macrophages. Engulfed Lp(a) particles transform macrophages into foam cells, while at the same time they trigger the production of IL-8. The latter activates MMPs which destabilize the plaque by degradation of extracellular matrix. An additional contribution of Lp(a) to the atherosclerotic process is mediated by the upregulation of adhesion and chemotactic molecules such as VCAM-1, E-selectin, and MCP-1, as well as the promotion of SMC migration and proliferation [71,72].

Prothrombotic Effects of Lp(a)

Elevated plasma levels of Lp(a) account for the promotion of a prothrombotic state involving three main mechanisms [1,71,72]:

1. Lp(a) binds and deactivates TF pathway inhibitor (TFP), which is produced by endothelial cells, monocytes and platelets in order to confine the coagulation cascade.
2. Lp(a)-induced upregulation of PAI-1.
3. Lp(a) displays a unique structural resemblance to plasminogen resulting in the competitive inhibition of the latter. Consequent inhibition of plasmin formation significantly impairs the fibrinolytic process.

Fibrinogen

Fibrinogen is both a procoagulant factor and an acute phase protein. It exerts both atherogenic and prothrombotic direct effects contributing to the development and aggravation of atherosclerotic disease.

Regarding its atherogenic potential, it has been demonstrated that fibrinogen upregulates adhesion molecules and triggers SMC and fibroblast migration and proliferation. Furthermore, it is believed to induce endothelial dysfunction and impair the endothelial barrier function. At the same time, it can bind to lipoproteins and facilitate lipid accumulation into the atheroma.

Concerning the prothrombotic properties of fibrinogen, it is already known that it constitutes the substrate for thrombin to form fibrin. Apart from that, fibrinogen is recognized as the strongest determinant of blood viscosity and it is also believed to promote platelet aggregation via interaction with glycoprotein IIb/IIIa receptor [72].

Infection

The attractive infectious theory of atherosclerosis was first proposed by Saikku et al., in 1988. Since then, interesting in vitro experimental findings have supported a link between different infectious agents, such as *Chlamydia pneumoniae*, cytomegalovirus, *Helicobacter pylori*, dental flora—and atherosclerotic disease. Chronic low-grade infection or repeated infection with *C. pneumoniae* has exhibited the strongest association. Nevertheless, in vivo data it did not manage to establish a direct causal relationship [1,73].

C. pneumoniae is transferred from the site of infection in the respiratory system to the vessel wall by circulating macrophages and lymphocytes. Endothelial injury is induced either directly by the infecting organism or indirectly by the binding of the bacterial Lipopolysaccharide to LDL. Modified LDL particles become extremely atherogenic. Furthermore, chlamydial heat shock protein-60 (HSP-60) has demonstrated distinct atherogenic properties. *C. pneumoniae* has the ability to infect and replicate into macrophages, endothelial cells and SMCs. By maintaining a chronic infection, it can induce inflammatory pathways involving the expression of adhesion molecules and chemotactic cytokines, the activation and migration of T cells and the proliferation of SMCs. In the setting of chronic inflammation, increased production of ROS can accelerate the oxidation of LDL and the apoptotic death of macrophages. Apoptotic process results in the release of MMPs which degrade the extracellular matrix and destabilize the plaque, whereas at the same time released TF promotes a prothrombotic environment. Finally, chlamydial infection increases the plasma levels of CRP and fibrinogen, each showing distinct and well-established atherogenic properties [71,73–75].

REFERENCES

[1] Frohlich J, Dobiasova M, Lear S, Lee KW. The role of risk factors in the development of atherosclerosis. Crit Rev Clin Lab Sci 2001;38:401–40.

[2] Safar ME. Arterial stiffness: a simplified overview in vascular medicine. Adv Cardiol 2007;44:1–18.

[3] Safar ME, Frohlich ED, editors. Atherosclerosis, large arteries and cardiovascular risk. Statewide agricultural land use baseline, vol. 1; 2015. p. 2007.

[4] Cecelja M, Chowienczyk P. Role of arterial stiffness in cardiovascular disease. JRSM Cardiovasc Dis 2012;1:11.

[5] Tegos TJ, Kalodiki E, Sabetai MM, Nicolaides AN. The genesis of atherosclerosis and risk factors: a review. Angiology 2001;52:89–98.

[6] LaRosa JC. Atherogenesis, I. relationship to coronary risk factors. Clin Cornerstone Hyperlipid 1998; 1:3–14.
[7] Fairweather D. Sex differences in inflammation during atherosclerosis. Inflamm Atheroscler Coron Artery Dis 2014;8:19–22.
[8] Kannel WB, McGee D, Gordon T. A general cardiovascular risk profile: the Framingham Study. Am J Cardiol 1976;38:46–51.
[9] Chambless LE, Folsom AR, Sharrett AR, Sorlie P, Couper D, Szklo M, et al. Coronary heart disease risk prediction in the Atherosclerosis Risk in Communities (ARIC) study. J Clin Epidemiol 2003;56: 880–90.
[10] Kagan A, Gordon T, Rhoads GG, Schiffman JC. Some factors related to coronary heart disease incidence in Honolulu Japanese men: the Honolulu Heart Study. Int J Epidemiol 1975;4:271–9.
[11] Welty TK, Lee ET, Yeh J, Cowan LD, Go O, Fabsitz RR, et al. Cardiovascular disease risk factors among American Indians. The strong heart study. Am J Epidemiol 1995;142:269–87.
[12] Neaton JD, Wentworth D. Serum cholesterol, blood pressure, cigarette smoking, and death from coronary heart disease. Overall findings and differences by age for 316,099 white men. Multiple Risk Factor Intervention Trial Research Group. Arch Intern Med 1992;152:56–64.
[13] van den Hoogen PC, Feskens EJ, Nagelkerke NJ, Menotti A, Nissinen A, Kromhout D. The relation between blood pressure and mortality due to coronary heart disease among men in different parts of the world. Seven Countries Study Research Group. N Engl J Med January 6, 2000;342(1):1–8 [Internet].
[14] Yannoutsos A, Levy BI, Safar ME, Slama G, Blacher J. Pathophysiology of hypertension: interactions between macro and microvascular alterations through endothelial dysfunction. J Hypertens Feburary 2014;32(2):216–24 [Internet].
[15] Hayoz D, Mazzolai L. Endothelial function, mechanical stress and atherosclerosis. Adv Cardiol 2007; 44:62–75 [Internet].
[16] Tabas I, García-Cardeña G, Owens GK. Recent insights into the cellular biology of atherosclerosis. J Cell Biol 2015;209(1):13–22.
[17] Ponticos M. Extracellular matrix synthesis in vascular disease: hypertension, and atherosclerosis. J Biomed Res 2014;28(1):25–39 [Internet].
[18] Gimbrone MA, García-Cardeña G. Endothelial cell dysfunction and the pathobiology of atherosclerosis. Circ Res 2016;118(4):620–36.
[19] Libby P. Changing concepts of atherogenesis. J Intern Med 2000;247(3):349–58.
[20] Touyz RM. Reactive oxygen species in vascular biology: role in arterial hypertension. Expert Rev Cardiovasc Ther 2003;1(1):91–106.
[21] Amar J, Chamontin B. Cardiovascular risk factors, atherosclerosis and pulse pressure. Adv Cardiol 2007;44:212–22.
[22] Abramson JL, Vaccarino V. Pulse pressure and inflammatory process in atherosclerosis. Adv Cardiol 2007;44:223–33.
[23] McEniery CM, Cockcroft JR. Does arterial stiffness predict atherosclerotic coronary events? Adv Cardiol 2007;44:160–72.
[24] Berliner JA, Navab M, Fogelman AM, Frank JS, Demer LL, Edwards PA, et al. Atherosclerosis: basic mechanisms. Oxidation, inflammation, and genetics. Circulation 1995;91:2488–96.
[25] O'Rourke MF, Safar ME. Relationship between aortic stiffening and microvascular disease in brain and kidney: cause and logic of therapy. Hypertens 1979;46:200–4.
[26] Touyz RM. Reactive oxygen species, vascular oxidative stress, and redox signaling in hypertension: what is the clinical significance? Hypertension 2004;44(3):248–52.
[27] Ritman EL, Lerman A. The dynamic vasa vasorum. Cardiovasc Res 2007;75:649–58.
[28] Schulman IH, Zhou M-S, Raij L. Interaction between nitric oxide and angiotensin II in the endothelium: role in atherosclerosis and hypertension. J Hypertens 2006;24:45–50.
[29] Paravicini TM, Touyz RM. Redox signaling in hypertension. Cardiovasc Res 2006;71:247–58.
[30] Crabtree MJ, Tatham AL, Hale AB, Alp NJ, Channon KM. Critical role for tetrahydrobiopterin recycling by dihydrofolate reductase in regulation of endothelial nitric-oxide synthase coupling: relative importance of the de novo biopterin synthesis versus salvage pathways. J Biol Chem 2009;284: 28128–36 [Internet]. American Society for Biochemistry and Molecular Biology.

[31] Schulz E, Gori T, Münzel T. Oxidative stress and endothelial dysfunction in hypertension. Hypertens Res 2011;34(6):665–73 [Internet].

[32] Smith SC, Milani RV, Arnett DK, Crouse JR, McDermott MM, Ridker PM, et al. Atherosclerosis vascular disease conference. Writing group II: risk factors. Circulation 2004;109:2613–6.

[33] Raupach T, Schäfer K, Konstantinides S, Andreas S. Secondhand smoke as an acute threat for the cardiovascular system: a change in paradigm. Eur Heart J 2006;27:386–92.

[34] Word Health Organization. Mortality attributable of tobacco: WHO global report. WHO Libr Cat Data 2012;4.

[35] Messner B, Bernhard D. Smoking and cardiovascular disease: mechanisms of endothelial dysfunction and early atherogenesis. Arterioscler Thromb Vasc Biol 2014;34:509–15.

[36] Katsiki N, Hatzitolios AI, Mikhailidis DP. Passive smoking: the democratic right of nonsmokers to survive. Angiology 2011;62:520–2.

[37] Barua RS, Ambrose JA. Mechanisms of coronary thrombosis in cigarette smoke exposure. Arterioscler Thromb Vasc Biol 2013;33:1460–7.

[38] Ambrose JA, Barua RS. The pathophysiology of cigarette smoking and cardiovascular disease: an update. J Am Coll Cardiol 2004;43:1731–7.

[39] Csordas A, Bernhard D. The biology behind the atherothrombotic effects of cigarette smoke. Nat Rev Cardiol 2013;10:219–30.

[40] Rahman MM, Laher I. Structural and functional alteration of blood vessels caused by cigarette smoking: an overview of molecular mechanisms. Curr Vasc Pharmacol 2007;5:276–92.

[41] Tsiara S, Elisaf M, Mikhailidis DP. Influence of smoking on predictors of vascular disease. Angiology 2003:507–30.

[42] Benowitz NL. The role of nicotine in smoking-related cardiovascular disease. Prev Med 1997;26: 412–7.

[43] Morris PB, Ference BA, Jahangir E, Feldman DN, Ryan JJ, Bahrami H, et al. Cardiovascular effects of exposure to cigarette smoke and electronic cigarettes: clinical perspectives from the prevention of cardiovascular disease section Leadership Council and early Career Councils of the American College of Cardiology. J Am Coll Cardiol 2015;66:1378–91.

[44] Arunachalam G, Yao H, Sundar IK, Caito S, Rahman I. SIRT1 regulates oxidant- and cigarette smoke-induced eNOS acetylation in endothelial cells: role of resveratrol. Biochem Biophys Res Commun 2010;393:66–72.

[45] Barua RS, Ambrose JA, Srivastava S, DeVoe MC, Eales-Reynolds LJ. Reactive oxygen species are involved in smoking-induced dysfunction of nitric oxide biosynthesis and upregulation of endothelial nitric oxide synthase: an in vitro demonstration in human coronary artery endothelial cells. Circulation 2003;107:2342–7.

[46] Jaimes EA, DeMaster EG, Tian RX, Raij L. Stable compounds of cigarette smoke induce endothelial superoxide anion production via NADPH oxidase activation. Arterioscler Thromb Vasc Biol 2004; 24(6):1031–6.

[47] Pistoia F, Cipollone F, Ferri C, Sarà M, Sudano I, Desideri GB. Cyclooxygenase and atherosclerosis: a smoking area. Curr Pharm Des 2010;16:2567–71 [Internet].

[48] D'Souza A, Hussain M, Howarth FC, Woods NM, Bidasee K, Singh J. Pathogenesis and pathophysiology of accelerated atherosclerosis in the diabetic heart. Mol Cell Biochem 2009;331:89–116.

[49] Du XL, Edelstein D, Rossetti L, Fantus IG, Goldberg H, Ziyadeh F, et al. Hyperglycemia-induced mitochondrial superoxide overproduction activates the hexosamine pathway and induces plasminogen activator inhibitor-1 expression by increasing Sp1 glycosylation. Proc Natl Acad Sci USA 2000;97: 12222–6.

[50] Itabe H. Oxidative modification of LDL: its pathological role in atherosclerosis. Clin Rev Allergy Immunol 2009;37:4–11.

[51] Calabresi L, Gomaraschi M, Simonelli S, Bernini F, Franceschini G. HDL and atherosclerosis: insights from inherited HDL disorders. Biochim Biophys Acta 2015;1851:13–8. Mol Cell Biol Lipids [Internet]. Elsevier B.V.

[52] Kita T, Kume N, Minami M, Hayashida K, Murayama T, Sano H, et al. Role of oxidized LDL in atherosclerosis. Ann N Y Acad Sci 2001;947:199–205. Discussion 205–206.

[53] Welty FK. How do elevated triglycerides and low HDL-cholesterol affect inflammation and atherothrombosis? Curr Cardiol Rep September 24, 2013;15:400 [Internet].
[54] Shao B, Heinecke JW. HDL, lipid peroxidation, and atherosclerosis. J Lipid Res 2009;50:599–601.
[55] Tenenbaum A, Klempfner R, Fisman EZ. Hypertriglyceridemia: a too long unfairly neglected major cardiovascular risk factor. Cardiovasc Diabetol 2014;13(1):159.
[56] Talayero B, Sacks F. The role of triglycerides in atherosclerosis. Curr Cardiol Rep 2012;13:544–52.
[57] Toth PP. Triglyceride-rich lipoproteins as a causal factor for cardiovascular disease. Vasc Health Risk Manag 2016;12:171–83.
[58] McCully KS. Vascular pathology of homocysteinemia: implications for the pathogenesis of arteriosclerosis. Am J Pathol 1969;56:111–28 [Internet]. American Society for Investigative Pathology.
[59] Temple ME, Luzier AB, Kazierad DJ. Homocysteine as a risk factor for atherosclerosis. Ann Pharmacother 2000;34:57–65.
[60] Papatheodorou L, Weiss N. Vascular oxidant stress and inflammation in hyperhomocysteinemia. Antioxid Redox Signal 2007;9:1941–58.
[61] Zhou S, Zhang Z, Xu G. Notable epigenetic role of hyperhomocysteinemia in atherogenesis. Lipids Health Dis 2014;13:134.
[62] Lentz SR. Mechanisms of homocysteine-induced atherothrombosis. J Thromb Haemost 2005;3:1646–54.
[63] Austin RC, Lentz SR, Werstuck GH. Role of hyperhomocysteinemia in endothelial dysfunction and atherothrombotic disease. Cell Death Differ 2004;11(Suppl. 1):S56–64.
[64] Lawrence De Koning AB, Werstuck GH, Zhou J, Austin RC. Hyperhomocysteinemia and its role in the development of atherosclerosis. Clin Biochem 2003;36:431–41.
[65] Splaver A, Lamas GA, Hennekens CH. Homocysteine and cardiovascular disease: biological mechanisms, observational epidemiology, and the need for randomized trials. Am Heart J 2004;148:34–40.
[66] McCully KS. Homocysteine and the pathogenesis of atherosclerosis. Expert Rev Clin Pharmacol 2015;8:1–9 [Internet].
[67] Paffen E. deMaat MPM. C-reactive protein in atherosclerosis: a causal factor? Cardiovasc Res 2006;71:30–9.
[68] Rudolf J, Lewandrowski KB. Cholesterol, lipoproteins, high-sensitivity c-reactive protein, and other risk factors for atherosclerosis. Clin Lab Med 2014;34(1):113–27.
[69] Ferri C, Croce G, Cofini V, De Berardinis G, Grassi D, Casale R, et al. C-reactive protein: interaction with the vascular endothelium and possible role in human atherosclerosis. Curr Pharm Des 2007;13:1631–45.
[70] Kullo IJ, Ballantyne CM. Conditional risk factors for atherosclerosis. Mayo Clin Proc 2005;80:219–30.
[71] Mallika V, Goswami B, Rajappa M. Atherosclerosis pathophysiology and the role of novel risk factors: a clinicobiochemical perspective. Angiology 2007;58:513–22.
[72] Catena C, Novello M, Lapenna R, Baroselli S, Colussi G, Nadalini E, et al. New risk factors for atherosclerosis in hypertension: focus on the prothrombotic state and lipoprotein(a). J Hypertens 2005;23:1617–31.
[73] Hoymans VY, Bosmans JM, Ieven MM, Vrints CJ. Chlamydia pneumoniae-based atherosclerosis: a smoking gun. Acta Cardiol 2007;62:565–71.
[74] McKechnie RS, Rubenfire M. The role of inflammation and infection in coronary artery disease: a clinical perspective. Acc Curr J Rev 2002;11:32–4.
[75] Mehta JL, Saldeen TG, Rand K. Interactive role of infection, inflammation and traditional risk factors in atherosclerosis and coronary artery disease. J Am Coll Cardiol 1998;31:1217–25.

CHAPTER 1.5

The Role of Inflammation

Sotirios Tsalamandris[1], Evangelos Oikonomou[1], Nikolaos Papageorgiou[1,2], Dimitris Tousoulis[1]

[1]1st Cardiology Clinic, 'Hippokration' General Hospital, National and Kapodistrian University of Athens, School of Medicine, Athens, Greece; [2]St Bartholomew's Hospital, London, United Kingdom

Contents

INTRODUCTION

Atherosclerosis is a complex mechanism that involves the interaction of inflammatory processes and genetic factors. Traditional risk factors (dyslipidemia, hypertension, diabetes mellitus, smoking, sedentary lifestyle, aging) have been found to be present in the vast majority ($\geq$50%) of patients with coronary artery disease (CAD) [1,2], whereas their elimination has been helpful in the prevention of developing coronary heart disease [3].

Coronary Artery Disease
ISBN 978-0-12-811908-2, http://dx.doi.org/10.1016/B978-0-12-811908-2.00005-2

Cardiovascular risk factors are responsible for the promotion of endothelial dysfunction in injured vessels by leading in augmented expression of adhesion molecules and secretion of proinflammatory molecules [4]. This agitated inflammatory sequence results in increased adherence, migration and accumulation of leukocytes in the sites of the development of atherosclerotic plaques.

During the last decade, researchers have discovered many promising inflammatory biomarkers for the prediction of cardiovascular events. The combination of already established inflammatory biomarkers with novel ones has significantly contributed to the estimation of inflammation in atherosclerosis [4–7]. Nevertheless, there is increasing need to discover an ideal biomarker, which will be able to facilitate the clinical diagnosis of the cardiovascular disease (CVD) from the physicians. The ideal biomarker should have the following characteristics: highly sensitive, specific, reliable, accessible, standardized, dependable, cost effective, and easily interpretable by clinicians.

In this chapter, we will review the pathophysiology of atherosclerosis, the main risk factors that are implicated in the development and progression of atherosclerosis and the role of inflammation in the atherosclerotic process.

PATHOPHYSIOLOGY OF ATHEROSCLEROSIS

Atherosclerosis is a disease of large and medium-sized muscular arteries and is characterized by endothelial dysfunction, vascular inflammation, and the build-up of lipids, cholesterol, calcium, and cellular debris within the intima of the vessel wall. This buildup results in plaque formation, vascular remodeling, acute and chronic luminal obstruction, abnormalities of blood flow, and diminished oxygen supply to target organs. The initial lesion in atherosclerosis involves the intima of the artery and begins in childhood with the development of fatty streaks [8].

Multiple factors contribute to the pathogenesis of atherosclerosis including endothelial dysfunction, dyslipidemia, inflammatory and immunologic factors, plaque rupture, and smoking.

In the complex process of atherosclerosis, there are some major key events that include the development of endothelial dysfunction [9,10], the oxidation of low-density lipoproteins (ox-LDL) [11–15] and the activation of numerous different cell types and chemoattractant agents [16–20]. More analytically, endothelial dysfunction is caused by the production of reactive oxygen species (ROS) and the uncoupling of endothelial nitric oxide synthase (eNOS) which leads in the reduced production of nitric oxide (NO) [9]. Furthermore, the increased expression of P-selectin and vascular cell adhesion molecule-1 (VCAM-1) result in the attachment of circulating monocytes and lymphocytes [10]. The latter in cooperation with macrophages and platelets, promote the migration and proliferation of smooth muscle cells (SMCs).

As far as lipoproteins are concerned, the intra-arterial inflammatory environment promotes their modification in ox-LDL [11]. ox-LDLs appear to play a significant role in the faster progression and development of atherosclerotic plaques [12,13], in contrary with high-density lipoproteins (HDL) which exert a protective anti-inflammatory role by lowering the risk of cardiovascular events [14,15]. Moreover, several molecules, such as chemokines, facilitate the migration of monocytes through the vessel's endothelial wall [16,17] with the latter being involved in the infiltration of leukocytes on the endothelium [18]. In consequence to these events and under the effect of macrophage colony stimulating factor (CSF), monocytes turn firstly into macrophages [11] and secondly into foam cells. Another cell category that seems to interact in the process of atherosclerosis is T-cells and especially Th1 [19]. Finally, in the areas where foam cells are accumulated matrix metalloproteinases destabilize the collagen matrix and cause the rupture of the atheromatous plaque [20].

It is obvious that the evolution of atherosclerosis depends from a cluster of inflammatory processes and if these processes could be controlled future cardiovascular events could have been predicted.

Main Risk Factors and Their Involvement in Inflammatory Process of Atherosclerosis

A number of factors commonly characterized as "risk factors" for atherosclerosis have been identified to facilitate the development of atherosclerosis by decreasing NO bioavailability in the vascular endothelium.

The Role of Hyperlipidemia in Inflammation

Increased cholesterol levels are a major risk factor for atherosclerosis and CAD. It has been evaluated that atherosclerosis begins at the first years of birth and continues through the adulthood, whereas increased cholesterol levels and atherosclerosis are closely related [21]. LDL and specifically ox-LDL triggers a number of inflammatory processes in the vasculature, thus promoting the local production of ROS, upregulating mechanisms, responsible for impaired ejection fraction (EF). Notably, EF is impaired even in children, exhibiting familiar hypercholesterolemia, as this has been assessed by ultrasound techniques including the flow-mediated dilation (FMD) [21,22].

The Role of Diabetes Mellitus in Inflammation

The endothelium of diabetic patients is more vulnerable in developing atherosclerotic plaques in comparison to the endothelium of nondiabetic individuals. A variety of mechanisms may contribute to the increased risk of developing atherosclerotic plaques in patients with diabetes mellitus (DM). The abnormal cluster of hyperglycemia, elevated free fatty acids, and insulin resistance which characterizes DM, act in concert to target the endothelial cell, resulting in oxidative stress and endothelial dysfunction [23].

In DM, impaired vasodilation results from reduced NO production and increased NO inactivation [24]. Endothelial dysfunction with impaired NO release is present both in patients with type 2 DM [25] and those with insulin-dependent DM [26,27]. The presence of insulin resistance alone may be associated with coronary endothelial dysfunction [28,29]. In addition, elevated levels of free fatty acids also contribute to reduce NO production, further impairing vascular relaxation [30,31]. These elevated levels of free fatty acids in DM, also promote the formation of ox-LDL, which consists of important initiating event for atherosclerosis. ox-LDL can damage endothelial cells and induce the expression of adhesion molecules (including P-selectin) [32] and chemotactic factors such as monocyte chemoattractant protein-1 and macrophage CSF [33,34] and therefore contribute to endothelial dysfunction in DM.

It has been shown that patients with diabetes mellitus are characterized by enhanced leukocytes-endothelial cells aggregation and increased vascular permeability, which are responsible for endothelial dysfunction, mediated by mechanisms such as activation of protein kinase C, decrease of cellular nicotinamide adenine dinucleotide phosphate (NADPH), insulin resistance and ROS [35]. It should be mentioned that asymmetrical dimethylarginine has a crucial role in endothelial dysfunction existing in states of diabetes mellitus [36]. Although, the extension of vascular imbalance is related both to diabetes duration and to LDL levels, changes during adolescent affect significantly the development of vascular dysfunction in young individuals with diabetes mellitus [35].

The Role of Hypertension in Inflammation

Endothelium-mediated vasodilation was found blunted in hypertensive compared to normotensive subjects as estimated by means of forearm blood flow measurements, whereas endothelium independent vasodilation resulted similar in the same setting [37]. That evidence was confirmed in the peripheral [38] and coronary [39] circulation and then extended to patients with secondary forms of hypertension, suggesting that microvascular endothelial dysfunction constitutes a generalized phenomenon associated with different hypertensive conditions. Additionally, a significantly blunted response of the baseline NO production was detected in the vessels of hypertensive patients compared to those of normotensive controls when a methylated analogous of L-arginine (i.e., NG-monomethyl-L-arginine) was infused in the brachial artery to test the baseline capacity of reduction in the forearm blood flow [40]. Thus, essential hypertension (EH) is accompanied by both increased basal vascular resistance and abnormal induced vasodilatory response to acetylcholine, reflecting an endothelium derived NO system defect.

The defect of endothelium-derived NO system in EH theoretically may be located in various domains of its metabolic pathway, comprised by both anabolic and catabolic biochemical processes. Regarding the anabolic pathway, decreased availability of L-arginine could beget reduced NO bioactivity; to test the latter hypothesis L-arginine deficit was proved not critical for the decreased NO bioactivity in hypertensives [41]. Other experiments, although found a preserved muscarinic receptor function in hypertensives, suggested that a generalized dysfunction is more likely to occur beyond the NO muscarinic receptor-mediated effects [42]. The signal transduction through G protein-related proved intact in hypertensives [43], a finding further supported by the evidence that the beta2 adrenergic-dependent NO activity operating via cyclic-adenosine monophosphate is not deleterious [44]. Taken together, all the above data suggest that endothelial dysfunction in EH is due to a selective abnormality of NO synthesis, probably related to impaired intracellular phosphatidyl inositol/Ca2+ pathway [45]. Last but not least, an intrinsic inhibitor of NO synthesis, namely asymmetric dimethyl arginine, is overexpressed in essential hypertensives compared to their normotensive counterparts and it was associated with endothelial dysfunction as assessed by forearm blood flow measurements [46]. We have previously demonstrated that the levels of asymmetric dimethylarginine could be adversely modulated in the setting of EH in subjects with sleep apnea which constitutes a condition of increased oxidative stress [47].

In experimental animal models of hypertension the excess of superoxide anions is linked to increased vascular tone and impaired endothelial dependent vasodilatation. An important intracellular source of superoxide radicals is the xanthine oxidase system, potentially inhibited by oxypurinol; oxypurinol infusion did not reverse the relatively impaired response to acetylcholine in hypertensives [48], suggesting that the increased intracellular oxidative load could be not accounted for the endothelial dysfunction associated with EH.

Cyclooxygenase cascade in the vascular endothelium may lead to the genesis of relevant vasodilatory and vasoconstrictive factors; however, diverse studies shown that these products do not play a major role in the endothelial dysfunction in hypertensives [49]. Conversely, it was demonstrated that endothelin-1 may play a role in the abnormal endothelial dependent dilation. Indeed, the combined endothelium A (ET_A) and endothelium B (ET_B) receptor blockade resulted in a significant potentiation of the vasodilatory response to acetylcholine, thereby, suggesting that an increased endothelin 1 (ET-1) activity may play a role in the abnormal endothelium dependent dilation in EH [50].

Upregulated renin–angiotensin system produced increased levels of angiotensin II, a potent vasoconstrictor that increases oxidative stress and breakdown of NO [51]. Furthermore, elevated angiotensin conversing enzyme concentrations antagonize NO activity not only by increasing angiotensin II generation but also by decreasing concentrations of bradychinin. Remarkable production of angiotensin II is accomplished by the tissues and there is evidence that complete suppression of the renin-angiotensin system could not be achieved by blocking the angiotensin converting enzyme-related production of angiotensin II alone [52]. Although, the relationship between angiotensin receptors and NO is not fully clear, angiotensin II converting enzyme inhibitors and angiotensin II receptor 1 blockers exerted protective effects on angiotensin II-mediated inflammatory response [53,54].

Pioneering studies have shown that the pulsatile flow triggers the inflammatory cascade by stimulating vascular adhesion molecule expression and leukocyte adhesive properties. Accordingly, higher levels of C reactive protein augment blood pressure (BP) by reducing NO availability, increasing endothelin-1 levels, upregulating angiotensin 1 receptor expression, thus contributing to the pathogenesis of EH and progression of subclinical organ damage, including the kidney dysfunctional vasculature expressed by microalbuminuria [55,56]. In hypertensives, we have previously demonstrated that increasing levels of microalbuminuria (i.e., an index of generalized endothelial dysfunction beyond the glomerulus) in tandem with decreasing levels of adiponectin are associated with aortic stiffness [57], and that molecules participating to the perpetuation of subclinical inflammation such as C reactive protein, interleukin-18 and CD-40 ligand are interrelated with microalbuminuria [58].

In gluteal artery biopsies of normotensives, it was shown that flow rate-dependent shear stress induced progressive vasodilation, which was blunted during the inhibition of NO synthesis and abolished after mechanical removal of the endothelium. In contrast, arteries from hypertensives showed significant impairment of FMD compared with those of normal controls, highlighting an abnormal flow-dependent dilation of the resistance arteries in EH [59]. Consequently, impaired shear stress dependent NO release, may contribute to the development and maintenance of increased peripheral resistance in EH and by preventing relaxation may lead to inappropriate elevations of BP [60].

The Role of Smoking in Endothelial Dysfunction Inflammation

The majority of clinical and experimental studies investigating cigarette smoke-induced endothelial dysfunction have monitored impairment of FMD. Many studies report that cigarette smoke reduces eNOS expression and the associated NO production [61,62], whereas one study shows that whole cigarette smoke, but not nicotine itself, causes endothelial dysfunction [63]. However, it is important to stress that the majority of investigations demonstrate that both cigarette smoke and nicotine causing endothelial dysfunction. Regardless of exposure time, cigarette smoke induced impaired vasodilation due to reduced stimulated release or attenuated basal production of endothelium-derived NO is generally regarded as a useful marker of early changes in endothelial cells that can precede morphologic changes [64]. Additionally, cigarette smoking reduces the number of endothelial progenitor cells and impedes their differentiation and functional activity [65]. A recent study reports that cigarette smoke impedes the endothelial repair process [66]. Recently, Akt/eNOS/NO pathway has been shown to be important for vascular endothelial growth factor (VEGF)-induced endothelial cell migration. Cigarette smoke extract inhibits this pathway and thus impairs VEGF-induced endothelial cell migration resulting poor endothelial function and vessel growth [67].

Moreover, cigarette smoking increases ET-1 (one of the most potent endogenous vasoconstrictors). Nicotine, tar, carbon monoxide and other smoke components increase ET-1 rapidly after exposure to smoke [68,69]. Intriguingly, chronic cigarette smoking does not lead to increased ET-1 mRNA expression even though plasma levels of ET-1 are often increased [70]. It seems that long-term smoking upregulates expression of endothelin receptors ET-A and ET-B [71], with a concomitant desensitization of the vasodilatory mechanism [72]. Irrespectively of decreased NO bioavailability and ET-1 effects contractile dysfunction observed in smokers are secondarily due thromboxane A2 production [73], upregulation of voltage-gated Ca2+ channels and downregulation of Ca2+-activated K+ channels [74]. Many of the vascular effects of chronic smoking are attributed to nicotine, a tertiary amine composed of a pyridine ring. Nicotine induces endothelial dysfunction [75—77], which in animal models depends on duration and dose of nicotine treatment [77]. Nicotine induced endothelial dysfunction has been associated with increased plasma levels of asymmetric dimethylarginine (ADMA), an endogenous NOS inhibitor and also a risk factor for cardiovascular disease. Nicotine induced ADMA accumulation is related to a decrease in activity of dimethylaminohydrolase (DDAH), a major hydrolase of ADMA [75].

Decreased NO bioavailability is central to smoking induced endothelial dysfunction. This phenomenon is attributed to decreased cyclic guanosine monophosphate production [78], decreased protein content [79] and activity [80] despite increased mRNA levels (probably due eNOS uncoupling) and increased oxidative stress. High concentrations of ROS, NO, peroxynitrite, peroxynitrate, and free radicals of organic compounds are found in the two phases of cigarette smoke: tar and gas-phase smoke. In addition to these short-lived, highly reactive substances, the gas phase of cigarette smoke contains varying amounts of more stable substances that also have the potential to increase intracellular production of ROS. These include a series of a, b-unsaturated aldehydes, such as acrolein and rotonaldehyde, a, b-unsaturated ketones, and a number of saturated aldehydes [81]. These stable compounds have been shown to react nonenzymatically with thiol groups that may be involved in the regulation of enzymes, including NADPH oxidase [81]. Cigarette smoking increases vascular superoxide production, resulting in decreased NO bioavailability and decreased production of prostacyclin [82]. Human coronary arterial endothelial cells incubated with smokers' serum have significantly lower NO production and eNOS activity but have increased eNOS expression, due to an increased production of H_2O_2 [83]. Furthermore, thiol-reactive stable compounds such as acrolein in cigarette smoke activate NADPH oxidase and increase endothelial O_2—production, thereby reducing NO bioavailability, resulting in endothelial dysfunction [84]. Finally, cigarette smoke-derived O_2—reacts with NO giving rise to peroxynitrite (ONOO−), which depletes tetrahydrobiopterin (an essential cofactor for eNOS) reserve by oxidation [85].

Interestingly, it has been observed that the amino acid L-arginine, improved endothelial function both in patients with cardiovascular risk factors and in healthy subjects. More specifically, in a randomized, placebo-controlled trial, endothelial function (FMD), aortic stiffness and an inflammatory marker were measured after oral administration of L-arginine in healthy subjects. L-arginine improved endothelial function and elastic properties during the acute phase of smoking, an effect accompanied by reduced inflammation in these subjects [86]. Also, L-arginine administration may be useful in several cardiovascular states, such as hypercholesterolemia, hypertension and premature atherosclerosis [87].

The Role of Infections in Inflammation

Recent evidence has shown that infections along with systemic inflammation play a key role in the initiation and progression of CAD [88]. The immune cells present in atherosclerotic lesions are predominantly macrophages and T lymphocytes, although mast cells, neutrophils and dendritic cells have also been recently suggested to play a role [89]. The initiating factor in the development of atherosclerotic lesions is endothelial activation and dysfunction, which facilitates the adhesion of immune cells and platelets to the endothelium and the entry of lipids in the vessel wall. $CD4^+$ helper T lymphocytes are predominant in the atherosclerotic plaques of CAD patients and they have been suggested to contribute to the growth and activity of the plaques [90]. Similar to data from murine models of atherosclerosis, most of the $CD4^+$ T cells in CAD patients secrete interferon-γ (IFN-γ), the signature proinflammatory cytokine of T helper 1 (Th1) cells [91]. IFN-γ has been shown to induce the expression of major histocompatibility complex class II by monocytes and other antigen presenting cells such as dendritic cells, and therefore allows these cells to present antigens derived from the plaques and amplify the local immune response [92]. In addition, IFN-γ stimulates the production of matrix metalloproteinases by macrophages, causing SMC apoptosis and the thinning of the fibrous cap, that culminates in plaque rupture and coronary thrombosis [93] (Fig. 1.5.1).

In addition, numerous different mechanisms have been implicated for the inflammatory response in the vasculature. It has been found that viral and bacterial infections are involved in the development and progress of atherosclerosis by triggering processes which in turn induce the expression of proinflammatory cytokines and adhesion molecules, migration and differentiation of SMCs, while inhibiting apoptosis and increasing the accumulation of lipids [94].

The Role of Hyperhomocysteinemia in Inflammation

Hyperhomocysteinemia represents a novel cardiovascular risk factor leading to endothelial dysfunction, whereas folate supplementation decreases homocysteine levels and results in endothelial dysfunction improvement. Studies have shown that homocysteine reduces

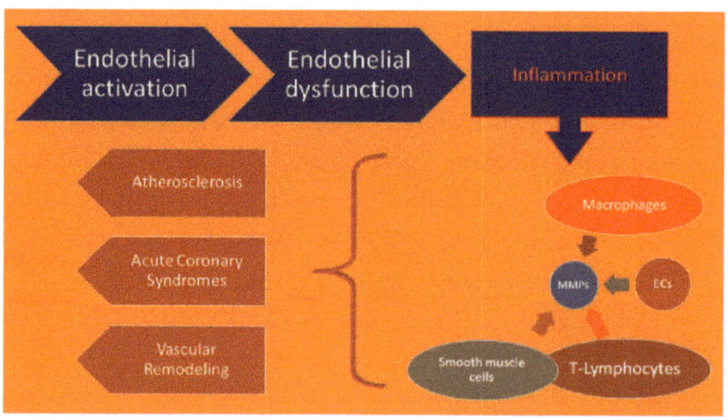

Figure 1.5.1 Role of inflammation and matrix metalloproteinases in atherosclerosis and vascular damage.

NO bioavailability by increasing the oxidative status [95]. It has been well evaluated that homocysteine leads to ADMA accumulation by its inhibitory effect of the enzyme dimethylarginine DDAH [96]. Moreover, experimental studies have evaluated that hyperhomocysteinemia may result in endothelial dysfunction through the accumulation of ADMA [97]. However, there are scarce data referring on the impact of hyperhomocysteinemia on EF [98]. Several mechanisms have been proposed to explain the role of increased homocysteine levels and some of them such as the previous could reliably explain the increased risk of CAD in patients exhibiting hyperhomocysteinemia [99].

The Role of Genes in Inflammation

Novel data suggest that atherosclerosis is a gene-related disease in which polymorphisms of genes regulating endothelial biology, appear to have a strong impact on vascular function under the effect of several environmental factors [100]. Numerous polymorphisms have been determined to be associated with EF and CAD including those of the renin-angiotensin system, proinflammatory cytokines and chemokines, acute phase proteins such as C-reactive protein and fibrinogen, NADPH oxidase and bradykinin receptor among many important polymorphisms [101].

Innate and Adaptive Immunity in Atherosclerosis

The concept of the link between atherogenesis and the immune system has recently gain ground. As our knowledge in atherosclerosis evolves, it is now well documented that the immune system has a fundamental role in the initiation and progression of atherosclerosis. The immune cells are present in the atherosclerotic lesions, expressing several proinflammatory cytokines through specific gene activation. Moreover, all these proinflammatory cytokines are secreted through the atherosclerotic lesions to the systemic circulation. The overhead functions are expressed in a synergetic way by the innate and adaptive immunity.

The innate immunity is a nonspecific response of the immune system, as the first line defense against pathogens, lasting temporarily compared to the adaptive immunity. It consists of endothelial and circulating immune cells. One component of this system is the macrophage foam cells located in the atherosclerotic plaque. These cells, except than accumulating the excess lipid in the atherosclerotic lesion, they are producing proinflammatory cytokines, chemokines and other lipid mediators. They also eradicate large amounts of oxidant species and the produced cocktail of inflammatory mediators can promote inflammation in the plaque, further stimulating the progression of the atherosclerotic plaque, independently from antigen stimulation. An interesting part of the abovementioned system is the Toll-like receptors on the immune cells membrane. These are a family of proteins that by recognizing and binding viral and bacterial components, propagates the signal in the cell and activates the NF-κB and interferon pathways.

The adaptive immunity is an antigen specific system. In the atherosclerotic lesions, antigen presenting cells (mononuclear phagocytes, dendritic cells, epithelial cells, B cells) are presenting antigens to the T cells and by triggering their activation, motivate them to secrete large amount of cytokines that can modify atherogenetic process. There are several types of T cells participating in this procedure. More specifically, helper T cells fall into two general subtypes (type 1 and 2), that have different effect on the atherosclerotic plaque, either destabilizing and heightening thrombogenicity or on the other hand inhibiting inflammation in the context of atherogenesis. Cytolytic T cells, express cytotoxic factors that can stimulate apoptosis of target cells (e.g., SMCs, ECs, and macrophages) contributing probably in this way to plaque progression and complication. Regulatory T cells on the other hand, can exert anti-inflammatory effects, whereas several experimental preparations suggest an antiatherosclerotic function of these cells in vivo. Finally, humoral immunity, expressed by B cells and antibodies, may have either atheroprotective or atherogenic properties, depending on the circumstances.

Current Prospective and Diagnostic Tools in the Assessment of Inflammation in Atherosclerosis

The serious clinical manifestations of atherosclerosis (including coronary heart disease, stroke, and peripheral vascular disease) augment the need of performing the appropriate diagnostic methods to the patients. The most important diagnostic methods include the usage of biochemical markers and the invasive and noninvasive imaging techniques assessing endothelial function (Table 1.5.1).

Table 1.5.1 Diagnostic tools in the assessment of inflammation in atherosclerosis

Biomarker	Structure	Role in prognosis
Adhesion molecules (ICAM−VCAM)	Transmembrane glycoproteins	Future CAD, IMT [102−104]
CRP	Homopentameric structure	CAD, MI, stroke [105−108]
Fibrinogen	Glycoprotein	CAD, MI, stroke cardiac death [109−113]
Proinflammatory cytokines (IL-6, TNF-a)	Glycoproteins	ACS death and heart failure [114−119]
Lp-PLA2	Calcium-independent member of the phospholipase A2 family	CVD, atherosclerotic plaque progression [120−123]
Chemokines	Chemotactic cytokines (CC, CXC, CX3C, XC)	CAD, coronary calcification, adverse cardiac events [124−128]
MRP8/14	Heterodimer of two calcium binding proteins	CVD [129−131]

ACS, acute coronary syndromes; *CAD*, coronary artery disease; *CRP*, C-reactive protein; *CVD*, cardiovascular disease; *ICAM-1*, intercellular adhesion molecule-1; *IL*, interleukin; *IMT*, intima-media thickness; *Lp-PLA2*, lipoprotein-associated phospholipase A2; *MI*, myocardial infarction; *MRP8/14*, myeloid-related protein 8/14 complex; *VCAM-1*, vascular cell adhesion molecule.

Traditional Inflammatory Biomarkers

The Role of Adhesion Molecules in Inflammatory Process

The cell adhesion molecule (CAMs) family includes integrins, the immunoglobulin superfamily and selectins [132,133]. They are proteins located on the cell surface [132] involved with the binding with other cells or with the extracellular matrix (ECM) in the process called cell adhesion. These proteins are typically transmembrane receptors and are composed of three domains: an intracellular domain that interacts with the cytoskeleton, a transmembrane domain, and an extracellular domain that interacts either with other CAMs of the same kind (homophilic binding) or with other CAMs or the ECM (heterophilic binding) [133]. The family molecules that have been mainly reported to be involved in atherosclerosis, are VCAM-1, P-selectin, intercellular adhesion molecule-1 (ICAM-1), platelet/endothelial cell adhesion molecule-1 (PECAM-l), and E-selectin [132—134]. More specifically, VCAM-1, ICAM-1, and PECAM-l interact in the adhesion of circulating leukocytes on the endothelial surface and their migration through the vessel wall [134].

The role of the most important adhesion molecules (ICAM-1 and VCAM-1) concerning the future development of cardiovascular events has been evaluated in several studies. The ICAM-1 and E-selectin have been found to play a significant predictive role of future cardiovascular events and may serve as molecular markers for atherosclerosis [the Atherosclerosis Risk In Communities (ARIC) study] [102]. Also in the same study, an association between levels of sICAM-1 and increased intima-media thickness as assessed by carotid ultrasound was established. Furthermore, an association was observed between increased levels of ICAM-1 and subsequent risk of first myocardial infarction in a nested case-control study from the Physicians' Health Study [103,104,133]. The association was present in the study overall as well as among nonsmokers and persisted after control for lipid and nonlipid risk factors. As shown, the risk of myocardial infarction associated with raised concentrations of sICAM-1 increased with the length of follow-up. Nevertheless, no association was found between sVCAM-1 levels and the ability of predicting future cardiovascular events. Other studies have defined the use of adhesion molecules as indexes of the extent of atherosclerotic disease [135,136]. However, as far as peripheral artery disease is concerned ICAM-1 seems to have a superior predictive role in comparison to VCAM-1 [137].

By aspects of secondary prevention, several studies [138—141] have pinpointed the role of adhesion molecules. In the Heart Outcomes Prevention Evaluation study it was demonstrated that sVCAM-1 levels are significantly related to future CAD events. In an interesting study [142] of patients with isolated coronary artery ectasia (CAE) it was shown that patients with isolated CAE have raised levels of plasma soluble ICAM-1, VCAM-1, and E-selectin in comparison with patients with obstructive CAD without CAE and control subjects with normal coronary arteries, suggesting the presence of a more severe and extensive chronic inflammation in the coronary circulation in these patients accompanied with poorer prognosis. Finally, in patients with acute coronary events (non-Q-wave myocardial infarction and unstable angina) elevated VCAM-1 levels may predict future cardiovascular events [133,143].

The Role of C-Reactive Protein in Inflammatory Process

C-reactive protein (CRP) was first described by Tillett and Frances in 1930, as a serum component present in acutely ill patients that reacted with a specific *Streptococcus pneumoniae* extract they termed Fraction C [144]. CRP is synthesized as a 206 amino acid polypeptide and secreted by hepatocytes as an approximately 23 kDa, nonglycosylated monomer, which noncovalently associates to form the homopentameric ring structure characteristic of pentraxin family members [145]. The human CRP gene is located on chromosome 1q23, and consists of two exons and one intron [145]. During the acute phase of inflammation, interleukin-6 (IL-6) is the major regulator of CRP production [145,146] and its structure requires conversion from a pentameric to a monomeric form during the atherosclerotic procedure [147].

CRP binds to a large number of autologous and extrinsic ligands, including native and modified plasma lipoproteins, phospholipids, and apoptotic cells, which are present in the atherosclerotic lesions [114,133,147]. When bound to ligands, CRP activates the classic pathway of complement, a major player in the immune and inflammatory response, and reacts with Fcγ receptors on phagocytic cells. In endothelial cells, CRP stimulates the production of ET-1 [148,149], downregulates eNOS [133,150] increases superoxide production, decreases NO production and alters eNOS phosphorylation [133,151]. CRP may also facilitate leukocyte adhesion and internalization into the arterial wall by stimulating the expression of VCAM-1, ICAM-1, E-selectin, and monocyte chemoattractant protein-1 [152].

C - reactive protein has been recommended as the marker of choice to monitor cardiovascular risk. Numerous studies have examined the predictive role of CRP in healthy populations [153], where increased levels have been associated with increased rates of myocardial infarction and stroke [105,107,133] both in men and women [106]. Moreover, in patients with known cardiovascular disease (chronic stable ischemic heart disease) a significant correlation between concentrations of C reactive protein within the normal range and long term cardiovascular risk was found [133,154] (European Concerted Action on Thrombosis and Disabilities Angina Pectoris Study). In a similar study of patients with stable coronary heart disease elevated CRP was associated with increased risk of unstable angina, myocardial infarction and death [155]. In patients with unstable angina (UA), higher concentrations of CRP levels were associated with increased risk of death [156]. In addition, in a recent interesting study [157], CRP was associated with the severity of CAD in the top to the bottom third comparison of CRP levels.

In patients with myocardial infarction there are conflicting results concerning the predictive role of CRP. For example, in Canakinumab Anti-inflammatory Thrombosis

Outcomes Study (CANTOS) no association was found between CRP levels and recurrent events in patients with myocardial infarction. On the contrary, in a study population consisted of patients who were consecutively hospitalized with acute myocardial infarction (AMI) within 12 h after symptom onset, admission CRP was a powerful parameter for risk stratification. Multivariable analysis showed that both troponin positivity and CRP positivity were associated with a twofold increased risk of dying within 28 days after the acute event for all patients with AMI [108].

Furthermore, the predictive role of CRP has been evaluated in the group of patients that undergo percutaneous coronary intervention (PCI). In an interesting study, inflammatory response to PCI was calculated as the difference between the peak postprocedural hsCRP level and the preprocedural hsCRP level (DeltaCRP). Twelve months follow-up data were obtained and clinical outcomes were compared with DeltaCRP. In receiver operating characteristics analyses, the cutoff point of DeltaCRP for major adverse cardiac events (MACE) was 3.0 mg/L, which yielded sensitivity of 61.7% and specificity of 69.7%. The patients with DeltaCRP >3 mg/L revealed higher incidence of myocardial infarction (37.7% vs. 14.6%, $P < .001$), and American College of Cardiology/American Heart Association (ACC/AHA) type B2/C lesion (81.5% vs. 68.7%, $P = .006$) than in patients with low DeltaCRP. There was significant correlation between DeltaCRP and the changes in troponin T after PCI ($r = 0.210$, $P < .001$). An increase in hsCRP >3 mg/L after PCI had a higher predictive value for the occurrence of MACE than low hsCRP elevation. In multivariate analysis, DeltaCRP and peak troponin T were independent predictors of MACE. In conclusion, postprocedural hsCRP elevation >3 mg/L was associated with higher incidence of MACE in patients with ACS. DeltaCRP determinations may be of value for risk stratification after PCI [158]. Recently, a study conducted in patients who underwent PCI using drug eluting stent revealed that poor renal function and high hsCRP were additively associated with a higher risk of hard endpoints and were independent predictors of mortality and recurrent MI even after correction for other factors [159]. Another similar study highlighted the role of higher CRP values which had a significantly higher rate of all-cause mortality and myocardial infarction during the follow-up of patients undergoing PCI [160]. Finally, a recent study using virtual histology-intravascular ultrasound (VH-IVUS) evaluated the relationship between high-sensitivity hsCRP levels and plaque components in acute ACS patients. VH-IVUS analysis has demonstrated that ACS patients with elevated hsCRP have more vulnerable plaque component, compared with ACS patients with normal hsCRP and so they consist a group of more high risk patients for future cardiovascular events [161].

The Role of Fibrinogen in Inflammatory Process

Fibrinogen is a soluble plasma glycoprotein synthesized by the liver that is converted by thrombin into fibrin during blood coagulation. Fibrinogen molecules are comprised of two sets of disulfide-bridged alpha-, beta-, and gamma-chains [162]. Each molecule contains two outer D domains connected to a central E domain by a coiled-coil segment. Fibrin is formed after thrombin cleavage of fibrin peptide A from fibrinogen alpha-chains, thus initiating fibrin polymerization [163].

Fibrinogen is one of the major components of the coagulation cascade and in addition plays a key role in the formation of atherosclerotic plaques in many aspects. More analytically, it incorporates into the arterial wall and is transformed into fibrin [133], promotes the binding and production of HDL [163], facilitates the migration of adhesion molecules to the endothelial surface [164], initiates the proliferation and migration of SMCs [165–167], interacts with leukocytes through Macrophage-1 (MAC-1) and alpha X beta 2 receptors [168,169], mediates the binding of monocytes and endothelial cells through ICAM-1 and MAC-1 [170–172], increases ICAM-1 levels leading to elevated leukocyte and platelet population on endothelial surface [173,174], and increases phagocytosis, leukocyte toxicity and delay of apoptosis [175] by rising intracellular calcium and increasing the expression of neutrophil activation factors.

The prognostic role of fibrinogen in cardiovascular disease has been evaluated in numerous studies. Two large studies, the Framingham study [109,110] and the European Concerted Action on Thrombosis and Disabilities (ECAT) study [111], revealed that increased levels of fibrinogen increase the risk for future cardiovascular events independently of traditional risk factors. In addition, a meta-analysis of the Fibrinogen Studies Collaboration [176] proved that fibrinogen increase is associated with increased risk of future cardiovascular events and mortality. Similar results were reported in the AtheroGene study [112] and in the Darwin Region Urban Indigenous Diabetes (DRUID) study [113]. In patients with diabetes mellitus the risk of developing peripheral artery disease was evaluated and was associated with elevated levels of fibrinogen [177]. Moreover, in patients with heterozygous familial hypercholesterolemia it was shown that elevated fibrinogen levels can strongly predict future cardiovascular events [178]. Finally, during the last years, many investigations have been conducted concerning the relationship between fibrinogen gene variants and the risk of future cardiovascular events. Data from these studies appear to be controversial whether there is a strong association between gene polymorphism and CVD risk, and so further investigation should be performed toward this direction [179–184].

The Role of Proinflammatory Cytokines in Inflammatory Process

The cytokines are an important group of molecules involved in coordinating the many and varied components of the immune system. They mainly consist of glycoproteins of 150 amino acids [185]. There are more than 100 cytokines, which may act independently or synergistically with other cytokines. Among them, the two most important are IL-6 [186] and tumor necrosis factor alpha (TNF-α) [187]. In particular, cytokines appear to orchestrate the chronic development of atherosclerosis, eventually leading to the formation of complex atherosclerotic plaques, which can trigger acute thromboembolic complications, such as myocardial infarction or stroke. Indeed the rupture-prone plaque, characterized by a thin fibrous cap overlaying a voluminous lipid core, exhibits accumulation of various proinflammatory cytokines. These cytokines may control the clinical consequences of plaques by mediating infiltration and accumulation of immunocompetent cells, directing the turnover of fibrillar collagens (governing the fragility of the fibrous cap), or enhancing foam cell formation and thrombogenicity of the lipid core [188]. IL-6 is produced at the site of inflammation and plays a key role in the acute phase response and is directly associated with the production of CRP [189]. The other important cytokine, TNF-α, is known to induce smooth muscle proliferation and to increase adherence of leucocytes to endothelial cells by inducing the expression of cell adhesion molecules such as ICAM-1 and VCAM-1 [190]. Furthermore, TNF-γ induces the expression of a wide range of cytokines, including chemokines and IL-6 by endothelial cells [191].

The predictive role of IL-6 in healthy individuals has been highlighted in combination with CRP [114] or alone [115] (with a twofold greater risk of death in patients with elevated IL-6 levels). Similar results were observed in a study of healthy men [117] with the risk of future cardiovascular events being increased accordingly to the elevation of IL-6. As far as the predictive role of TNF-α in future cardiovascular events is concerned, it has been demonstrated that elevation of TNF-α may partly explain the increased risk of cardiovascular events in patients with low HDL-C levels [116].

In populations of patients with already established CAD, IL-6 has been shown to be an independent predictor for cardiovascular mortality [192]. The study was conducted in a group of patients with first acute ST-segment elevation myocardial infarction (STEMI) who were admitted to our Hospital within 6 h after the symptoms onset. Clinical data were recorded and serum admission levels of IL-6, soluble ICAM-1 (sICAM-1), soluble VCAM-1 (sVCAM-1), and soluble P-selectin (sP-selectin) were determined. The patients were then followed up for 3 years to document cardiovascular mortality. During the follow-up, 34 patients died from cardiovascular causes. The admission levels of IL-6 were significantly higher in these patients, whereas sICAM-1, sVCAM-1, and sP-selectin were comparable between these and the survived patients. The Kaplan–Meier plots revealed a significant increase in cardiovascular mortality with increasing levels of IL-6. The logistic regression analysis indicated that IL-6 was an independent predictor for cardiovascular mortality. In conclusion, findings indicated that elevated admission levels of IL-6, but not soluble adhesion molecules, provide valuable information for risk assessment of long-term cardiovascular mortality in patients with STEMI [192]. Furthermore, in patients undergoing PCI, highest IL-6 quartile was associated with stent thrombosis, especially in clopidogrel users regardless of the time of ST, suggesting the establishment of IL-6 as independent predictor of stent thrombosis [118]. Lately, the role of the −174G>C polymorphism of interleukin-6 (IL-6) as a predisposing factor to angina pectoris was investigated and it was found that there is a high incidence of the −174G>C polymorphism of the IL-6 gene in patients with angina pectoris compared with those carrying the G allele, reinforcing the contribution of genetic factors to the symptoms of angina pectoris [119].

The predictive role of TNF-α in patients with increased risk of manifesting recurrent coronary events was shown in the Cholesterol and Recurrent Events (CARE) study, where elevated levels of TNF-α were associated with higher risk or cardiovascular death or recurrent fatal myocardial infarction [193]. Recently, in patients with chronic heart failure ongoing myocardial damage the up regulation of monocyte proinflammatory cytokine (mainly TNF-α) production by CRP was proved to be significantly related to future cardiac events [194].

Novel Inflammatory Biomarkers

The Role of Lipoprotein-Associated Phospholipase A2 in Inflammatory Process

Phospholipases A2 (PLA2s) are enzymes that release fatty acids from the second carbon group of glycerol. They are characterized by a low molecular weight (13–15 kDa), histidine in the catalytic site, $Ca2^{+}$ bound in the active site, as well as six conserved disulfide bonds with one or two variable disulfide bonds [133,195]. These enzymes all catalyze the hydrolysis through the same mechanism of abstraction of a proton from a water molecule followed by a nucleophilic attack on the sn-2 bond. Lipoprotein-associated PLA2 (Lp-PLA2) also known as platelet-activating factor acetyl hydrolase is a phospholipase A2 enzyme that in humans is encoded by the PLA2G7 gene [133,195].

Lp-PLA2 is a 45-kDa protein of 441 amino acids [133,195]. Lp-PLA2 is involved in the development of atherosclerosis. In human atherosclerotic lesions, two main sources of Lp-PLA2 can be identified, including that which is brought into the intima bound to LDL (from the circulation), and that which is synthesized de novo by plaque inflammatory cells (macrophages, T cells, mast cells) [195]. In the blood it travels mainly with LDL. Less than 20% is associated with high-density lipoprotein HDL [196]. It is an enzyme produced by inflammatory cells and hydrolyzes oxidized phospholipids in LDL. When LDLs are oxidized, Lp-PLA2 mediates their conversion in lysophosphatidylcholine and oxidized fatty acid [133].

The predictive role of Lp-PLA2 in primary and secondary level has been investigated in several studies. Folsom and colleagues studied the influence of inflammatory markers, in addition to traditional risk factors, in the prediction of CAD [120]. The traditional risk-factor model showed CAD prediction with a receiver operating characteristic area under curve of 0.8, but the addition of Lp-PLA2 showed a statistically significant increase in the area under curve. Notably, hsCRP (an acute-phase reactant that increases both in obesity and in acute and chronic systemic inflammation) did not significantly increase the area under curve. Similarly, the West of Scotland Coronary Prevention Study (WOSCOPS) showed that in multivariate analysis, Lp-PLA2 was significantly associated with risk, compared with hsCRP. However, further trials are needed to confirm an incremental prognostic value of Lp-PLA2 over hsCRP [121]. Although Lp-PLA2 at this time could be speculated to have better specificity than CRP, evidence of better risk prediction is increasing when the two complement one another [122]. The ARIC trial showed that low-to-moderate levels of both markers in combination significantly increase the risk, in comparison with a combination either of high Lp-PLA2 and low CRP or of high CRP and low Lp-PLA2 [123]. The MONICA study also showed better risk prediction when high Lp-PLA2 and high CRP were combined, when compared with each used alone [197]. The IHCS trial revealed better prediction of CAD death with combinations of high CRP/low Lp-PLA2 and high CRP/high Lp-PLA2, but not with combinations of low CRP/low Lp-PLA2 and low CRP/high Lp-PLA2 [198]. Oldgren and colleagues in Fragmin and/or Early Revascularization during Instability in Coronary Artery Disease (FRISC) II and Growing Up in Singapore Towards healthy Outcomes (GUSTO) IV, failed to find an association, in ACS patients, between Lp-PLA2 mass measurement and future CV events, including death, within 6 months−1 year of follow-up [199]. Similar results were found when measured early after an ACS event in the Pravastatin or Atorvastatin Evaluation and Infection Therapy-Thrombolysis in Myocardial Infarction (PROVE IT-TIMI) 22 trial, but at 30 days after the event, Lp-PLA2 activity in the highest quintile was associated with increased risk of additional CV events, when compared with the lowest quintile [200].

The Role of Chemokines in the Inflammatory Process

Chemokines are a family of small cytokines, or proteins secreted by cells. Their name is derived from their ability to induce directed chemotaxis in nearby responsive cells; they are chemotactic cytokines [133,201]. Proteins are classified as chemokines according to shared structural characteristics such as small size (they are all approximately 8−10 kDa in size), and the presence of four cysteine residues in conserved locations that are key to forming their 3-dimensional shape [201]. They are divided into four families (CC, CXC, CX3C, XC) based on the position of the first two cysteine residues [133]. Chemokines are made and secreted by endothelial cells and immune cells and act to attract immune cells that express the appropriate receptors on their surface [122]. By this mechanism they direct leukocytes to sites of atherosclerotic lesion and inflammation and play a pivotal role in the progression of atherosclerosis. Among the numerous chemokines the following have been found to be mainly involved in atherosclerosis: IL-8, monocyte chemoattractant protein-1, growth-related oncogene-α and interferon-γ inducible protein-10 [202−206]. These chemokines promote atherosclerosis by producing ROS, secreting enzymes, interfering in foam cell formation, platelet differentiation and activation, and by facilitating SMC migration [207,208].

The prognostic role of chemokines in cardiovascular prognosis has been mainly examined in IL-8 and MCP-1 [124,125]. In apparently healthy individuals European Prospective Investigation into Cancer (EPIC)-Norfolk population study [126] concluded that elevated plasma concentrations of IL-8 are associated with an increased risk of future CAD. This relationship appears to be stronger in men than in women and is independent of traditional cardiovascular risk factors and CRP. MCP-1 has been found to be associated with coronary artery calcium score when adjusted for traditional coronary risk factors [127]. In the etude PRospective de l'infarctus du MyocardE (the etymology is in French) (PRIME) study, higher systemic levels of regulated on activation, normal T cell expressed and secreted (RANTES) and IP-10 are independent predictors of ischemic stroke but not MCP-1 [128]. Moreover, in patients with CAD and/or peripheral artery disease MCP-1 levels were higher compared to controls [209]. In the population of patients with acute coronary syndromes, increased MCP-1 levels were associated with almost twofold increased risk of death or recurrent myocardial infarction [210]. Additionally, increased plasma levels of MCP-1, as well as the T-cell marker soluble IL-2 receptor, may be useful in predicting new coronary events independently of other inflammatory markers [211]. In conclusion, although chemokines are associated with coronary artery, further studies need to be conducted in order to be established as a clinical biomarker.

The Role of Myeloid-Related Protein 8/14 Complex in Inflammatory Process

Myeloid-related protein 8/14 complex (MRP-8/14) is a heterodimer of two myeloid calcium-binding proteins associated with different types of acute inflammatory processes (including intracellular calcium dependent signaling, cell differentiation, cell cycle progression, and cytoskeleton—membrane interactions) [133,212]. Monocytes and neutrophils are the cells that mainly secret myeloid related protein 8/14 complex [133,213]. In inflamed tissues, the MRP-8/14 complex is deposited onto the endothelium of venules associated with extravagating leukocytes and constitutes a marker of phagocyte activation [133]. MRP-8/14 has been reported to aid uptake of arachidonic acid by binding CD36, which is now recognized as a fatty acid transporter protein [214], helping in this way in the formation of the atheromatus plaque. In human carotid endarterectomy specimens, we previously described an association among high Mrp-8, Mrp-14, and Mrp-8/14 levels and the features of the rupture-prone lesions [129], suggesting a role for high levels of MRP proteins in plaque destabilization and disruption.

The predictive role of MRP-8/14 has been indicated in a study where levels of systemic Mrp-8/14 appear to increase before markers of myocardial necrosis (myoglobin, creatine kinase-MB, and troponin), and high levels are associated with increased risk of recurrent cardiovascular events [130]. Additional data from a recent study in a population of patients who underwent carotid endarterectomy, suggested that high Mrp-8/14 plasma and plaque levels are related to an increased risk of adverse cardiovascular events after a carotid endarterectomy, independent of traditional cardiovascular risk factors [131]. For imaging techniques, which aim to target plaque proteins and use them to stratify patients at risk of recurrent events, plaque Mrp-8/14 might be a potential target.

CONCLUSIONS

Our understanding of atherosclerosis has evolved beyond the view that these lesions consist of a lifeless collection of lipid debris. Current evidence supports a central role for inflammation in all phases of the atherosclerotic process. Substantial biological data implicate inflammatory pathways in early atherogenesis, in the progression of lesions, and finally in the thrombotic complications of this disease. Risk assessment of atherosclerosis can be improved through integration of bioimaging techniques with a combination of biochemical markers. The effectiveness of some biomarkers in the major cardiovascular diseases caused by the development of atherosclerotic plaques, is already well established, but for many of them additional research and testing are needed. Finally, the quest to identify proximal stimuli for inflammation, as one pathogenic process in atherogenesis or trigger to lesion complication, may yield novel therapeutic targets in years to come.

REFERENCES

[1] Canto JG, Iskandrian AE. Major risk factors for cardiovascular disease: debunking the "only 50%" myth. JAMA 2003;290(7):947–9.

[2] Hackam DG, Anand SS. Emerging risk factors for atherosclerotic vascular disease: a critical review of the evidence. JAMA 2003;290(7):932–40.

[3] Yusuf S, et al. Effect of potentially modifiable risk factors associated with myocardial infarction in 52 countries (the INTERHEART study): case-control study. Lancet 2004;364(9438):937–52.

[4] Tousoulis D, Charakida M, Stefanadis C. Endothelial function and inflammation in coronary artery disease. Heart 2006;92(4):441–4.

[5] Hansson GK. Inflammation, atherosclerosis, and coronary artery disease. N Engl J Med 2005;352(16): 1685–95.

[6] Friedman GD, Klatsky AL, Siegelaub AB. The leukocyte count as a predictor of myocardial infarction. N Engl J Med 1974;290(23):1275–8.

[7] Tousoulis D, et al. New biochemical markers in acute coronary syndromes. Curr Med Chem 2008; 15(13):1288–96.

[8] Charakida M, et al. Inflammatory and thrombotic processes are associated with vascular dysfunction in children with familial hypercholesterolemia. Atherosclerosis 2009;204(2):532–7.

[9] Tousoulis D, et al. Role of inflammation and oxidative stress in endothelial progenitor cell function and mobilization: therapeutic implications for cardiovascular diseases. Atherosclerosis 2008;201(2): 236–47.

[10] van der Giessen AG, et al. Plaque and shear stress distribution in human coronary bifurcations: a multislice computed tomography study. EuroIntervention 2009;4(5):654–61.

[11] Libby P, Ridker PM, Maseri A. Inflammation and atherosclerosis. Circulation 2002;105(9):1135–43.

[12] Zeibig S, et al. Effect of the oxLDL binding protein Fc-CD68 on plaque extension and vulnerability in atherosclerosis. Circ Res 2011;108(6):695–703.

[13] Parthasarathy S, Santanam N, Auge N. Oxidized low-density lipoprotein, a two-faced Janus in coronary artery disease? Biochem Pharmacol 1998;56(3):279–84.

[14] Johnsen SH, et al. Elevated high-density lipoprotein cholesterol levels are protective against plaque progression: a follow-up study of 1952 persons with carotid atherosclerosis the Tromso study. Circulation 2005;112(4):498–504.

[15] Chrysohoou C, et al. The emerging anti-inflammatory role of HDL-cholesterol, illustrated in cardiovascular disease free population; the ATTICA study. Int J Cardiol 2007;122(1):29–33.

[16] Ikeda U, et al. Monocyte chemoattractant protein-1 and coronary artery disease. Clin Cardiol 2002; 25(4):143–7.

[17] Tousoulis D, Charakida M, Stefanadis C. Endothelial function and inflammation in coronary artery disease. Postgrad Med J 2008;84(993):368–71.

[18] George SJ. Therapeutic potential of matrix metalloproteinase inhibitors in atherosclerosis. Expert Opin Invest Drugs 2000;9(5):993–1007.

[19] Aukrust P, et al. Chemokines and cardiovascular risk. Arterioscler Thromb Vasc Biol 2008;28(11):1909–19.
[20] Kaperonis EA, et al. Inflammation and atherosclerosis. Eur J Vasc Endovasc Surg 2006;31(4):386–93.
[21] The multiple risk factor intervention trial (MRFIT). A national study of primary prevention of coronary heart disease. JAMA 1976;235(8):825–7.
[22] de Jongh S, et al. Family history of cardiovascular events and endothelial dysfunction in children with familial hypercholesterolemia. Atherosclerosis 2002;163(1):193–7.
[23] Aljada A. Endothelium, inflammation, and diabetes. Metab Syndr Relat Disord 2003;1(1):3–21.
[24] Hsueh WA, Quinones MJ. Role of endothelial dysfunction in insulin resistance. Am J Cardiol 2003;92(4a):10j–7j.
[25] Balletshofer BM, et al. Endothelial dysfunction is detectable in young normotensive first-degree relatives of subjects with type 2 diabetes in association with insulin resistance. Circulation 2000;101(15):1780–4.
[26] Clarkson P, et al. Impaired vascular reactivity in insulin-dependent diabetes mellitus is related to disease duration and low density lipoprotein cholesterol levels. J Am Coll Cardiol 1996;28(3):573–9.
[27] Makimattila S, et al. Chronic hyperglycemia impairs endothelial function and insulin sensitivity via different mechanisms in insulin-dependent diabetes mellitus. Circulation 1996;94(6):1276–82.
[28] Quinones MJ, et al. Coronary vasomotor abnormalities in insulin-resistant individuals. Ann Intern Med 2004;140(9):700–8.
[29] Hsueh WA, Lyon CJ, Quinones MJ. Insulin resistance and the endothelium. Am J Med 2004;117(2):109–17.
[30] Navab M, et al. The Yin and Yang of oxidation in the development of the fatty streak. A review based on the 1994 George Lyman Duff Memorial Lecture. Arterioscler Thromb Vasc Biol 1996;16(7):831–42.
[31] Ross R. Atherosclerosis—an inflammatory disease. N Engl J Med 1999;340(2):115–26.
[32] Vora DK, et al. Induction of P-selectin by oxidized lipoproteins. Separate effects on synthesis and surface expression. Circ Res 1997;80(6):810–8.
[33] Cushing SD, et al. Minimally modified low density lipoprotein induces monocyte chemotactic protein 1 in human endothelial cells and smooth muscle cells. Proc Natl Acad Sci USA 1990;87(13):5134–8.
[34] Rajavashisth TB, et al. Induction of endothelial cell expression of granulocyte and macrophage colony-stimulating factors by modified low-density lipoproteins. Nature 1990;344(6263):254–7.
[35] Varughese GI. The impact of diabetes mellitus on endothelial dysfunction. South Med J 2007;100(2):128–9.
[36] Chan NN, Chan JC. Asymmetric dimethylarginine (ADMA): a potential link between endothelial dysfunction and cardiovascular diseases in insulin resistance syndrome? Diabetologia 2002;45(12):1609–16.
[37] Panza JA, et al. Abnormal endothelium-dependent vascular relaxation in patients with essential hypertension. N Engl J Med 1990;323(1):22–7.
[38] Linder L, et al. Indirect evidence for release of endothelium-derived relaxing factor in human forearm circulation in vivo. Blunted response in essential hypertension. Circulation 1990;81(6):1762–7.
[39] Brush Jr JE, et al. Abnormal endothelium-dependent coronary vasomotion in hypertensive patients. J Am Coll Cardiol 1992;19(4):809–15.
[40] Panza JA, et al. Role of endothelium-derived nitric oxide in the abnormal endothelium-dependent vascular relaxation of patients with essential hypertension. Circulation 1993;87(5):1468–74.
[41] Panza JA, et al. Effect of increased availability of endothelium-derived nitric oxide precursor on endothelium-dependent vascular relaxation in normal subjects and in patients with essential hypertension. Circulation 1993;87(5):1475–81.
[42] Panza JA, et al. Impaired endothelium-dependent vasodilation in patients with essential hypertension: evidence that the abnormality is not at the muscarinic receptor level. J Am Coll Cardiol 1994;23(7):1610–6.
[43] Panza JA, et al. Impaired endothelium-dependent vasodilation in patients with essential hypertension. Evidence that nitric oxide abnormality is not localized to a single signal transduction pathway. Circulation 1995;91(6):1732–8.

[44] Cardillo C, et al. Decreased vasodilator response to isoproterenol during nitric oxide inhibition in humans. Hypertension 1997;30(4):918–21.
[45] Cardillo C, et al. Selective defect in nitric oxide synthesis may explain the impaired endothelium-dependent vasodilation in patients with essential hypertension. Circulation 1998;97(9):851–6.
[46] Perticone F, et al. Asymmetric dimethylarginine, L-arginine, and endothelial dysfunction in essential hypertension. J Am Coll Cardiol 2005;46(3):518–23.
[47] Thomopoulos C, et al. Obstructive sleep apnoea syndrome is associated with enhanced sub-clinical inflammation and asymmetric dimethyl-arginine levels in hypertensives. J Hum Hypertens 2009; 23(1):65–7.
[48] Cardillo C, et al. Xanthine oxidase inhibition with oxypurinol improves endothelial vasodilator function in hypercholesterolemic but not in hypertensive patients. Hypertension 1997;30(1 Pt 1):57–63.
[49] Campia U, et al. Role of cyclooxygenase products in the regulation of vascular tone and in the endothelial vasodilator function of normal, hypertensive, and hypercholesterolemic humans. Am J Cardiol 2002;89(3):286–90.
[50] Cardillo C, et al. Improved endothelium-dependent vasodilation after blockade of endothelin receptors in patients with essential hypertension. Circulation 2002;105(4):452–6.
[51] Annuk M, et al. Endothelial function, CRP and oxidative stress in chronic kidney disease. J Nephrol 2005;18(6):721–6.
[52] Petrie MC, et al. Angiotensin converting enzyme (ACE) and non-ACE dependent angiotensin II generation in resistance arteries from patients with heart failure and coronary heart disease. J Am Coll Cardiol 2001;37(4):1056–61.
[53] Tamarat R, et al. Angiotensin II angiogenic effect in vivo involves vascular endothelial growth factor- and inflammation-related pathways. Lab Invest 2002;82(6):747–56.
[54] Chen R, et al. Important role of nitric oxide in the effect of angiotensin-converting enzyme inhibitor imidapril on vascular injury. Hypertension 2003;42(4):542–7.
[55] Sesso HD, et al. C-reactive protein and the risk of developing hypertension. JAMA 2003;290(22): 2945–51.
[56] Tsioufis C, et al. Inter-relationships of microalbuminuria with the other surrogates of the atherosclerotic cardiovascular disease in hypertensive subjects. Am J Hypertens 2004;17(5 Pt 1):470–6.
[57] Tsioufis C, et al. Low-grade inflammation and hypoadiponectinaemia have an additive detrimental effect on aortic stiffness in essential hypertensive patients. Eur Heart J 2007;28(9):1162–9.
[58] Tsioufis C, et al. Diverse associations of microalbuminuria with C-reactive protein, interleukin-18 and soluble CD 40 ligand in male essential hypertensive subjects. Am J Hypertens 2006;19(5):462–6.
[59] Paniagua OA, Bryant MB, Panza JA. Role of endothelial nitric oxide in shear stress-induced vasodilation of human microvasculature: diminished activity in hypertensive and hypercholesterolemic patients. Circulation 2001;103(13):1752–8.
[60] Ueno H, et al. Blood flow regulates the development of vascular hypertrophy, smooth muscle cell proliferation, and endothelial cell nitric oxide synthase in hypertension. Hypertension 2000;36(1): 89–96.
[61] Tentolouris C, et al. Effects of smoking on nitric oxide synthesis in epicardial normal and atheromatous coronary arteries. Int J Cardiol 2004;95(1):69–73.
[62] Barua RS, et al. Dysfunctional endothelial nitric oxide biosynthesis in healthy smokers with impaired endothelium-dependent vasodilatation. Circulation 2001;104(16):1905–10.
[63] Zhang JY, et al. Lipid-soluble smoke particles damage endothelial cells and reduce endothelium-dependent dilatation in rat and man. BMC Cardiovasc Disord 2006;6:3.
[64] Cohen RA, et al. Loss of selective endothelial cell vasoactive functions caused by hypercholesterolemia in pig coronary arteries. Circ Res 1988;63(5):903–10.
[65] Michaud SE, et al. Circulating endothelial progenitor cells from healthy smokers exhibit impaired functional activities. Atherosclerosis 2006;187(2):423–32.
[66] Su Y, et al. Cigarette smoke extract inhibits angiogenesis of pulmonary artery endothelial cells: the role of calpain. Am J Physiol Lung Cell Mol Physiol 2004;287(4):L794–800.
[67] Michaud SE, et al. Cigarette smoke exposure impairs VEGF-induced endothelial cell migration: role of NO and reactive oxygen species. J Mol Cell Cardiol 2006;41(2):275–84.

[68] Goerre S, et al. Effect of cigarette smoking and nicotine on plasma endothelin-1 levels. J Cardiovasc Pharmacol 1995;26(Suppl. 3):S236–8.
[69] Lee WO, Wright SM. Production of endothelin by cultured human endothelial cells following exposure to nicotine or caffeine. Metabolism 1999;48(7):845–8.
[70] Rahman MM, et al. Increased vascular contractility in isolated vessels from cigarette smoking rats is mediated by basal endothelin release. Vasc Pharmacol 2007;46(1):35–42.
[71] Granstrom BW, et al. Smoking particles enhance endothelin A and endothelin B receptor-mediated contractions by enhancing translation in rat bronchi. BMC Pulm Med 2006;6:6.
[72] Migneault A, et al. Chronically elevated endothelin levels reduce pulmonary vascular reactivity to nitric oxide. Am J Respir Crit Care Med 2005;171(5):506–13.
[73] Iida M, et al. Mechanisms underlying cerebrovascular effects of cigarette smoking in rats in vivo. Stroke 1998;29(8):1656–65.
[74] Gerzanich V, et al. Chronic nicotine alters NO signaling of Ca(2+) channels in cerebral arterioles. Circ Res 2001;88(3):359–65.
[75] Jiang DJ, et al. Involvement of DDAH/ADMA/NOS pathway in nicotine-induced endothelial dysfunction. Biochem Biophys Res Commun 2006;349(2):683–93.
[76] Xiao D, et al. Direct effects of nicotine on contractility of the uterine artery in pregnancy. J Pharmacol Exp Ther 2007;322(1):180–5.
[77] Miller VM, et al. Time and dose effect of transdermal nicotine on endothelial function. Am J Physiol Heart Circ Physiol 2000;279(4):H1913–21.
[78] Nagy J, et al. Induction of endothelial cell injury by cigarette smoke. Endothelium 1997;5(4): 251–63.
[79] Wang XL, et al. Genotype dependent and cigarette specific effects on endothelial nitric oxide synthase gene expression and enzyme activity. FEBS Lett 2000;471(1):45–50.
[80] Su Y, et al. Effect of cigarette smoke extract on nitric oxide synthase in pulmonary artery endothelial cells. Am J Respir Cell Mol Biol 1998;19(5):819–25.
[81] Cooper KO, Witz G, Witmer C. The effects of alpha, beta-unsaturated aldehydes on hepatic thiols and thiol-containing enzymes. Fundam Appl Toxicol 1992;19(3):343–9.
[82] Raij L, DeMaster EG, Jaimes EA. Cigarette smoke-induced endothelium dysfunction: role of superoxide anion. J Hypertens 2001;19(5):891–7.
[83] Barua RS, et al. Reactive oxygen species are involved in smoking-induced dysfunction of nitric oxide biosynthesis and upregulation of endothelial nitric oxide synthase: an in vitro demonstration in human coronary artery endothelial cells. Circulation 2003;107(18):2342–7.
[84] Jaimes EA, et al. Stable compounds of cigarette smoke induce endothelial superoxide anion production via NADPH oxidase activation. Arterioscler Thromb Vasc Biol 2004;24(6):1031–6.
[85] Heitzer T, et al. Tetrahydrobiopterin improves endothelium-dependent vasodilation in chronic smokers: evidence for a dysfunctional nitric oxide synthase. Circ Res 2000;86(2):E36–41.
[86] Siasos G, et al. The impact of oral L-arginine supplementation on acute smoking-induced endothelial injury and arterial performance. Am J Hypertens 2009;22(6):586–92.
[87] Siasos G, et al. L-Arginine, the substrate for NO synthesis: an alternative treatment for premature atherosclerosis? Int J Cardiol 2007;116(3):300–8.
[88] Prasad A, et al. Predisposition to atherosclerosis by infections: role of endothelial dysfunction. Circulation 2002;106(2):184–90.
[89] Weber C, Zernecke A, Libby P. The multifaceted contributions of leukocyte subsets to atherosclerosis: lessons from mouse models. Nat Rev Immunol 2008;8(10):802–15.
[90] Daugherty A, Rateri DL. T lymphocytes in atherosclerosis: the yin-yang of Th1 and Th2 influence on lesion formation. Circ Res 2002;90(10):1039–40.
[91] Methe H, et al. Enhanced T-helper-1 lymphocyte activation patterns in acute coronary syndromes. J Am Coll Cardiol 2005;45(12):1939–45.
[92] Leon ML, Zuckerman SH. Gamma interferon: a central mediator in atherosclerosis. Inflamm Res 2005;54(10):395–411.
[93] Johnson JL. Matrix metalloproteinases: influence on smooth muscle cells and atherosclerotic plaque stability. Expert Rev Cardiovasc Ther 2007;5(2):265–82.

[94] Jacob HS, et al. Herpes virus infection of endothelium: new insights into atherosclerosis. Trans Am Clin Climatol Assoc 1992;103:95–104.
[95] Virdis A, et al. Mechanisms responsible for endothelial dysfunction induced by fasting hyperhomocystinemia in normotensive subjects and patients with essential hypertension. J Am Coll Cardiol 2001;38(4):1106–15.
[96] Stuhlinger MC, et al. Homocysteine impairs the nitric oxide synthase pathway: role of asymmetric dimethylarginine. Circulation 2001;104(21):2569–75.
[97] Stuhlinger MC, et al. Endothelial dysfunction induced by hyperhomocyst(e)inemia: role of asymmetric dimethylarginine. Circulation 2003;108(8):933–8.
[98] Wanby P, et al. Asymmetric dimethylarginine and total homocysteine in plasma after oral methionine loading. Scand J Clin Lab Invest 2003;63(5):347–53.
[99] Mallamaci F, et al. Hyperhomocysteinemia predicts cardiovascular outcomes in hemodialysis patients. Kidney Int 2002;61(2):609–14.
[100] Lee WH, et al. Genetic factors associated with endothelial dysfunction affect the early onset of coronary artery disease in Korean males. Vasc Med 2001;6(2):103–8.
[101] Imamura A, et al. Endothelial nitric oxide synthase and methylenetetrahydrofolate reductase gene polymorphisms are associated with endothelial dysfunction in young, healthy men. Can J Cardiol 2004;20(12):1229–34.
[102] Hwang SJ, et al. Circulating adhesion molecules VCAM-1, ICAM-1, and E-selectin in carotid atherosclerosis and incident coronary heart disease cases: the Atherosclerosis Risk in Communities (ARIC) study. Circulation 1997;96(12):4219–25.
[103] de Lemos JA, Hennekens CH, Ridker PM. Plasma concentration of soluble vascular cell adhesion molecule-1 and subsequent cardiovascular risk. J Am Coll Cardiol 2000;36(2):423–6.
[104] Ridker PM, et al. Plasma concentration of soluble intercellular adhesion molecule 1 and risks of future myocardial infarction in apparently healthy men. Lancet 1998;351(9096):88–92.
[105] Cao JJ, et al. C-reactive protein, carotid intima-media thickness, and incidence of ischemic stroke in the elderly: the Cardiovascular Health Study. Circulation 2003;108(2):166–70.
[106] Ridker PM, et al. Inflammation, aspirin, and the risk of cardiovascular disease in apparently healthy men. N Engl J Med 1997;336(14):973–9.
[107] Curb JD, et al. C-reactive protein and the future risk of thromboembolic stroke in healthy men. Circulation 2003;107(15):2016–20.
[108] Kuch B, et al. Differential impact of admission C-reactive protein levels on 28-day mortality risk in patients with ST-elevation versus non-ST-elevation myocardial infarction (from the monitoring trends and determinants on cardiovascular diseases [MONICA]/Cooperative health research in the region of Augsburg [KORA] Augsburg myocardial infarction Registry). Am J Cardiol 2008; 102(9):1125–30.
[109] Kannel WB, et al. Diabetes, fibrinogen, and risk of cardiovascular disease: the Framingham experience. Am Heart J 1990;120(3):672–6.
[110] Kannel WB, et al. Fibrinogen and risk of cardiovascular disease. The Framingham Study. JAMA 1987;258(9):1183–6.
[111] Juhan-Vague I, Thompson SG, Jespersen J. Involvement of the hemostatic system in the insulin resistance syndrome. A study of 1500 patients with angina pectoris. The ECAT Angina Pectoris Study Group. Arterioscler Thromb 1993;13(12):1865–73.
[112] Sinning JM, et al. Impact of C-reactive protein and fibrinogen on cardiovascular prognosis in patients with stable angina pectoris: the AtheroGene study. Eur Heart J 2006;27(24):2962–8.
[113] Maple-Brown LJ, et al. Fibrinogen and associated risk factors in a high-risk population: urban Indigenous Australians, the DRUID Study. Cardiovasc Diabetol 2010;9:69.
[114] Ridker PM, et al. C-reactive protein and other markers of inflammation in the prediction of cardiovascular disease in women. N Engl J Med 2000;342(12):836–43.
[115] Harris TB, et al. Associations of elevated interleukin-6 and C-reactive protein levels with mortality in the elderly. Am J Med 1999;106(5):506–12.
[116] Yamagishi S, et al. Decreased high-density lipoprotein cholesterol level is an independent correlate of circulating tumor necrosis factor-alpha in a general population. Clin Cardiol 2009;32(9):E29–32.

[117] Ridker PM, et al. Plasma concentration of interleukin-6 and the risk of future myocardial infarction among apparently healthy men. Circulation 2000;101(15):1767–72.
[118] Hwang SJ, et al. High plasma interleukin-6 is associated with drug-eluting stent thrombosis: possible role of inflammatory cytokines in the development of stent thrombosis from the Korea Stent Thrombosis Registry. Circ J 2011;75(6):1350–7.
[119] Lee KW, et al. Circulating endothelial cells, von Willebrand factor, interleukin-6, and prognosis in patients with acute coronary syndromes. Blood 2005;105(2):526–32.
[120] Folsom AR, et al. An assessment of incremental coronary risk prediction using C-reactive protein and other novel risk markers: the atherosclerosis risk in communities study. Arch Intern Med 2006; 166(13):1368–73.
[121] Packard CJ, et al. Lipoprotein-associated phospholipase A2 as an independent predictor of coronary heart disease. West of Scotland Coronary Prevention Study Group. N Engl J Med 2000;343(16): 1148–55.
[122] Madjid M, Ali M, Willerson JT. Lipoprotein-associated phospholipase A2 as a novel risk marker for cardiovascular disease: a systematic review of the literature. Tex Heart Inst J 2010;37(1):25–39.
[123] Ballantyne CM, et al. Lipoprotein-associated phospholipase A2, high-sensitivity C-reactive protein, and risk for incident coronary heart disease in middle-aged men and women in the Atherosclerosis Risk in Communities (ARIC) study. Circulation 2004;109(7):837–42.
[124] Boring L, et al. Decreased lesion formation in CCR2-/- mice reveals a role for chemokines in the initiation of atherosclerosis. Nature 1998;394(6696):894–7.
[125] Gu L, et al. Absence of monocyte chemoattractant protein-1 reduces atherosclerosis in low density lipoprotein receptor-deficient mice. Mol Cell 1998;2(2):275–81.
[126] Boekholdt SM, et al. IL-8 plasma concentrations and the risk of future coronary artery disease in apparently healthy men and women: the EPIC-Norfolk prospective population study. Arterioscler Thromb Vasc Biol 2004;24(8):1503–8.
[127] Deo R, et al. Association among plasma levels of monocyte chemoattractant protein-1, traditional cardiovascular risk factors, and subclinical atherosclerosis. J Am Coll Cardiol 2004;44(9):1812–8.
[128] Canoui-Poitrine F, et al. Systemic chemokine levels, coronary heart disease, and ischemic stroke events: the PRIME study. Neurology 2011;77(12):1165–73.
[129] Ionita MG, et al. High levels of myeloid-related protein 14 in human atherosclerotic plaques correlate with the characteristics of rupture-prone lesions. Arterioscler Thromb Vasc Biol 2009;29(8): 1220–7.
[130] Morrow DA, et al. Myeloid-related protein 8/14 and the risk of cardiovascular death or myocardial infarction after an acute coronary syndrome in the Pravastatin or Atorvastatin Evaluation and Infection Therapy: Thrombolysis in Myocardial Infarction (PROVE IT-TIMI 22) trial. Am Heart J 2008; 155(1):49–55.
[131] Ionita MG, et al. High myeloid-related protein: 8/14 levels are related to an increased risk of cardiovascular events after carotid endarterectomy. Stroke 2010;41(9):2010–5.
[132] Krieglstein CF, Granger DN. Adhesion molecules and their role in vascular disease. Am J Hypertens 2001;14(6 Pt 2):44s–54s.
[133] Stefanadi E, et al. Inflammatory biomarkers predicting events in atherosclerosis. Curr Med Chem 2010;17(16):1690–707.
[134] Chia MC. The role of adhesion molecules in atherosclerosis. Crit Rev Clin Lab Sci 1998;35(6): 573–602.
[135] De Caterina R, et al. Soluble vascular cell adhesion molecule-1 as a biohumoral correlate of atherosclerosis. Arterioscler Thromb Vasc Biol 1997;17(11):2646–54.
[136] Rohde LE, et al. Circulating cell adhesion molecules are correlated with ultrasound-based assessment of carotid atherosclerosis. Arterioscler Thromb Vasc Biol 1998;18(11):1765–70.
[137] Blann AD, et al. Circulating ICAM-1 and VCAM-1 in peripheral artery disease and hypercholesterolaemia: relationship to the location of atherosclerotic disease, smoking, and in the prediction of adverse events. Thromb Haemost 1998;79(6):1080–5.
[138] Smieja M, et al. Multiple infections and subsequent cardiovascular events in the heart outcomes prevention evaluation (HOPE) study. Circulation 2003;107(2):251–7.

[139] Mulvihill NT, et al. Enhanced endothelial activation in diabetic patients with unstable angina and non-Q-wave myocardial infarction. Diabet Med 2001;18(12):979–83.
[140] Parker 3rd C, Vita JA, Freedman JE. Soluble adhesion molecules and unstable coronary artery disease. Atherosclerosis 2001;156(2):417–24.
[141] Mulvihill NT, et al. Evidence of prolonged inflammation in unstable angina and non-Q wave myocardial infarction. J Am Coll Cardiol 2000;36(4):1210–6.
[142] Turhan H, et al. Plasma soluble adhesion molecules; intercellular adhesion molecule-1, vascular cell adhesion molecule-1 and E-selectin levels in patients with isolated coronary artery ectasia. Coron Artery Dis 2005;16(1):45–50.
[143] Mulvihill NT, et al. Risk stratification in unstable angina and non-Q wave myocardial infarction using soluble cell adhesion molecules. Heart 2001;85(6):623–7.
[144] Volanakis JE. Human C-reactive protein: expression, structure, and function. Mol Immunol 2001; 38(2–3):189–97.
[145] Srinivasan N, et al. Comparative analyses of pentraxins: implications for protomer assembly and ligand binding. Structure 1994;2(11):1017–27.
[146] Lefkou E, Fragakis N, Varlamis G. The inflammatory process during childhood and the risk of developing atheromatous disease in adult life: the role of C-reactive protein. Hell J Cardiol 2006;47(3):164–9.
[147] Verma S, Szmitko PE, Yeh ET. C-reactive protein: structure affects function. Circulation 2004; 109(16):1914–7.
[148] Cermak J, et al. C-reactive protein induces human peripheral blood monocytes to synthesize tissue factor. Blood 1993;82(2):513–20.
[149] Pearson TA, et al. Markers of inflammation and cardiovascular disease: application to clinical and public health practice: a statement for healthcare professionals from the Centers for Disease Control and Prevention and the American Heart Association. Circulation 2003;107(3):499–511.
[150] Hein TW, et al. Human C-reactive protein induces endothelial dysfunction and uncoupling of eNOS in vivo. Atherosclerosis 2009;206(1):61–8.
[151] Singh U, et al. C-reactive protein decreases endothelial nitric oxide synthase activity via uncoupling. J Mol Cell Cardiol 2007;43(6):780–91.
[152] Takeya M, et al. Detection of monocyte chemoattractant protein-1 in human atherosclerotic lesions by an anti-monocyte chemoattractant protein-1 monoclonal antibody. Hum Pathol 1993;24(5): 534–9.
[153] Tuomisto K, et al. C-reactive protein, interleukin-6 and tumor necrosis factor alpha as predictors of incident coronary and cardiovascular events and total mortality. A population-based, prospective study. Thromb Haemost 2006;95(3):511–8.
[154] Haverkate F, et al. Production of C-reactive protein and risk of coronary events in stable and unstable angina. European concerted action on thrombosis and Disabilities angina pectoris study group. Lancet 1997;349(9050):462–6.
[155] Eldrup N, et al. Prognosis by C-reactive protein and matrix metalloproteinase-9 levels in stable coronary heart disease during 15 years of follow-up. Nutr Metab Cardiovasc Dis 2012;22(8):677–83.
[156] Toss H, et al. Prognostic influence of increased fibrinogen and C-reactive protein levels in unstable coronary artery disease. FRISC Study Group. Fragmin during Instability in Coronary Artery Disease. Circulation 1997;96(12):4204–10.
[157] Liu HH, et al. C-reactive protein predicts the severity of coronary artery disease beyond low-density lipoprotein cholesterol. Angiology 2012;63(3):218–22.
[158] Yun KH, et al. Response of high-sensitivity C-reactive protein to percutaneous coronary intervention in patients with acute coronary syndrome. Heart Vessels 2009;24(3):175–80.
[159] Choi DH, et al. Renal dysfunction and high levels of hsCRP are additively associated with hard endpoints after percutaneous coronary intervention with drug eluting stents. Int J Cardiol 2011;149(2):174–81.
[160] Gao F, et al. C-reactive protein and the risk of contrast-induced acute kidney injury in patients undergoing percutaneous coronary intervention. Am J Nephrol 2011;34(3):203–10.
[161] Hong YJ, et al. Relation between high-sensitivity C-reactive protein and coronary plaque components in patients with acute coronary syndrome: virtual histology-intravascular ultrasound analysis. Korean Circ J 2011;41(8):440–6.

[162] Kamath S, Lip GY. Fibrinogen: biochemistry, epidemiology and determinants. QJM 2003;96(10): 711–29.
[163] Ernst E, Resch KL. Fibrinogen as a cardiovascular risk factor: a meta-analysis and review of the literature. Ann Intern Med 1993;118(12):956–63.
[164] Miyao Y, et al. Elevated plasma interleukin-6 levels in patients with acute myocardial infarction. Am Heart J 1993;126(6):1299–304.
[165] Smith EB, et al. Fate of fibrinogen in human arterial intima. Arteriosclerosis 1990;10(2):263–75.
[166] Stroncek DF, Shankar RA, Skubitz KM. The subcellular distribution of myeloid-related protein 8 (MRP8) and MRP14 in human neutrophils. J Transl Med 2005;3:36.
[167] Thompson WD, Smith EB. Atherosclerosis and the coagulation system. J Pathol 1989;159(2): 97–106.
[168] Altieri DC, et al. Oligospecificity of the cellular adhesion receptor Mac-1 encompasses an inducible recognition specificity for fibrinogen. J Cell Biol 1988;107(5):1893–900.
[169] Colman RW. Interactions between the contact system, neutrophils and fibrinogen. Adv Exp Med Biol 1990;281:105–20.
[170] van de Stolpe A, et al. Fibrinogen binding to ICAM-1 on EA.hy 926 endothelial cells is dependent on an intact cytoskeleton. Thromb Haemost 1996;75(1):182–9.
[171] Duperray A, et al. Molecular identification of a novel fibrinogen binding site on the first domain of ICAM-1 regulating leukocyte-endothelium bridging. J Biol Chem 1997;272(1):435–41.
[172] Harley SL, Sturge J, Powell JT. Regulation by fibrinogen and its products of intercellular adhesion molecule-1 expression in human saphenous vein endothelial cells. Arterioscler Thromb Vasc Biol 2000;20(3):652–8.
[173] Gardiner EE, D'Souza SE. A mitogenic action for fibrinogen mediated through intercellular adhesion molecule-1. J Biol Chem 1997;272(24):15474–80.
[174] Forsyth CB, et al. Integrin alpha(M)beta(2)-mediated cell migration to fibrinogen and its recognition peptides. J Exp Med 2001;193(10):1123–33.
[175] Rubel C, et al. Fibrinogen promotes neutrophil activation and delays apoptosis. J Immunol 2001; 166(3):2002–10.
[176] Danesh J, et al. Plasma fibrinogen level and the risk of major cardiovascular diseases and nonvascular mortality: an individual participant meta-analysis. JAMA 2005;294(14):1799–809.
[177] Wattanakit K, et al. Risk factors for peripheral arterial disease incidence in persons with diabetes: the Atherosclerosis Risk in Communities (ARIC) Study. Atherosclerosis 2005;180(2):389–97.
[178] Panagiotakos DB, et al. Importance of LDL/HDL cholesterol ratio as a predictor for coronary heart disease events in patients with heterozygous familial hypercholesterolaemia: a 15-year follow-up (1987–2002). Curr Med Res Opin 2003;19(2):89–94.
[179] Reiner AP, et al. Association between patterns of nucleotide variation across the three fibrinogen genes and plasma fibrinogen levels: the Coronary Artery Risk Development in Young Adults (CARDIA) study. J Thromb Haemost 2006;4(6):1279–87.
[180] Kardys I, et al. Fibrinogen gene haplotypes in relation to risk of coronary events and coronary and extracoronary atherosclerosis: the Rotterdam Study. Thromb Haemost 2007;97(2):288–95.
[181] Carty CL, et al. Associations between common fibrinogen gene polymorphisms and cardiovascular disease in older adults. The Cardiovascular Health Study. Thromb Haemost 2008;99(2):388–95.
[182] Siegerink B, Rosendaal FR, Algra A. Genetic variation in fibrinogen; its relationship to fibrinogen levels and the risk of myocardial infarction and ischemic stroke. J Thromb Haemost 2009;7(3): 385–90.
[183] Rallidis LS, et al. G-455A polymorphism of beta-fibrinogen gene and the risk of premature myocardial infarction in Greece. Thromb Res 2010;125(1):34–7.
[184] Wassel CL, et al. Association of genomic loci from a cardiovascular gene SNP array with fibrinogen levels in European Americans and African-Americans from six cohort studies: the Candidate Gene Association Resource (CARe). Blood 2011;117(1):268–75.
[185] Kelso A. Cytokines: structure, function and synthesis. Curr Opin Immunol 1989;2(2):215–25.
[186] Somers W, Stahl M, Seehra JS. 1.9 A crystal structure of interleukin 6: implications for a novel mode of receptor dimerization and signaling. Embo J 1997;16(5):989–97.

[187] Perez C, et al. A nonsecretable cell surface mutant of tumor necrosis factor (TNF) kills by cell-to-cell contact. Cell 1990;63(2):251–8.
[188] Young JL, Libby P, Schonbeck U. Cytokines in the pathogenesis of atherosclerosis. Thromb Haemost 2002;88(4):554–67.
[189] Tzoulaki I, et al. C-reactive protein, interleukin-6, and soluble adhesion molecules as predictors of progressive peripheral atherosclerosis in the general population: Edinburgh Artery Study. Circulation 2005;112(7):976–83.
[190] Szekanecz Z. Pro-inflammatory cytokines in atherosclerosis. Isr Med Assoc J 2008;10(7):529–30.
[191] Popa C, et al. The role of TNF-alpha in chronic inflammatory conditions, intermediary metabolism, and cardiovascular risk. J Lipid Res 2007;48(4):751–62.
[192] Fan ZX, et al. Interleukin-6, but not soluble adhesion molecules, predicts a subsequent mortality from cardiovascular disease in patients with acute ST-segment elevation myocardial infarction. Cell Biochem Biophys 2011;61(2):443–8.
[193] Ridker PM, et al. Elevation of tumor necrosis factor-alpha and increased risk of recurrent coronary events after myocardial infarction. Circulation 2000;101(18):2149–53.
[194] Nakagomi A, et al. Upregulation of monocyte proinflammatory cytokine production by C-reactive protein is significantly related to ongoing myocardial damage and future cardiac events in patients with chronic heart failure. J Card Fail 2010;16(7):562–71.
[195] Zalewski A, Macphee C. Role of lipoprotein-associated phospholipase A2 in atherosclerosis: biology, epidemiology, and possible therapeutic target. Arterioscler Thromb Vasc Biol 2005;25(5):923–31.
[196] Caslake MJ, et al. Lipoprotein-associated phospholipase A(2), platelet-activating factor acetylhydrolase: a potential new risk factor for coronary artery disease. Atherosclerosis 2000;150(2):413–9.
[197] Koenig W, et al. Lipoprotein-associated phospholipase A2 adds to risk prediction of incident coronary events by C-reactive protein in apparently healthy middle-aged men from the general population: results from the 14-year follow-up of a large cohort from southern Germany. Circulation 2004; 110(14):1903–8.
[198] Anderson JL. Lipoprotein-associated phospholipase A2: an independent predictor of coronary artery disease events in primary and secondary prevention. Am J Cardiol 2008;101(12a):23f–33f.
[199] Oldgren J, et al. Lipoprotein-associated phospholipase A2 does not predict mortality or new ischaemic events in acute coronary syndrome patients. Eur Heart J 2007;28(6):699–704.
[200] Serruys PW, et al. Effects of the direct lipoprotein-associated phospholipase A(2) inhibitor darapladib on human coronary atherosclerotic plaque. Circulation 2008;118(11):1172–82.
[201] Braunersreuther V, Mach F, Steffens S. The specific role of chemokines in atherosclerosis. Thromb Haemost 2007;97(5):714–21.
[202] Breland UM, et al. A potential role of the CXC chemokine GROalpha in atherosclerosis and plaque destabilization: downregulatory effects of statins. Arterioscler Thromb Vasc Biol 2008;28(5):1005–11.
[203] Cheng C, et al. Shear stress-induced changes in atherosclerotic plaque composition are modulated by chemokines. J Clin Invest 2007;117(3):616–26.
[204] Damas JK, et al. Enhanced expression of the homeostatic chemokines CCL19 and CCL21 in clinical and experimental atherosclerosis: possible pathogenic role in plaque destabilization. Arterioscler Thromb Vasc Biol 2007;27(3):614–20.
[205] Lee WH, et al. A novel chemokine, Leukotactin-1, induces chemotaxis, pro-atherogenic cytokines, and tissue factor expression in atherosclerosis. Atherosclerosis 2002;161(2):255–60.
[206] Smith C, et al. Increased levels of neutrophil-activating peptide-2 in acute coronary syndromes: possible role of platelet-mediated vascular inflammation. J Am Coll Cardiol 2006;48(8):1591–9.
[207] Kodali RB, et al. CCL11 (Eotaxin) induces CCR3-dependent smooth muscle cell migration. Arterioscler Thromb Vasc Biol 2004;24(7):1211–6.
[208] Weber C. Platelets and chemokines in atherosclerosis: partners in crime. Circ Res 2005;96(6):612–6.
[209] Hoogeveen RC, et al. Plasma MCP-1 level and risk for peripheral arterial disease and incident coronary heart disease: atherosclerosis Risk in Communities study. Atherosclerosis 2005;183(2):301–7.
[210] de Lemos JA, et al. Association between plasma levels of monocyte chemoattractant protein-1 and long-term clinical outcomes in patients with acute coronary syndromes. Circulation 2003;107(5): 690–5.

[211] Kervinen H, et al. Prognostic usefulness of plasma monocyte/macrophage and T-lymphocyte activation markers in patients with acute coronary syndromes. Am J Cardiol 2004;94(8):993–6.
[212] Schafer BW, Heizmann CW. The S100 family of EF-hand calcium-binding proteins: functions and pathology. Trends Biochem Sci 1996;21(4):134–40.
[213] Hessian PA, Edgeworth J, Hogg N. MRP-8 and MRP-14, two abundant Ca(2+)-binding proteins of neutrophils and monocytes. J Leukoc Biol 1993;53(2):197–204.
[214] Abumrad NA, et al. Cloning of a rat adipocyte membrane protein implicated in binding or transport of long-chain fatty acids that is induced during preadipocyte differentiation. Homology with human CD36. J Biol Chem 1993;268(24):17665–8.

FURTHER READING

[1] Virchow R. Cellular pathology. As based upon physiological and pathological histology. Lecture XVI—Atheromatous affection of arteries. 1858. Nutr Rev 1989;47(1):23–5.
[2] Kriszbacher I, Koppan M, Bodis J. Inflammation, atherosclerosis, and coronary artery disease. N Engl J Med 2005;353(4):429–30. Author reply 429–430.
[3] Mogensen TH. Pathogen recognition and inflammatory signaling in innate immune defenses. Clin Microbiol Rev 2009;22(2):240–73.
[4] Niessner A, Weyand CM. Dendritic cells in atherosclerotic disease. Clin Immunol 2010;134(1):25–32.
[5] Mallat Z, et al. The role of adaptive T cell immunity in atherosclerosis. J Lipid Res 2009;(Suppl. 50):S364–9.
[6] Libby P, Lichtman AH, Hansson GK. Immune effector mechanisms implicated in atherosclerosis: from mice to humans. Immunity 2013;38(6):1092–104.
[7] Andersson J, Libby P, Hansson GK. Adaptive immunity and atherosclerosis. Clin Immunol 2010;134(1):33–46.
[8] Tsimikas S, Hall JL. Lipoprotein(a) as a potential causal genetic risk factor of cardiovascular disease: a rationale for increased efforts to understand its pathophysiology and develop targeted therapies. J Am Coll Cardiol 2012;60(8):716–21.
[9] Chou MY, et al. Oxidation-specific epitopes are important targets of innate immunity. J Intern Med 2008;263(5):479–88.
[10] Hermansson A, et al. Inhibition of T cell response to native low-density lipoprotein reduces atherosclerosis. J Exp Med 2010;207(5):1081–93.
[11] Taleb S, Tedgui A, Mallat Z. Regulatory T-cell immunity and its relevance to atherosclerosis. J Intern Med 2008;263(5):489–99.
[12] Lichtman AH, et al. Adaptive immunity in atherogenesis: new insights and therapeutic approaches. J Clin Invest 2013;123(1):27–36.

CHAPTER 1.6

The Role of Oxidative Stress

Sotirios Tsalamandris, Georgia Vogiatzi, Alexios S. Antonopoulos, Dimitris Tousoulis
1st Cardiology Clinic, 'Hippokration' General Hospital, National and Kapodistrian University of Athens, School of Medicine, Athens, Greece

Contents

INTRODUCTION

The vascular endothelium consists of a thin semipermeable layer of cells upon the basal membrane covering the internal surface of vessels and forming a boundary between the vessel wall and the blood flow. Its structural and functional integrity is vital for the protection of the vessel wall and circulatory function [1]. Over the last decade, the vascular endothelium has emerged mostly as a paracrine organ responsible for the secretion of several beneficial substances with antiatherogenic effects [2]. One of these chemical molecules, nitric oxide (NO) is widely known for its antithrombotic, antiapoptotic, anti-inflammatory, and antioxidant effects. However, in pathophysiological conditions NO production is reduced, reactive oxygen species (ROS) and adhesion molecules are accumulated, insulin resistance is observed, and oxidative status is impaired while oxidative stress constitutes an essential link from endothelial dysfunction to progression of coronary artery disease (CAD) [3].

Several pro-oxidant pathways related to endothelial injury have been identified, and these oxidation mechanisms may be important in the structural and functional changes that occur during the initiation and progression of CAD. This consistency has provided the rationale for the study of antioxidant strategies in clinical trials, with the currently available data being rather controversial. In general, secondary prevention trials have shown more consistent benefit than primary prevention trials.

Therefore, in this chapter we present current information regarding molecular and biochemical aspects of oxidative pathways and focus on pharmacokinetic and pharmacodynamic aspects of antioxidants and its therapeutic potential in clinical practice.

Coronary Artery Disease
ISBN 978-0-12-811908-2, http://dx.doi.org/10.1016/B978-0-12-811908-2.00006-4

PHYSIOLOGY OF REACTIVE OXYGEN SPECIES METABOLISM

ROS, are highly reactive oxygen-based metabolic products with unpaired electrons. Physiologically ROS are signaling molecules that influence many processes, by serving as an integral component of cellular signaling pathways [4]. Under basal conditions, a healthy endothelium maintains its functional and structural vascular properties and keeps the balance between vasoconstriction and vasodilation, growth promotion and inhibition, prothrombotic and antithrombotic mechanisms, proinflammatory and anti-inflammatory status as well as oxidative and antioxidative mechanisms [5,6]. To protect against cellular oxidant stress and its adverse effects, adaptive enzymatic mechanisms have evolved to metabolize ROS, into less reactive forms and minimize their potential to produce oxidative damage and cellular dysfunction. The balance between ROS production and their removal by antioxidant systems is called "redox state". Oxidative stress is defined as an excess production of ROS relative to the levels of antioxidants which can modify endothelial function, cell and organism phenotype [7].

The forms of ROS that are relevant in biological systems include free radicals, such as superoxide $\left(O_2^-\right)$ and hydroxyl radical (OH•), and nonradicals capable of generating free radicals, such as hydrogen peroxide (H_2O_2). ROS are produced by various oxidase enzymes, including nicotinamide-adenine dinucleotide phosphate (NADPH) oxidase, xanthine oxidase, uncoupled endothelial NO synthase (eNOS), cyclooxygenase, glucose oxidase, and lipooxygenase, and mitochondrial electron transport [8,9].

On the other hand, a number of key antioxidant enzymes, such as superoxide dismutases (SODs), catalases, glutathione peroxidases (GSHPxs), hemeoxygenase, and glucose-6-phosphate dehydrogenase, limit intracellular and extracellular accumulation of deleterious ROS [10,11].

O_2^- is a primary radical that could lead to the formation of other ROS, such as OH• and H_2O_2. A small amount of O_2^- is normally produced as a byproduct of molecular oxygen during mitochondrial oxidative phosphorylation. O_2^- gets inactivated by either NO or SOD. SOD enzymes, rapidly converts O_2^- to H_2O_2, which is itself broken down by GSHPx and catalase to water. Under pathological conditions, the single-electron reduction of H_2O_2 may lead to formation of highly reactive OH radicals, either via the Fenton reaction in the presence of endogenous iron or via Haber—Weiss reaction by reacting with O_2^- (Fig. 1.6.1).

REACTIVE OXYGEN SPECIES AND ENDOTHELIAL DYSFUNCTION

A major mechanism by which ROS contribute to vascular disease is via the superoxide-mediated inactivation of NO. This reaction, which is essentially diffusion limited, not only results in reduced NO bioavailability, and thus a loss of its vasoprotective actions, but also in the formation of peroxynitrite ($ONOO^-$). Through its

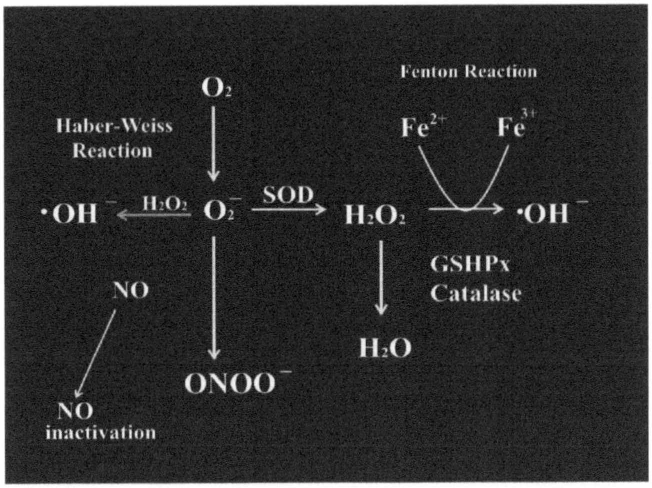

Figure 1.6.1 Reactive oxygen species metabolism. O_2^- is a primary radical that could lead to the formation of other ROS, such as OH• and H_2O_2. O_2^- gets inactivated by either nitric oxide or SOD. Glutathione peroxidase catalyze H_2O_2 to H_2O. *GSHPx*, glutathione peroxidase catalyse; *H_2O_2*, hydrogen peroxide; *NO*, nitric oxide; *O_2^-*, superoxide; *OH•*, hydroxyl radical; *ROS*, reactive oxygen species; *SOD*, superoxide dismutase.

powerful oxidizing properties, peroxynitrite causes irreversible damage to macromolecules including proteins, lipids and DNA, thereby disrupting crucial cell signaling pathways and promoting cell death. Peroxynitrite can also react with smaller cellular molecules, including the key eNOS cofactor, tetrahydropterin. In the absence of tetrahydropterin, eNOS becomes enzymatically uncoupled and generates superoxide instead of NO, thus exacerbating oxidative stress within the cell [7,12]. Superoxide may also cause lipid peroxidation (LOOH), whereas myeloperoxidase, in turn, catalyzes the reaction of hydrogen peroxide and chloride (Cl^-) to form hypochlorous acid (HOCl) [11].

In addition cardiovascular risk factors increase the expression and/or activity of NADPH oxidases (NOX) in the vascular wall [12]. NOXs are the only enzymes with the primary function of generating ROS [13,14]. Other enzymes (cyclooxygenases, cytochrome P450, enzymes of the mitochondrial electron transport chain) can produce ROS, but only as a byproduct of their normal function. Moreover, abnormal or nonphysiological sources of ROS exist, including "uncoupled" eNOS and xanthine oxidase, which are dysfunctional variants of eNOS and xanthine dehydrogenase, respectively [15–17]. Of note, these latter enzymes only become ROS-generators in the presence of an external source of ROS (such as NOXs) and can thus be considered as secondary sources of ROS. NOXs catalyze the transfer of electrons from cytosolic NADPH to molecular O_2^- via their membrane-bound catalytic NOX or DUOX subunit to generate the ROS, superoxide or H_2O_2 [18]. NOX1, NOX2, NOX4, and NOX5 have emerged as major sources of ROS production in the endothelium. Under physiological conditions these enzymes are expressed at relatively low levels on membranes of the endoplasmic reticulum and nucleus where they contribute to intracellular redox signaling processes. However, in cardiovascular risk/disease settings such as hyperlipidemia, hypertension, diabetes, and atherosclerosis, endothelial expression of NOX2, and to a lesser extent of NOX1, appears to be enhanced, resulting in excessive superoxide generation—especially in the extracellular compartment. This ultimately leads to loss of vasoprotective NO, damage to macromolecules including extracellular matrix proteins, and activation of proinflammatory signaling pathways, all of which contribute to vessel disease [12].

Xanthine oxidase is another potential source of ROS associate with vascular disease [11]. It is an iron–sulfur molybdenum flavoprotein enzyme that catalyzes the last steps of purine metabolism, the transformation of hypoxanthine and xanthine to uric acid, with O_2^- or H_2O_2 generation as byproducts. In addition to being a housekeeping enzyme, xanthine oxidase is known as the rate-limiting enzyme in purine catabolism [19]. The enzyme readily donates electrons to molecular oxygen, via a one-electron reduction pathway producing O_2^- and a two-electron reduction pathway generating H_2O_2. Both ex vivo experiments and patient studies show beneficial effects of xanthine oxidase inhibitors in postmyocardial infarction ventricular function and endothelial function, respectively [20,21].

THE IMPACT OF OXIDATIVE STRESS IN ATHEROSCLEROSIS

ROS leads to NO inactivation and impaired endothelium-dependent vasorelaxation. They also cause vascular smooth muscle cell hypertrophy, inflammatory molecule expression, platelet activation and/or adhesion [12], endothelial cells apoptosis [22], increased monocyte adhesion [23], and play an important role in angiogenesis [24].

Risk factors for atherosclerosis induce coronary atherogenesis, partly through the increased free radicals production [25,26]. Hypertension is associated with enhanced angiotensin II-induced O_2^- production and increased oxidative stress in humans [27], whereas diabetes mellitus increases free radicals production through the increased formation of advanced glycation end products [25]. More specifically, hyperglycemia stimulates cellular ROS production by different sources, with the mitochondrial electron transport chain acting as the initial O_2^- producer. Overproduction of mitochondrial derived O_2^- leads to activation of protein kinase C (PKC) and formation of advanced glycation end products [28]. PKC and advanced glycation end-products can, on the one hand, activate NADPH oxidase and, on the other hand, inhibit eNOS activity through posttranslational modifications [29].

Smoking also leads to endothelial dysfunction, possibly by a nonspecific attenuation of the vessel at the level of vascular smooth muscle and by an enhanced degradation of nitric oxide as a result of endothelial and/or smooth muscle production of oxygen derived free radicals. O_2^- from cigarette smoking may reach the vascular endothelium and react with NO [30], reducing the vasoactive levels of NO, and diminishing the response to endothelium-dependent vasodilators via formation of peroxynitrite anion ($ONOO^-$) a highly reactive intermediate with strong cytotoxic potency [30,31]. The damaging free radicals in cigarette smoke may cause either direct arterial wall injury or may initiate and/or accelerate secondary processes including depletion of antioxidants, lipid peroxidation [32] and activation of phagocyte-platelet endothelial cell interactions. Oxygen free radicals in the vascular wall may initiate a vicious cycle by oxidizing lipids or lipoproteins, which in fact may inhibit endothelium-dependent vasodilation directly or indirectly by the formation of superoxide anions [33].

Another major involvement of free radicals in atherogenesis is the oxidation of low-density lipoprotein (LDL), which leads to the formation of oxidized-LDL, a highly reactive free radical with direct cytotoxic effects on endothelial cells. It is a chemoattractant for circulating monocytes and induces vascular smooth muscle cells proliferation and cytokine production [34]. Furthermore, oxidized-LDL is taken up by monocytes to form foam cells, triggering atherogenesis [25].

REFERENCES

[1] Tousoulis D, et al. Oxidative stress and early atherosclerosis: novel antioxidant treatment. Cardiovasc Drugs Ther 2015;29(1):75−88.
[2] Ignarro LJ. Biosynthesis and metabolism of endothelium-derived nitric oxide. Annu Rev Pharmacol Toxicol 1990;30:535−60.
[3] Esper RJ, et al. Endothelial dysfunction: a comprehensive appraisal. Cardiovasc Diabetol 2006;5:4.
[4] Montezano AC, Touyz RM. Oxidative stress, Noxs, and hypertension: experimental evidence and clinical controversies. Ann Med 2012;44(Suppl. 1):S2−16.
[5] Vallance P, Collier J, Moncada S. Effects of endothelium-derived nitric oxide on peripheral arteriolar tone in man. Lancet 1989;2(8670):997−1000.
[6] Vanhoutte PM. Endothelium and control of vascular function. State of the Art lecture. Hypertension 1989;13(6 Pt 2):658−67.
[7] Pacher P, Beckman JS, Liaudet L. Nitric oxide and peroxynitrite in health and disease. Physiol Rev 2007;87(1):315−424.
[8] Papageorgiou N, et al. Antioxidant treatment and endothelial dysfunction: is it time for flavonoids? Recent Pat Cardiovasc Drug Discov 2013;8(2):81−92.
[9] Tousoulis D, et al. Oxidative stress and endothelial function: therapeutic interventions. Recent Pat Cardiovasc Drug Discov 2011;6(2):103−14.
[10] Griendling KK, FitzGerald GA. Oxidative stress and cardiovascular injury: part II: animal and human studies. Circulation 2003;108(17):2034−40.
[11] Mueller CF, et al. ATVB in focus: redox mechanisms in blood vessels. Arterioscler Thromb Vasc Biol 2005;25(2):274−8.
[12] Drummond GR, Sobey CG. Endothelial NADPH oxidases: which NOX to target in vascular disease? Trends Endocrinol Metab 2014;25(9):452−63.

[13] Drummond GR, et al. Combating oxidative stress in vascular disease: NADPH oxidases as therapeutic targets. Nat Rev Drug Discov 2011;10(6):453–71.
[14] Lassegue B, San Martin A, Griendling KK. Biochemistry, physiology, and pathophysiology of NADPH oxidases in the cardiovascular system. Circ Res 2012;110(10):1364–90.
[15] Nishino T, et al. Mammalian xanthine oxidoreductase – mechanism of transition from xanthine dehydrogenase to xanthine oxidase. FEBS J 2008;275(13):3278–89.
[16] Luo S, et al. Molecular mechanisms of endothelial NO synthase uncoupling. Curr Pharm Des 2014; 20(22):3548–53.
[17] Montezano AC, Touyz RM. Reactive oxygen species and endothelial function–role of nitric oxide synthase uncoupling and Nox family nicotinamide adenine dinucleotide phosphate oxidases. Basic Clin Pharmacol Toxicol 2012;110(1):87–94.
[18] de Oliveira-Junior EB, et al. The human NADPH oxidase: primary and secondary defects impairing the respiratory burst function and the microbicidal ability of phagocytes. Scand J Immunol 2011;73(5): 420–7.
[19] Battelli MG, Bolognesi A, Polito L. Pathophysiology of circulating xanthine oxidoreductase: new emerging roles for a multi-tasking enzyme. Biochim Biophys Acta 2014;1842(9):1502–17.
[20] Ohara Y, Peterson TE, Harrison DG. Hypercholesterolemia increases endothelial superoxide anion production. J Clin Invest 1993;91(6):2546–51.
[21] O'Driscoll JG, et al. Nitric oxide-dependent endothelial function is unaffected by allopurinol in hypercholesterolaemic subjects. Clin Exp Pharmacol Physiol 1999;26(10):779–83.
[22] Li AE, et al. A role for reactive oxygen species in endothelial cell anoikis. Circ Res 1999;85(4): 304–10.
[23] Marui N, et al. Vascular cell adhesion molecule-1 (VCAM-1) gene transcription and expression are regulated through an antioxidant-sensitive mechanism in human vascular endothelial cells. J Clin Invest 1993;92(4):1866–74.
[24] Maulik N, Das DK. Redox signaling in vascular angiogenesis. Free Radic Biol Med 2002;33(8): 1047–60.
[25] Antoniades C, et al. Oxidative stress, antioxidant vitamins, and atherosclerosis. From basic research to clinical practice. Herz 2003;28(7):628–38.
[26] Tousoulis D, et al. Asymmetric dimethylarginine (ADMA): is really a biomarker for cardiovascular prognosis? Int J Cardiol 2011;153(2):123–5.
[27] Landmesser U, et al. Role of p47(phox) in vascular oxidative stress and hypertension caused by angiotensin II. Hypertension 2002;40(4):511–5.
[28] Giacco F, Brownlee M. Oxidative stress and diabetic complications. Circ Res 2010;107(9):1058–70.
[29] Funk SD, Yurdagul Jr A, Orr AW. Hyperglycemia and endothelial dysfunction in atherosclerosis: lessons from type 1 diabetes. Int J Vasc Med 2012;2012:569654.
[30] Radi R, et al. Peroxynitrite-induced membrane lipid peroxidation: the cytotoxic potential of superoxide and nitric oxide. Arch Biochem Biophys 1991;288(2):481–7.
[31] Zeiher AM, Schachinger V, Minners J. Long-term cigarette smoking impairs endothelium-dependent coronary arterial vasodilator function. Circulation 1995;92(5):1094–100.
[32] Reznick AZ, et al. Modification of plasma proteins by cigarette smoke as measured by protein carbonyl formation. Biochem J 1992;286(Pt 2):607–11.
[33] Galle J, et al. Impairment of endothelium-dependent dilation in rabbit renal arteries by oxidized lipoprotein(a). Role of oxygen-derived radicals. Circulation 1995;92(6):1582–9.
[34] Noguchi N. Novel insights into the molecular mechanisms of the antiatherosclerotic properties of antioxidants: the alternatives to radical scavenging. Free Radic Biol Med 2002;33(11):1480–9.

CHAPTER 1.7

Genetics of Atherosclerosis

Nikolaos Papageorgiou[1,2], Evangelos Diamantis[1], Effimia Zacharia[1], Dimitris Tousoulis[1]

[1]1st Cardiology Clinic, 'Hippokration' General Hospital, National and Kapodistrian University of Athens, School of Medicine, Athens, Greece; [2]St Bartholomew's Hospital, London, United Kingdom

Contents

INTRODUCTION

Atherosclerosis is a chronic progressive inflammatory disorder leading to coronary artery disease (CAD) [1,2]. The modification of known risk factors for CAD such as hypercholesterolemia, hypertension, and smoking has been assessed in multiple randomized placebo-controlled clinical trials and is associated with 30%—40% reduction in clinical events, such as death and myocardial infarction (MI) [3]. Despite tremendous efforts in the field of its prevention and treatment, CAD remains a major cause of morbidity and mortality worldwide [4—6]. It is generally accepted that environmental factors like smoking and lifestyle are involved in the pathogenesis of CAD. However, epidemiological and family studies like Framingham and Interheart have repeatedly shown that genetic predisposition accounts for 40%—60% of the cumulative risk for (CAD), pointing out the predominant role of genes (Table 1.7.1), [7—12]. Nowadays, CAD is considered to be a complex and polygenic disorder, because several genes contribute to its predisposition, each having only mild to moderate influence. The attempts to identify these genes have been numerous. Still, due to lack of required technology and small sample sizes, only single gene disorders—primarily those involving cholesterol—have been studied so far [13].

Coronary Artery Disease
ISBN 978-0-12-811908-2, http://dx.doi.org/10.1016/B978-0-12-811908-2.00007-6

Table 1.7.1 Heritability estimates of atherosclerosis and its risk factors, and percentage of genetic variability

Risk factor	Heritability (%)	Genetic variability (%)	References
Coronary artery disease/ myocardial infarction	40–50	6	[92,102]
Blood pressure	30	0.9	[103,104]
Diabetes mellitus	26	10	[105–107]
LDL cholesterol	45–60	25–30	[108–110]
HDL cholesterol	47–62	25–30	[108–110]
Lp(a)	65–94	20	[111–113]
Smoking	50	1	[114–116]

HDL, high-density lipoprotein; *LDL*, low-density; *Lp*, lipoprotein.
Adapted and modified from Holdt LM, Teupser D. Genetic background of atherosclerosis and its risk factors. In: The ESC textbook of preventive cardiology. European Society of Cardiology; 2015.

During the last decade, the rapid progress in molecular genetics led to the identification of an increasingly large number of candidate genes, genetic polymorphisms, and susceptibility loci, which are associated with atherosclerosis. Two main approaches have been used so far in the interpretation of the genetic background of CAD: linkage analysis (genome-wide and fine-mapping) and association analysis studies (candidate-gene or genome-wide).

CANDIDATE GENES

Over the last decade, epidemiological studies have identified several risk factors for CAD, such as elevated lipid levels, diabetes mellitus, smoking, hypertension, inflammation, oxidative stress, and increased coagulation (Table 1.7.2). Thus, it was assumed that the identification of the genes potentially involved in the aforementioned processes would unravel the underlying pathophysiologic pathways of atherosclerosis (Fig. 1.7.1). However, the results of the studies were not consistent, probably due to underestimation of the polymorphisms' frequency in the control population or due to small sample sizes. Indicatively, Morgan et al., in their study of 1461 participants, failed to confirm the results of previous studies, which had shown that 85 genetic variants are significant susceptibility factors for atherosclerosis or acute coronary syndrome (ACS). This emphasizes the need for robust replication of putative genetic risk factors before their introduction into clinical care [14]. Moreover, caution must be taken regarding the analysis of the data. Indeed, a positive association does not necessarily establish the causality of the allele. Therefore, further analysis is needed for all identified candidate genes, so as to reduce the impact of potential confounding parameters (such as the specific study population, the sample size and presence of genetic polymorphisms) and to establish definitive evidence of selection, mechanism of selection, and functional effects of the allelic variants under selection. In the following paragraphs we will discuss some of the most important candidate genes.

Table 1.7.2 Examples of single nucleotide polymorphism (SNP) or haplotype association studies of coronary heart disease

Gene	Gene name	SNPs or haplotype	References
ALOX5AP	Arachidonate 5-lipoxgenase-activating protein	Haplotype A	[117]
APOE	Apolipoprotein E	ε4 allele rs4986978	[118]
GJA4	Gap junction protein α1	rs1764391	[119,120]
CX3CR1	Chemokine (C–X3–C motif) receptor 1	rs3732378	[121,122]
SERPINE1	Serpin peptidase inhibitor, clade E member 1	rs1799889	[119,123]
PSMA6	Proteasome subunit α type 6	rs1048990	[124]
MMP3	Matrix metallopeptidase 3	rs3025058	[125,126]
PDE4D	Phosphodiesterase 4D, cAMP specific	*haplotypes*	[127–129]
LTA4H	Leukotriene A4 hydrolase	Haplotype K	[130]
USF1	Upstream transcription factor 1	Haplotype across USF1	[131]
GATA2	GATA binding protein 2	rs2713604; rs3803λ	[132]
TNFSF4	Tumor necrosis factor (ligand) superfamily, member 4	rs3850641	[133]

Adapted and modified from Kullo, et al. Mechanisms of disease: the genetic basis of coronary heart disease. Nat Clin Pract Cardiovasc Med October 2007;4(10):558–69.

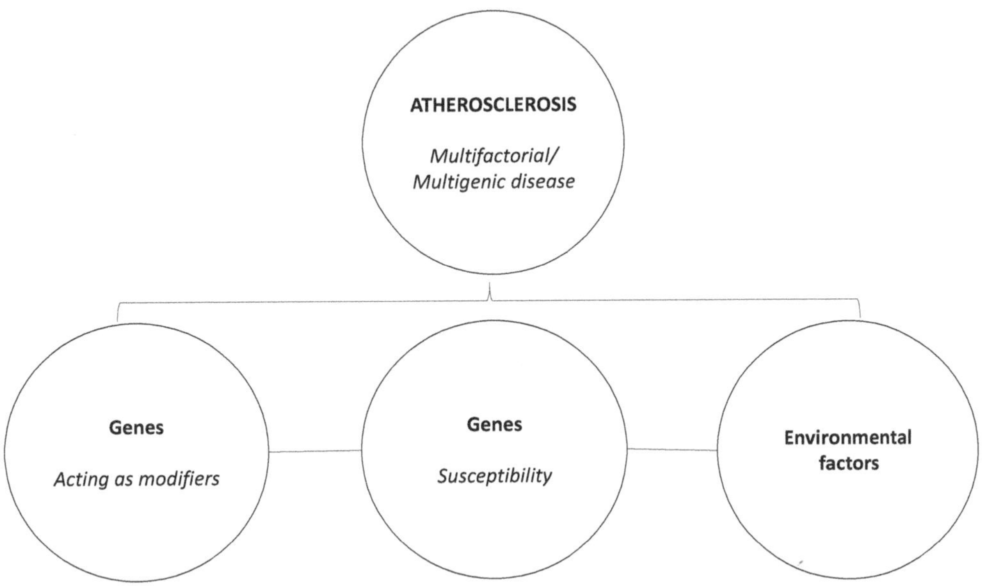

Figure 1.7.1 Atherosclerosis is the combined result of the interaction between genetic and environmental factors.

Endothelial eNOS

Endothelial-derived nitric oxide (NO), having a dominant role in vascular regulation, is involved in a variety of different important physiological functions and has been implicated in the pathogenesis of atherosclerosis and CAD. It is derived from L-arginine and is regulated by the NOS3 gene. The latter is located on chromosome 7q35-36 and consists of 26 exons with a total size of 21 kb [15—17]. At least 100 polymorphisms have been identified in the eNOS gene, some of which have functional consequences. The first point mutation was described by Hingorani et al.; guanine (G) was replaced by thymine (T) at nucleotide 1917 in exon 7 of the NOS3 gene, leading to substitution of glutamic acid by aspartic acid at codon 298 (Glu298Asp, also known as G894-T, rs1799983) [18]. The exact effect of this single nucleotide polymorphism (SNP) on NO production has not been clarified yet. Although it has been shown that eNOS Asp298 decreases vascular NO generation, McDonald et al. failed to show a significant effect on its production in vivo [19—23].

Apart from its effect on intracellular NO production, this allele has been associated with increased red blood cell aggregation, which in turn increases the susceptibility to acute coronary events, particularly those of thrombotic origin [24].

This variant has been shown to be associated with MI and coronary artery spasm [25—27]. However, correlation of this SNP with CAD remains debatable [2—4,13,16—23], as some studies have also reported its lack of association with the disease probably due to geo-ethnic differences and environmental factors.

A study that investigated a possible correlation between eNOS Glu298Asp and NO-mediated endothelial function in CAD patients showed inconsistent NO-mediated endothelial vasorelaxations among patients. At the same time, reduced values of NO were associated with increased number of clinical risk factors for atherosclerosis [28].

A recent study assessed the possible correlation between eNOS G894T gene polymorphism with target organ damage markers, oxidative stress and metabolic and inflammatory parameters in 126 children with newly diagnosed primary hypertension and 83 healthy children. The results did not show prevalence of the eNOS T allele among patients, but did show, an association with early vascular damage, which was independent of metabolic abnormalities [29]. On the other hand, Antoniades et al. indicated that G894-T correlated with impaired endothelial function and higher levels of von Willebrand factor (vWF) in a cohort of 56 patients with a history of premature MI. This suggested an association with endothelial function and with markers of endothelial cells (ECs) injury and activation [30]. Moreover, the G894T eNOS allele (rs1799983) has been shown to be a genetic marker of MI among the Moroccan population [31].

Another common variant of the eNOS gene is the result of a cytosine replacing a thymine at nucleotide 2786 (T786-C, rs2070744) at the 59-flanking region [32]. Several studies have pointed out that this SNP in the NOS3 promoter region (T786-C) reduces both the enzyme expression and its efficiency up to 50%, leading to increased CAD risk [33–36]. It has been associated with an increased risk for developing hypertension and coronary artery spasm [32,33,37]. However, its role in the presence of CAD remains debatable [38]. In fact, a recent study found no impact of −786 C/T or 4 a/b allele on the long-term outcomes of 532 patients with angiographically confirmed CAD [38].

Recently, Kumar et al. (2016) studied the genotype distribution, allele frequencies and haplotypes of the eNOS −786 T/C, 4 b/4a and 894 G/T among Indian patients with CHD (n = 300), their first degree relatives (FDRS) (n = 100) and healthy age/sex-matched controls (n = 300) and noticed significant differences among study participants. Further analysis revealed a strong positive association between eNOS gene haplotypes GCB, TCB (G-allele of 894 G/T, C-allele −786 T/C, B-allele of Intron 4 b/4a respectively) and nitrite/nitrate levels in both controls and CHD patients compared to controls ($P < .01$). This finding underlied the importance of eNOS polymorphism not only in the pathogenesis of CHD but also in the prediction of future coronary events in asymptomatic healthy FDRS [39].

C-reactive Protein

C-reactive protein (CRP) is one of the most widely examined markers of inflammation and is involved in the pathophysiology of atherosclerosis through several mechanisms. CRP enhances the binding of LDL-cholesterol from macrophages that subsequently form foam cells and contributes to the development of vulnerable plaques. The latter stimulate the secretion of proinflammatory cytokines and tissue factor. Thus, CRP contributes to vascular remodeling [40,41]. CRP levels have been inversely correlated with endothelial dysfunction (ED) and with the prediction of future or recurrent cardiovascular (CV) events [42].

CRP gene is located on chromosome 1 (region 1q21–q23). Its SNPs have been associated with circulating CRP levels. Thus, they predict the risk of developing inflammatory disease and the severity of the latter [42–44].

A meta-analysis of 18,637 individuals showed that the rs1130864 (on the 3′ untranslated region of CRP) was associated with increased serum CRP levels, whereas increased serum CRP levels were, in tandem, associated with coronary heart disease. However, no direct association was found between the SNP and coronary heart disease [45]. Schulz et al. investigated the importance of CRP levels and genetic variants as prognostic markers for further CV events (3-year follow-up) in a cohort of 939 patients with angiographic proven cardiovascular disease (CVD). Both CRP levels ≥5 mg/L (log-rank test: $P = .001$, Cox regression: hazard ratio = 1.77, 95% confidence interval [CI]: 1.2–2.7) and SNP rs1800947 (log–rank test: $P = .01$, Cox regression: hazard ratio = 1.99, 95% CI: 1.1–3.6) were found to be independent risk factors for further adverse CV events among patients with CVD within a 3-year follow-up [46].

Cytokines

Cytokines are glycoproteins with a monomeric molecular mass of 15–25 kDa and are the main mediators of the inflammatory response. Genetic variants of the IL-6 gene, which are located at chromosome 7p21-24, have been associated with its plasma concentrations and lead in altered biological responses. Thus, IL-6 SNPs are considered to be crucial modulators in the pathogenesis of several diseases [47–50].

Nevertheless, the results of studies that investigated the involvement of IL-6 polymorphisms in the pathogenesis of CAD were not consistent [51–57]. Mao et al. recruited 224 CAD patients and 260 control subjects in order to investigate the role of IL-6 polymorphism (−174 G/C and −572 C/G) in the development of CAD. The investigators concluded that IL-6 −174 G/C polymorphisms are associated with the pathogenesis of CAD because carriers of the CC genotype at IL-6 −174 G/C had a 2.32 (95%CI = 1.33–4.06) fold risk of developing CAD compared to carriers of the GG genotype [58].

Moreover, according to experimental studies which examined the functional effect of four polymorphisms in the IL-6 promoter (G-597A,G-572 GC, −373A(n)T(n), G-174 GC), more than one of the polymorphic sites is functional. These polymorphic sites may influence IL-6 transcription not by a simple additive mechanism but rather through complex interactions determined by the haplotype [59]. Furthermore, the HIFMECH study (hypercoagulability and impaired fibrinolytic function mechanisms predisposing to MI) investigated a possible association between two interleukin 6 (IL-6) promoter polymorphisms, IL-6-572G>C and IL-6-174G>C, and MI. The study showed that the North American carriers of −174C allele had lower risk of acute MI compared to −174GG homozygotes, whereas no association with MI was observed for the −572G>C polymorphism (odds ratio [OR] 0.53; CI 0.32–0.86). In contrast, the south European homozygotes had no increased risk for MI [60]. Similarly, Malarstig et al. reported no association of the −572G>C polymorphism with the clinical outcomes of 3027 patients with non-ST-elevation ACS, but did report a more distinct inflammatory response [61]. It is also worth noting that, according to preliminary data, there is a significant and independent association between the IL-6 G-572C gene polymorphism (presence of C allele) and adolescents with a family history of premature atherosclerosis [62].

In addition, IL-6 −174CC and nt565AA genotypes, as well as IL-10 ATA haplotypes were correlated with ED in patients with peripheral artery disease, as indicated by decreased values of flow-mediated dilation (FMD) and by a high short-term risk of acute postoperative CV events [63].

Moreover, in the setting of non-ST-elevation ACS, genetic variations at the IL-1 gene locus correlate with changes in the levels of soluble endothelial activation markers (vWF and E-selectin) [64].

A population-based case-control study investigated the variations in inflammation-related genes—interleukin (IL)-1beta, IL-6, CRP, IL-10, IL-18, and the tumor necrosis factor (TNF) superfamily [lymphotoxin (LT)-alpha, TNF-alpha, LT-beta]—with respect to nonfatal incident MI or ischemic stroke risk in postmenopausal and/or hypertensive patients aged 30–79 years. Bis et al. showed that only the gene-wide IL-18 variation was associated with higher MI risk (OR 0.74; 95% CI 0.64–0.87), in contrast to IL-6, CRP, IL-10, and TNF superfamily gene [65].

Fibrinogen

Fibrinogen is a soluble glycoprotein that facilitates platelet aggregation and plays a key role in the initiation and progression of atherogenesis [66–68]. Several prospective epidemiological studies and clinical observations have provided evidence that plasma fibrinogen levels correlate with risk for CAD [69–73]. Indeed, the Northwick Park Heart and the PROspective CArdiovascular Munster Study, showed an association between fibrinogen levels and incidence of coronary events, which was independent of other risk factors [74,75].

Plasma fibrinogen levels have a high degree of heterogeneity, which could be attributed to the alternative processing of fibrinogen mRNAs, posttranslational modifications and proteolytic degradation of the protein. Research on fibrinogen polymorphisms has mainly focused on the promoter of the b gene [76]. There are several polymorphisms on fibrinogen chain genes which affect its levels, however only few of the genetic variants are associated with increased CV risk. Data have shown that, although the −455G/A and −148C/T fibrinogen promoter region variants were strongly correlated with fibrinogen levels, they were not associated with increased risk of CAD [77,78]. In line with these observations, Papageorgiou et al., examined the combined effects of the G455A and the G58A fibrinogen genetic polymorphisms on prothrombotic profile, endothelial function and the risk of CAD in 422 patients with angiographically documented CAD and 277 sex/age-matched controls. The investigators reported that, although the two polymorphisms had no significant effect on the risk for CAD, the 455AA homozygosity was associated with increased fibrinogen levels in the control ($P = .035$) and the CAD group ($P < .001$) compared to the G allele carriers. Moreover, both the 58AA ($P = .016$) and 455AA homozygotes ($P = .022$) presented with higher levels of D-Dimers in the CAD group, which suggested that these genetic polymorphisms may modify the atherosclerotic process via effects on the coagulation cascade [79].

A recent study examined the effect of rs180070 (allele A) and rs2070011 (allele G) polymorphism of the fibrinogen gene on the risk of CAD in 744 patients (332 hypertensive) who underwent coronary angiography due to symptoms of stable angina. The presence of allele A homozygosity (rs180070) was associated with significantly higher levels of fibrinogen ($P = .05$), interleukin-6 ($P = .046$) and D-dimers compared to the presence of G allele ($P = .048$) in hypertensive patients. Importantly, fibrinogen levels (>443 mg/dL) but not A homozygosity (odds ratio = 3.00; 95% confidence interval, 0.78–11.90; $P = .110$) were shown to be independent predictors (odds ratio = 3.50; 95% confidence interval, 1.14–10.90; $P = 0.029$), of the angiographic presence of CAD in hypertensive patients [80].

Other Endothelium-Derived Mediators

Endothelin-1 (ET-1), an endothelium-derived peptide with multiple functions, is shown to be a powerful vasoconstrictor with mitogenic activity on vascular smooth muscle and fibroblasts. Genetic variants of the ET-1 gene (3A/4A, Lys198Asn) have been suggested to alter its expression and play a role in endothelial function. More specifically, a recent study concerning patients with idiopathic pulmonary hypertension showed that ED was more pronounced in individuals with ET-1 variants, which also were shown to play a significant role in susceptibility to the disease and its clinical progression [83].

Several other molecules produced by vascular endothelium, like PAI-1, tissue factor, and vWF—which are regulated by the production of NO—prostacyclin, tPA, and thrombomodulin, have the ability to affect blood fluidity and thrombosis.

Furthermore, investigators have pointed out the role of CD40 and its ligand CD40Ls, which are members of the TNF superfamily and are expressed in the surface of many cells involved in atherosclerosis—such as ECs, SMCs, macrophages, T lymphocytes and platelets. Several inflammatory processes are initiated as a consequence of the CD40L binding to its receptor. Noticeably, both are molecules with a dual prothrombotic and proinflammatory role [23,82]. However, it is still unknown whether genetic variations in CD40/CD40L genes contribute to atherogenesis. It is worth noting that genetic polymorphisms related to plasma CD40L levels could be used to examine whether this molecule constitutes a causal factor. Accordingly, a study evaluated the role of genetic variations in CD40 and CD40L genes in subclinical atherosclerosis, which was assessed by coronary artery calcification and IMT in subjects from families included in the Diabetes Heart Study. The results of the study showed that the two SNPs in the CD40 gene (rs1535045 and rs3765459) were significantly associated with decreased CAD in this population. However, they were not associated with other parameters of the study [82,86].

GENOME-WIDE ASSOCIATION STUDIES

A major breakthrough was achieved in 2005 with the development of a microarray of 500,000 DNA markers followed by arrays containing 1 million DNA markers [81,82]. The Human Genome Project, the Hap Map Project [83], resulted in the identification of many of the common SNPs, which were used as DNA markers in order to perform unbiased genome-wide association studies (GWAS). This provided the opportunity to map and identify the susceptibility genes, not only for single-gene (Mendelian) disorders, but also for complex polygenic (non-Mendelian) traits.

The GWAS studies have one severe limitation: the exceedingly high threshold of significance required —a P value $< .05/1{,}000{,}000$ or $<5 \times 10^{-8}$ [84]. Therefore, in order to avoid false positive results, it has become the standard to require replication of these markers showing genomewide significance in an independent large sample size.

In the Framingham Heart Study, investigators studied associations with FMD and were only able to demonstrate one SNP that reached a reasonable level of significance for inclusion as a loci of relevance. The locus was on chromosome 7 with a candidate gene close to the CFTR gene, which raises the possibility that chloride channel function may be relevant to the flow-mediated response. Baseline flow was also related to a distinct loci. These associations have not been repeated in other cohorts; thus, it has not been confirmed whether these findings are replicable. However, they raise some interesting ideas for further study [85].

The largest international GWAS for CAD formation involved the collaboration of 14 GWAS is the Coronary Artery Disease Genome-wide Replication and Meta-Analysis (CARDIoGRAM) [86] which included 87,000 subjects; 22,000 CAD patients and 65,000 controls. This study led to the discovery of 13 new risk variants for CAD and confirmation of 10 others previously described [87] (Table 1.7.3). Interestingly, investigators reported that only 10 risk variants were associated with hypertension or lipids, whereas 23 of them mediate their risk via unknown mechanisms. This underlines the fact that there are several mechanisms contributing to the pathogenesis of CAD, which are yet to be elucidated.

In an another study, Reilly et al. in a cohort of 12,393 patients with CAD and 7383 healthy controls identified 2 novel loci: ADAMTS7 (a disintegrin and metalloproteinase with thrombospondin motifs 7), which predisposes for CAD, and the ABO blood group locus, which predisposes for MI [88].

Moreover, the CAD C4D Genetics Consortium studied 575,000 genotyped SNPs in 15,420 individuals with CAD (8424 Europeans and 6996 South Asians) and 15,062 controls [39] and identified five additional genetic risk variants for CAD [89]. The IBC 50K CAD Consortium, using a customized gene array, examined 49,094 genetic variants in 2100 genes of CV relevance in 15,596 European and South Asian patients with confirmed CAD cases and 34,992 controls and identified three additional risk variants for CAD [90].

Wang et al., identified a new association between rs6903956 in a putative gene denoted as C6orf105 on chromosome 6p24.1 and CAD in the Chinese Han population; however, the study found no effect in the Caucasian population [91].

A recent meta-analysis combined data from the CARDIoGRAM and C4D studies and increased the sample size up to 240,000. The researchers confirmed 46 loci susceptible to with CAD and discovered further 104 independent variants ($r < 0.2$) strongly

Table 1.7.3 Risk loci for coronary artery disease discovered by genome-wide association studies

Chromosomal location	Single nucleotide polymorphism (SNP)	Genes	Odds ratio (95% confidence interval)	References
Loci that associate with cholesterol				
1p32.3	rs11206510	*PCSK9*	1.15 (1.10−1.21)	[134]
1p13.3	rs599839	*SORT1*	1.29 (1.18−1.40)	[84]
2p21	rs4299376	*ABCG8*	1.08 (1.04−1.13)	[90]
6q25.3	rs3798220	*LPA*	1.92 (1.48−2.49)	[135,136]
8q24.13	rs17321515	*TRIB1*	1.10 (1.06−1.15)	[90]
9q34.2	rs579459	*ABO*	1.10 (1.07−1.13)	[87]
11q23.3	rs964184	*ZNF259, APOA5-A4-C3-A1*	1.13 (1.10−1.16)	[87]
19p13.2	rs1122608	*LDLR*	1.14 (1.09−1.19)	[134]
Loci that associate with hypertension				
10q24.32	rs12413409	*CYP17A1, CNNM2, NT5C2*	1.12 (1.08−1.16)	[87]
12q24.12	rs3184504	*SH2B3*	1.13 (1.08−1.18)	[137,138]
Loci that do not associate with known risk factors				
1p32.2	rs17114036	*PPAP2B*	1.17 (1.13−1.22)	[87]
1q41	rs17465637	*MIA3*	1.20 (1.12−1.30)	[84]
2q33.1	rs6725887	*WDR12*	1.16 (1.10−1.22)	[134]
3q22.3	rs2306374	*MRAS*	1.15 (1.11−1.19)	[139]
5q31.1	rs2706399	*IL5*	1.07 (1.03−1.11)	[90]
6p24.1	rs6903956	*C6orf105*	1.65 (1.44−1.90)	[91]
6p24.1	rs12526453	*PHACTR1*	1.13 (1.09−1.17)	[134]
6p21.3	rs3869109	HCG27 and HLA-C	1.15	[87]
6q23.2	rs12190287	*TCF21*	1.08 (1.06−1.10)	[87]
7q22.3	rs10953541	*BCAP29*	1.08 (1.05−1.11)	[89]
7q32.2	rs11556924	*ZC3HC1*	1.09 (1.07−1.12)	[87]
9p21.3	rs4977574	*CDKN2A* *CDKN2B* *ANRIL* 0.46	1.25 (1.18−1.31) 1.37 (1.26 −1.48)	[89,93]
10p11.23	rs2505083	*KIAA1462*	1.07 (1.04−1.09)	[89,140]
10q11.21	rs1746048	*CXCL12*	1.33 (1.20−1.48)	[84]
10q23.31	rs1412444	*LIPA*	1.09 (1.07−1.12)	[89,90]
11q22.3	rs974819	*PDGF*	1.07 (1.04−1.09)	[89]
13q34	rs4773144	*COL4A1, COL4A2*	1.07 (1.05−1.09)	[87]
14q32.2	rs2895811	*HHIPL1*	1.07 (1.05−1.10)	[87]
15q25.1	rs3825807	*ADAMTS7*	1.08 (1.06−1.10)	[87,89]
17p13.3	rs216172	*SMG6, SRR*	1.07 (1.05−1.09)	[87]
17p11.2	rs12936587	*RASD1, SMCR3, PEMT*	1.07 (1.05−1.09)	[87]
17q21.32	rs46522	*UBE2Z, GIP, ATP5G1, SNF8*	1.06 (1.04−1.08)	[87]
21q22.11	rs9982601	*MRPS6*	1.19 (1.13−1.27)	[134]

Adapted and modified from Roberts R. Genetics of coronary artery disease. Circ Res June 6, 2014;114(12):1890−903.

associated with CAD at a 5% false discovery rate. If put together, these variants explain approximately 10.6% of CAD heritability [92].

Currently, the strongest and most robust susceptibility locus identified by four genome-wide association studies for CAD is located on chromosome 9p21 [84,93–95]. Although this risk variant occurs in 75% of the European population (50% heterozygous and 25% homozygous risk), the mechanisms involving it to atherosclerosis have not been fully clarified. The risk for CAD is increased by 25% in the case of one copy and by 50% in the case of two copies of the 9p21.3 risk allele. 9p21.3 has also been associated with a twofold increase in risk for premature CAD [94]. A recent meta-analysis by Samani et al. concluded that this risk locus was significantly associated with early age of onset of heart disease [84]. Interestingly, this association has not been shown in African patients [96–100]. Furthermore, a recent population-based, prospective study concluded that the rs1333049 polymorphism located on chromosome 9p21 enhances atherosclerosis development and progression in a cohort of 769 individuals, aged 40–79 years in the years between1990 and 1995 (OR: 1.73 [95% CI: 1.36 to 2.21] and OR: 1.87 [95% CI: 1.44 to 2.42] [101] respectively).

CONCLUSIONS

It is widely accepted that atherosclerosis is a multifactorial inflammatory disease with several environmental and genetic interactions involved. The genetic predisposition to CAD has been studied with respect to several candidate genes encoding components of the inflammatory cascade, including cytokines, adhesion molecules and other biomarkers. Nevertheless, results have not always been consistent and data were limited to single studies. Therefore, it has been difficult to draw firm conclusions about these genes. On the other hand, genome-wide association studies have contributed to a great extend to the identification of genetic determinants of CAD. Some of these uncertainties could be resolved by systematic reviews and meta-analyses of the relevant genotype–disease association studies. However, future large studies are required to identify genes that regulate vascular function.

REFERENCES

[1] Lusis AJ. Atherosclerosis. Nature 2000;407(6801):233–41.
[2] Ross R. Atherosclerosis—an inflammatory disease. N Engl J Med 1999;340(2):115–26.
[3] Wald NJ, Law MR. A strategy to reduce cardiovascular disease by more than 80%. BMJ 2003; 326(7404):1419.
[4] Murray CJ, Lopez AD. Measuring the global burden of disease. N Engl J Med 2013;369(5):448–57.
[5] Lopez AD, Murray CC. The global burden of disease, 1990–2020. Nat Med 1998;4(11):1241–3.
[6] Juvonen J, Surcel HM, Satta J, Teppo AM, Bloigu A, Syrjala H, et al. Elevated circulating levels of inflammatory cytokines in patients with abdominal aortic aneurysm. Arterioscler Thromb Vasc Biol 1997;17(11):2843–7.

[7] Arnett DK, Baird AE, Barkley RA, Basson CT, Boerwinkle E, Ganesh SK, et al. Relevance of genetics and genomics for prevention and treatment of cardiovascular disease: a scientific statement from the American Heart Association Council on Epidemiology and Prevention, the Stroke Council, and the Functional Genomics and Translational Biology Interdisciplinary Working group. Circulation 2007;115(22):2878–901.
[8] Roberts R. Personalized medicine: an idea whose time is approaching. Methodist DeBakey Cardiovasc J 2007;3(1):3–7.
[9] Schildkraut JM, Myers RH, Cupples LA, Kiely DK, Kannel WB. Coronary risk associated with age and sex of parental heart disease in the Framingham Study. Am J Cardiol 1989;64(10):555–9.
[10] Hopkins PN, Williams RR, Kuida H, Stults BM, Hunt SC, Barlow GK, et al. Family history as an independent risk factor for incident coronary artery disease in a high-risk cohort in Utah. Am J Cardiol 1988;62(10 Pt 1):703–7.
[11] Rissanen AM. Familial occurrence of coronary heart disease: effect of age at diagnosis. Am J Cardiol 1979;44(1):60–6.
[12] Cooper JA, Miller GJ, Humphries SE. A comparison of the PROCAM and Framingham point-scoring systems for estimation of individual risk of coronary heart disease in the Second Northwick Park Heart Study. Atherosclerosis 2005;181(1):93–100.
[13] Brown MS, Goldstein JL. Familial hypercholesterolemia: a genetic defect in the low-density lipoprotein receptor. N Engl J Med 1976;294(25):1386–90.
[14] Morgan TM, Krumholz HM, Lifton RP, Spertus JA. Nonvalidation of reported genetic risk factors for acute coronary syndrome in a large-scale replication study. JAMA 2007;297(14):1551–61.
[15] Glasser SP, Selwyn AP, Ganz P. Atherosclerosis: risk factors and the vascular endothelium. Am Heart J 1996;131(2):379–84.
[16] Tousoulis D. Biomarkers in cardiovascular disease. Curr Med Chem 2012;19(16):2483–4.
[17] Marsden PA, Schappert KT, Chen HS, Flowers M, Sundell CL, Wilcox JN, et al. Molecular cloning and characterization of human endothelial nitric oxide synthase. FEBS Lett 1992;307(3):287–93.
[18] Hingorani AD, Liang CF, Fatibene J, Lyon A, Monteith S, Parsons A, et al. A common variant of the endothelial nitric oxide synthase (Glu298->Asp) is a major risk factor for coronary artery disease in the UK. Circulation 1999;100(14):1515–20.
[19] Hingorani AD. Endothelial nitric oxide synthase polymorphisms and hypertension. Curr Hypertens Rep 2003;5(1):19–25.
[20] Persu A, Stoenoiu MS, Messiaen T, Davila S, Robino C, El-Khattabi O, et al. Modifier effect of ENOS in autosomal dominant polycystic kidney disease. Hum Mol Genet 2002;11(3):229–41.
[21] McDonald DM, Alp NJ, Channon KM. Functional comparison of the endothelial nitric oxide synthase Glu298Asp polymorphic variants in human endothelial cells. Pharmacogenetics 2004;14(12):831–9.
[22] Philip I, Plantefeve G, Vuillaumier-Barrot S, Vicaut E, LeMarie C, Henrion D, et al. G894T polymorphism in the endothelial nitric oxide synthase gene is associated with an enhanced vascular responsiveness to phenylephrine. Circulation 1999;99(24):3096–8.
[23] Angeline T, Isabel W, Tsongalis GJ. Endothelial nitric oxide gene polymorphisms, nitric oxide production and coronary artery disease risk in a South Indian population. Exp Mol Pathol 2010;89(3):205–8.
[24] Bor-Kucukatay M, Demir S, Akbay R, Dursunoglu D, Akdag B, Semiz E. Relationship between hemorheology and Glu(298)Asp polymorphism of endothelial nitric oxide synthase gene in patients with coronary artery disease. Mol Biol Rep 2010;37(1):171–8.
[25] Shimasaki Y, Yasue H, Yoshimura M, Nakayama M, Kugiyama K, Ogawa H, et al. Association of the missense Glu298Asp variant of the endothelial nitric oxide synthase gene with myocardial infarction. J Am Coll Cardiol 1998;31(7):1506–10.
[26] Hibi K, Ishigami T, Tamura K, Mizushima S, Nyui N, Fujita T, et al. Endothelial nitric oxide synthase gene polymorphism and acute myocardial infarction. Hypertension 1998;32(3):521–6.
[27] Dafni C, Drakoulis N, Landt O, Panidis D, Reczko M, Cokkinos DV. Association of the eNOS E298D polymorphism and the risk of myocardial infarction in the Greek population. BMC Med Genet 2010;11:133.

[28] Guzik TJ, Black E, West NE, McDonald D, Ratnatunga C, Pillai R, et al. Relationship between the G894T polymorphism (Glu298Asp variant) in endothelial nitric oxide synthase and nitric oxide-mediated endothelial function in human atherosclerosis. Am J Med Genet 2001;100(2):130–7.
[29] Sladowska-Kozlowska J, Litwin M, Niemirska A, Wierzbicka A, Roszczynko M, Szperl M. Associations of the eNOS G894T gene polymorphism with target organ damage in children with newly diagnosed primary hypertension. Pediatr Nephrol 2015;30(12):2189–97.
[30] Antoniades C, Tousoulis D, Vasiliadou C, Pitsavos C, Toutouza M, Tentolouris C, et al. Genetic polymorphisms G894T on the eNOS gene is associated with endothelial function and vWF levels in premature myocardial infarction survivors. Int J Cardiol 2006;107(1):95–100.
[31] Hassani Idrissi H, Hmimech W, Diakite B, Korchi F, Baghdadi D, Habbal R, et al. Association of G894T eNOS, 4G/5G PAI and T1131C APOA5 polymorphisms with susceptibility to myocardial infarction in Morocco. Meta Gene 2016;9:56–61.
[32] Nakayama M, Yasue H, Yoshimura M, Shimasaki Y, Kugiyama K, Ogawa H, et al. T-786->C mutation in the 5'-flanking region of the endothelial nitric oxide synthase gene is associated with coronary spasm. Circulation 1999;99(22):2864–70.
[33] Sandrim VC, de Syllos RW, Lisboa HR, Tres GS, Tanus-Santos JE. Endothelial nitric oxide synthase haplotypes affect the susceptibility to hypertension in patients with type 2 diabetes mellitus. Atherosclerosis 2006;189(1):241–6.
[34] Miyamoto Y, Saito Y, Nakayama M, Shimasaki Y, Yoshimura T, Yoshimura M, et al. Replication protein A1 reduces transcription of the endothelial nitric oxide synthase gene containing a -786T->C mutation associated with coronary spastic angina. Hum Mol Genet 2000;9(18):2629–37.
[35] Alvarez R, Gonzalez P, Batalla A, Reguero JR, Iglesias-Cubero G, Hevia S, et al. Association between the NOS3 (-786 T/C) and the ACE (I/D) DNA genotypes and early coronary artery disease. Nitric Oxide 2001;5(4):343–8.
[36] Rios DL, Callegari-Jacques SM, Hutz MH. Endothelial nitric oxide synthase and fractalkine chemokine receptor polymorphisms on angiographically assessed coronary atherosclerosis. Clin Chim Acta 2005;362(1–2):138–46.
[37] Hyndman ME, Parsons HG, Verma S, Bridge PJ, Edworthy S, Jones C, et al. The T-786->C mutation in endothelial nitric oxide synthase is associated with hypertension. Hypertension 2002; 39(4):919–22.
[38] Ragia G, Nikolaidis E, Tavridou A, Arvanitidis KI, Kanoni S, Dedoussis GV, et al. Endothelial nitric oxide synthase gene polymorphisms -786T > C and 894G > T in coronary artery bypass graft surgery patients. Hum Genomics 2010;4(6):375–83.
[39] Kumar GR, Spurthi KM, Kumar GK, Aiyengar TM, Padala C, Shyamala N, et al. Genetic polymorphisms of eNOS (-786T/C, Intron 4b/4a & 894G/T) and its association with asymptomatic first degree relatives of coronary heart disease patients. Nitric Oxide 2016;60.
[40] Wilson AM, Ryan MC, Boyle AJ. The novel role of C-reactive protein in cardiovascular disease: risk marker or pathogen. Int J Cardiol 2006;106(3):291–7.
[41] Singh U, Devaraj S, Vasquez-Vivar J, Jialal I. C-reactive protein decreases endothelial nitric oxide synthase activity via uncoupling. J Mol Cell Cardiol 2007;43(6):780–91.
[42] Fichtlscherer S, Rosenberger G, Walter DH, Breuer S, Dimmeler S, Zeiher AM. Elevated C-reactive protein levels and impaired endothelial vasoreactivity in patients with coronary artery disease. Circulation 2000;102(9):1000–6.
[43] Hage FG, Szalai AJ. The role of C-reactive protein polymorphisms in inflammation and cardiovascular risk. Curr Atheroscler Rep 2009;11(2):124–30.
[44] Hage FG, Szalai AJ. C-reactive protein gene polymorphisms, C-reactive protein blood levels, and cardiovascular disease risk. J Am Coll Cardiol 2007;50(12):1115–22.
[45] Lawlor DA, Harbord RM, Timpson NJ, Lowe GD, Rumley A, Gaunt TR, et al. The association of C-reactive protein and CRP genotype with coronary heart disease: findings from five studies with 4,610 cases amongst 18,637 participants. PLoS One 2008;3(8):e3011.
[46] Schulz S, Ludike H, Lierath M, Schlitt A, Werdan K, Hofmann B, et al. C-reactive protein levels and genetic variants of CRP as prognostic markers for combined cardiovascular endpoint (cardiovascular death, death from stroke, myocardial infarction, and stroke/TIA). Cytokine 2016;88:71–6.

[47] Fernandes MT, Fernandes KB, Marquez AS, Colus IM, Souza MF, Santos JP, et al. Association of interleukin-6 gene polymorphism (rs1800796) with severity and functional status of osteoarthritis in elderly individuals. Cytokine 2015;75(2):316–20.

[48] de Albuquerque Maranhao RM, Martins Esteves FA, Crovella S, Segat L, Eleuterio Souza PR. Tumor necrosis factor-alpha and interleukin-6 gene polymorphism association with susceptibility to celiac disease in Italian patients. Genet Mol Res 2015;14(4):16343–52.

[49] Xie XM, Ke R, Zhang YH, Wang GZ, Zhu YT, Hussian T, et al. Interleukin-6 gene -174G>C polymorphism and chronic obstructive pulmonary disease risk: a meta-analysis. Genet Mol Res 2015;14(3):8516–25.

[50] Ramos Dos Santos L, Belcavello L, Camporez D, Iamonde Maciel de Magalhaes C, Zandonade E, Lirio Morelato R, et al. Association study of the BIN1 and IL-6 genes on Alzheimer's disease. Neurosci Lett 2016;614:65–9.

[51] Brull DJ, Montgomery HE, Sanders J, Dhamrait S, Luong L, Rumley A, et al. Interleukin-6 gene -174g>c and -572g>c promoter polymorphisms are strong predictors of plasma interleukin-6 levels after coronary artery bypass surgery. Arterioscler Thromb Vasc Biol 2001; 21(9):1458–63.

[52] Humphries SE, Luong LA, Ogg MS, Hawe E, Miller GJ. The interleukin-6-174 G/C promoter polymorphism is associated with risk of coronary heart disease and systolic blood pressure in healthy men. Eur Heart J 2001;22(24):2243–52.

[53] Li L, Li E, Zhang LH, Jian LG, Liu HP, Wang T. IL-6-174G/C and IL-6-572C/G polymorphisms are associated with increased risk of coronary artery disease. Genet Mol Res 2015;14(3):8451–7.

[54] Liu SL, Yin YW, Sun QQ, Hu AM, Zhang SJ. Genetic polymorphisms of interleukin-6 gene and susceptibility to coronary artery disease in Chinese population: evidence based on 4582 subjects. Hum Immunol 2015;76(7):505–10.

[55] Wang K, Dong PS, Zhang HF, Li ZJ, Yang XM, Liu H. Role of interleukin-6 gene polymorphisms in the risk of coronary artery disease. Genet Mol Res 2015;14(2):3177–83.

[56] Yang HT, Wang SL, Yan LJ, Qian P, Duan HY. Association of interleukin gene polymorphisms with the risk of coronary artery disease. Genet Mol Res 2015;14(4):12489–96.

[57] Tousoulis D, Kampoli AM, Papageorgiou N, Androulakis E, Antoniades C, Toutouzas K, et al. Pathophysiology of atherosclerosis: the role of inflammation. Curr Pharm Des 2011;17(37): 4089–110.

[58] Mao L, Geng GY, Han WJ, Zhao MH, Wu L, Liu HL. Interleukin-6 (IL-6) -174G/C genomic polymorphism contribution to the risk of coronary artery disease in a Chinese population. Genet Mol Res 2016;15(2).

[59] Terry CF, Loukaci V, Green FR. Cooperative influence of genetic polymorphisms on interleukin 6 transcriptional regulation. J Biol Chem 2000;275(24):18138–44.

[60] Kelberman D, Hawe E, Luong LA, Mohamed-Ali V, Lundman P, Tornvall P, et al. Effect of interleukin-6 promoter polymorphisms in survivors of myocardial infarction and matched controls in the North and South of Europe. The HIFMECH Study. Thromb Haemost 2004; 92(5):1122–8.

[61] Malarstig A, Wallentin L, Siegbahn A. Genetic variation in the interleukin-6 gene in relation to risk and outcomes in acute coronary syndrome. Thromb Res 2007;119(4):467–73.

[62] Celik A, Ozcetin M, Ates O, Altunkas F, Karaman K, Akar I, et al. Analyses of C-reactive protein, endothelial nitric oxide synthase and interleukin-6 gene polymorphisms in adolescents with a family history of premature coronary artery disease: a pilot study. Balkan Med J 2015;32(4):397–402.

[63] Stoica AL, Stoica E, Constantinescu I, Uscatescu V, Ginghina C. Interleukin-6 and interleukin-10 gene polymorphism, endothelial dysfunction, and postoperative prognosis in patients with peripheral arterial disease. J Vasc Surg 2010;52(1):103–9.

[64] Ray KK, Camp NJ, Bennett CE, Francis SE, Crossman DC. Genetic variation at the interleukin-1 locus is a determinant of changes in soluble endothelial factors in patients with acute coronary syndromes. Clin Sci 2002;103(3):303–10.

[65] Bis JC, Heckbert SR, Smith NL, Reiner AP, Rice K, Lumley T, et al. Variation in inflammation-related genes and risk of incident nonfatal myocardial infarction or ischemic stroke. Atherosclerosis 2008;198(1):166–73.

[66] Fibrinogen Studies C, Kaptoge S, White IR, Thompson SG, Wood AM, Lewington S, et al. Associations of plasma fibrinogen levels with established cardiovascular disease risk factors, inflammatory markers, and other characteristics: individual participant meta-analysis of 154,211 adults in 31 prospective studies: the fibrinogen studies collaboration. Am J Epidemiol 2007;166(8):867–79.
[67] Forsyth CB, Solovjov DA, Ugarova TP, Plow EF. Integrin alpha(M)beta(2)-mediated cell migration to fibrinogen and its recognition peptides. J Exp Med 2001;193(10):1123–33.
[68] Schneider DJ, Taatjes DJ, Howard DB, Sobel BE. Increased reactivity of platelets induced by fibrinogen independent of its binding to the IIb-IIIa surface glycoprotein: a potential contributor to cardiovascular risk. J Am Coll Cardiol 1999;33(1):261–6.
[69] Rana JS, Cote M, Despres JP, Sandhu MS, Talmud PJ, Ninio E, et al. Inflammatory biomarkers and the prediction of coronary events among people at intermediate risk: the EPIC-Norfolk prospective population study. Heart 2009;95(20):1682–7.
[70] Chuang SY, Bai CH, Chen WH, Lien LM, Pan WH. Fibrinogen independently predicts the development of ischemic stroke in a Taiwanese population: CVDFACTS study. Stroke 2009;40(5): 1578–84.
[71] Rizzo M, Corrado E, Coppola G, Muratori I, Novo G, Novo S. Markers of inflammation are strong predictors of subclinical and clinical atherosclerosis in women with hypertension. Coron Artery Dis 2009;20(1):15–20.
[72] Green D, Chan C, Kang J, Liu K, Schreiner P, Jenny NS, et al. Longitudinal assessment of fibrinogen in relation to subclinical cardiovascular disease: the CARDIA study. J Thromb Haemost 2010;8(3): 489–95.
[73] Fibrinogen Studies C, Danesh J, Lewington S, Thompson SG, Lowe GD, Collins R, et al. Plasma fibrinogen level and the risk of major cardiovascular diseases and nonvascular mortality: an individual participant meta-analysis. JAMA 2005;294(14):1799–809.
[74] Meade TW, Mellows S, Brozovic M, Miller GJ, Chakrabarti RR, North WR, et al. Haemostatic function and ischaemic heart disease: principal results of the Northwick Park Heart Study. Lancet 1986;2(8506):533–7.
[75] Assmann G, Cullen P, Heinrich J, Schulte H. Hemostatic variables in the prediction of coronary risk: results of the 8 year follow-up of healthy men in the Munster Heart Study (PROCAM). Prospective Cardiovascular Munster Study. Isr J Med Sci 1996;32(6):364–70.
[76] Kathiresan S, Yang Q, Larson MG, Camargo AL, Tofler GH, Hirschhorn JN, et al. Common genetic variation in five thrombosis genes and relations to plasma hemostatic protein level and cardiovascular disease risk. Arterioscler Thromb Vasc Biol 2006;26(6):1405–12.
[77] Smith GD, Harbord R, Milton J, Ebrahim S, Sterne JA. Does elevated plasma fibrinogen increase the risk of coronary heart disease? Evidence from a meta-analysis of genetic association studies. Arterioscler Thromb Vasc Biol 2005;25(10):2228–33.
[78] Keavney B, Danesh J, Parish S, Palmer A, Clark S, Youngman L, et al. Fibrinogen and coronary heart disease: test of causality by 'Mendelian randomization'. Int J Epidemiol 2006;35(4):935–43.
[79] Papageorgiou N, Tousoulis D, Miliou A, Hatzis G, Kozanitou M, Androulakis E, et al. Combined effects of fibrinogen genetic variability on atherosclerosis in patients with or without stable angina pectoris: focus on the coagulation cascade and endothelial function. Int J Cardiol 2013;168(5): 4602–7.
[80] Papageorgiou N, Briasoulis A, Hatzis G, Androulakis E, Kozanitou M, Miliou A, et al. Coronary artery atherosclerosis in hypertensive patients: the role of fibrinogen genetic variability. Rev Esp Cardiol 2016;70.
[81] LaFramboise T. Single nucleotide polymorphism arrays: a decade of biological, computational and technological advances. Nucleic Acids Res 2009;37(13):4181–93.
[82] Lander ES. Initial impact of the sequencing of the human genome. Nature 2011;470(7333): 187–97.
[83] International HapMap C. The International HapMap Project. Nature 2003;426(6968):789–96.
[84] Samani NJ, Erdmann J, Hall AS, Hengstenberg C, Mangino M, Mayer B, et al. Genomewide association analysis of coronary artery disease. N Engl J Med 2007;357(5):443–53.

[85] Vasan RS, Larson MG, Aragam J, Wang TJ, Mitchell GF, Kathiresan S, et al. Genome-wide association of echocardiographic dimensions, brachial artery endothelial function and treadmill exercise responses in the Framingham Heart Study. BMC Med Genet 2007; 8(Suppl. 1):S2.
[86] Preuss M, Konig IR, Thompson JR, Erdmann J, Absher D, Assimes TL, et al. Design of the coronary ARtery DIsease genome-wide replication and meta-analysis (CARDIoGRAM) study: a genome-wide association meta-analysis involving more than 22,000 cases and 60,000 controls. Circ Cardiovasc Genet 2010;3(5):475–83.
[87] Schunkert H, Konig IR, Kathiresan S, Reilly MP, Assimes TL, Holm H, et al. Large-scale association analysis identifies 13 new susceptibility loci for coronary artery disease. Nat Genet 2011;43(4):333–8.
[88] Reilly MP, Li M, He J, Ferguson JF, Stylianou IM, Mehta NN, et al. Identification of ADAMTS7 as a novel locus for coronary atherosclerosis and association of ABO with myocardial infarction in the presence of coronary atherosclerosis: two genome-wide association studies. Lancet 2011;377(9763): 383–92.
[89] Coronary Artery Disease Genetics C. A genome-wide association study in Europeans and South Asians identifies five new loci for coronary artery disease. Nat Genet 2011;43(4):339–44.
[90] Consortium IKC. Large-scale gene-centric analysis identifies novel variants for coronary artery disease. PLoS Genet 2011;7(9):e1002260.
[91] Wang F, Xu CQ, He Q, Cai JP, Li XC, Wang D, et al. Genome-wide association identifies a susceptibility locus for coronary artery disease in the Chinese Han population. Nat Genet 2011; 43(4):345–9.
[92] Consortium CAD, Deloukas P, Kanoni S, Willenborg C, Farrall M, Assimes TL, et al. Large-scale association analysis identifies new risk loci for coronary artery disease. Nat Genet 2013;45(1):25–33.
[93] Helgadottir A, Thorleifsson G, Manolescu A, Gretarsdottir S, Blondal T, Jonasdottir A, et al. A common variant on chromosome 9p21 affects the risk of myocardial infarction. Science 2007; 316(5830):1491–3.
[94] McPherson R, Pertsemlidis A, Kavaslar N, Stewart A, Roberts R, Cox DR, et al. A common allele on chromosome 9 associated with coronary heart disease. Science 2007;316(5830):1488–91.
[95] Wellcome Trust Case Control C. Genome-wide association study of 14,000 cases of seven common diseases and 3,000 shared controls. Nature 2007;447(7145):661–78.
[96] Ding H, Xu Y, Wang X, Wang Q, Zhang L, Tu Y, et al. 9p21 is a shared susceptibility locus strongly for coronary artery disease and weakly for ischemic stroke in Chinese Han population. Circ Cardiovasc Genet 2009;2(4):338–46.
[97] Kumar J, Yumnam S, Basu T, Ghosh A, Garg G, Karthikeyan G, et al. Association of polymorphisms in 9p21 region with CAD in North Indian population: replication of SNPs identified through GWAS. Clin Genet 2011;79(6):588–93.
[98] Saleheen D, Alexander M, Rasheed A, Wormser D, Soranzo N, Hammond N, et al. Association of the 9p21.3 locus with risk of first-ever myocardial infarction in Pakistanis: case-control study in South Asia and updated meta-analysis of Europeans. Arterioscler Thromb Vasc Biol 2010;30(7): 1467–73.
[99] Kral BG, Mathias RA, Suktitipat B, Ruczinski I, Vaidya D, Yanek LR, et al. A common variant in the CDKN2B gene on chromosome 9p21 protects against coronary artery disease in Americans of African ancestry. J Hum Genet 2011;56(3):224–9.
[100] Shen GQ, Li L, Rao S, Abdullah KG, Ban JM, Lee BS, et al. Four SNPs on chromosome 9p21 in a South Korean population implicate a genetic locus that confers high cross-race risk for development of coronary artery disease. Arterioscler Thromb Vasc Biol 2008;28(2):360–5.
[101] Ye S, Willeit J, Kronenberg F, Xu Q, Kiechl S. Association of genetic variation on chromosome 9p21 with susceptibility and progression of atherosclerosis: a population-based, prospective study. J Am Coll Cardiol 2008;52(5):378–84.
[102] Zdravkovic S, Wienke A, Pedersen NL, Marenberg ME, Yashin AI, De Faire U. Heritability of death from coronary heart disease: a 36-year follow-up of 20 966 Swedish twins. J Intern Medicine 2002; 252(3):247–54.

[103] International Consortium for Blood Pressure Genome-Wide Association S, Ehret GB, Munroe PB, Rice KM, Bochud M, Johnson AD, et al. Genetic variants in novel pathways influence blood pressure and cardiovascular disease risk. Nature 2011;478(7367):103–9.
[104] Miall WE, Oldham PD. The hereditary factor in arterial blood-pressure. Br Med J 1963;1(5323): 75–80.
[105] Morris AP, Voight BF, Teslovich TM, Ferreira T, Segre AV, Steinthorsdottir V, et al. Large-scale association analysis provides insights into the genetic architecture and pathophysiology of type 2 diabetes. Nat Genet 2012;44(9):981–90.
[106] Poulsen P, Kyvik KO, Vaag A, Beck-Nielsen H. Heritability of type II (non-insulin-dependent) diabetes mellitus and abnormal glucose tolerance—a population-based twin study. Diabetologia 1999; 42(2):139–45.
[107] Voight BF, Scott LJ, Steinthorsdottir V, Morris AP, Dina C, Welch RP, et al. Twelve type 2 diabetes susceptibility loci identified through large-scale association analysis. Nat Genet 2010; 42(7):579–89.
[108] Friedlander Y, Kark JD, Stein Y. Biological and environmental sources of variation in plasma lipids and lipoproteins: the Jerusalem Lipid Research Clinic. Hum Hered 1986;36(3):143–53.
[109] Perusse L, Despres JP, Tremblay A, Leblanc C, Talbot J, Allard C, et al. Genetic and environmental determinants of serum lipids and lipoproteins in French Canadian families. Arteriosclerosis 1989;9(3): 308–18.
[110] Teslovich TM, Musunuru K, Smith AV, Edmondson AC, Stylianou IM, Koseki M, et al. Biological, clinical and population relevance of 95 loci for blood lipids. Nature 2010;466(7307):707–13.
[111] Austin MA, Sandholzer C, Selby JV, Newman B, Krauss RM, Utermann G. Lipoprotein(a) in women twins: heritability and relationship to apolipoprotein(a) phenotypes. Am J Hum Genet 1992;51(4):829–40.
[112] Scholz M, Kraft HG, Lingenhel A, Delport R, Vorster EH, Bickeboller H, et al. Genetic control of lipoprotein(a) concentrations is different in Africans and Caucasians. Eur J Hum Genet 1999;7(2): 169–78.
[113] Qi Q, Workalemahu T, Zhang C, Hu FB, Qi L. Genetic variants, plasma lipoprotein(a) levels, and risk of cardiovascular morbidity and mortality among two prospective cohorts of type 2 diabetes. Eur Heart J 2012;33(3):325–34.
[114] Koopmans JR, Slutske WS, Heath AC, Neale MC, Boomsma DI. The genetics of smoking initiation and quantity smoked in Dutch adolescent and young adult twins. Behav Genet 1999;29(6):383–93.
[115] Li MD, Cheng R, Ma JZ, Swan GE. A meta-analysis of estimated genetic and environmental effects on smoking behavior in male and female adult twins. Addiction 2003;98(1):23–31.
[116] Tobacco GC. Genome-wide meta-analyses identify multiple loci associated with smoking behavior. Nat Genet 2010;42(5):441–7.
[117] Helgadottir A, Gretarsdottir S, St Clair D, Manolescu A, Cheung J, Thorleifsson G, et al. Association between the gene encoding 5-lipoxygenase-activating protein and stroke replicated in a Scottish population. Am J Hum Genet 2005;76(3):505–9.
[118] Song Y, Stampfer MJ, Liu S. Meta-analysis: apolipoprotein E genotypes and risk for coronary heart disease. Ann Intern Med 2004;141(2):137–47.
[119] Yamada Y, Izawa H, Ichihara S, Takatsu F, Ishihara H, Hirayama H, et al. Prediction of the risk of myocardial infarction from polymorphisms in candidate genes. N Engl J Med 2002;347(24): 1916–23.
[120] Hirashiki A, Yamada Y, Murase Y, Suzuki Y, Kataoka H, Morimoto Y, et al. Association of gene polymorphisms with coronary artery disease in low- or high-risk subjects defined by conventional risk factors. J Am Coll Cardiol 2003;42(8):1429–37.
[121] Moatti D, Faure S, Fumeron F, Amara Mel W, Seknadji P, McDermott DH, et al. Polymorphism in the fractalkine receptor CX3CR1 as a genetic risk factor for coronary artery disease. Blood 2001; 97(7):1925–8.
[122] McDermott DH, Fong AM, Yang Q, Sechler JM, Cupples LA, Merrell MN, et al. Chemokine receptor mutant CX3CR1-M280 has impaired adhesive function and correlates with protection from cardiovascular disease in humans. J Clin Invest 2003;111(8):1241–50.

[123] Margaglione M, Cappucci G, Colaizzo D, Giuliani N, Vecchione G, Grandone E, et al. The PAI-1 gene locus 4G/5G polymorphism is associated with a family history of coronary artery disease. Arterioscler Thromb Vasc Biol 1998;18(2):152–6.

[124] Ozaki K, Sato H, Iida A, Mizuno H, Nakamura T, Miyamoto Y, et al. A functional SNP in PSMA6 confers risk of myocardial infarction in the Japanese population. Nat Genet 2006;38(8):921–5.

[125] Terashima M, Akita H, Kanazawa K, Inoue N, Yamada S, Ito K, et al. Stromelysin promoter 5A/6A polymorphism is associated with acute myocardial infarction. Circulation 1999;99(21):2717–9.

[126] Rauramaa R, Vaisanen SB, Luong LA, Schmidt-Trucksass A, Penttila IM, Bouchard C, et al. Stromelysin-1 and interleukin-6 gene promoter polymorphisms are determinants of asymptomatic carotid artery atherosclerosis. Arterioscler Thromb Vasc Biol 2000;20(12):2657–62.

[127] Gretarsdottir S, Thorleifsson G, Reynisdottir ST, Manolescu A, Jonsdottir S, Jonsdottir T, et al. The gene encoding phosphodiesterase 4D confers risk of ischemic stroke. Nat Genet 2003;35(2):131–8.

[128] Zee RY, Brophy VH, Cheng S, Hegener HH, Erlich HA, Ridker PM. Polymorphisms of the phosphodiesterase 4D, cAMP-specific (PDE4D) gene and risk of ischemic stroke: a prospective, nested case-control evaluation. Stroke 2006;37(8):2012–7.

[129] Brophy VH, Ro SK, Rhees BK, Lui LY, Lee JM, Umblas N, et al. Association of phosphodiesterase 4D polymorphisms with ischemic stroke in a US population stratified by hypertension status. Stroke 2006;37(6):1385–90.

[130] Helgadottir A, Manolescu A, Helgason A, Thorleifsson G, Thorsteinsdottir U, Gudbjartsson DF, et al. A variant of the gene encoding leukotriene A4 hydrolase confers ethnicity-specific risk of myocardial infarction. Nat Genet 2006;38(1):68–74.

[131] Komulainen K, Alanne M, Auro K, Kilpikari R, Pajukanta P, Saarela J, et al. Risk alleles of USF1 gene predict cardiovascular disease of women in two prospective studies. PLoS Genet 2006;2(5):e69.

[132] Connelly JJ, Wang T, Cox JE, Haynes C, Wang L, Shah SH, et al. GATA2 is associated with familial early-onset coronary artery disease. PLoS Genet 2006;2(8):e139.

[133] Wang X, Ria M, Kelmenson PM, Eriksson P, Higgins DC, Samnegard A, et al. Positional identification of TNFSF4, encoding OX40 ligand, as a gene that influences atherosclerosis susceptibility. Nat Genet 2005;37(4):365–72.

[134] Myocardial Infarction Genetics C, Kathiresan S, Voight BF, Purcell S, Musunuru K, Ardissino D, et al. Genome-wide association of early-onset myocardial infarction with single nucleotide polymorphisms and copy number variants. Nat Genet 2009;41(3):334–41.

[135] Clarke R, Peden JF, Hopewell JC, Kyriakou T, Goel A, Heath SC, et al. Genetic variants associated with Lp(a) lipoprotein level and coronary disease. N Engl J Med 2009;361(26):2518–28.

[136] Tregouet DA, Konig IR, Erdmann J, Munteanu A, Braund PS, Hall AS, et al. Genome-wide haplotype association study identifies the SLC22A3-LPAL2-LPA gene cluster as a risk locus for coronary artery disease. Nat Genet 2009;41(3):283–5.

[137] Gudbjartsson DF, Bjornsdottir US, Halapi E, Helgadottir A, Sulem P, Jonsdottir GM, et al. Sequence variants affecting eosinophil numbers associate with asthma and myocardial infarction. Nat Genet 2009;41(3):342–7.

[138] Soranzo N, Spector TD, Mangino M, Kuhnel B, Rendon A, Teumer A, et al. A genome-wide meta-analysis identifies 22 loci associated with eight hematological parameters in the HaemGen consortium. Nat Genet 2009;41(11):1182–90.

[139] Erdmann J, Grosshennig A, Braund PS, Konig IR, Hengstenberg C, Hall AS, et al. New susceptibility locus for coronary artery disease on chromosome 3q22.3. Nat Genet 2009;41(3):280–2.

[140] Erdmann J, Willenborg C, Nahrstaedt J, Preuss M, Konig IR, Baumert J, et al. Genome-wide association study identifies a new locus for coronary artery disease on chromosome 10p11.23. Eur Heart J 2011;32(2):158–68.

CHAPTER 2

Coronary Artery Circulation: Basic Principles

CHAPTER 2.1

Functional Anatomy

Alexios S. Antonopoulos[1], Gerasimos Siasos[1], Charalambos Antoniades[2], Dimitris Tousoulis[1]
[1]1st Cardiology Clinic, 'Hippokration' General Hospital, National and Kapodistrian University of Athens, School of Medicine, Athens, Greece; [2]University of Oxford, Oxford, United Kingdom

Contents

ANATOMY AND FUNCTION OF CORONARY ARTERIES

The coronary artery system, which is responsible for the perfusion of the heart, normally consists of the left and right coronary arteries.

Left Coronary Artery

The left coronary artery arises from the left sinus of Valsalva, located just above the aortic root, at the beginning of the ascending aorta. The initial part of the left coronary artery is called the left main stem or trunk, which is a short branch (mean length 1 cm and diameter 4 mm) that passes between the pulmonary trunk and the left atrial appendage. It is then divided into two major branches; the left anterior descending (LAD) and the left circumflex artery (LCx).

Left Anterior Descending Artery

The LAD branch proceeds downwards into the anterior interventricular groove. Throughout its route it supplies branches for the perfusion of both ventricles and the

Coronary Artery Disease
ISBN 978-0-12-811908-2, http://dx.doi.org/10.1016/B978-0-12-811908-2.00008-8

interventricular septum (diagonals and septal perforators). More specifically, from the LAD artery emerges the anterior septal perforating arteries, which supply the anterior two-thirds of the interventricular septum. In addition, other LAD branches supply the anterior, lateral, and apical wall of the left ventricle, the right and upper-ascending left branch of the bundle of His and the anterolateral papillary muscle of the left ventricle. The LAD artery also provides collateral circulation to the anterior right ventricle through the circle of Vieusens, the posterior part of the interventricular septum, and the posterior descending artery. The LAD artery is accompanied by nerves, lymphatic vessels, and the great cardiac vein. It moves toward the apex of the heart following its curved anatomy. Occasionally, a long left anterior descending artery may wrap over the apex and be supplied with blood the inferior wall of the left ventricle. The LAD is divided into proximal, mid, and distal segments. The first septal perforator generally divides the proximal and mid-segments of the LAD. The diagonals are varied in number and size and are labeled (proximally to distally) D1, D2, D3, and so on.

Left Circumflex Artery

The LCx is located at the left part of the coronary groove and reaches the posterior longitudinal groove (left atrial—ventricular sulcus). It is close to the mitral orifice. Its branches supply the greatest part of the left atrium, the posterior and lateral walls of the left ventricle and part of the anterior papillary muscle. Additionally, the LCx provides perfusion to the sinoatrial node in 40%—50% of hearts. The left circumflex branches are labeled obtuse marginal (OMs) branches. The OMs are labeled from proximal to distal, OM1, OM2, OM3, and so on. The major branch of the LCx is the left marginal artery. In around 10%—15% of the population, the left circumflex joins to the right coronary artery (RCA) to give rise to the posterior descending artery. Ostium refers to the segment of origin of the artery.

Right Coronary Artery

The RCA emerges from the right sinus of Valsalva, proceeds behind the pulmonary artery, and descends through the atrioventricular groove then curves posteriorly, making a bend at the crux of the heart and continuing downward in the posterior interventricular sulcus. The RCA gives two major branches; the conus (arteriosus) artery and the sinoatrial artery. The conus artery is the first provided branch by the RCA; it originates from the ostium of the RCA or separately from the right coronary sinus and has a superior and anterior course. It supplies the right ventricle. The atrial branch is generally the second artery to be visualized; it stems from the from the proximal RCA and has a posterior course. From the atrial branch originates the sinoatrial artery, from which the artery that supplies the sinoatrial node in 50%—70% of hearts emerges, running along the anterior right atrium to the superior vena cava and encircling it before reaching the sinoatrial node. The RCA gives off multiple branches for the perfusion of the right atrium and ventricle. The greatest branch is the right acute marginal branch that supplies the right ventricle. At the course of its downward route to the posterior surface of the heart, the RCA gives off two or three branches. The artery of the atrioventricular node (in 50%–60% of hearts), which emerges from the RCA at the crux of the heart and proceeds anteriorly along the base of the atrial septum, also supplies proximal parts of the bundles of His, and the parts of the posterior interventricular septum. Other acute marginal (AM) branches, may also originate from the RCA, varying in size and number and are referred to as (from proximal to distal) AM1, AM2, AM3, etc. Throughout its route, the RCA is accompanied by branches of the cardiac nerve plexus, lymphatic vessels and small cardiac veins. The RCA is divided in proximal, mid, and distal segments. The segment of the RCA from the ostium to the origin of the first acute marginal artery is usually referred to as the proximal RCA.

Dominance refers to whether the posterior descending artery (PDA) originates from the RCA (right dominant), LCx (left dominant), or both (codominant). Approximately 80% of humans are right dominant. In right dominance, the distal RCA at the level of the crux of the heart typically bifurcates into the PDA and a posterolateral branch. The PDA courses in the posterior ventricular septum giving origin to the SA nodal artery and posterior ventricular branch. In left dominance, the PDA originates from the distal LCx. In codominance, there are right and left PDAs originating from the RCA and LCx. In right predominance (20%–25% of population) the RCA across the posterior interventricular septum and reach as far as the left marginal artery to supply the diaphragmatic surface of the left ventricle (Fig. 2.1.1).

ANATOMY AND FUNCTION OF CARDIAC VEINS

Coronary Sinus Ostium

The coronary sinus ostium is an opening of the coronary venous system into the right atrium. It is found between the inferior vena cava and the tricuspid valve on the inferior part of the interatrial septum, forming a corner of Koch's triangle. Its valve is called "the Thebesian valve" [1].

Coronary Sinus

The coronary sinus, the length of which varies from 15 to 65 mm, is found at the posterior part of the coronary sulcus on the diaphragmatic or posterior surface of the heart and is the principal collector of the venous blood of the heart. It empties the blood of the coronary veins right into the right atrium, proximal to the conjunction of the posterior interventricular sulcus and the coronary sulcus. The coronary sinus receives drainage from multiple veins including the great cardiac vein, the left marginal vein, the posterior

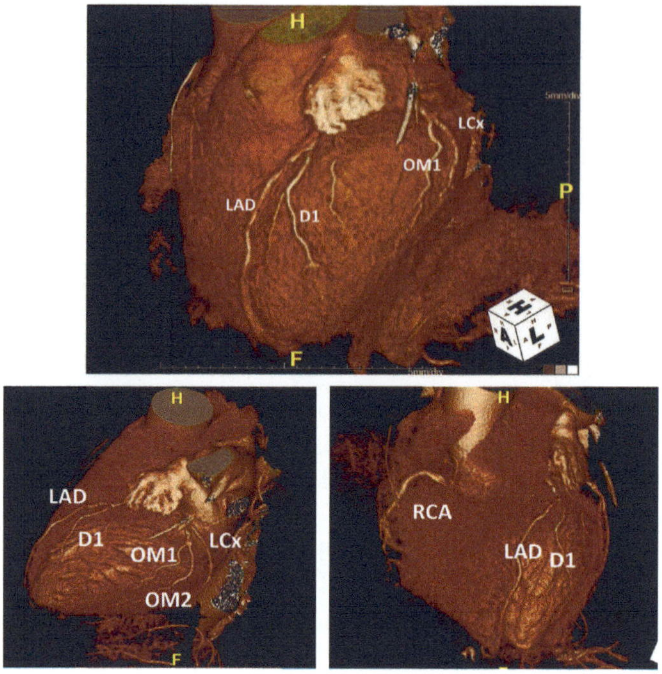

Figure 2.1.1 Coronary circulation as shown on multiple three-dimensional reconstruction projections of a coronary computerized angiography scan. *D1*, first diagonal branch; *LAD*, left anterior descending artery; *LCx*, left circumflex artery; *OM*, obtuse marginal branch; *RCA*, right coronary artery.

vein of the left ventricle, the posterior interventricular vein, the oblique vein of the left atrium and other veins of the left and right atrium.

Great Cardiac Vein

In most hearts, the distal part of the coronary sinus narrows to form the great cardiac vein. The great cardiac vein receives drainage from the anterior surfaces of the left ventricle. The ostium of the great cardiac vein is covered by valves of Vieusen in most cases.

Left Marginal Vein

The ostium of the left marginal vein is typically located at the distal end of the coronary sinus. This vein receives deoxygenated blood from the area of the lateral wall of the left ventricle.

Posterior Vein of the Left Ventricle

Located close to the left marginal vein, the posterior vein of the left ventricle also receives drainage from the lateral wall of the left ventricle.

Posterior Interventricular Vein

The ostium of this vein is typically located close to the ostium of the coronary sinus. A valve associated with the ostium is found in most hearts as this drainage enters the coronary sinus. This vein receives deoxygenated blood from the posterior areas of both right and left ventricles and follows a path along the posterior ventricular septal border.

VARIATIONS OF THE CORONARY CIRCULATION

The arteries of the coronary circulation considerably vary in size, course and bifurcations. Understanding of the coronary anatomy is essential to know which branch supplies blood to specific myocardial segments as well as to guide decision making during percutaneous coronary interventions in patients with coronary artery disease [2]. Before listing the most common congenital anomalies of the coronary circulation, there are also some variations of the coronary course that one should be aware of.

A *coronary fistula* is a communication between a coronary artery and another cardiac chamber or vessel or vascular structure (e.g., coronary sinus). Coronary fistulas are found accidentally in about 0.3%—1.3% of patients undergoing angiography [3]. More than 90% of coronary fistulas drain into the venous circulation into low-pressure structures and usually involve the RCA. Fistulas involving the left anterior descending and the left ventricle are among the rarest type of coronary fistulas [3]. Elective closure at an early age is recommended for symptomatic patients due to the high prevalence of late symptoms and complications. Hemodynamically significant fistulas may lead to myocardial ischemia by steal phenomenon or even to heart failure or pulmonary artery hypertension by progression of the left-to-right shunt. Multiplanar CT coronary angiography allows identification of coronary anomalies and excellent characterization of the vascular anatomy due its high temporal and spatial resolution and offers a good noninterventional alternative modality to angiogram [4].

Myocardial bridges are segments of coronary arteries with an intramyocardial course and are among the most common course variations of coronary circulation. Myocardial bridges are most frequently observed in the left anterior descending artery. They are usually benign and patients can be totally asymptomatic, but there is also the possibility of strangling the coronary artery during systole and cause ischemia symptoms.

Duplication is the second common anomaly of the course of coronary arteries. Duplication usually occurs in the left anterior descending artery (even though cases of right coronary duplication have been also reported). In the case of duplication there is usually a long and a short left anterior descending artery. Duplication is a benign coronary anomaly, with no pathophysiological importance.

Congenital Anomalies

The incidence of congenital coronary anomalies is unknown, since they are usually found in subjects referred for diagnostic coronary angiography due to other reasons or at necropsy. One of the most common coronary anomalies is the anomalous origin of the left main coronary artery from the right aortic sinus of Valsalva. It is benign unless if the left coronary artery has an interarterial course between the aorta and pulmonary trunk, which may predispose to angina and/or sudden cardiac death, caused by compression of the left main coronary artery [5].

Coronary Collaterals

The initial documentation of coronary collaterals in humans was not made until 1964, although their presence was considered as far back as the 16th century. There are few and small collateral vessels and they may develop into a major vascular network in patients with obstructive coronary artery disease. It is generally believed that the collateral vessels, or network, develop only when the occlusive disease or stenosis is severe enough to produce a substantial transstenotic pressure drop. When collaterals are present in patients with severe or significant occlusive CAD, the part of the myocardium supplied by the stenotic vessel seems to have better contractile function than in patients without collaterals [6]. Further, it is hypothesized that preexisting collaterals may also play an important role by decreasing the extent and degree of myocardial damage at the time or during an acute coronary episode. In general, the coronary collaterals are capable of preserving the function and structure in a resting myocardium, however, the coronary flow per gram in a collateral-dependent myocardium is reduced when compared with the normally perfused myocardium. One reason for this could be the change in pressure gradient across the collateral vessels (arterial pressure is lower than aortic pressure in the collateralized segment) [7].

REFERENCES

[1] Mori S, Spicer DE, Anderson RH. Revisiting the anatomy of the living heart. Circ J 2016;80(1):24–33.
[2] Reyes E, Underwood SR. Coronary anatomy and function: a story of Yin and Yang. Eur Heart J Cardiovasc Imaging 2015;16(8):831–3.
[3] Zenooz NA, et al. Coronary artery fistulas: CT findings. Radiographics 2009;29(3):781–9.
[4] Pursnani A, et al. Coronary CTA assessment of coronary anomalies. J Cardiovasc Comput Tomogr 2012;6(1):48–59.
[5] Roberts WC. Major anomalies of coronary arterial origin seen in adulthood. Am Heart J 1986;111(5): 941–63.
[6] Baroldi G, Mantero O, Scomazzoni G. The collaterals of the coronary arteries in normal and pathologic hearts. Circ Res 1956;4(2):223–9.
[7] David DS. Coronary collaterals. Circulation 2003;108(23):e161.

CHAPTER 2.2

Myocardial Oxygen Consumption

Alexios S. Antonopoulos, Athina Goliopoulou, Georgia Vogiatzi, Dimitris Tousoulis
1st Cardiology Clinic, 'Hippokration' General Hospital, National and Kapodistrian University of Athens, School of Medicine, Athens, Greece

Contents

OVERVIEW OF MYOCARDIAL ENERGETICS

The heart's energy is a product of aerobic metabolism. Hence, the oxygen consumption of the myocardium is directly related to myocardial energy use. It is, therefore, very important to understand the factors that determine myocardial oxygen consumption (MVO_2), which requires understanding the basic concepts of cardiomyocyte energetics.

Adequate supply of oxygen is quintessential for the cellular processes that require ATP and a decrease in oxygen demand may be of critical importance in clinical situations in which the oxygen requirements of the myocardium exceed the oxygen delivery levels of which the coronary circulation is capable (such as in cases of obstructive coronary artery disease resulting in myocardial ischemia).

Cardiomyocytes, in contrast to other tissues, e.g., skeletal muscles, contain only a limited glycogen pool, which lasts to sustain cardiac work only for a few minutes. This is the reason why cardiac cells depend on constant supply of oxygen to cover their needs and sustain aerobic cellular metabolism. Within cardiomyocytes myoglobin serves as the protein molecule that is responsible for storage of oxygen, similar to oxy-hemoglobin. In order to support myocardial oxygen needs, particularly during times

Coronary Artery Disease
ISBN 978-0-12-811908-2, http://dx.doi.org/10.1016/B978-0-12-811908-2.00009-X

of increased oxygen demand (e.g., during exercise), the heart must extract oxygen from the arterial blood supplying the myocardium.

The primary energy fuel of the healthy myocardium is fatty acids which are metabolized in the rich-in-mitochondria cardiomyocytes via oxidative phosphorylation resulting in ATP production and consumption of oxygen. In disease states, e.g., in heart failure, the failing myocardium may shift to the use of glucose or lactic acid as an energy source. ATP production is used by cells for events such as myocyte contraction and relaxation (e.g., myosin ATPase) and membrane transport mechanisms (e.g., $Na+/K+$-ATPase pump). Approximately a quarter of myocardial ATP, and consequently oxygen consumption, is used for the basal metabolism of the heart tissue in a resting individual. This energy is consumed in various cellular activities other than contraction. Thus, it cannot be reduced for the clinical benefit of the patient. Approximately 75% of myocardial energy is used at the processes required for the contraction of the myocardial muscle. This mostly reflects ATP splitting associated with cross-bridge splitting during the isovolumetric contraction and ejection phases of the cardiac cycle. A smaller amount of ATP is also consumed for calcium confiscation at the end of each contraction. The largest part (approximately 50%) of the total MVO_2 occurs during the isovolumetric contraction phase of the cardiac cycle, which has high energy demands. The energy required for the isovolumetric contraction is dependent on the cardiac afterload, that is, on the pressure that develops inside the ventricles during this phase. It is then obvious that cardiac afterload is an important determinant of MVO_2. As a result, a reduction in cardiac afterload can prove to be very helpful in the decrease of myocardial energy demands and, hence, MVO_2.

DETERMINANTS OF MYOCARDIAL OXYGEN CONSUMPTION

The study of the determinants of MVO_2 begins with the work of pioneers of cardiac physiology, as early as the beginning of the 20th century. In 1912 Rohde demonstrated that MVO_2 varies as a function of the product of developed pressure and heart rate in the cat isovolumetric left ventricular preparation. Evans and Matsuoka added later on that "*there is a relation between the tension set up on contraction and the metabolism of the contractile tissue*". We now know that MVO_2 actually depends on myocardial oxygen supply and demand. *Myocardial oxygen supply* depends on: (1) the oxygen carrying capacity of blood (i.e., hemoglobin) and (2) the coronary blood flow or the regulation of vascular resistance; the latter is determined by factors such as extravascular compressive forces, humoral factors, neural control, autoregulation, metabolic control and the diastolic phase. On the other hand *myocardial oxygen demand* is determined mainly by three factors the heart rate, cardiac contractility (or velocity of contraction) and systolic wall tension plus the myocardial mass (which is constant for a given heart).

At the course of the ejection phase of the cardiac cycle, the myocardium performs external work and the energy demands of the heart depend on the amount of this work. Because the blood circulated is a fluid system, work (force $\times$ distance) is equal

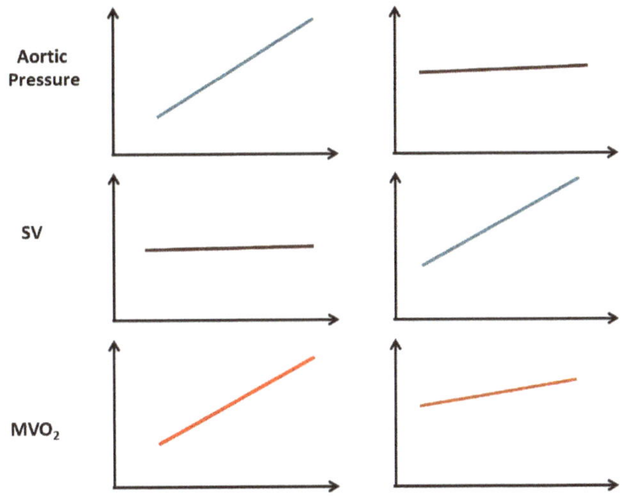

Figure 2.2.1 Relationship between aortic pressure, stroke volume, and myocardial oxygen consumption (MVO_2) in experimental models. MVO_2 rises more steeply with an increase in afterload (aortic pressure). On the other hand, an increase in stroke volume results in a much smaller increase in MVO_2.

to pressure (force/surface) multiplied by volume (distance^3). The external work produced by the left ventricle in one heartbeat is called *stroke work* and is equal to the area enclosed by the left ventricular pressure-volume loop. Increases in either pressure (increased afterload) or stroke volume induce an increase in stroke work. Pressure work increase causes greater ATP expenses and oxygen consumption compared to volume work increase [1]. Hence, a reduction in cardiac afterload would be very beneficial in reducing myocardial oxygen demands during external work (Fig. 2.2.1).

In experimental models it has been demonstrated that there is no simple linear relationship between minute work and MVO_2, i.e., MVO_2. MVO_2 is directly proportional to the *tension-time index*. The latter is a measure of ventricular work and oxygen demand that is found by multiplying the average pressure in the ventricle during the period in which it ejects blood by the time it takes to do this i.e., the area beneath the left ventricular pressure pulse per minute. These two variables change in the same direction at all levels of aortic pressure and cardiac output. MVO_2 is thus directly proportional to the cardiac output at a constant arterial pressure and heart rate, as well as to the heart rate at a constant arterial pressure and cardiac output. Nevertheless, in other conditions such as during physical activity or positive inotropes infusion, indirect indices of myocardial workload, (such as the tension-time index) are poorly correlated with MVO_2. Studies have explored this indiscrepancy and have demonstrated that the *velocity of contraction* is an additional important determinant of MVO_2. Factors that change the velocity of cardiac contraction, e.g., calcium, paired electrical stimulation, norepinephrine, lead to increases of MVO_2. In these circumstances, the tension-time index is disassociated from MVO_2 consumption because the former declines (as a result of the shortening of ejection duration) while the latter increases. Below we report the exact role of each of the major variables affecting MVO_2.

Myocardial Mass

The mass of perfused myocardium is another determinant of MVO_2 (even though it is not subjected to dynamic alterations but is constant for a given heart at a given time point). Myocardial mass is mainly dependent on the mass of the left ventricle. The latter can be calculated by use of imaging techniques such as 2D/3D echocardiography or cardiac magnetic resonance imaging. Geometrical assumptions are often required for left ventricle shape and myocardial mass calculations by using echocardiography. MVO_2 greatly varies between resting and exercise conditions, according to the state of mechanical activity. It has been calculated that MVO2 during resting heart rate is 3–15 mL O_2/min/100 g of tissue, whereas it exceeds 70 mL O_2/min/100 g of tissue during heavy exercise. On the contrast the arrested human heart utilizes only 1.2–1.4 mL O_2/min/100 g of tissue, which corresponds to basal ATP requirements, primary for membrane transport. Any additional increase above that MVO_2 level is directed toward the support of cardiac contraction and relaxation. At resting conditions, myocardial mass is linearly associated with MVO_2.

Heart Rate

The heart rate is one of the factors that significantly influences MVO_2 because the energy cost per minute is affected by the energy cost per heartbeat and the number of beats per minute. Clinical experience has shown that, for a given cardiac output, lower oxygen demands are achieved by a lower heart rate and high stroke volume rather than by a higher heart rate and low stroke volume. This is a result of less pressure (wall tension) developed less often with a low heart rate, which is especially beneficial because the pressure development phase of the cardiac cycle requires high energy supply.

Even though the exact amount of oxygen consumption per heart beat cannot be reliably measured, it is well established that the heart rate is the most important determinant of total MVO_2. Tachycardia affects myocardial oxygen supply–demand balance by shortening diastole duration, i.e., the period of coronary arteries' filling, and thus affects primarily myocardial oxygen supply. Nevertheless tachycardia also increases contractility due to the accumulation of intracellular calcium in myocytes (*Treppe phenomenon*, staircase or *Bowditch effect*). Therefore increased heart rate also raises MVO_2 (see section below for the effects of contractility on MVO_2). Extreme tachycardia can lead to ischemia even in normal hearts, a phenomenon typically observed in patients presenting with atrial fibrillation with rapid ventricular response or other forms of supraventricular tachycardias and manifests with angina symptoms and/or a release of myocardial necrosis enzymes in the circulation (type II myocardial infarction [MI] according to third universal definition of MI). Nevertheless ischemia with increased heart rate is more easily observed in patients with preexisting myocardial hypertrophy or severe aortic stenosis.

The close relationship between heart rate and myocardial O_2 needs suggests that the former can be accordingly modified with pharmacological therapeutic interventions to lower MVO_2, e.g., in patients with angina.

Systolic Wall Tension

Myocardial wall tension or more correctly meridional wall stress (stress = tension divided by area) is one of the most critical determinants of MVO_2 and is determined by the Laplace relation, i.e., depends on ventricular systolic pressure, ventricular radius, and is inversely related to ventricular wall thickness [2].

$$\text{Stress} = Pr/2h$$

(where P, pressure; r, radius, and h, wall thickness).

According to the Laplace relation, preload leads to an increase in myocardial wall tension and oxygen demand by increasing ventricular radius, whereas afterload, affects the magnitude of systolic pressure generation. Systolic wall tension depends on the wall hypertrophy, and decreases with increasing wall thickness, suggesting that cardiac concentric hypertrophy is a compensatory mechanism to reduce wall tension and oxygen demand. A twofold increase in myocardial wall tension doubles MVO_2. Even though wall tension cannot be directly quantified, it is important to understand the directional changes that occur in myocardial oxygen demand with respective changes in wall tension (or each of its determinants). Thus when the balance of oxygen supply and demand is disturbed, interventions can be attempted to induce those directional changes in the above parameters to lower myocardial oxygen demand [2].

Calculation of average myocardial wall stress is feasible by measurement of *left ventricular pressure*, the *radius* or *diameter* of the left ventricular cavity short-axis cross-section (and sometimes the long axis dimension), and *left ventricular wall thickness* (postero-lateral wall, septum, or an average). Nevertheless, geometric assumptions are required for left ventricle (LV) cavity and the final formula is only an approximation and not totally reliably. One of the commonly used formulas for myocardial wall stress is the following [3].

$$\text{Wall stress} = \frac{Pr}{2h}\,\frac{1.33}{1+\dfrac{h}{2r}}$$

(where: h, LV wall thickness, r, the internal LV radius, and P, peak systolic pressure. The factor of 1.33 converts mm Hg to kdynes/cm^2. The normal values in humans are

The above formula is one the best validated in humans. Normal values are 123–179 kdynes/cm^2). Various measurements during the cardiac cycle are needed because the maxima of the various components of the formula may correspond to different time points.

Another formula, a variation of the first is the following:

$$\text{Wall stress} = 0.86 \times 0.33 \times \frac{P \times 2r}{h\left(1 + \frac{h}{2r}\right)} - 27$$

Which corrects peak left ventricular systolic pressure (*P*) to peak systolic stress. Normal values are 137 ± 28 kdynes/cm^2.

Contractility (or Velocity of Contraction)

Any alterations in myocardial contractility can have a severe impact on oxygen requirements for basal metabolism, isovolumic wall tension generation and external work. The myocardial muscle cells consume more energy in the rapid development of a certain tension and shortening by a given amount than in performing the same activity more slowly. Additionally, increases in myocardial contractility result in more energy expenses in active calcium transport. This is often called the "energy wasting" effect of increased contractility [4].

Contractility is affected by autonomic nervous system, heart rate, blood calcium, temperature etc. In heart failure, positive inotropes, e.g., catecholamines, lead to an increase in oxygen consumption. Nevertheless, the improved systolic function and decrease in end diastolic volume and wall tension also favorably affect oxygen demand. Thus catecholamines and positive inotropes increase contractility, so that the left ventricular end-diastolic pressure and volume fall substantially. This decrease in ventricular volume leads (according to the Laplace relation) to a decline in intramyocardial tension, which tends to reduce MVO_2. Thus, the change in MVO_2 following administration of cardiac inotropes depends on the degree to which intramyocardial tension is reduced in relation to the extent to which the contractile state is augmented. This suggests that although in the failing heart with increased end diastolic volumes, inotropes may lead to favorable changes in MVO_2 by reducing wall tension, in the nonfailing heart they will consistently lead to an increase in MVO_2.

On the other hand reductions in contractility lead to respective declines in MVO_2. In animal experiments administration of cardiac depressant drugs, such as propranolol, procaine amide or pronethalol led to reductions in MVO_2 when wall tension is held constant. In these experiments, the change in MVO_2 was related to the changes in dp/dt, or the velocity of contraction [5].

In studies performed by Buckberg and Ross in anesthetized dogs, increased contractility by infusion of isoproterenol led to increased MVO_2 in parallel with increases in the time-tension index. In other experiments, a formula linking Vmax of contraction and wall tension with MVO_2 was derived [3]:

$$MVO_2\ (\text{mL/beat/100 g}) = k + 0.25\ \text{peak developed tension}\ \left(\text{g/cm}^2\right) + 1.43\ \text{Vmax}\ (\text{cm/sec})$$

Even though both tension and contractility are critical determinants of MVO_2, it is challenging to assess the relative importance of each one. In experimental models employing a dog model on cardiopulmonary bypass and a latex balloon inserted in the left ventricle, it was feasible to alter wall tension by keeping cardiac contractility constant as well as alter cardiac contractility (inotropes infusion) by keeping developed tension at a constant level. By increasing ventricular volume (via balloon inflation) the peak developed tension was progressively increased too, leading to respective rises in MVO_2. During norepinephrine infusion, MVO_2 was further increased by an increase in cardiac contractility (at a constant tension). The model showed that papillary muscle's oxygen consumption increased in an almost linear fashion with increases in afterload up to approximately 50% of isometric force. From that level onwards, further additions of afterload are associated with only small increases in MVO_2 [4].

These experimental observations suggested that

1. at any given level of contractile state and initial muscle length, MVO_2 does not simply correlate directly with the external work performed by the muscle and
2. even at a level of constant contractility, MVO_2 is not a simple linear function of tension development [4].

Moreover in other experiments it has been shown that further to contractility and tension development, MVO_2 is influenced by external shortening against a load (Fenn effect) [6]. Shortening of the skeletal muscle against an external load results in additional heat production which depends on the work performed by the muscle. This is the work performed by the muscle to overcome its elastic component or "internal contractile element work" [7]. During isometric contraction, when the external work of the muscle is zero, the work required for the shortening of the contracting elements can be calculated as the product of load and extent of shortening of the contractile apparatus. Indeed the oxygen cost required for the internal contractile element work is higher than that required for the external work of the muscle. The sum of the two elements, i.e., (1) internal contractile element work and (2) external muscle work is nonlinearly associated with the total contractile element work. Nevertheless the MVO_2 required for the internal work is approximately twice that required for the external fiber shortening work. In other terms, external work performed against high pressure loads requires a much higher MVO_2 than similar levels of external work performed at high stroke volumes [4].

These observations have also been validated in experimental models of valvular regurgitation (mitral and aortic). Experimentally provoked mitral or aortic regurgitation resulting in a doubling of the stroke volume, were associated with small increases in MVO_2, mainly due to the increased oxygen cost required for the internal contractile element work at a higher wall tension. Conversely, when wall tension was kept constant, there was no increase in MVO_2 by experimentally induced valvular regurgitation [4].

MVO_2 in Heart Disease States

In patients with coronary artery stenosis, angina symptoms are provoked by myocardial ischemia due to inadequate myocardial oxygenation. Understanding of the mechanisms involved in MVO_2 in these patients is important in dictating therapeutic interventions. When increasing myocardial blood supply is not feasible, then reductions in myocardial oxygen demand could help to alleviate angina symptoms. Therapeutic strategies then could be directed toward reducing heart rate (i.e., the number of contractions per minute), and reducing wall tension development and cardiac contractility, as a means to re-establish an equilibrium between myocardial oxygen supply and demand. Next to drug interventions, other methods that have been employed to reduce angina symptoms in these patients are the stimulation of carotid sinus nerves which suppresses sympathetic outflow discharge leading to reductions in heart rate and myocardial contractility. In experimental studies this appears to be an efficient antiangina strategy relieving angina symptoms in these patients. Importantly, the reductions in MVO_2 achieved by carotid sinus nerve stimulation, have helped these patients to increase the levels of angina-free physical activity performed, improving thus their quality of life.

Other states related to ischemic heart disease that can dramatically alter myocardial oxygen needs are myocardial stunning and the hibernating myocardium.

Myocardial stunning is a state of reversible decreased myocardial contraction. It is a situation that occurs right after the occlusion of an artery of the heart or after the reperfusion of an ischemic area. In the first case, the first minutes after total ischemia occurs, the myocardium uses anaerobic glycolysis instead of aerobic for energy production, resulting in reduced ability to produce ATP to cover its metabolic needs. The lack of energy and the accumulation of lactate result in reduced myocardial contraction. Thus, the period of "myocardial stunning" begins, during which reversible ischemic damage occurs to myocardial cells. Approximately 30 min after the total occlusion takes place, the damage is irreversible, ending the phase of myocardial stunning. In situations of reperfusion, myocardial stunning occurs because for a period of about 2 weeks following primary coronary intervention or coronary artery bypass surgery, myocardial blood flow has returned to normal but contractile function is still reduced.

Hibernating myocardium is a state of the heart in which some segments of scarred myocardium demonstrate contractile dysfunction, with a myocardial wall that does not move or has very reduced mobility. In the case of stunned myocardium this can be partly improved with inotropic agents such as dobutamine. Stunned myocardium is usually the result of chronic ischemia. For the prevention of a supply—demand imbalance in a hibernating myocardium, oxygen consumption and function is downregulated as an adaptive response to the limited coronary flow.

Methods to Estimate of Myocardial Oxygen Consumption

Fick Principle for Estimation of Myocardial Oxygen Consumption

MVO_2 relies upon the extraction of oxygen from the supplied coronary blood. Thus a way to estimate MVO_2 is to calculate the arterial venous oxygen difference (CaO_2-CvO_2) and apply the **Fick principle**, allowing the estimation of MVO_2 a function of:

$$MVO_2 = CBF \times (CaO_2 - CvO_2)$$

(where CBF, coronary blood flow (mL/min); and (CaO_2-CvO_2) is the arterial-venous oxygen content difference (mL O_2/mL blood)).

Nevertheless, although CaO_2 can be easily calculated, measurement of CBF and CvO_2 require coronary catheterization and blood sampling from the coronary sinus.

Pressure-Rate Product

Another way to estimate the *relative changes in MVO_2*, which is important, e.g., when deciding about the therapeutic interventions to reduce MVO_2. The relative changes in MVO_2 can be estimated indirectly by the pressure-rate product [8].

A simple variation of the pressure-rate product index is the following

$$\text{Pressure-rate product} = \text{aortic SP} \times \text{HR}$$

(where aortic SP, aortic systolic pressure; HR, heart rate).

Hemodynamic response	RPP (rate pressure product)
High	more than 30,000
High intermediate	25,000–29,999
Intermediate	20,000–24,999
Low intermediate	15,000–19,999
Low	10,000–14,999

Pressure-rate product is used to estimate the myocardial workload. It is a measure of the stress put on the cardiac muscle per contraction against the afterload the muscle pumps against (i.e., aortic systolic pressure). Multiplied by the heart rate it is an index of the workload per minute and a surrogate marker for MVO_2.

Diastolic Pressure-Time Index/Systolic Pressure-Time Index Ratio

Another index that can be used to estimate the balance of myocardial oxygen supply–demand is the diastolic pressure-time index/systolic pressure-time index (DPTI/SPTI) ratio.

DPTI is the area under the aortic diastolic pressure curve and represents the oxygen supply to the heart.

The SPTI is the area under the left ventricle pressure curve and represents the oxygen demand by the heart. This index estimates the oxygen needs of the working myocardium but is an approximation of MVO_2 because the latter is more closely related to wall stress rather than pressure and does not take into account muscle mass or contractility [3].

In experimental models with anesthetized dogs, cut-off values for subendocardial ischemia for the DPTI/SPTI ratio have been calculated (i.e., DPTI/SPTI < 0.8). Even when corrected for myocardial mass, this threshold value seems to have good discriminating value for subendocardial ischemia. In humans, in subjects undergoing exercise treadmill testing, it has been found that a DPTI/SPTI ratio <0.45 is strongly predictive of ischemic changes in electrocardiogram. Threshold values for DPTI/SPTI ratio have been also calculated in human heart disease, but these were carried out in clinical studies with small populations, e.g., in patients with severe aortic stenosis and normal coronary arteries in an effort to monitor for an imbalance in myocardial supply:demand and ischemia triggering. Also in patients with heart disease, a prolongation of the systolic period is observed, due to a decrease in cardiac contractility and prolonged time for fall in LV pressure. This leads to an increase in SPTI and reduced time for diastolic coronary blood supply (i.e., decrease in DPTI). It has been estimated that a rise in left ventricular end-diastolic pressure from 10 to 20 mm Hg leads to a decrease in DPTI/SPTI by 20% [3].

As an alternative to cardiac catheterization, similar tracing and calculation of the DPTI/SPTI ratio can be achieved by analysis of the brachial intra-arterial pressure waveforms and estimation of the left ventricular diastolic pressure from the pulmonary arterial wedge pressure (e.g., during intensive care unit stay). It should be noted though that this method for DPTI/SPTI calculation by use of brachial artery pressure waveforms, tends to overestimate SPTI due to the augmentation effect on peripheral systolic blood pressure.

REFERENCES

[1] Braunwald E, et al. Hemodynamic determinants of coronary flow: effect of changes in aortic pressure and cardiac output on the relationship between myocardial oxygen consumption and coronary flow. Am J Physiol 1958;192(1):157–63.

[2] Rodbard S, et al. Myocardial tension and oxygen uptake. Circ Res 1964;14:139–49.

[3] Hoffman JI, Buckberg GD. The myocardial oxygen supply:demand index revisited. J Am Heart Assoc 2014;3(1):e000285.

[4] Braunwald E. Thirteenth Bowditch lecture. The determinants of myocardial oxygen consumption. Physiologist 1969;12(2):65–93.

[5] Coleman 3rd HN. Role of acetylstrophanthidin in augmenting myocardial oxygen consumption. Relation of increased O-2 consumption to changes in velocity of contraction. Circ Res 1967;21(4):487–95.

[6] Gibbs CL. Cardiac energetics and the Fenn effect. Basic Res Cardiol 1987;82(Suppl. 2):61–8.

[7] Britman NA, Levine HJ. Contractile element work: a major determinant of myocardial oxygen consumption. J Clin Invest 1964;43:1397–408.

[8] Gobel FL, et al. The rate-pressure product as an index of myocardial oxygen consumption during exercise in patients with angina pectoris. Circulation 1978;57(3):549–56.

CHAPTER 2.3

Regulation of Oxygen Transport and Coronary Blood Flow

Konstantinos Aznaouridis, Konstantina Masoura, Dimitris Tousoulis
1st Cardiology Clinic, 'Hippokration' General Hospital, National and Kapodistrian University of Athens, School of Medicine, Athens, Greece

Contents

CORONARY ANATOMY

The major epicardial arteries of the heart are the left main coronary, which bifurcates into the left anterior descending artery (LAD) and the circumflex artery, and the right coronary artery. The two main coronary arteries originate from the coronary ostia, which are located at the proximal aorta, just above the cusps of the aortic valve (Fig. 2.3.1). The LAD supplies the anteroseptal, anterolateral, and anteroapical segments of the left ventricle. The circumflex artery supplies the lateral and posterior segments, whereas the right coronary artery supplies the inferior, inferoseptal, and posterior segments. The blood supply of the inferior, posterior, inferolateral, and posterolateral segments varies considerably, with varying contribution of the right coronary and circumflex arteries. More frequently, most of those myocardial areas are supplied by the right coronary artery (right dominance).

Normal large epicardial coronary arteries have low vascular resistance. In contrast, the smaller coronary arteries and arterioles in the microcirculation of the myocardium are the

Coronary Artery Disease
ISBN 978-0-12-811908-2, http://dx.doi.org/10.1016/B978-0-12-811908-2.00010-6

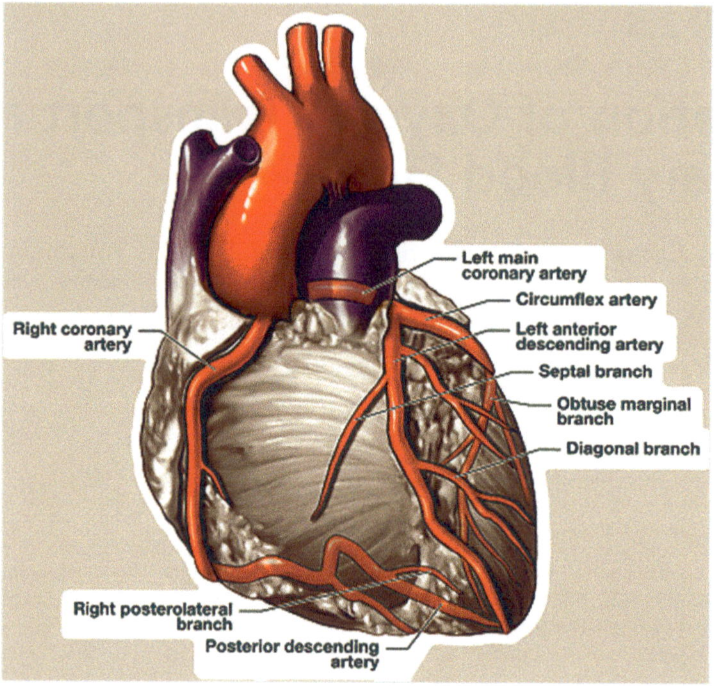

Figure 2.3.1 Normal epicardial coronary arteries. *(Adapted from http://www.cardiachealth.org/how-does-your-heart-work-anatomy-heart-coronary-arteries.)*

primary sites of vascular resistance, and comprise a major site for the regulation of myocardial blood flow. The coronary arterioles provide numerous capillaries that lie adjacent to the cardiac myocytes. The myocardial circulation has a high number of capillaries for each cardiomyocyte, which ensures a short diffusion distance and adequate oxygen delivery to meet the high oxygen demand of the cardiomyocytes. The capillary blood then enters venules, which merge to form the cardiac veins, which eventually drain

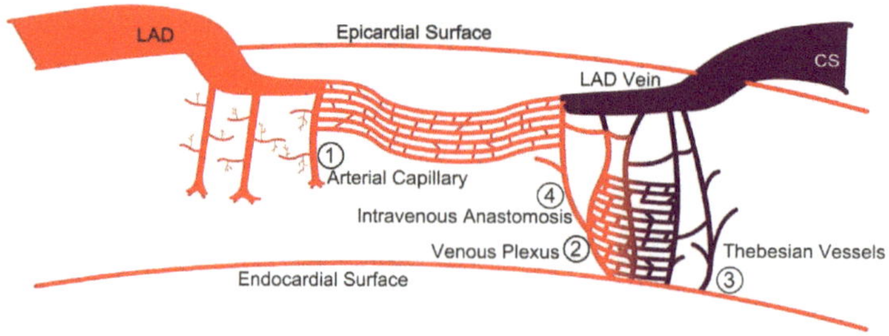

Figure 2.3.2 Anatomy of the medium-sized and small arteries/capillaries and venous system in normal myocardium. The large epicardial left anterior descending artery (LAD) is divided into several medium-sized arteries and arterioles, which provide a rich capillary bed. Blood from capillaries drains into the coronary veins. In addition to this in-series vascular system, the myocardial capillaries may contribute to a second parallel microvascular bed (thebesian network), which is usually a remnant from the embryonic development. *(Adapted from Kassab GS, et al. J Appl Physiol 2008;104:1266–72.)*

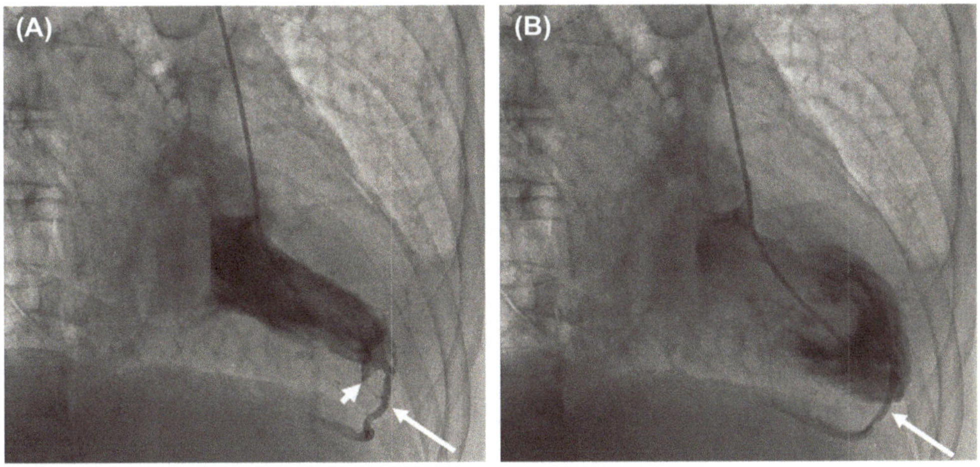

Figure 2.3.3 Left ventriculography showing an unusual connection of left ventricular cavity with a large posterior interventricular vein (*arrows*), possibly through a small thebesian vein; (A) systole and (B) diastole. *(Adapted from Aznaouridis K, et al. Inadvertent cardiac phlebography. World J Cardiol 2017;9:558–61.)*

into the coronary sinus (Fig. 2.3.2). The coronary sinus is located in the left side of the atrioventricular groove and drains into the right atrium. In addition to the coronary veins, the thebesian veins (venae cordis minimae) drain directly into the cardiac chambers. The thebesian veins are small valveless venous conduits that connect the coronary arteries, veins, or capillaries with the cardiac chambers. Most thebesian veins of the ventricles are connected to the cardiac venous system (Figs. 2.3.2 and 2.3.3).

REGULATION OF CORONARY BLOOD FLOW

The coronary circulation is the oxygen supplier to the myocardium. The blood flow in the coronary arteries is primarily determined by the perfusion pressure and coronary vascular resistance. The coronary vascular resistance is regulated by dynamic changes (vasodilation vs. vasoconstriction; external compression during the cardiac cycle) and fixed changes (atherosclerotic stenosis) of the diameter of coronary arteries. Several factors described in the following discussion contribute to the dynamic changes of coronary resistance.

Flow across a blood vessel is governed by a fundamental law of physics, Ohm's Law, which states that flow (F) equals the pressure difference or pressure gradient across the vessel (ΔP) divided by vascular resistance (R).

$$F = \Delta P/R$$

As far as the blood flow of an organ is concerned, the pressure gradient is calculated as the difference between the mean pressure in the artery supplying the organ minus the mean pressure of the vein or veins that drain blood from this organ.

Therefore flow and resistance are inversely related. In particular, when the pressure gradient across a vascular segment remains constant, flow is reduced proportionally to an increase in resistance, and vice versa. Similarly, when the blood flow along a blood vessel is constant, an increase in resistance is associated with a respective increase of the pressure gradient across this vascular segment.

Perfusion of the coronaries happens mainly during diastole. Myocardial contraction during systole causes systolic extravascular compression of the coronaries due to systolic increase of both intraventricular and intramyocardial pressures (Fig. 2.3.4), which diminishes coronary blood flow (CBF) during systole. Hence the coronary flow is higher during the diastolic phase of the cardiac cycle (Fig. 2.3.5). Therefore, coronary perfusion is mainly a diastolic phenomenon. The extravascular compression is more prominent to the small-caliber arteries travelling from the epicardial coronaries to the subendocardium (Fig. 2.3.4), and therefore the subendocardial layers of the myocardium are more

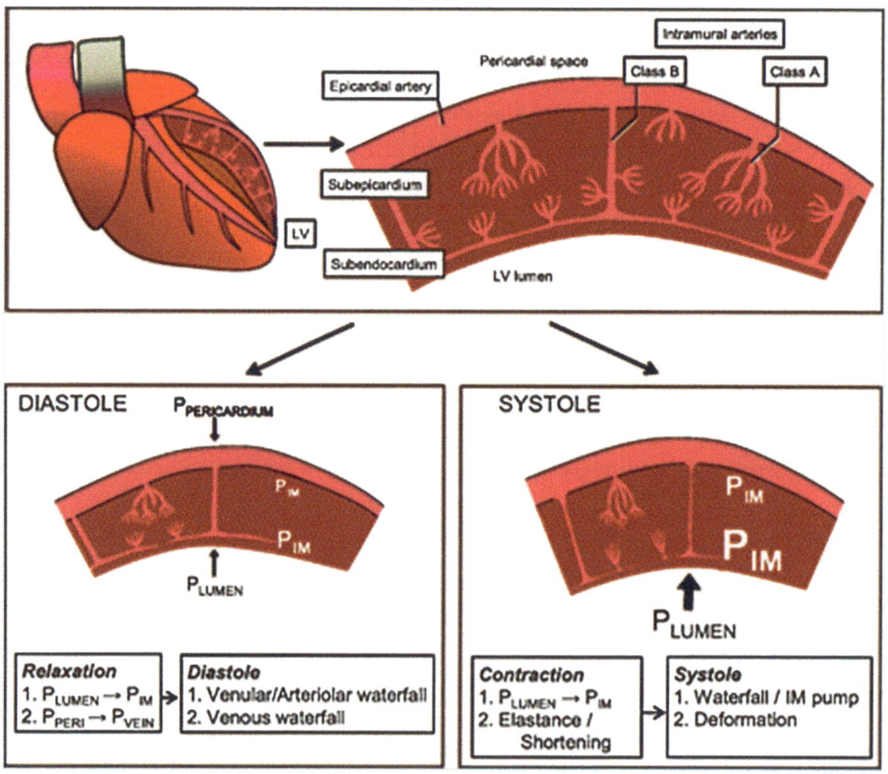

Figure 2.3.4 Schematic drawing of the intramyocardial vasculature (top) and the effect of extravascular forces (intracardiac chamber pressure P_{LUMEN}, intrapericardial pressure P_{PERI}, and intramyocardial pressure P_{IM}) acting on the coronary vessels during the phases of the cardiac cycle (bottom). *LV*, left ventricle. *(Adapted from Duncker DJ, et al. Physiol Rev 2008;88:1009–86.)*

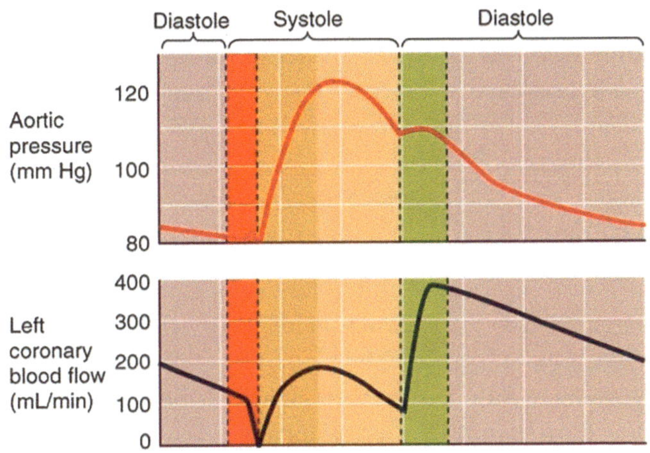

Figure 2.3.5 Blood flow in a coronary artery. Normal coronary flow occurs mainly during diastole. *(Adapted from http://www.cvphysiology.com/Blood%20Flow/BF001.)*

susceptible to ischemia. This subendocardial ischemia may be aggravated (1) when the perfusion pressure is low; (2) when the heart rate is increased, resulting in short diastolic duration and less time available for coronary flow during diastole; or (3) when there is significant coronary artery disease with stenosis of the epicardial coronaries and impaired flow reserve of the coronaries.

As in most organs, blood flow in the myocardium is closely related to the metabolic activity of the heart muscle. In conditions of high demand and increased metabolic turnover, as happens during physical activity, an increase in myocardial blood flow normally occurs. There are several mechanisms that contribute to the regulation of blood flow in the myocardium. Autoregulation, metabolic regulation through locally produced substances, as well as effects of the autonomic nervous system contribute to this regulatory process. The net result of these mechanisms is to supply the myocardium with the amount of oxygen it needs to meet its requirements, and to remove waste metabolic products from the myocardial tissue.

MECHANISMS OF CORONARY FLOW REGULATION

Hypoxia

Low partial pressure of oxygen at tissue level due to impaired oxygen supply or increased oxygen needs causes vasodilation, which is mainly mediated by locally produced metabolites and molecules (discussed in detail later). The coronary vascular response to hypoxia usually depends on several metabolic and endothelial vasodilatory factors that are activated in proportion to the degree of hypoxemia, and these responses depend mainly on K^+(ATP [adenosine triphosphate]) channels. Interestingly, hypoxia may also induce a degree of direct vasodilation, as the vascular smooth muscle may not be able to maintain its vascular tone in a hypoxic environment. In general, hypoxia induces a degree of vasodilation in most organs, except the pulmonary vasculature, which reacts with vasoconstriction in case of hypoxia.

Autonomic Nervous System

The autonomic nervous system is an important regulator of CBF. Several regulatory mechanisms are triggered each time there is increased mechanical and metabolic activity of the heart. The activation of sympathetic nerve fibers innervating the coronary vasculature causes only transient vasoconstriction, which is mediated by α1-adrenoceptors. This brief and mild vasoconstrictive effect is usually followed by vasodilation, which is mediated by the production of substances with vasodilator properties. These vasodilators lead to active hyperemia, which is also mediated by activation of the β1-adrenoceptors of the myocardium. Therefore the net outcome of the sympathetic activation to the heart is coronary vasodilation and increased coronary flow to the myocardium, although sympathetic activation induces a direct—albeit mild and transient—vasoconstrictive effect on the coronaries.

The stimulation of the parasympathetic nervous system induces a direct mild vasodilatory effect on the coronary vasculature, which is mediated by acetylcholine. Nevertheless, when the parasympathetic activation of the heart is accompanied by a significant reduction in heart rate and a respective decrease in myocardial oxygen requirements, then coronary vasoconstriction and increase of vascular resistance of the coronary vasculature ensues.

Autoregulation of Coronary Blood Flow

Autoregulation of the CBF is the ability of the heart to maintain a constant blood flow when the perfusion (driving) pressure changes. This beneficial response does not depend on systemic neurogenic or humoral effects, but is an intrinsic regulatory mechanism of the heart mediated by a myogenic response mainly in the small-resistance arteries and also involves local metabolic and endothelial responses in the coronary circulation.

When the perfusion pressure in the coronaries falls (for example, due to a fall in the systemic aortic pressure due to acute blood loss or systemic vasodilation), the CBF tends to decrease initially. This is because the blood flow is proportional to the perfusion pressure for a given vascular resistance. Shortly after the CBF falls due to the lowered perfusion pressure, the small coronary arteries and arterioles vasodilate and the vascular resistance of the myocardium also falls. This low resistance contributes to maintain a constant coronary flow despite a lower perfusion pressure.

In a theoretical model without the ability of autoregulation, a sudden decrease in perfusion pressure would lead to an acute and sustained reduction of blood flow and to a slight increase in peripheral vascular resistances due to passive vasoconstriction secondary to lower blood flow. On the other hand, in the coronary circulation and in other organs able to autoregulate blood flow (such as brain), a sudden decrease in perfusion pressure leads to an initial fall in blood flow, but then the blood flow increases and is gradually restored almost to the previous levels, because of vasodilation and reduction of vascular resistance in the coronary vasculature (Fig. 2.3.6). The coronary vascular system has a high degree of autoregulation, which means that blood flow can be restored to almost normal levels despite a significant change (increase or reduction) in the perfusion pressure.

The coronary circulation may be autoregulated for a wide range of perfusion pressures between ~50–60 mm Hg and ~180–200 mm Hg. Below a perfusion pressure of ~50–60 mm Hg the coronary flow is reduced, as the autoregulatory mechanisms do not suffice to keep the coronary flow constant, given that the vasculature is already vasodilated maximally (Fig. 2.3.7). In normal systemic perfusion pressures, failure of autoregulation may be clinically significant in the presence of a significant coronary stenosis. In this case, the vasculature distal to this stenosis may be already vasodilated, and therefore a reduction in the systemic perfusion pressure may lead to a significant reduction in the flow of the respective myocardial segment, as locally there will be no significant autoregulatory reserve.

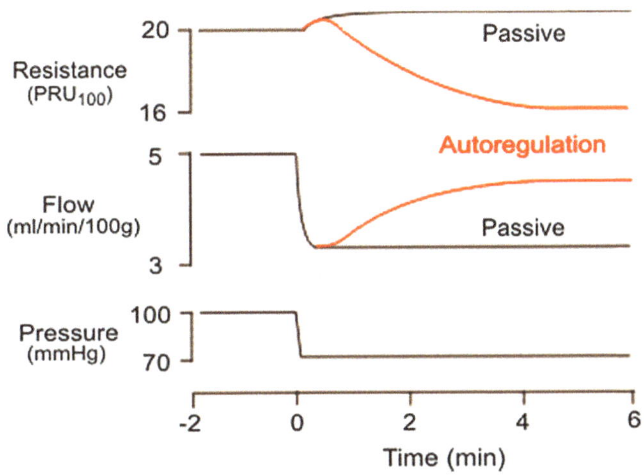

Figure 2.3.6 Model of the effects of acute reduction of perfusion pressure by 30 mm Hg. In the absence of autoregulation (passive model), there will be a fast and sustained fall in blood flow, and a mild passive increase of vascular resistance. In a normal coronary vascular bed with autoregulatory capabilities, an initial fall in blood flow will be followed by a gradual increase in blood flow coupled with a simultaneous vasodilation and a decrease in vascular resistance. *(Adapted from http://www.cvphysiology.com/Blood%20Flow/BF004.)*

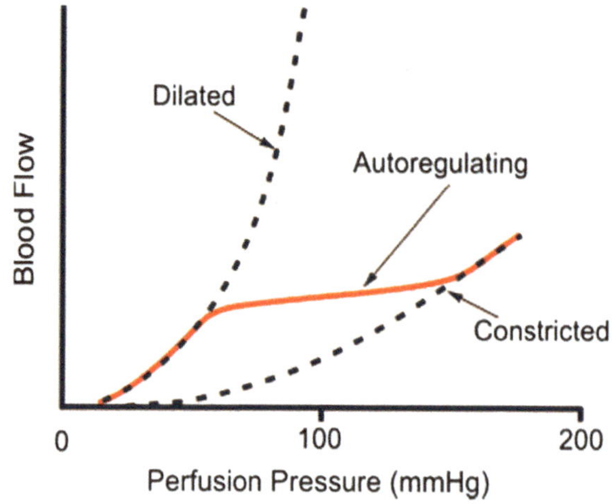

Figure 2.3.7 Schematic diagram of pressure–flow relationships from an experiment showing responses of blood flow to changes in perfusion pressure. In a model of maximal vasodilation and absence of autoregulation (curve "Dilated"), a small increase in perfusion pressure leads to a substantial increase in blood flow. The "Constricted" curve shows the pressure–flow relationships in a model of maximal vasoconstriction and absence of autoregulation. The normal coronary circulation is similar to the "Autoregulating" model, where the vascular resistance is in normal range and autoregulatory mechanisms are also functioning. In this model, coronary flow changes relatively little despite a large change in perfusion pressure. Even in a normal coronary circulation, a very low perfusion pressure outside the autoregulatory spectrum (below 50 mm Hg) would be associated with a significant reduction in coronary flow. *(Adapted from http://www.cvphysiology.com/Blood%20Flow/BF004.)*

Tissue Metabolites, Ions, and Endothelial Derivatives

A variety of tissue metabolites, ions, and endothelium-derived substances play a crucial role in the regulation of CBF in normal conditions (Fig. 2.3.8).

Adenosine is an important molecule that mediates active hyperemia and contributes to autoregulation of coronary flow. Adenosine is synthesized from adenosine monophosphate, which is produced from the hydrolysis of intracellular ATP and adenosine diphosphate. The synthesis of adenosine increases during conditions of hypoxia and increased oxygen demands (Fig. 2.3.9). Adenosine serves as a metabolic mediator, which "senses" changes in myocardial oxygen consumption and subsequently induces respective changes in CBF.

Nitric oxide (NO) is an endothelial-derived molecule that serves as a regulator of CBF (Fig. 2.3.8). The production of NO is not that important for the vasodilation in coronary arterioles during high myocardial oxygen demand as in this case vasodilation is primarily mediated by metabolic or myogenic mechanisms. However, the endothelium-dependent production of NO maintains the basal vasomotor tone both in larger conduit and smaller resistance vessels in the coronary circulation. NO release may be impaired in the large epicardial coronaries of subjects with atherosclerotic risk factors without

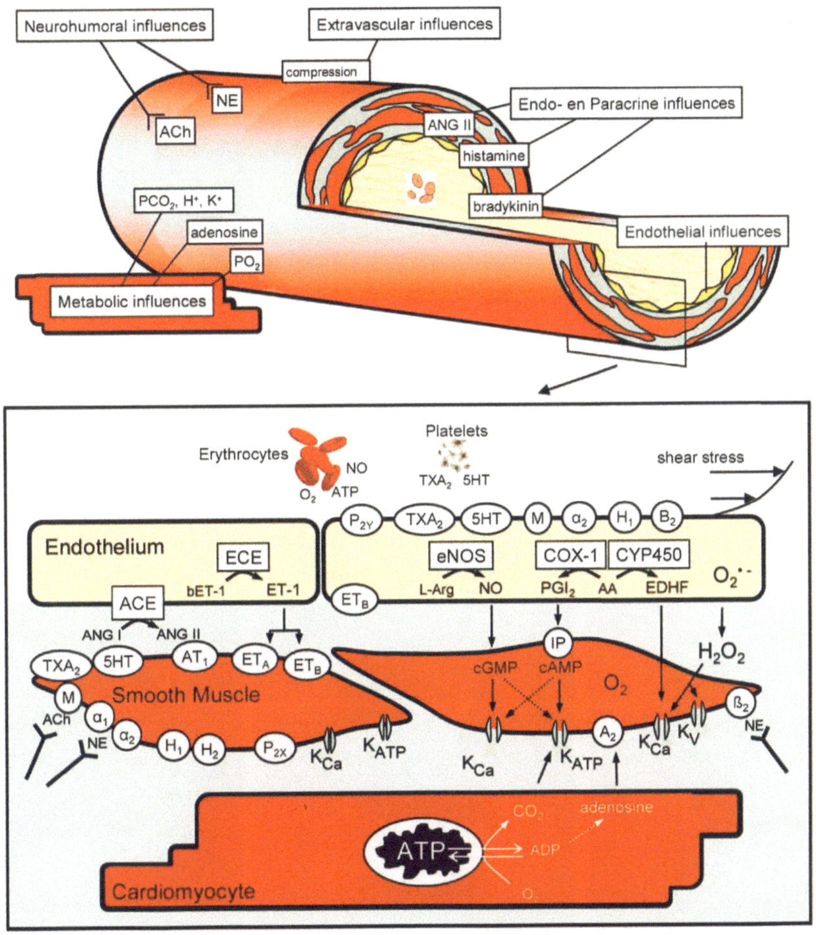

Figure 2.3.8 Schematic drawing of the complex interplay between cardiomyocytes, endothelial cells, and smooth muscle cells of coronary arterioles for regulation of coronary myocardial blood flow. *(Adapted from Duncker DJ, et al. Physiol Rev 2008;88:1009–86.)*

evidence of established coronary artery disease. This latter effect may be clinically significant in patients with episodes of angina-like chest pain in the absence of significant fixed coronary stenoses (Fig. 2.3.10).

Potassium ions are released in the myocardial tissue during the contraction of the heart, as potassium leaves the myocardial fibers with each action potential. In case of tachycardia (increased frequency of action potentials), more potassium ions leave the myocardial fibers and are transferred to the extracellular space of the myocardium. This ionic gradient is normally restored from the Na^+/K^+-ATPase pump. However, the ability of this pump may be saturated with very fast depolarizations (rapid heart rate) and K^+ may accumulate in the extracellular space. A small increase in the extracellular tissue

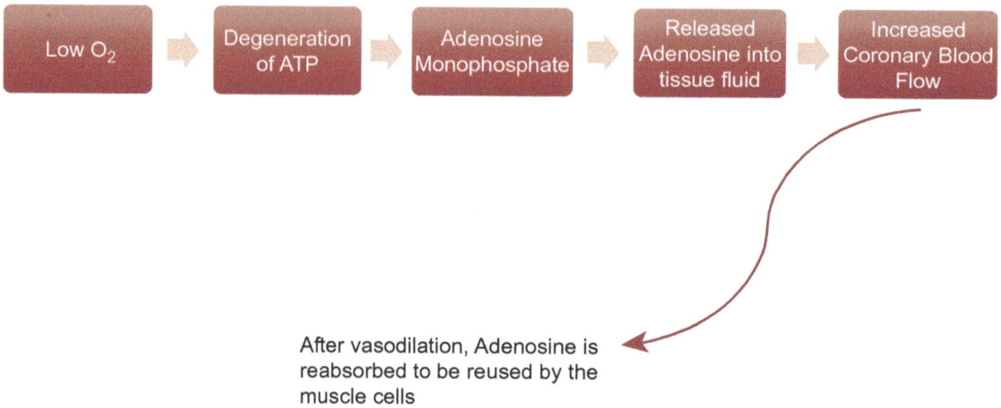

Figure 2.3.9 Role of adenosine in the regulation of coronary blood flow in conditions of impaired myocardial oxygen supply. *ATP*, adensosine triphosphate. *(Adapted from http://www.wikilectures.eu/index.php/Coronary_Blood_Flow.)*

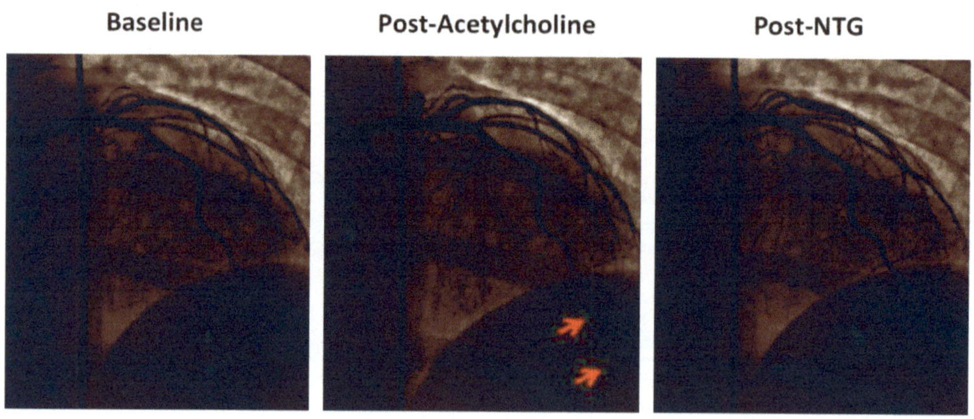

Figure 2.3.10 The role of nitric oxide (NO) in coronary vasomotion; coronary angiogram and coronary reactivity testing with intracoronary administration of acetylcholine in a patient with recurrent episodes of angina and unobstructed coronaries. When the coronary endothelium is normal, acetylcholine induces vasodilation, as her weak direct vasoconstrictive effect on vascular smooth muscle is counterbalanced by a potent vasodilatory effect due to NO release from endothelial cells. However, in this patient with coronary endothelial dysfunction, acetylcholine causes vasoconstriction (panel in the middle), as her vasoconstrictive effect is unopposed. *(Adapted from Laksanakorn W, et al. Int J Case Rep Images 2015;6:239–44.)*

levels of potassium in the myocardium may initiate a variety of regulatory effects. Potassium ions produce a repolarization of the vascular smooth muscle of the coronaries, leading to vasorelaxation, reduction of coronary resistance, and active hyperemia through activation of the Na+/K+-ATPase pump and an increase of the membrane conductance to potassium. This role of potassium ion contributes to the regulation of coronary flow to some extent, but in the heart it is not as important as it is for the contracting skeletal muscle. Recent investigations suggest that voltage-dependent K^+ channels represent an important mechanism that modulates CBF not only at rest but also during conditions of myocardial oxygen consumption and periods of cardiac ischemia.

Carbon dioxide level is increased during conditions of increased oxidative metabolism in the myocardium. It reaches the vascular smooth muscle cells of the coronary resistance arterioles and contributes to coronary vasodilation. The role of CO_2-mediated vasodilation seems not to be as important as it is in the regulation of the blood flow in the brain.

Lactic acid is mainly produced through anaerobic metabolism, in the presence of myocardial ischemia. It is a vasodilator of the coronary circulation, but its importance is rather limited.

CORONARY FLOW AND OXYGEN TRANSPORT AND DELIVERY TO THE MYOCARDIUM

The oxygen turnover in the cardiac muscle depends on the balance between the supply of oxygen through the coronaries and the oxygen demand from the myocardium. The increased oxygen demands of the myocardium (for example, in case of tachycardia and increased contractility during physical exercise) are met mainly through an increase in the CBF, given that the arterial oxygen content remains relatively constant. Increasing CBF helps maintaining adequate tissue oxygen levels during periods of high oxygen demand (Fig. 2.3.11). This increase in the CBF is mediated mainly by the local regulatory mechanisms, which were discussed earlier in this chapter. This close relationship between high myocardial oxygen demands and increased CBF is impaired when significant coronary stenosis of an epicardial coronary artery is present, or less commonly, in the presence of diffuse microvascular coronary dysfunction despite the absence of significant epicardial coronary disease. In such cases, the increased myocardial needs for oxygen do not induce an adequate increase of CBF, and therefore the myocardial oxygen demand is not met and tissue hypoxia ensues.

Therefore, an abnormal oxygen supply/demand balance due to significant obstructive coronary artery disease causes myocardial hypoxia, ischemia, and eventually angina.

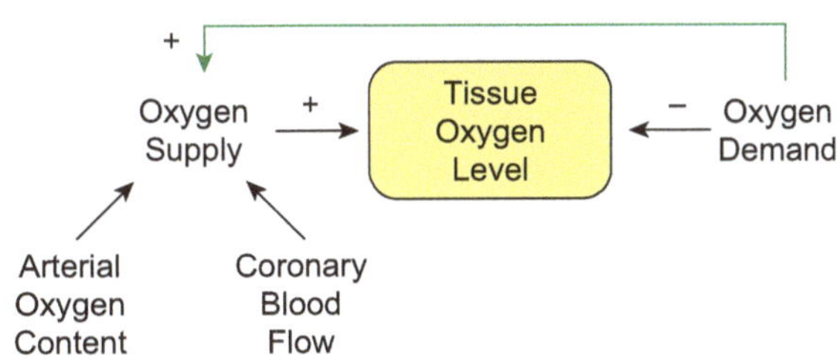

Figure 2.3.11 Determinants of myocardial tissue oxygen level. *(Adapted from http://www.cvphysiology.com/CAD/CAD006.)*

Antianginal medications aim at inducing a favorable modification of this balance. In particular, calcium channel blockers and nitrates cause vasodilation and increase myocardial blood flow and oxygen supply, whereas β-blockers decrease the heart rate and myocardial contractility and decrease myocardial oxygen needs.

The myocardium has a high oxygen consumption (8–10 mL O_2/min/100 g myocardium) and a very high arteriovenous oxygen difference (10–12 mL O_2/100 mL blood). This high arteriovenous oxygen difference is due to the high extraction of oxygen form blood in the capillaries of coronary circulation, as the oxygen consumption from the heart is significantly higher than the respective consumption in other organs. For example, the oxygen consumption in the brain and in the kidneys is 3–4 mL O_2/min/100 g and 4–6/min/100 g, respectively. In the skeletal muscle, the oxygen consumption is 0.8–1.2 mL O_2/min/100 g at rest and 40–55 mL O_2/min/100 g during heavy exercise. On the other hand, during heavy exercise, the oxygen consumption from the myocardium may increase to very high levels, up to 70 mL O_2/min/100 g myocardium. In normal coronary arteries, the increase in oxygen consumption in conditions of increased cardiac activity is accompanied by a proportional increase in the CBF.

The supply of oxygen to the myocardium is determined by two factors: (1) the oxygen concentration of the arterial blood (O_2 conc) and (2) CBF.

$$O_2 \text{ supply (mL } O_2\text{/min)} = \text{CBF (in mL blood/min)} \times O_2 \text{ conc (mL } O_2\text{/mL blood)}$$

Considering that (1) the usual oxygen concentration of the arterial blood (O_2 conc) is about 20 mL O_2/100 mL blood or 0.2 mL O_2/mL blood or 20 vol %. and that (2) CBF is usually approximately 80 mL/min per 100 g of myocardial tissue at rest and at a heart rate of ~60–70 beats per minute, then the myocardial oxygen supply at rest is approximately 16 mL O_2/min per 100 g of myocardial tissue. Given that the average weight of the heart in a normal weight adult is about 300–310 g, the oxygen supply to the heart at rest is approximately 48–50 mL O_2/min.

To maintain a normal oxidative metabolism, the myocardium extracts oxygen from the coronary circulation. Oxygen extraction is defined as the difference between the arterial oxygen content (coronary artery) and the venous oxygen content (coronary veins). The myocardium's oxygen extraction is typically high, as approximately 50% of the oxygen available in the arterial blood is extracted when the blood passes through the myocardial capillaries.

The oxygen extraction in the heart as expressed by the arteriovenous oxygen difference is very high (arteriovenous difference of 10–12 mL O_2/100 mL blood). This difference is much lower in other organs in the body. For example, the arteriovenous difference is 2.5–3.5 mL O_2/100 mL blood in the kidneys, 4.0–5.5 mL O_2/100 mL blood in the intestine, and 2.5–5.0 mL O_2/100 mL blood in the skeletal muscle at rest.

The extraction of oxygen in the myocardium can be calculated using Fick's principle, as:

$$O_2 \text{ extraction (mL } O_2\text{/min)} = \text{CBF (in mL blood/min)} \times \text{Arteriovenous } O_2 \text{ difference (mL } O_2\text{/mL blood)}$$

When the CBF is 80 mL/min per 100 g of myocardial tissue and the arteriovenous difference is 10 mL O_2/100 mL blood, the O_2 extraction would be = 8 mL O_2/min per 100 g of myocardial tissue.

The maximal amount of oxygen that can be extracted is theoretically 20 mL O_2/100 mL blood (20 vol%). However, the actual maximal oxygen extraction in real conditions is up to 16 mL O_2/100 mL blood. This means that the myocardium extracts at least one-half and up to two-thirds of the physiologically available oxygen at rest. In other words, the oxygen supply and demand are very tightly coupled in the myocardium to keep myocardial oxygenation at physiological levels. It is obvious that when the oxygen demands increase, increasing the oxygen extraction from the blood can meet the increased requirements only in part, and therefore the heart muscle almost exclusively relies on a respective increase of blood flow. When the coronary circulation is normal (no significant coronary stenosis), the CBF increases proportionately to meet the higher myocardial oxygen needs and consumption, and to prevent myocardial hypoxia. This increase in myocardial flow is mediated by the complex local and neurogenic regulatory mechanisms, which have been described earlier.

When significant obstructive coronary artery disease is present (significant fixed stenosis in a major epicardial coronary artery), the available blood supply usually meets the oxygen demand of the myocardium in resting conditions. In contrast, in circumstances of increased needs (for example, during physical activity), when the heart rate increases and the contractility increases as well, the CBF may not be able to increase proportionately and the high myocardial needs for oxygen may not be met. In these conditions, the ischemic myocardium will increase the oxygen extraction from the supplied blood, and the arteriovenous oxygen difference in the heart will also increase. However, as explained before, the oxygen extraction in the heart is already high enough even at resting conditions, and therefore the capacity of the myocardium to meet its increased oxygen needs from increasing oxygen extraction is limited. Therefore there will be a lack of adequate oxygen to meet the high metabolic demands of the contracting heart. This leads to hypoxia of the myocardium, and when this hypoxia is severe the patient develops angina. In such circumstances, the myocardial blood flow in the diseased coronary artery can be improved within either antianginals (nitrates or calcium channel antagonists, which improve myocardial blood flow) or with revascularization (placement of a coronary stent to expand the coronary artery's lumen, or coronary artery bypass surgery to bypass the diseased coronary artery with a vascular graft).

The oxygen concentration of the arterial blood (O_2 conc) is mainly determined by the oxygen that is carried by hemoglobin in the red blood cells, as only a very small fraction of oxygen is dissolved in the blood's plasma. Therefore diseases or conditions that reduce the hematocrit or the content of hemoglobin in the erythrocytes (anemia) have an impact on the oxygen-carrying capacity of the blood and the oxygen supply to tissues, including the myocardium. In addition, the myocardial oxygen supply may be reduced in cases of hypoxemia (for example, in the presence of chronic severe lung disease). However, in the absence of conditions causing hypoxemia, the oxygen content of arterial blood does not change significantly, and therefore, as already stated, the main determinant of oxygen supply to the myocardium is CBF.

MYOCARDIAL METABOLISM

The heart muscle utilizes ATP as a primary source of energy for most functions, such as the function of membrane Na^+/K^+-ATPase. The other major functions of the myocardial fiber, such as the contraction and relaxation of the sarcomere, which are mediated by myosin ATPase and ATP-dependent transport of calcium, also consume ATP. An increase in the activity of the heart by increasing the heart rate and contractility increases the metabolism of the myocardium, and the respective needs for energy (ATP) and oxygen. Most cardiac functions are served through aerobic production of ATP, as the anaerobic capacity of the heart is rather limited (Fig. 2.3.12). Therefore the intracellular levels of ATP decrease if there is no adequate O_2 available to the myocardial fiber to produce ATP aerobically, or if there is an excessive increase in ATP utilization that is not accompanied by a respective increase in ATP synthesis.

The myocardium is able to utilize several substrates for oxidation to generate ATP and meet its energetic requirements. The heart uses mainly fatty acids as well as carbohydrates as a substrate. The cardiomyocytes have a considerable adaptive ability and can utilize almost exclusively carbohydrates such as glucose after a meal rich in carbohydrates. During periods of strenuous exercise, lactate can serve as an important substrate for the myocardium (Fig. 2.3.13). In certain scenarios such as acidosis, amino acids and ketones can be utilized. These alternative pathways are particularly important during myocardial

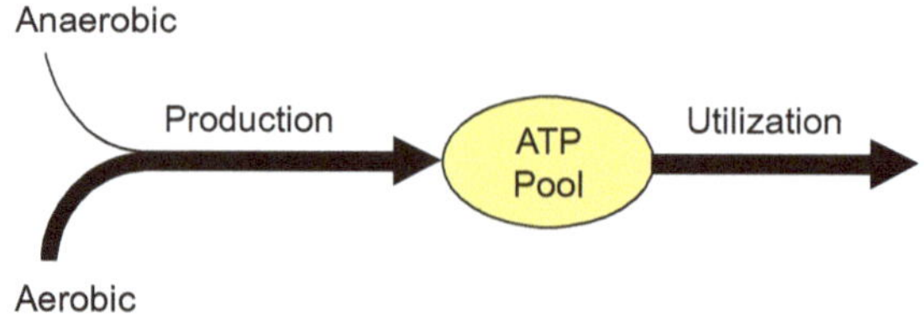

Figure 2.3.12 Aerobic production of adenosine triphosphate (ATP) is a central process in myocardial energy supply. *(Adapted from http://www.cvphysiology.com/CAD/CAD009.)*

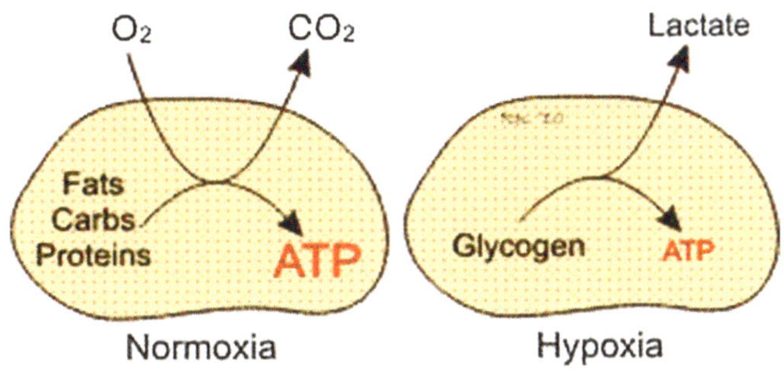

Figure 2.3.13 In hypoxic conditions, the myocardium shifts to anaerobic metabolism and uses glycogen-derived carbohydrates as a substrate for the production of ATP. *(Adapted from http://www.cvphysiology.com/CAD/CAD009.)*

ischemia, when the CBF cannot efficiently supply oxygen and deliver metabolic substrates to the myocardium, and therefore the mechanisms of aerobic metabolism are altered. In such conditions, the metabolism may shift to anaerobic and the heart muscle uses glycogen-derived carbohydrates as a substrate for anaerobic metabolism, production of ATP, and formation of lactic acid. Importantly, the heart is able to produce only a limited amount of ATP using this pathway, as it has a limited amount of glycogen, which is consumed rapidly under conditions of severe hypoxia.

BLOOD FLOW IN CORONARY STENOSIS

Coronary stenosis refers to an abnormally narrowed coronary artery. The stenosis is usually discrete, but may sometimes be diffuse and involve a long segment of an artery. In coronaries, stenoses are usually located in the large epicardial arteries with diameters of more than 2 mm, but they may also occur in smaller arteries. In the majority of cases, a stenosis is caused by atherosclerosis, but other causes such as systemic inflammatory diseases may be involved. Coronary stenoses are usually chronic fixed changes, but sometimes a transient, dynamic stenosis due to spasm of the artery may happen.

From a pathophysiological point of view, a coronary stenosis impairs flow to the subtending myocardium, as the perfusion pressure is decreased distal to the stenosis (Fig. 2.3.14). The flow dynamics in a vessel is governed by Poiseuille's law, which states that resistance in a vessel has an inverse relationship with the radius of the vessel to the fourth power. For example, the diameter (and therefore the radius as well) of a coronary artery may be reduced by two-thirds (for example, an artery with a diameter of 3 mm with a fixed stenosis at a discrete segment may have a residual diameter of only 1 mm in that segment). In that case, the remaining diameter (and the radius) in the diseased segment will be one-third of the initial diameter/radius, and according to Poiseuille's law the resistance within that stenotic segment will increase by 81-fold. If the perfusion

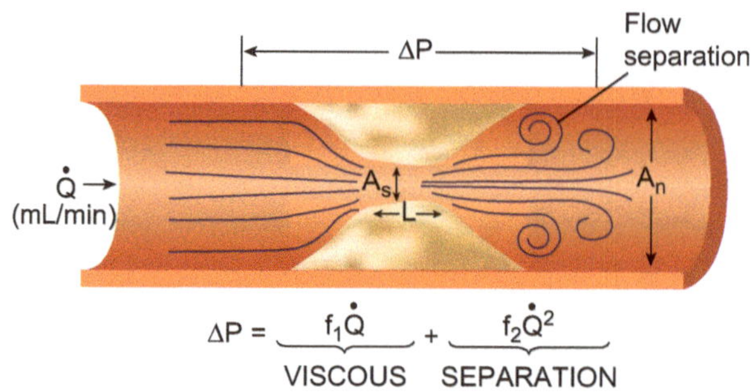

Figure 2.3.14 The drop of perfusion pressure across a coronary stenosis may be due to turbulent flow, separation losses, and viscous losses.

pressure in that coronary artery remained constant, then the blood flow distal to this stenosis would decrease by 81-fold. However, in vivo, the major epicardial coronary arteries contribute only minimally to the overall resistance of the coronary circulation, and the major determinant of the resistance of the coronary system are the small coronary arteries and arterioles, which are coupled in series with the large epicardial coronaries. Therefore what actually happens is that the smaller arteries of the coronary tree decrease their resistance through autoregulatory mechanisms in such a way that the overall resistance of the coronary vasculature distal to the stenosis remains more or less unchanged, and the coronary flow stays unimpaired. These mechanisms are effective when the stenosis of the epicardial coronary is of mild or moderate severity.

However, if the coronary stenosis develops further and becomes significant ("obstructive" coronary artery disease), the autoregulatory mechanisms in the small arteries and arterioles may become exhausted and reach their limit. At rest, the myocardial blood flow to the subtending myocardial tissue may remain normal if the distal arterial bed vasodilates to the maximum and the respective distal resistances are kept to the minimum. Collateral blood flow from other sites of the myocardium may also contribute to reduced microvascular resistances and preservation of normal tissue flow. However, in case of increased needs (during physical activity), these regulatory and counterbalancing mechanisms may become exhausted and insufficient, resulting in a significant reduction in blood flow in a distal vascular bed, with resultant tissue ischemia, hypoxia, and angina. From a clinical point of view, a coronary stenosis is considered to be significant when there is a 70%–75% reduction in the diameter of a large epicardial artery by angiography or intracoronary imaging (or ~50% reduction in the cross-sectional area of a large epicardial artery measured with invasive intracoronary imaging methods). Therefore invasive revascularization with either transcatheter methods (angioplasty) or open heart surgery (bypass grafting) is usually not considered, unless at least one stenosis of more than ~70%–75% is present in a large epicardial coronary artery.

HEMODYNAMICS OF CORONARY STEAL PHENOMENON

Coronary steal is the phenomenon where physiological or pharmacological vasodilation of a myocardial segment's vasculature is associated with "steal" of blood from another myocardial segment, which is already significantly vasodilated due to the presence of a significant stenosis in a large epicardial artery. The pathophysiological background for the phenomenon of coronary steal is that, as explained before, in the presence of a significant coronary stenosis, the blood flow to the myocardial tissue may remain normal at rest as the distal arterial bed vasodilates to the maximum and the respective distal resistances are kept to the minimum.

The following example demonstrates the anatomical and physiological conditions that may allow the coronary "steal" phenomenon to appear. Fig. 2.3.15 shows a main artery (let us say a left main coronary artery) with a moderate stenosis, which bifurcates to a LAD with a significant proximal stenosis and a circumflex artery that is free of disease. As explained before, the vascular bed supplied by the diseased LAD is likely maximally vasodilated with minimal resistances due to activation of autoregulatory mechanisms, to keep the myocardial flow to this bed constant. This is because the combined effect of the left main stenosis and LAD stenosis decrease the perfusion pressure distal to the LAD stenosis significantly, and therefore the blood supply to the distal LAD vasculature would be reduced if this distal LAD vasculature had not been maximally vasodilated.

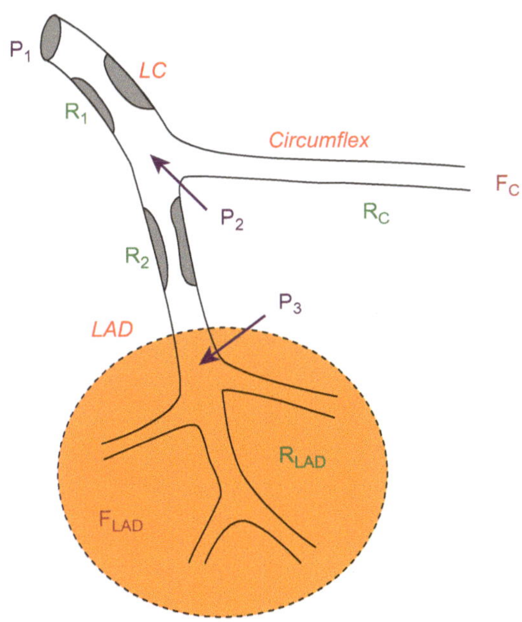

Figure 2.3.15 Hemodynamics of coronary steal phenomenon. *LAD*, left anterior descending artery; *LC*, left main coronary artery. *(Adapted from http://www.cvphysiology.com/Hemodynamics/H009.)*

On the other hand, the moderate left main stenosis may not alter substantially the perfusion pressure to the circumflex, and therefore a normal blood flow to the distal circumflex vasculature may be maintained in the presence of normal resistances (no need for vasodilation in the vasculature distal to the circumflex).

In this scenario, in the presence of vasodilatory stimulus such as physical exercise or a vasodilatory medication, the resistances of the vascular bed of the circumflex artery would decrease and the respective blood flow would increase. The resistance in the LAD territory could not decrease further, as this vascular bed is already at maximal vasodilation. However, the increased blood flow and decreased resistance in the circumflex territory would decrease the perfusion pressure of the circumflex and LAD (P_2) distal to the left main stenosis, as the overall resistance of the vascular bed supplied by the left main would decrease (due to the vasodilation of the circumflex vascular bed mediated by the autoregulatory mechanisms due to physical exercise of drugs with vasodilatory effects). This reduction in the perfusion pressure of LAD (P_2) would induce a decrease in the perfusion pressure of the LAD vascular bed distal to the LAD stenosis (P_3). This would inevitably lead to a further decrease of the blood flow to the distal LAD vascular bed, as the perfusion pressure to the already maximally vasodilated vascular bed falls and further vasodilation is not possible. Therefore, in this scenario, there is redistribution of blood flow with an increase of flow in the circumflex territory, coupled by an erroneous decrease of the blood flow in the LAD vascular bed, resulting in a deterioration of ischemia and hypoxia of the LAD vascular bed. The aforementioned phenomena have clinical importance when the conditions of this scenario are met, and also comprise the pathophysiological basis for noninvasive tests of ischemia with the use of drugs generating pharmacological vasodilatory stress (such as dipyridamole).

ANGINA

Angina is the clinical manifestation of severe myocardial ischemia. It appears clinically with pain or discomfort in the central chest area, usually in the retrosternal area, which is caused when the myocardial oxygen needs are not met. This is due to an abnormal balance between oxygen supply and the oxygen demand in the heart muscle, resulting in myocardial tissue hypoxia. Impaired oxygen supply may manifest during increased oxygen demand (physical activity) in the presence of a fixed significant coronary stenosis, or during an acute complete or incomplete coronary occlusion due to plaque rupture and coronary thrombosis during an acute coronary syndrome. For didactic, clinical, and pathophysiological purposes, angina is classified into the following types, all of which are characterized by an abnormal balance between oxygen supply and demand.

Stable Angina

The pathophysiological background of stable angina is the presence of a significant fixed stenosis in a large epicardial coronary artery, which is usually caused by atherosclerosis.

As explained earlier, if the coronary artery is severely stenosed above a certain level (~70%–75% stenosis in diameter), there will be no physiological increase of CBF during increased needs (physical activity, sexual intercourse, etc.) and myocardial ischemia and hypoxia will appear. So the oxygen supply/demand balance is altered when oxygen demands increase. Clinically, this cascade of events manifests as angina during physical activities. This anginal pain usually occurs predictably and reproducibly each time the patient exerts above a certain level. Stable angina may also appear under other conditions of increased oxygen demands, for example, during intense emotional stress or after a large meal. Stable angina is treated with antianginal medications aiming at improving blood flow during conditions of high oxygen demands (nitrates, calcium channel antagonists) or reducing oxygen demands (β-blockers).

Unstable Angina

Unstable angina is traditionally not caused by increased oxygen demands, but usually appears when an atherosclerotic plaque of an epicardial coronary artery ruptures, resulting in thrombus formation, impairment of blood flow, and inability of the coronary system to meet the basal flow and oxygen requirements at rest in a particular part of the myocardium. So in unstable angina, the oxygen supply/demand balance is altered due to a change in oxygen (blood) supply. Depending on the severity and duration of the ischemia, necrosis of myocardial tissue (myocardial infarction) may or may not happen. The basis of treatment for unstable angina is antiplatelet and antithrombotic medications in addition to antianginal therapy. Invasive revascularization also plays a major role in a substantial subset of patients with unstable angina.

Unstable angina may sometimes occur in the presence of chronic, fixed coronary artery stenoses, and may be triggered by a combination of impaired oxygen supply and increased oxygen needs. For example, a patient with a severe coronary stenosis and history of predictable stable angina may develop unstable angina in the presence of acute arrhythmia with high heart rate (such as paroxysmal atrial fibrillation) or an acute febrile systemic illness accompanied by significant sinus tachycardia.

Variant Angina (Prinzmetal Angina)

Prinzmetal angina is the consequence of spasm of a major epicardial coronary artery, with significant transient reduction of CBF, reduced oxygen supply, and hypoxia in a myocardial segment. Both coronaries with preexisting fixed stenoses and "normal" coronaries without significant atherosclerosis may be prone to coronary vasospasm. Conditions such as extreme emotional or physical stress and endothelial dysfunction may elicit Prinzmetal angina. Medications that release or prevent coronary spasm, such as nitrates and calcium channel antagonists, are the cornerstone of treatment for this form of angina.

FURTHER READING

[1] Nishikawa Y, Ogawa S. Importance of nitric oxide in the coronary artery at rest and during pacing in humans. J Am Coll Cardiol January 1997;29(1):85–92.

[2] Duncker DJ, Bache RJ. Regulation of coronary blood flow during exercise. Physiol Rev July 2008; 88(3):1009–86.

[3] Kassab GS, Navia JA, March K, Choy JS. Coronary venous retroperfusion: an old concept, a new approach. J Appl Physiol May 2008;104(5):1266–72.

[4] Aznaouridis K, Masoura C, Kastellanos S, Alahmar A. Inadvertent cardiac phlebography. World J Cardiol June 26, 2017;9(6):558–61.

[5] http://www.wikilectures.eu/index.php/Coronary_Blood_Flow.

[6] Tune JD. Control of coronary blood flow during hypoxemia. Adv Exp Med Biol 2007;618:25–39.

[7] Laksanakorn W, Laprattanagul T, Wei J, Shufelt C, Minissian M, Mehta PK, Merz CNB. Cardiac rehabilitation for cardiac syndrome X and microvascular angina: a case report. Int J Case Rep Images 2015; 6(4):239–44.

[8] http://www.cvphysiology.com.

[9] Berwick ZC, Dick GM, Moberly SP, Kohr MC, Sturek M, Tune JD. Contribution of voltage-dependent K^+ channels to metabolic control of coronary blood flow. J Mol Cell Cardiol April 2012; 52(4):912–9.

[10] van de Hoef TP, Nolte F, Rolandi MC, Piek JJ, van den Wijngaard JP, Spaan JA, Siebes M. Coronary pressure-flow relations as basis for the understanding of coronary physiology. J Mol Cell Cardiol April 2012;52(4):786–93.

[11] Deussen A, Ohanyan V, Jannasch A, Yin L, Chilian W. Mechanisms of metabolic coronary flow regulation. J Mol Cell Cardiol April 2012;52(4):794–801.

[12] Beyer AM, Gutterman DD. Regulation of the human coronary microcirculation. J Mol Cell Cardiol April 2012;52(4):814–21.

[13] Duncker DJ, Koller A, Merkus D, Canty Jr JM. Regulation of coronary blood flow in health and ischemic heart disease. Prog Cardiovasc Dis 2015 ;57(5):409–22.

[14] Tune JD, Gorman MW, Feigl EO. Matching coronary blood flow to myocardial oxygen consumption. J Appl Physiol (1985) July 2004;97(1):404–15.

[15] Goodwill AG, Dick GM, Kiel AM, Tune JD. Regulation of coronary blood flow. Compr Physiol March 16, 2017;7(2):321–82.

[16] Schremmer B, Dhainaut JF. Regulation of myocardial oxygen delivery. Intensive Care Med 1990; 16(Suppl. 2):S157–63.

[17] Jacob M, Chappell D, Becker BF. Regulation of blood flow and volume exchange across the microcirculation. Crit Care October 21, 2016;20(1):319.

CHAPTER 2.4

Stable Angina Pectoris

Charalambos Vlachopoulos, Christos Georgakopoulos, Dimitris Pollalis, Dimitris Tousoulis
1st Cardiology Clinic, 'Hippokration' General Hospital, National and Kapodistrian University of Athens, School of Medicine, Athens, Greece

Contents

Coronary Artery Disease
ISBN 978-0-12-811908-2, http://dx.doi.org/10.1016/B978-0-12-811908-2.00011-8

DEFINITION AND DIFFERENTIAL DIAGNOSIS

Definition—Medical History

Angina pectoris is a discomfort in the chest or adjacent areas caused by myocardial ischemia. It is usually brought on by exertion and associated with a disturbance in myocardial function, but without myocardial necrosis [1] Heberden's initial description of the chest discomfort as conveying a sense of "strangling and anxiety" is still remarkably pertinent, although adjectives frequently used to describe this distress include "viselike", "constricting", "suffocating", "crushing", "heavy", and "squeezing". In other patients, the quality of the sensation is more vague and described as a mild pressure-like discomfort, an uncomfortable numb sensation, or a burning sensation. The site of the discomfort is usually retrosternal, but radiation is common and usually occurs down the ulnar surface of the left arm; the right arm and the outer surfaces of both arms may also be involved [1]. Epigastric discomfort alone or in association with chest pressure is not uncommon. Anginal discomfort above the mandible, below the epigastrium, or confined to the ear is rare. Anginal "equivalents" (i.e., symptoms of myocardial ischemia other than angina), such as dyspnea, faintness, fatigue, and eructations, are common, particularly in the elderly. A history of abnormal exertional dyspnea may be an early indicator of coronary artery disease (CAD) even when angina is absent or no electrocardiographic (ECG) evidence of ischemic heart disease can be found [2]. Dyspnea at rest or with exertion may be a manifestation of severe ischemia and lead to increases in left ventricular filling pressure. Nocturnal angina should raise the suspicion of sleep apnea [3].

A careful clinical history is key to making the correct diagnosis and is particularly important in this era of cost-conscious practice of medicine because it may obviate more expensive testing. If the quality of the pain and its duration, precipitating factors, and associated symptoms are taken into consideration, it is usually possible to arrive at a correct diagnosis. The typical episode of angina pectoris usually begins gradually and reaches its maximum intensity over a period of minutes before dissipating. It is unusual for angina pectoris to reach its maximum severity within seconds, and it is characteristic that patients with angina usually prefer to rest, sit, or stop walking during episodes [1]. Chest discomfort while walking in the cold, uphill, or after a meal is suggestive of angina. Features suggesting the absence of angina pectoris include pleuritic pain, pain localized to the tip of one finger, pain reproduced by movement or palpation of the chest wall or arms, and constant pain lasting many hours or, alternatively, very brief episodes of pain lasting seconds. Pain radiating into the lower extremities is also a highly unusual manifestation of angina pectoris.

Typical angina pectoris is relieved within minutes by rest or by the use of nitroglycerin. The response to the latter is often a useful diagnostic tool, although it should be remembered that esophageal pain and other syndromes may also respond to nitroglycerin. A delay of more than 5–10 min before relief is obtained by rest and nitroglycerin suggests that the symptoms are either not due to ischemia or, alternatively, are due to severe ischemia, i.e., acute myocardial infarction or unstable angina. The phenomenon of "first effort" or "warm-up" angina is used to describe the ability of some patients in whom angina develops with exertion to subsequently continue at the same level of exertion without symptoms after an intervening period of rest. This attenuation of myocardial ischemia observed with repeated exertion has been postulated to be due to ischemic preconditioning [4].

Mechanism of Cardiac Pain

The mechanisms of cardiac pain and the neural pathways involved are poorly understood [5]. It is presumed that angina pectoris results from ischemic episodes that excite chemosensitive and mechanoreceptive receptors in the heart. Stimulation of these receptors results in the release of adenosine, bradykinin, and other substances that excite the sensory ends of the sympathetic and vagal afferent fibers. The afferent fibers traverse the nerves that connect to the upper five thoracic sympathetic ganglia and upper five distal thoracic roots of the spinal cord. Impulses are transmitted by the spinal cord to the thalamus and hence to the neocortex.

Within the spinal cord, cardiac sympathetic afferent impulses may converge with impulses from somatic thoracic structures, which may be the basis for referred cardiac pain, for example, to the chest. In comparison, cardiac vagal afferent fibers synapse in the nucleus tractus solitarius of the medulla and then descend to excite the upper cervical spino-thalamic tract cells, which may contribute to the anginal pain experienced in the neck and jaw [6]. On the basis of positron emission tomographic (PET) findings on changes in regional cerebral blood flow associated with angina pectoris, it has been proposed that cortical activation is necessary for pain sensation and the thalamus acts as a gate for afferent pain signals [7].

Differential Diagnosis of Chest Pain

Differentiation of various disorders from CAD is challenging because the severity of the chest pain and the seriousness of the underlying disorder are not necessarily related. Compounding the difficulty in differential diagnosis is the common myth that pain in the left arm or left side of the chest is an ominous sign signifying the presence of CAD. However, a host of other disorders can also cause discomfort in these locations (Fig. 2.4.1).

Esophageal Disorders

The common esophageal disorders that may simulate or coexist with angina pectoris are gastro-esophageal reflux and disorders of esophageal motility, including diffuse spasm as well as "nutcracker" esophagus, which is characterized by high-amplitude peristaltic

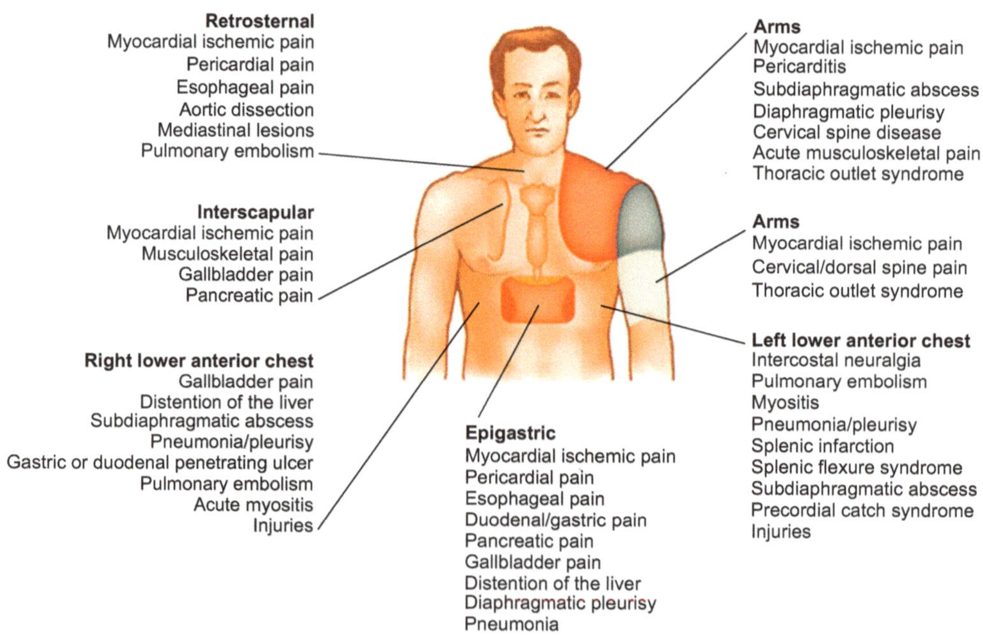

Figure 2.4.1 Location of discomfort and cause of chest symptoms. The location of angina is usually retrosternal, but radiation is common. An epigastric location of angina may also occur. *(Modified from Braunwald E. The history. In: Zipes D, Libby P, Bonow RO, Braunwald E, editors. Braunwald's heart disease. 7th ed. Philadelphia: WB Saunders; 2005.)*

contractions and vigorous achalasia [8]. Symptomatic esophageal reflux is common and estimated to occur in 7%–14% of an otherwise "healthy" US population. In a comparative study of patients with chest pain and normal coronary angiograms and controls with confirmed CAD, esophageal function testing (including manometry, provocation tests, and 24-h ambulatory pH monitoring) commonly implicated the esophagus as a cause of pain in patients with normal coronary angiograms. Nonetheless, a similar high frequency of esophageal abnormalities among patients with angina pectoris suggests that the esophagus may be an unrecognized source of pain in both groups of patients [8]. Further evidence of a relationship between esophageal abnormalities and angina pectoris was provided by a prospective study demonstrating that esophageal acid stimulation can cause anginal attacks in association with a significant reduction in coronary blood flow in patients with CAD [9]. A lack of any significant effect in this study in heart transplant recipients with cardiac denervation suggests a neural origin.

The classic manifestation of esophageal pain is "heartburn", particularly in connection with changes in posture and meals and in association with dysphagia. Esophageal spasm may also cause constant retrosternal discomfort of uniform intensity or severe spasmodic pain during or after swallowing. To further compound the difficulty in distinguishing between angina and esophageal pain, both may be relieved by nitroglycerin. However, esophageal pain is often relieved by milk, antacids, foods, or, occasionally, warm liquids.

Gastroesophageal Reflux

The esophageal acid perfusion, or Bernstein test, may be helpful in its use of alternate infusions of dilute acid and normal saline by a nasal gastric catheter with the tip placed at the level of the midesophagus [8]. Infusion of acid produces pain in over 90% of patients with subjective and objective evidence of gastroesophageal acid reflux, but it is particularly useful if the patient's symptoms are reproduced. Acid reflux into the esophagus can also be recognized by recording the pH from an electrode at the tip of a catheter inserted into the distal portion of the esophagus.

Esophageal Motility Disorders

Esophageal motility disorders are not uncommon in patients with retrosternal chest pain of unclear cause and should be specifically excluded or confirmed, if possible [10]. In addition to chest pain, the majority of such patients have dysphagia. Although barium studies may reveal motility problems, esophageal manometry may show diffuse esophageal spasm, increased pressure at the lower esophageal sphincter, and other motility disorders. Provocative pharmacological agents such as metacholine may provoke esophageal pain and manometric signs of spasm.

A more complex problem is determining whether part or all of the symptoms in patients with known CAD are due to esophageal disease. Both CAD and esophageal disease are common clinical entities that may coexist. Diagnostic evaluation for an esophageal disorder may be indicated in patients with CAD who have a poor symptomatic response to antianginal therapy in the absence of documentation of severe ischemia or in patients with persistent symptoms despite adequate coronary revascularization.

Biliary Colic

Although visceral symptoms are a common association of myocardial ischemia, cholecystitis, and related hepatobiliary disorders may also mimic ischemia and should always be considered in patients with atypical chest discomfort, particularly those with diabetes [11]. The pain is steady, usually lasts 2—4 h, and subsides spontaneously without any symptoms between attacks. It is generally most intense in the right upper abdominal area but may also be felt in the epigastrium or precordium. This discomfort is often referred to the scapula, may radiate around the costal margin to the back, or may in rare cases be felt in the shoulder and suggest diaphragmatic irritation. Ultrasonography is accurate in diagnosing gallstones and allows determination of gallbladder size and thickness and whether the bile ducts are dilated.

Costosternal Syndrome

In 1921, Tietze first described a syndrome of local pain and tenderness, usually limited to the anterior chest wall and associated with swelling of costal cartilage. This condition causes pain that can resemble angina pectoris. The full-blown Tietze syndrome, i.e., pain associated with tender swelling of the costochondral junctions, is uncommon, whereas costochondritis causing tenderness of the costochondral junctions (without swelling) is relatively common [12]. Pain on palpation of these joints is a useful clinical sign. Local pressure should be applied routinely to the anterior chest wall during examination of a patient with suspected angina pectoris. In addition, costochondritis is usually well localized. Although palpation of the chest wall often reproduces pain in patients with various musculoskeletal conditions, it should be appreciated that chest wall tenderness may also be associated with and does not exclude symptomatic CAD [13].

Other Musculoskeletal Disorders

Cervical radiculitis may be confused with angina. This condition may occur as a constant ache, sometimes resulting in a sensory deficit. The pain may be related to motion of the neck, just as motion of the shoulder triggers attacks of pain from bursitis. A hyperalgesic area noted by running the finger down the back and exerting pressure may lead to a suspicion of thoracic root pain. Occasionally, pain mimicking angina can be due to compression of the brachial plexus by the cervical ribs, and tendinitis or bursitis involving the left shoulder may also cause angina-like pain. Physical examination may also detect pain brought about by movement of an arthritic shoulder or a calcified shoulder tendon.

Other Causes of Angina-Like Pain

Acute myocardial infarction is usually associated with prolonged (>30 min), severe pain occurring at rest that apart from duration and intensity, may be similar to angina pectoris. It is associated with characteristic ECG changes and the release of cardiac markers. Unstable angina is a severe form of angina that may also occur at rest and may not be relieved by nitroglycerin.

Severe pulmonary hypertension may be associated with exertional chest pain with the characteristics of angina pectoris, and indeed, this pain is thought to be due to right ventricular ischemia that develops during exertion. Other associated symptoms include exertional dyspnea, dizziness, and syncope.

Pulmonary embolism is initially characterized by dyspnea as the cardinal symptom, but chest pain may also be present. Pleuritic pain suggests pulmonary infarction, and a history of exacerbation of the pain with inspiration, along with a pleural friction rub, usually helps distinguish it from angina pectoris.

The pain of acute pericarditis may at times be difficult to distinguish from angina pectoris. However, pericarditis tends to occur in younger patients than does angina, and the diagnosis depends on the combination of chest pain not relieved by rest or nitroglycerin, a pericardial friction rub, and ECG changes.

Chronic CAD can and frequently does coexist with any of the other disorders mentioned above, and noncardiac disease can trigger a true angina attack in a patient with CAD. An additional component of the history is an evaluation of risk factors for CAD because such risk factors in turn have an effect on both the probability of significant obstructive CAD and the overall prognosis [14].

PHYSICAL EXAMINATION

General Examination

Inspection of the eyes may reveal a corneal arcus, and examination of the skin may show xanthomas. Among patients with heterozygous familial hypercholesterolemia (in whom CAD is common), the presence of a corneal arcus increases with age and, in some studies, correlates positively with levels of cholesterol and low-density lipoprotein (LDL) and also with the prognosis [15,16]. Xanthelasma, in which lipid deposits are intracellular, appears to be promoted by increased levels of triglycerides and a relative deficiency of high-density lipoprotein (HDL). The presence of xanthelasma is a strong marker of dyslipidemia and, often, a family history of cardiovascular disease, and should provide a strong impetus for performing a comprehensive lipid profile [17]. Retinal arteriolar changes are common in patients with CAD and diabetes mellitus or hypertension. Moreover, diabetes-associated visual impairments, including retinopathy, are independent predictors of increased mortality from all causes, including CAD [18].

Blood pressure may be chronically elevated or may rise acutely (along with the heart rate) during an angina attack. Changes in blood pressure may precede (and precipitate) or follow (and be caused by) angina.

Other important features of the general physical examination are abnormalities in arterial pulses and the venous system. A rapid pulse may be a clue to cardiac decompensation or to a systemic condition such as thyrotoxicosis or anemia, which can exacerbate angina pectoris. The association between peripheral vascular disease (PVD) and CAD is strong and well documented [19,20]. This association is not confined to patients with symptomatic or clinically overt PVD or CAD but is also seen in asymptomatic subjects with a reduced ankle-brachial blood pressure index or evidence of early carotid disease on ultrasonography [21]. The presence of carotid and peripheral arterial disease on palpation and auscultation increases the likelihood that chest discomfort of unclear origin is caused by CAD. Evaluation of the patient's venous system, particularly in the legs, may have an important bearing on the type of grafting procedure used in subsequent coronary bypass surgery.

Cardiac Examination

The physical findings of hypertrophic cardiomyopathy or aortic valve disease suggest that angina may be due to conditions other than (or in addition to) CAD [22]. Paradoxical splitting of the second heart sound may occur transiently during angina and appears to be related to asynergy and prolongation of left ventricular contraction, which results in delayed closure of the aortic valve. If other obvious cardiac diseases are absent, a third or loud fourth heart sound suggests ischemia as the basis for the chest pain. These sounds are common in patients with angina at rest, and their frequency is increased during handgrip exercise [23]. A sustained apical cardiac impulse is common in patients with moderate or severe left ventricular dysfunction. A displaced ventricular impulse, particularly if dyskinetic, is a sign of significant left ventricular systolic dysfunction, especially in a patient who previously had a myocardial infarction.

Transient apical systolic murmurs are a common finding in stable coronary artery disease (SCAD) and have been attributed to reversible papillary muscle dysfunction secondary to transient myocardial ischemia. When persistent, such murmurs may be due to papillary muscle fibrosis, which is often a manifestation of subendocardial infarction or a regional wall motion abnormality altering the alignment of the papillary muscles in relation to other components of the mitral valve apparatus. These murmurs are more prevalent in patients with extensive CAD, especially those with prior myocardial infarction and left ventricular dysfunction, and may indicate an adverse prognosis [24]. Systolic murmurs may assume a variety of configurations (early, late, or holosystolic) and may be accentuated by exertion or during angina. A mid-systolic click, often followed by a late systolic murmur produced by mitral valve prolapse, also occurs in patients with CAD. A diastolic murmur or a continuous murmur is a rare finding in CAD and has been attributed to turbulent flow across a proximal coronary artery stenosis [25].

PATHOPHYSIOLOGY

Angina pectoris results from myocardial ischemia, which is caused by an imbalance between myocardial O_2 requirements and myocardial O_2 supply [5]. The former may be elevated by increases in heart rate, left ventricular wall stress, and contractility; the latter is determined by coronary blood flow and coronary arterial O_2 content (Fig. 2.4.2).

Angina Caused by Increased Myocardial O_2 Requirements

In this condition, sometimes termed "demand angina", the myocardial O_2 requirement increases in the face of a constant and usually restricted O_2 supply. The increased requirement commonly stems from norepinephrine release by adrenergic nerve endings in the heart and vascular bed, a physiological response to exertion, emotion, or mental stress. Of great importance to the myocardial O_2 requirement is the rate at which any task is carried out. Hurrying is particularly likely to precipitate angina, as are efforts involving motion of the hands over the head. Mental stress may also precipitate angina, presumably by increased hemodynamic and catecholamine responses to stress, increased adrenergic tone, reduced vagal activity, and stiffening of the large arteries [26,27].

The combination of physical exertion and emotion in association with sexual activity commonly precipitates angina pectoris, but sexual activity seldom triggers myocardial

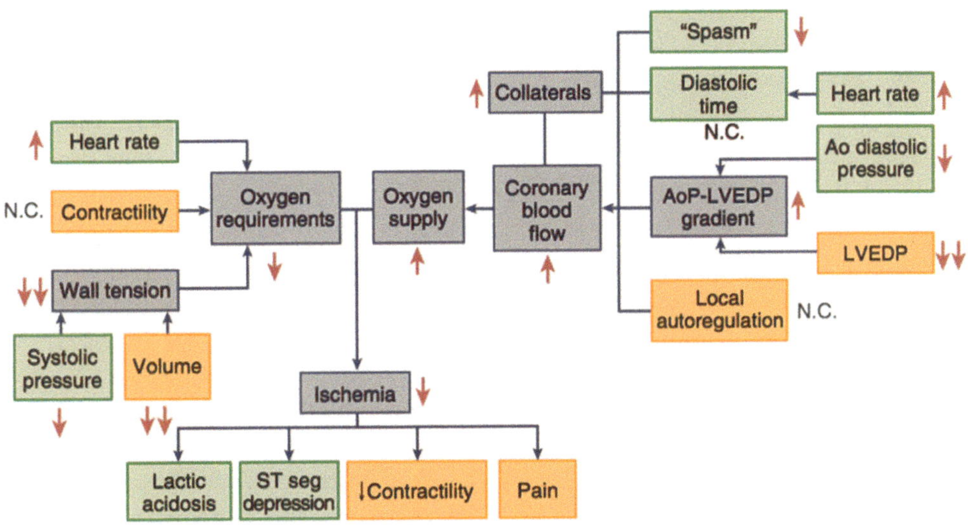

Figure 2.4.2 Factors influencing the balance between myocardial O_2 requirement (left) and supply (right). Arrows indicate effects of nitrates. In relieving angina pectoris, nitrates exert favorable effects by reducing O_2 requirements and increasing supply. Although a reflex increase in heart rate would tend to reduce the time for coronary flow, dilation of collaterals and enhancement of the pressure gradient for flow to occur as left ventricular end-diastolic pressure (LVEDP) falls tend to increase coronary flow. *AoP-LVEDP*, aortic pressure minus LVEDP; *N.C.*, no change. *(From Frishman WH. Pharmacology of the nitrates in angina pectoris. Am J Cardiol 1985;56:81.)*

infarction [28]. Anger may produce constriction of coronary arteries with preexisting narrowing without necessarily affecting O_2 demand. Other factors causing angina secondary to an increase in myocardial O_2 requirement in patients with obstructive CAD include physical exertion after a heavy meal and the excessive metabolic demands imposed by chills, fever, thyrotoxicosis, tachycardia from any cause, and hypoglycemia. Among patients with stable, fixed obstructive CAD, several studies using ambulatory ECG monitoring have documented the importance of increases in myocardial O_2 ischemia [29].

In all these conditions, underlying coronary artery obstruction is usually present, and the other factors (e.g., exertion, emotion, or fever) precipitate ischemia and chest discomfort by stimulating myocardial O_2 need in the presence of a relatively fixed and limited myocardial O_2 supply.

Angina Caused by Transiently Decreased O_2 Supply

Increasing evidence suggests that not only unstable angina but also chronic stable angina may be caused by transient reductions in O_2 supply as a consequence of coronary vasoconstriction [5,30], a condition that is sometimes termed "supply angina" and due to the entity of "dynamic stenosis". [31] The coronary arterial bed is well innervated, and a variety of stimuli alter coronary tone. Two main explanations have been offered for the association between coronary vasoconstriction and spasm in the presence of organic stenoses. First, platelet thrombi and leukocytes may elaborate vasoconstrictor substances such as serotonin and thromboxane A_2. Second, endothelial damage in atherosclerotic coronary arteries may result in decreased production of vasodilator substances and an abnormal vasoconstrictor response to exercise and other stimuli. A variable threshold of myocardial ischemia in patients with chronic stable angina may be due to dynamic changes in peristenotic smooth muscle tone and also to constriction of arteries distal to the stenosis [32]. In this setting, calcium antagonists and nitrates are less effective than in patients with variant angina, perhaps because of the nature of the constricting stimuli and the site of constriction [32].

Patients with angina precipitated by a transient reduction in myocardial O_2 supply may have a spectrum of signs and symptoms that depend on the severity of the underlying fixed defect and the degree of the dynamic change in coronary arterial tone. In a typical patient with chronic stable angina, the degree of fixed obstruction is sufficient to result in an inadequate coronary flow rate to cope with the increased O_2 demands of exercise. However, episodes of transient coronary vasoconstriction may be superimposed on this inadequate flow rate and cause additional limitations to coronary flow reserve in many patients.

In rare patients without organic obstructing lesions, severe dynamic obstruction occurring at rest alone can cause myocardial ischemia and result in angina (Prinzmetal angina).

On the other hand, in patients with severe fixed obstruction to coronary blood flow, only a minor increase in dynamic obstruction is necessary for blood flow to fall below a critical level and cause myocardial ischemia.

Fixed Compared With Variable-Threshold Angina

The threshold for angina differs widely among patients with chronic angina. In patients with fixed-threshold angina precipitated by increased O_2 demands with few if any dynamic (vasoconstrictor) components, the level of physical activity required to precipitate angina is relatively constant. Characteristically, these patients can predict the amount of physical activity that will precipitate angina, e.g., walking up exactly two flights of stairs at a customary pace. When these patients are tested on a treadmill or bicycle, the pressure-rate product (the so-called double product, a correlate of the myocardial O_2 requirement) that elicits angina and/or ECG evidence of ischemia is relatively constant.

In patients with fixed-threshold, demand angina, the specific threshold at which ischemia develops (as reflected in angina and/or ST segment depression) is a function of the myocardial O_2 requirement. As the activity of the left ventricle (LV) (and, therefore, its O_2 requirement) increases, a point is reached at which perfusion distal to a critical coronary arterial obstruction cannot supply sufficient O_2 to myocardium perfused by the obstructed artery; ischemia, and angina ensue. This relationship is, however, modified by the effects of coronary vasomotor tone on myocardial O_2 supply [33]. Coronary vascular reserve (or coronary vasodilator or flow reserve) is impaired in patients with significant obstructive CAD and also in those with microvascular disease or endothelial damage from conditions such as hypertension [34].

The majority of patients with variable-threshold angina have atherosclerotic coronary arterial narrowing, but dynamic obstruction caused by vasoconstriction plays an important role in causing myocardial ischemia. These patients typically have "good days", when they are capable of substantial physical activity, as well as "bad days", when even minimal activity can cause clinical and/or ECG evidence of myocardial ischemia or angina at rest. Often, even in the course of a single day, they may be capable of substantial physical activity at one time while minimal activity results in angina at another. Patients with variable-threshold angina often complain of a circadian variation in angina that is more common in the morning. Angina on exertion and sometimes even at rest may be precipitated by cold temperature [35], emotion and mental stress [36]. A cold environment has been shown to increase peripheral resistance, both at rest and during exercise [37]. The rise in arterial pressure, by augmenting myocardial O_2 requirements, lowers the threshold for the development of angina. An alternative or additional explanation is the development of cold-induced coronary vasoconstriction via activation of peripheral and reflex mechanisms [37].

The entity of postprandial angina has been recognized for about two centuries and may be a marker of severe multivessel CAD [38]. The mechanism has not been explained, but it may be due to redistribution of coronary blood flow away from the territory supplied by severely stenosed vessels [39]. Some evidence indicates that this phenomenon is more prominent after high-carbohydrate than high-fat meals [38].

EVALUATION

Grading of Angina Pectoris

A system of grading the severity of angina pectoris proposed by the Canadian Cardiovascular Society has gained widespread acceptance [40] (Table 2.4.1). The system is a modification of the New York Heart Association functional classification but allows patients to be categorized in more specific terms. Other grading systems include a specific activity scale developed by Goldman et al. [41] and an anginal "score" developed by Califf and colleagues [42]. The Goldman scale is based on the metabolic cost of specific activities and appears to be valid when used by both physicians and nonphysicians. The anginal score of Califf and coworkers integrates the clinical features and "tempo" of angina together with ECG.

ST and T wave changes and offers independent prognostic information above that provided by age, gender, left ventricular function, and coronary angiographic anatomy. A limitation of all these grading systems is their dependence on accurate patient observation and patients' widely varying tolerance for symptoms. Prospective evaluation of the reproducibility of the New York Heart Association estimates of functional class made by two physicians demonstrated a reproducibility of only 56%, and only 51% of the estimates agreed with treadmill exercise performance. Functional estimates based on the Canadian Cardiovascular Society criteria were more reproducible (73%) but still did not correlate well with objective measures of exercise performance [41].

Table 2.4.1 Grading of severity of angina according to the canadian cardiovascular society

Class	Description
I	Angina with strenuous, rapid, or prolonged exertion at work or during recreation. Ordinary physical activity, such as walking or climbing stairs, does not cause angina.
II	Slight limitation of ordinary activity. Angina occurs on walking more than two blocks on the level and climbing more than one flight of stairs at a normal pace and in normal conditions.
III	Marked limitations of ordinary physical activity. Angina occurs on walking one or two blocks on the level and climbing one flight of stairs in normal conditions and at a normal pace.
IV	Inability to carry on any physical activity without discomfort. Angina symptoms may be present at rest.

Noninvasive Testing

Biochemical Tests

In patients with SCAD, metabolic abnormalities that are risk factors for the development of CAD are frequently detected. Such abnormalities include hypercholesterolemia and other dyslipidemias, carbohydrate intolerance, and insulin resistance. Moreover, chronic kidney disease is strongly associated with risk for atherosclerotic vascular disease. All patients with established or suspected CAD warrant biochemical evaluation of total cholesterol, LDL cholesterol, HDL cholesterol, triglyceride, serum creatinine (estimated glomerular filtration rate [eGFR]), and fasting blood glucose levels.

Other biochemical markers have also been shown to be associated with higher risk for future cardiovascular events. Measurement of lipoprotein(a) and other lipid elements that are particularly atherogenic, such as apolipoprotein B and small dense LDL, appears to add prognostic information to the measurement of total cholesterol and LDL and may be considered a secondary target for therapy in patients who have achieved therapeutic targets for LDL. However, no consensus has been reached regarding routine measurement, and a simple approach based on calculation of non-HDL cholesterol (particularly in patients with triglyceride levels >200 mg/dL) may capture important information related to the presence of residual very LDL atherogenic remnants [43]. Although homocysteine has also been linked to atherogenesis and correlates with risk for CAD, prospective studies, in aggregate, have supported, at most, a modest increase in risk associated with elevated homocysteine levels and have not consistently demonstrated a relationship independent of traditional risk factors or other biochemical markers [44]. Moreover, placebo-controlled trials have failed to demonstrate clinical benefit associated with folate replacement therapy as an intervention to mitigate the adverse effects of increased homocysteine levels [45]. Therefore general screening for elevated homocysteine levels is not recommended.

Advances in understanding regarding the pathobiology of atherothrombosis have generated interest in inflammatory biomarkers as noninvasive indicators of underlying atherosclerosis and cardiovascular risk. Measurement of the acute-phase protein high-sensitivity C-reactive protein (hsCRP) has shown a consistent relationship to risk for incident cardiovascular events, as well as with imaging findings of vulnerable atherosclerosis [46]. The prognostic value of hsCRP is additive to traditional risk factors, including lipids [47]; however, its incremental clinical value for screening continues to be debated.

Other biomarkers of inflammation, such as interleukin-6, myeloperoxidase, growth factors, and metalloproteinases, remain under study as potential markers of underlying atherosclerosis [48].

Biomarkers of Myocyte Injury, Ischemia, and Hemodynamic Stress

Blood levels of cardiac markers of myocardial necrosis are typically used to differentiate patients with acute myocardial infarction (MI) from those with SCAD. However, with more sensitive assays for cardiac troponin, circulating biomarkers of myocyte injury have now been detected in patients with SCAD and shown to have a graded relationship with the subsequent risk for cardiovascular mortality and heart failure [49]. Although such evidence may lead to new applications of troponin in patients with SCAD, clinical use in this population is currently not recommended [50]. Biomarkers of myocardial ischemia are also under study. For example, the plasma concentration of brain natriuretic peptide (BNP) increases in response to spontaneous or provoked ischemia. Although BNP and N-terminal pro-BNP may not have sufficient specificity to aid in the diagnosis of SCAD, their concentration is associated with risk for cardiovascular events in those at risk for and with established CAD. Other novel biomarkers of hemodynamic stress, such as mid-regional pro-adrenomedullin (MR-pro-ADM) and mid-regional pro-atrial natriuretic peptide (MR-pro-ANP), also appear to provide incremental information regarding the risk for cardiovascular death in patients with SCAD and may be useful for guiding decisions regarding therapy [51].

Resting Electrocardiogram

Findings on the resting ECG are normal in approximately half of patients with stable ischemic heart disease (SIHD), and even patients with severe CAD may have a normal tracing at rest. The most common abnormalities on the ECG in patients with chronic CAD are nonspecific ST-T wave changes with or without abnormal Q waves. In patients with known CAD, however, the occurrence of ST-T wave abnormalities on the resting ECG (particularly if obtained during an episode of angina) can correlate with the severity of the underlying heart disease. This correlation explains the adverse association of ST-T wave changes with prognosis in these patients. In contrast, a normal resting ECG is a more favorable long-term prognostic sign in patients with suspected or definite CAD.

Interval ECGs may reveal the development of Q wave MIs that have gone unrecognized clinically. Various conduction disturbances, most frequently left bundle branch block and left anterior fascicular block, may occur in patients with SIHD. They are often associated with impairment of LV function, reflect multivessel CAD, and are an indicator of a relatively poor prognosis. Various arrhythmias, especially ventricular premature beats, may be present on the ECG, but they too have low sensitivity and specificity for accurately detecting CAD.

During an episode of angina pectoris, findings on the ECG become abnormal in 50% or more of patients with normal resting ECGs. The most common finding is ST-segment depression, although ST-segment elevation and normalization of previous resting ST-T wave depression or inversion (pseudonormalization) may develop. Ambulatory ECG monitoring provides a quantitative estimate of the frequency and duration of ischemic episodes during routine activities; however, its sensitivity for detecting CAD is less than that of exercise electrocardiography.

Noninvasive Stress Testing

Noninvasive stress testing can provide useful and often indispensable information to establish the diagnosis and estimate the prognosis in patients with chronic stable angina. However, indiscriminate use of such tests yields limited incremental information beyond that provided by the physician's detailed and thoughtful clinical assessment. Appropriate application of noninvasive tests requires consideration of Bayesian principles, which state that the reliability and predictive accuracy of any test are defined not only by its sensitivity and specificity but also by the prevalence of disease (or pretest probability) in the population under study. A reasonable estimate of the pretest probability of CAD may be made on clinical grounds (Table 2.4.2).

Table 2.4.2 Clinical pre-test probabilities[a] in patients with stable chest pain symptoms

	Typical angina		Atypical angina		Non-anginal pain	
Age	Men	Women	Men	Women	Men	Women
30–39	59	28	29	10	18	5
40–49	69	37	38	14	25	8
50–59	77	47	49	20	34	12
60–69	84	58	59	28	44	17
70–79	89	68	69	37	54	24
>80	93	76	78	47	65	32

ECG = electrocardiogram; PTP = pre-test probability; SCAD = stable coronary artery disease.

[a] Probabilities of obstructive coronary disease shown reflect the estimates for patients aged 35, 45, 55, 65, 75 and 85 years.

- Groups in white boxes have a PTP <15% and hence can be managed without further testing.
- Groups in blue boxes have a PTP of 15–65%. They could have an exercise ECG if feasible as the initial test. However, if local expertise and availability permit a non-invasive imaging based test for ischaemia this would be preferable given the superior diagnostic capabilities of such tests. In young patients radiation issues should be considered.
- Groups in light red boxes have PTPs between 66–85% and hence should have a non-invasive imaging functional test for making a diagnosis of SCAD.
- In groups in dark red boxes the PTP is >85% and one can assume that SCAD is present. They need risk stratification only.

Reproduced from 2013 ESC Guidelines on the management of stable coronary artery disease.

Noninvasive testing should be performed only if the incremental information provided by a test is likely to alter the planned management strategy. The value of noninvasive stress testing is greatest when the pretest likelihood is intermediate because the test result is likely to have the greatest effect on the posttest probability of CAD and hence on clinical decision making.

Exercise Electrocardiography

Diagnosis of Coronary Artery Disease The exercise ECG is particularly helpful in patients with chest pain syndromes who are considered to have a moderate probability of CAD and in whom the resting ECG is normal, provided that they are capable of achieving an adequate workload [52]. Although the incremental diagnostic value of exercise testing is limited in patients in whom the estimated prevalence of CAD is high or low, the test provides useful additional information about the degree of functional limitation in both groups of patients and about the severity of ischemia and prognosis in patients with a high pretest probability of CAD. Interpretation of the exercise test should include consideration of the patient's exercise capacity (duration and metabolic equivalents) and clinical, hemodynamic, and ECG responses.

Influence of Antianginal Therapy Antianginal pharmacologic therapy may reduce the sensitivity of exercise testing as a screening tool, and if the purpose of the exercise test is to diagnose ischemia, it should be performed, if possible, in the absence of antianginal medications, particularly long-acting beta-blocking agents, which should be omitted for 2–3 days before testing. For long-acting nitrates, calcium antagonists, and short-acting beta blockers, discontinuing use of the medications the day before testing usually suffices.

Nuclear Cardiology Techniques

Stress Myocardial Perfusion Imaging Exercise myocardial perfusion imaging (MPI) with simultaneous ECG recording is generally considered to be superior to an exercise ECG alone in detecting CAD, in identifying multivessel CAD, in localizing diseased vessels, and in determining the magnitude of ischemic and infarcted myocardium [53]. Exercise single-photon emission computed tomography (SPECT) yields an average sensitivity and specificity of 88% and 72%, respectively (ranges, 71%–98% and 36%–92%, respectively), as opposed to 68% sensitivity and 77% specificity for exercise electrocardiography alone [53,54]. Referral bias may account, in part, for the low specificity of many studies, and the few studies that have adjusted for referral bias report a specificity higher than 90%. Perfusion imaging is also valuable for detecting myocardial viability in patients with regional or global LV dysfunction, with or without Q waves, and provides important information in regard to prognosis in all patients.

Stress MPI is particularly helpful in the diagnosis of CAD in patients with abnormal resting ECGs and in those in whom ST-segment responses cannot be interpreted accurately, such as patients with repolarization abnormalities caused by LV hypertrophy, those with left bundle branch block, and those receiving digitalis. Because stress MPI is a relatively expensive test (three to four times the cost of an exercise ECG), stress MPI should not be used as a screening test in patients in whom the prevalence of CAD is low because most abnormal test findings will yield false-positive results, and a regular exercise ECG should always be considered first in patients with chest pain and a normal resting ECG for screening and detection of CAD [54].

Myocardial Perfusion Imaging With Pharmacologic Vasodilator Stress For patients unable to exercise adequately, especially older adults and those with PVD, pulmonary disease, arthritis, orthopedic problems, severe obesity, or a previous stroke, pharmacologic vasodilator stress with dipyridamole or adenosine derivatives may be used [55]. In most nuclear cardiology laboratories, such patients account for approximately 40%—50% of those referred for perfusion imaging. Although the diagnostic accuracy of pharmacologic vasodilator stress perfusion imaging is comparable to that achieved with exercise perfusion imaging, treadmill testing is preferred for patients who are capable of exercising because the exercise component of the test provides additional diagnostic and prognostic information, including ST-segment changes, effort tolerance and symptomatic response, and heart rate and blood pressure response. Vasodilator stress agents are also used with PET to diagnose CAD and determine its severity.

Stress Echocardiography

Two-dimensional echocardiography is useful for the evaluation of patients with chronic CAD because it can be used to assess global and regional LV function under basal conditions and during ischemia, as well as to detect LV hypertrophy and associated valve disease [56]. Stress echocardiography may be performed with exercise or pharmacologic stress and allows detection of regional ischemia by identifying new areas of wall motion disorders. Adequate images can be obtained in more than 85% of patients, and the test is highly reproducible. Numerous studies have shown that exercise echocardiography can detect the presence of CAD with an accuracy similar to that of stress MPI and is superior to exercise electrocardiography alone. Stress echocardiography is also valuable in localizing and quantifying ischemic myocardium. As with perfusion imaging, stress echocardiography also provides important prognostic information about patients with known or suspected CAD. Pharmacologic stress, such as with dobutamine, should be used in patients unable to exercise, those unable to achieve adequate heart rates with exercise, and those in whom the quality of the echocardiographic images during or immediately after exercise is poor.

Stress echocardiography is an excellent alternative to nuclear cardiology procedures [56]. Limitations imposed by poor visualization of endocardial borders in a sizable subset of patients have been reduced by newer techniques, including myocardial contrast perfusion imaging, three-dimensional imaging, and strain-rate echocardiography. Although less expensive than nuclear perfusion imaging, stress echocardiography is more expensive than and not as widely available as exercise electrocardiography.

Chest X-ray

The chest X-ray is generally within normal limits in patients with SIHD, particularly if they have normal findings on the resting ECG and have not experienced MI. If cardiomegaly is present, it is indicative of severe CAD with previous MI, preexisting hypertension, or an associated nonischemic condition such as concomitant valvular heart disease or cardiomyopathy.

Computed Tomography

Cardiac CT has made substantial advances as a noninvasive approach to imaging atherosclerosis and its consequences [57]. In addition to being a highly sensitive method for detecting coronary calcification, which is a good marker of the total coronary atherosclerotic burden, cardiac CT can also provide angiography of the coronary arterial tree, assessment of myocardial perfusion, and quantification of ventricular function and myocardial viability as an emerging application [57,58].

Although coronary calcification is a highly sensitive (~90%) finding in patients with CAD, its specificity for identifying patients with obstructive CAD is low (~50%). In view of this limitation, CT is currently *not recommended* as a routine approach to screening for obstructive CAD in individuals at low risk for IHD (<10% 10-year estimated risk for coronary events) [59]. However, selective screening of individuals at intermediate risk for CAD events may be reasonable to consider because a high calcium score may reclassify an individual as being at higher risk and thereby lead to more intense risk factor modification [59]. In patients with known CAD, exercise testing is preferable to CT in guiding decisions for coronary angiography.

CT technology has progressed such that in selected individuals, high-quality images of the coronary arteries may be obtained. Consequently, CT angiography may be reasonable in symptomatic patients at intermediate risk for CAD after initial evaluation, in particular, those with indeterminate results of stress testing [60,61]. In experienced centers with advanced technology, CT has also been used to characterize plaque composition and, when paired with PET in a hybrid PET/CT scanner, can offer an assessment of coronary anatomy concurrent with information regarding myocardial blood flow and metabolism [62,63]. Nevertheless, in most individuals at present, the temporal resolution of coronary CT angiography is lower than optimal for accurate, complete coronary artery depiction because of nonevaluable segments and limited accuracy in estimating the degree of luminal stenosis. The capacity of CT for determination of plaque composition is also currently not sufficient for routine application [60]. Therefore, currently, the clinical strength of CT angiography remains its ability to exclude significant CAD with a high negative predictive value [57]. The results of ongoing investigation and new technologic innovations will guide evolution of the role of cardiac CT in the future assessment and management of SIHD.

Cardiac Magnetic Resonance Imaging

Cardiac magnetic resonance (CMR) is established as a valuable clinical tool for imaging the aorta and cerebral and peripheral arterial vasculature and is evolving as a versatile noninvasive cardiac imaging modality that has multiple applications in patients with IHD [60,64]. Clinical use of CMR for assessment of myocardial viability has grown because of evidence demonstrating its ability to predict functional recovery after percutaneous or surgical revascularization and its very good correlation with PET. Pharmacologic stress perfusion imaging with CMR compares favorably with SPECT and also offers accurate characterization of LV function, as well as delineation of patterns of myocardial disease that are often useful in discriminating ischemic from nonischemic myocardial dysfunction.

Because of its ability to visualize arteries in three dimensions and differentiate tissue constituents, CMR has received interest as a method to characterize arterial atheroma and assess vulnerability to rupture on the basis of compositional analysis. Characterization of arterial plaque has been achieved in the aorta and carotid arteries in humans and has been shown to be predictive of subsequent vascular events. Moreover, CMR coronary angiography in humans is established as a modality to characterize congenital coronary anomalies and has shown promise in detecting stenoses in the proximal and middle segments of major epicardial vessels or surgical bypass grafts. Accordingly, CMR is continuing to develop as a single approach for assessment of cardiac function, structure, blood flow, and viability without exposing the patient to ionizing radiation.

Invasive Testing

Coronary Angiography

The clinical examination and noninvasive techniques described earlier are extremely valuable in establishing the diagnosis of CAD and are indispensable to the overall assessment of patients with this condition. Currently, however, definitive diagnosis of CAD and precise assessment of its anatomic severity still require cardiac catheterization and coronary arteriography. Nevertheless, it should not be forgotten that myocardial ischemia may occur in the absence of epicardial CAD. In patients with chronic stable angina referred for coronary arteriography, approximately 25% each have single-, double-, or triple-vessel anatomically significant CAD (i.e., >70% luminal diameter narrowing). Five percent to ten percent have obstruction of the left main coronary artery, and in at least 15% and as high as 30% in some series, no critical obstruction is detectable. Advanced invasive imaging techniques such as intravascular ultrasonography (IVUS) provide a cross-sectional view of the coronary artery and have substantially enhanced the detection and quantification of coronary atherosclerosis, as well as the potential to characterize the vulnerability of coronary atheroma [65]. Studies incorporating both coronary angiography and IVUS have demonstrated that the severity of CAD may be underestimated by angiography alone. Intravascular optical coherence tomography, angioscopy, and thermography are evolving as additional tools for more complete characterization of coronary atheroma [66–68]. Moreover, studies of coronary flow reserve (maximum flow divided by resting flow) and endothelial function frequently produce abnormal results in patients with CAD and can play an important role in determining the functional significance of a stenosis or in detecting microvascular dysfunction in those without obstructive epicardial disease.

Coronary angiographic findings differ between patients with acute MI and those with SIHD. Patients with unheralded MI have fewer diseased vessels, fewer stenosis, and chronic occlusions, and less diffuse disease than do patients with chronic stable angina, thus suggesting that the pathophysiologic substrate and propensity for thrombosis differ between these two groups of patients. In patients with SIHD who have a history of previous MI, total occlusion of at least one major coronary artery is more common than in those without such a history.

Natural History and Risk Stratification

In a registry of patients with a history of stable angina managed in general practices, 29% experienced angina one or more times per week along with associated greater physical limitation and worse quality of life. The frequency of reported angina varied substantially between clinics, thus suggesting significant heterogeneity in the success of identifying and managing angina [69]. Women have a similar incidence of stable angina as men, and angina in both sexes is associated with higher risk for mortality than in the general population. Data from the Framingham Study, obtained before the widespread use of aspirin, beta-blocking agents, and aggressive modification of risk factors, revealed an average annual mortality rate of 4% in patients with SCAD. The combination of these treatments has improved the prognosis, with an annual mortality rate of 1%–3% and a rate of major ischemic events of 1%–2%. For example, among 38,602 outpatients with SCAD enrolled in 2003–2004, the 1-year rate of cardiovascular death was 1.9% (95% confidence interval [CI], 1.7%–2.1%), that of all-cause mortality was 2.9% (95% CI, 2.6%–3.2%), and that of cardiovascular death, MI, or stroke was 4.5% (95% CI, 4.2%–4.8%) [70]. Clinical, noninvasive, and invasive tools are useful for refining the estimated risk in individual patients with SCAD. Moreover, noninvasively acquired information is valuable in identifying patients who are candidates for invasive evaluation with cardiac catheterization.

Clinical Criteria

Clinical characteristics, including age, male sex, diabetes mellitus, previous MI, and the presence of symptoms typical of angina, are predictive of the presence of CAD [48]. A number of studies have attested to the adverse prognostic implications of heart failure in patients with SCAD. The severity of angina, especially the tempo of intensification, and the presence of dyspnea are also important predictors of outcome [48].

Noninvasive Testing

Exercise Electrocardiography

The prognostic importance of the treadmill exercise test was determined by observational studies in the 1980s and early 1990s. One of the most important and consistent predictors is maximal exercise capacity, regardless of whether it is measured by exercise duration or by workload achieved or whether the test was terminated because of dyspnea, fatigue, or angina. After adjustment for age, the peak exercise capacity measured in metabolic equivalents is among the strongest predictors of mortality in men with cardiovascular disease [71].

Echocardiography

Assessment of LV function is one of the most valuable aspects of echocardiography. Such testing is not necessary for all patients with angina pectoris, and in patients with normal ECGs and no previous history of MI, the likelihood of preserved LV systolic function is high. In contrast, in patients with a history of MI, ST-T wave changes, or conduction defects or Q waves on the ECG, LV function should be measured via echocardiography or an equivalent technique. The presence or absence of inducible regional wall motion abnormalities and the response of the ejection fraction to exercise or pharmacologic stress appear to provide incremental prognostic information to that provided by the resting ECG. Moreover, a negative stress echocardiographic result portends a low risk for future events (<1% per person-year).

Angiographic Criteria

The independent impact of multivessel CAD and LV dysfunction and their interaction on the prognosis of patients with CAD is well established. However, manifestations of CAD in the absence of obstructive atherosclerosis are now recognized and have been associated with an adverse prognosis [72,73].

Extent of Coronary Artery Disease

Although several indices have been used to quantify the extent or severity of CAD, the simple classification of disease into single-, double-, or triple-vessel or left main CAD is the most widely used and is effective. Additional prognostic information is provided by the severity of the obstruction and its location, whether proximal or distal. The importance to survival of the quantity of myocardium that is jeopardized is reflected in the observation that an obstructive lesion proximal to the first septal perforating branch of the left anterior descending (LAD) coronary artery was associated with a 5-year survival rate of 90%, as opposed to 98% in patients with more distal lesions. Scoring systems to capture more detailed assessment of the full extent and severity of epicardial CAD, such as the Synergy between percutaneous coronary intervention (PCI) with Taxus and Cardiac Surgery score, have been developed and validated [74] (Fig. 2.4.3).

Steps	Variable assessed	Description
Step 1	Dominance	The weight of individual coronary segments varies according to coronary artery dominance (right or left). Co-dominance does not exist as an option in the SYNTAX score.
Step 2	Coronary segment	The diseased coronary segment directly affects the score as each coronary segment is assigned a weight, depending on its location, ranging from 0.5 (i.e. posterolateral branch) to 6 (i.e. left main in case of left dominance).

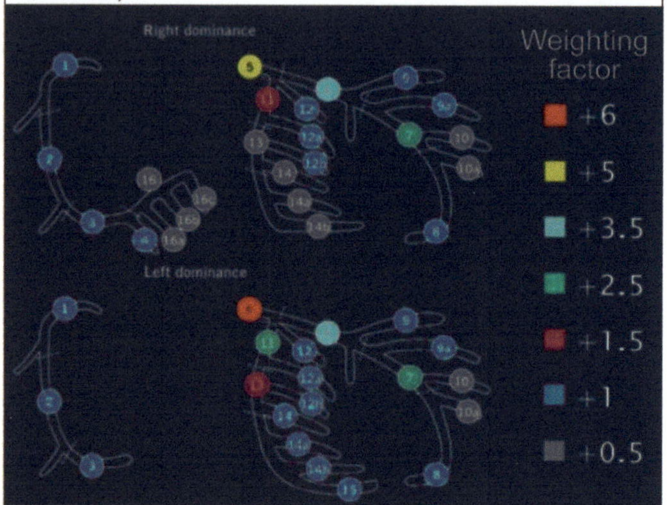

Steps	Variable assessed	Description
Step 3	Diameter stenosis	The score of each diseased coronary segment is multiplied by 2 in case of a stenosis 50–99% and by 5 in case of total occlusion. In case of total occlusion, additional points will be added as follows: - Age >3 months or unknown +1 - Blunt stump +1 - Bridging +1 - First segment visible distally +1 per non visible segment - Side branch at the occlusion +1 if <1.5 mm diameter +1 if both <1.5 and ≥1.5 mm diameter +0 if ≥1.5 mm diameter (i.e. bifurcation lesion)
Step 4	Trifurcation lesion	The presence of a trifurcation lesion adds additional points based on the number of diseased segments: - 1 segment +3 - 2 segments +4 - 3 segments +5 - 4 segments +6
Step 5	Bifurcation lesion	The presence of a bifurcation lesion adds additional points based on the type of bifurcation according to the Medina classification:[29] - Medina 1,0,0 or 0,1,0 or 1,1,0: add 1 additional point - Medina 1,1,1 or 0,0,1 or 1,0,1 or 0,1,1: add 2 additional point Additionally, the presence of a bifurcation angle <70° adds 1 additional point.
Step 6	Aorto-ostial lesion	The presence of aorto-ostial lesion segments adds 1 additional point.
Step 7	Severe tortuosity	The presence of severe tortuosity proximal of the diseased segment adds 2 additional points.
Step 8	Lesion length	Lesion length >20 mm adds 1 additional point.
Step 9	Calcification	The presence of heavy calcification adds 2 additional points.
Step 10	Thrombus	The presence of thrombus adds 1 additional point.
Step 11	Diffuse disease/small vessels	The presence of diffusely diseased and narrowed segments distal to the lesion (i.e. when at least 75% of the length of the segment distal to the lesion has a vessel diameter of <2 mm) adds 1 point per segment number.

Figure 2.4.3 Guide to calculate the Synergy between percutaneous coronary intervention (PCI) with Taxus and Cardiac Surgery score.

High-grade lesions of the left main coronary artery or its equivalent, as defined by severe proximal LAD and proximal left circumflex CAD, are particularly life-threatening. Mortality in medically treated patients has been reported to be 29% at 18 months and 43% at 5 years. Survival is better in patients with 50%–70% stenosis (1- and 3-year survival rates of 91% and 66%, respectively) than in patients with a left main coronary artery stenosis greater than 70% (1- and 3-year survival rates of 72% and 41%, respectively), especially in those with preserved LV systolic function.

Limitations of Angiography

Coronary angiography provides information principally about the degree of luminal stenosis of the coronary arteries. The pathophysiologic significance of coronary stenoses lies both in their impact on resting and exercise-induced blood flow and in their potential for plaque rupture with superimposed thrombotic occlusion. It is generally accepted that a stenosis involving more than 60% of the luminal diameter is hemodynamically significant in that it may be responsible for a reduction in exercise-induced myocardial blood flow that causes ischemia. The immediate functional significance of obstruction of intermediate severity (≈50% diameter stenosis) is less well established. Coronary angiography is not a reliable indicator of the functional significance of stenosis. Furthermore, coronary angiographic determinants of the severity of stenosis are based on a decrease in the caliber of the lumen at the site of the lesion relative to adjacent reference segments, which are considered, often erroneously, to be relatively free of disease. This approach may lead to significant underestimation of the severity and extent of atherosclerosis.

The most serious limitation to the routine use of coronary angiography for prognosis in patients with SIHD is its inability to identify which coronary lesions can be considered to be at high risk, or vulnerable, for future events, such as MI or sudden death. Although it is widely accepted that MI is the result of thrombotic occlusion at the site of plaque rupture or erosion, it is clear that it is not necessarily the plaque causing the most severe stenosis that subsequently ruptures. Lesions causing mild obstruction can rupture, thrombose, and occlude, thereby leading to MI and sudden death. In fact, two thirds to three quarters of all acute MIs emanate from antecedent coronary stenoses that involve less than 50% of the luminal diameter. Approaches to quantifying the extent of CAD, inclusive of nonobstructive lesions, appear to offer additional prognostic information [74].

In summary, angiographic documentation of the extent of CAD provides useful information for assessment of the patient's risk for death and future ischemic events and is an indispensable step in the selection of patients for coronary revascularization, particularly if the interaction between the anatomic extent of disease, LV function, and severity of ischemia is taken into account. However, angiography is not helpful in predicting sites of subsequent plaque rupture or erosion that can precipitate MI or sudden cardiac death. Additional tools that improve the imaging of coronary atheroma, such as IVUS, or functional assessment of a stenosis, such as determination of the fractional flow reserve can be helpful in deciding on the flow-limiting significance of a specific lesion and the need for coronary revascularization. Characterization of the atheroma and also coronary flow reserve with CT or CMR remains under evaluation but is not yet a routine clinical tool.

MANAGEMENT

Comprehensive management of SCAD has five aspects (Fig. 2.4.4): (1) identification and treatment of associated diseases that can precipitate or worsen angina and ischemia; (2) reduction of coronary risk factors; (3) application of pharmacologic and nonpharmacologic interventions for secondary prevention, with particular attention directed to adjustments in lifestyle; (4) pharmacologic management of angina; and (5) revascularization by catheter-based PCI or by coronary artery bypass grafting (CABG), when indicated. All five of these approaches must be considered, often simultaneously, in each patient. Of the medical therapies, three (aspirin, angiotensin-converting enzyme [ACE] inhibition, and effective lipid lowering) have been shown to reduce mortality and morbidity in patients with SIHD and preserved LV function. Other therapies such as nitrates, beta blockers, calcium antagonists, and ranolazine have been shown to improve symptomatology and exercise performance but their effect, if any, on survival in patients with SIHD has not been demonstrated.

Reduction of Coronary Disease Risk Factors

Hypertension

Epidemiologic links between increased blood pressure and CAD severity and mortality are well established. For individuals from 40 to 70 years of age, risk for IHD doubles

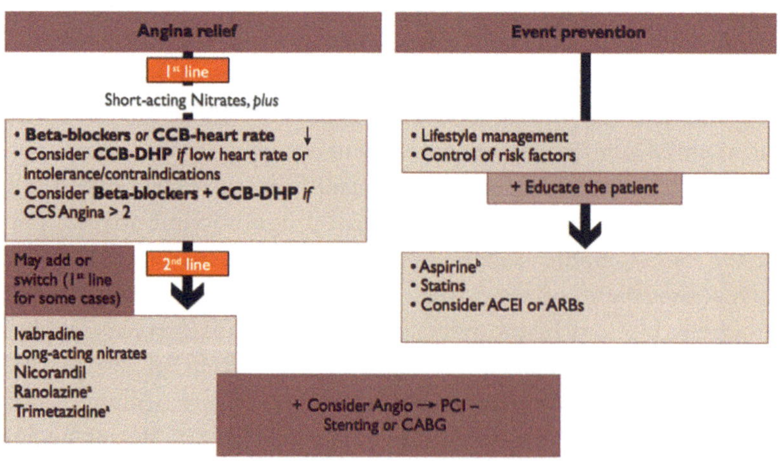

Figure 2.4.4 Management of stable angina. *(From 2013 ESC guidelines on the management of stable coronary artery disease.)*

for each 20 mm Hg increment in systolic blood pressure across the entire range of 115–185 mm Hg [75]. Hypertension predisposes to vascular injury, accelerates the development of atherosclerosis, increases myocardial O_2 demand, intensifies ischemia in patients with preexisting obstructive CAD, and predisposes to atrial fibrillation. Although the relationship between hypertension and CAD is linear, LV hypertrophy is a stronger predictor of MI and CAD death than is the actual degree of increase in blood pressure [75].

Physical Activity

Regular physical activity is associated with a decrease in CV morbidity and mortality in patients with established CAD and physical activity should be incorporated into daily activities. Aerobic exercise should be offered to patients with known CAD, usually as part of a structured cardiac rehabilitation program, with the need for an evaluation of both exercise capacity and exercise-associated risk. Patients with stable angina pectoris should undergo moderate-to-vigorous intensity aerobic exercise training ≥ 3 times a week and for 30 min per session. Sedentary patients should be strongly encouraged to start light-intensity exercise programs after adequate exercise-related risk stratification. In patients with significant CAD who are not candidates for revascularization, exercise training may offer an alternative means of symptom alleviation and improved prognosis.

Cigarette Smoking

Smoking remains one of the most powerful risk factors for the development of CAD in all age groups. In patients with angiographically documented CAD, cigarette smokers have a higher 5-year risk for sudden death, MI, and all-cause mortality than do those who have stopped smoking. Cigarette smoking may be responsible for aggravating angina other than through the progression of atherosclerosis. It may increase myocardial O_2 demand and reduce coronary blood flow by means of an alpha-adrenergically mediated increase in coronary artery tone and thereby cause acute ischemia. Moreover, passive exposure to cigarette smoke has adverse cardiovascular effects that are almost as large as those of active smoking. Smoking cessation lessens the risk for adverse coronary events in patients with established CAD and is one of the most effective and cost-saving approaches to prevention of disease progression in native vessels and bypass grafts [76,77].

Management of Dyslipidemia

Clinical trials in patients with established atherosclerotic vascular disease have demonstrated a significant reduction in subsequent cardiovascular events in patients with a wide range of serum cholesterol and LDL cholesterol levels treated with statins. In the aggregate, angiographic trials of cholesterol lowering in patients with chronic CAD have shown that its effects on coronary obstruction are modest in comparison to the substantive reduction in cardiovascular events, thus suggesting that regression of atherosclerosis is not the primary mechanism of benefit. Nonetheless, in trials using IVUS [78] intensive lipid-lowering therapy with preprotein convertase serine kexin (PCSK)-9 inhibitors has led to regression of coronary atherosclerotic burden in patients with angiographic CAD.

Patients with established CAD are regarded as being at very high risk for cardiovascular events and statin treatment should be considered, irrespective of LDL-C levels. According to the 2016 European Society of Cardiology Guidelines [86] regarding the management of dyslipidemias, the goals of treatment are LDL-C below 1.8 mmol/L (70 mg/dL) or a reduction of at least 50% if the baseline is between 1.8 and 3.5 mmol/L (70 and 135 mg/dL). In the majority of patients this is achievable through high-intensity statin monotherapy or combination therapy with high intensity statin plus ezetimibe.

Management of Diabetes Mellitus

Patients with diabetes mellitus are at significantly higher risk for atherosclerotic vascular disease. Diabetes mellitus is a strong risk factor for CV complications, increases the risk of progression of coronary disease and should be managed carefully, with good control of glycated hemoglobin (HbA1c) to 7.0% (53 mmol/mol) generally and 6.5%−6.9% (48−52 mmol/mol) on an individual basis. Glucose control should be based on individual considerations, depending on the patient's characteristics including age, presence of complications and diabetes duration. Although a favorable impact of control of glycemia on microvascular complications of diabetes has been established, the effect on macrovascular complications (including CAD) is unclear. Weight management, physical activity, blood pressure control, and lipid management are recommended for all patients with SCAD and diabetes [76].

Influenza Vaccination

An annual influenza vaccination is recommended for patients with CAD, especially the elderly [78,79].

Pharmacotherapy for Secondary Prevention

Aspirin

A meta-analysis of 140,000 patients in 300 studies has confirmed the prophylactic benefit of aspirin in men and women with angina pectoris, with previous MI or stroke, and after CABG. In a Swedish trial of men and women with chronic stable angina, 75 mg of aspirin in conjunction with the beta-blocking agent sotalol conferred a 34% reduction in acute MI and sudden death. In a smaller study confined to men with chronic stable angina but without a history of MI, 325 mg of aspirin on alternate days reduced the risk for MI by 87% during 5 years of follow-up. Therefore administration of aspirin daily is advisable in patients with SIHD and no contraindications to this drug. Dosing at 75−162 mg daily appears to have comparable effects on secondary prevention as dosing at 160–325 mg daily and seems to be associated with lower bleeding risk. Thus aspirin, 75–162 mg daily, is preferred for secondary prevention in the absence of recent intracoronary stenting [76].

P_2Y_{12} Inhibitors

P_2Y_{12} inhibitors (clopidogrel, prasugrel, ticagrelor) act as antagonists of the platelet adenosine diphosphate receptor P_2Y_{12}, thereby inhibiting platelet aggregation. The major study supporting the use of thienopyridine in stable coronary patients is the Clopidogrel versus Aspirin in Patients at Risk of Ischaemic Events trial, which showed an overall benefit of clopidogrel as compared with aspirin (with also a favorable safety profile) in preventing CV events in three categories of patients with previous MI, previous stroke or PVD [80]. The clopidogrel benefit was driven by the PVD subgroup and the dose of aspirin with which it was compared (325 mg/day) may not be the safest dose. Clopidogrel should thus be proposed as a second-line treatment, especially for aspirin-intolerant patients. Prasugrel and ticagrelor are new P_2Y_{12} antagonists that achieve greater platelet inhibition, compared with clopidogrel [81,82]. Prasugrel and ticagrelor are both associated with a significant reduction of CV outcomes as compared with clopidogrel in acute coronary syndrome (ACS) patients [83,84], but no clinical studies have evaluated the benefit of these drugs in SCAD patients.

Angiotensin-Converting Enzyme Inhibitors and Angiotensin Receptor Blockers

Although inhibitors of the renin–angiotensin–aldosterone system are not indicated for the treatment of angina, these drugs appear to have important benefits in reducing the risk for future ischemic events in some patients with cardiovascular disease [85]. Potentially beneficial effects of ACE inhibitors include reductions in LV hypertrophy, vascular hypertrophy, progression of atherosclerosis, plaque rupture, and thrombosis, in addition to a potentially favorable influence on myocardial O2 supply-and-demand relationships, cardiac hemodynamics, sympathetic activity, and coronary endothelial vaso-motor function. Two trials have provided strong evidence supporting the therapeutic benefit of ACE inhibitors in patients with normal LV function and absence of heart failure. In the Heart Outcomes Protection Evaluation study, ramipril significantly decreased the risk for major vascular events by a relative 22% in 9297 patients with atherosclerotic vascular disease or diabetes mellitus. European Trial on Reduction of Cardiac Events with Perindopril in Stable CAD similarly showed a 20% relative reduction in the risk for cardiovascular death, MI, or cardiac arrest in 13,655 patients with stable CAD in the absence of heart failure. ACE inhibitors are recommended for all patients with CAD and LV dysfunction and for those with hypertension, diabetes, or chronic kidney disease [76]. ACE inhibitors may be considered for optional use in all other patients with SIHD, a normal LV ejection fraction, and well-controlled cardiovascular risk factors in whom revascularization has been performed [76].

Lipid-Lowering Agents

Patients with established CAD are regarded as being at very high risk for cardiovascular events and statin treatment should be considered, irrespective of LDL-C levels. According to the 2016 European Society of Cardiology Guidelines [86] regarding the management of dyslipidemias, the goals of treatment are LDL-C below 1.8 mmol/L (70 mg/dL) or a reduction of at least 50% if the baseline is between 1.8 and 3.5 mmol/L (70 and 135 mg/dL).

Pharmacologic Management of Angina

Beta Adrenoceptor–Blocking Agents

B-blockers act directly on the heart to reduce heart rate, contractility, atrio-ventricular conduction and ectopic activity (Fig. 2.4.5). Additionally, they may increase perfusion of ischemic areas by prolonging the diastole and increasing vascular resistance in nonischemic areas. In post-MI patients, b-blockers achieved a 30% risk reduction for CV death and MI [87]. Thus b-blockers may also be protective in patients with SCAD, but without supportive evidence from placebo-controlled clinical trials. However, a recent retrospective analysis of the Reduction of Atherothrombosis for Continued Health (REACH) registry suggested that, in patients with either CAD risk factors only, known prior MI, or known CAD without MI, the use of b-blockers was not associated with a lower risk of cardiovascular events [88]. Although propensity score matching was used for the analysis, the

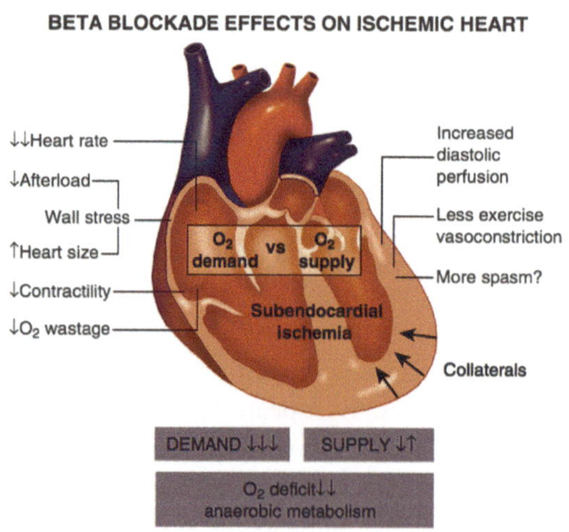

Figure 2.4.5 Effects of beta blockade on an ischemic heart. Beta blockade has a beneficial effect on ischemic myocardium unless (1) the preload rises substantially, as in left-sided heart failure, or (2) vasospastic angina is present, in which case spasm may be promoted in some patients. Note the suggestion that beta blockade diminishes exercise-induced vasoconstriction. *(Modified from Opie LH. Drugs for the heart. 4th ed. Philadelphia: WB Saunders; 1995. p. 6.)*

demonstration lacks the strength of a randomized evaluation. Among other limitations, most of the b-blocker trials in post-MI patients were performed before the implementation of other secondary prevention therapies, such as statins and ACE inhibitors, leaving uncertainty regarding their efficacy when added to modern therapeutic strategies. b-Blockers are clearly effective in controlling exercise-induced angina, improving exercise capacity and limiting both symptomatic as well as asymptomatic ischemic episodes. Regarding angina control, b-blockers and calcium channel blockers (CCBs) are similar [89–92]. B-blockers can be combined with dihydropyridines (DHPs) to control angina [93–97]. Combination therapy of b-blockers with verapamil and diltiazem should be avoided because of the risk of bradycardia or atrioventricular block. The most widely used b-blockers in Europe are those with pre-dominant b1-blockade, such as metoprolol, bisoprolol, atenolol, or nevibolol. Carvedilol, a nonselective b-a_1 blocker, is also often used. All of these reduce cardiac events in patients with heart failure [99–102]. In summary, there is evidence for prognostic benefits from the use of b-blockers in post-MI patients, or in heart failure. Extrapolation from these data suggests that b-blockers may be the first-line antianginal therapy in stable CAD patients without contraindications. Nevibolol and bisoprolol are partly secreted by the kidney, whereas carvedilol and metoprolol are metabolized by the liver, hence being safer in patients with renal compromise.

Short-Acting Nitrates for Acute Effort Angina

Sublingual nitroglycerin is the standard initial therapy for effort angina. When angina starts, the patient should rest sitting (standing promotes syncope, lying down enhances venous return and heart work) and take sublingual nitroglycerin (0.3–0.6 mg) every 5 min until the pain goes or a maximum of 1.2 mg has been taken within 15 min. Nitroglycerin spray acts more rapidly. Nitroglycerin can be used prophylactically when angina can be expected, such as activity after a meal, emotional stress, sexual activity and in colder weather. Isosorbide dinitrate (5 mg sublingually) helps to abort anginal attacks for about 1 h. Because the dinitrate requires hepatic conversion to the mononitrate, the onset of antianginal action (within 3–4 min) is slower than with nitroglycerin. After oral ingestion, hemodynamic and antianginal effects persist for several hours, conferring longer protection against angina than sublingual nitroglycerin.

Long-Acting Nitrates for Angina Prophylaxis

Long-acting nitrates are not continuously effective if regularly taken over a prolonged period without a nitrate-free or nitrate-low interval of about 8–10 h (tolerance). Worsening of endothelial dysfunction is a potential complication of long-acting nitrates, hence the common practice of the routine use of long-acting nitrates as first line therapy for patients with effort angina needs reevaluation. *Isosorbide dinitrate* (oral preparation) is frequently given for the prophylaxis of angina. In a crucial placebo-controlled study, exercise duration improved significantly for 6–8 h after single oral doses of

15–120 mg isosorbide dinitrate, but for only 2 h when the same doses were given repetitively four times daily, despite much higher plasma isosorbide dinitrate concentrations during sustained than during acute therapy. With the extended-release formulation of isosorbide dinitrate, eccentric twice-daily dosing, with 40 mg in the morning, repeated 7 h later, was not superior to placebo in a large multicenter study. Thus, prolonged therapy with isosorbide dinitrate is not evidence-based.

Mononitrates have similar dosage and effects to those of isosorbide dinitrate. Nitrate tolerance—likewise a potential problem—can be prevented by changes in dosing and timing of administration, as well as by using slow-release preparations. Thus only twice-daily rapid-release preparations or very high doses of slow-release mononitrate—also twice daily—give sustained antianginal benefit.

Transdermal nitroglycerin patches fail to cover 24 h during prolonged use. A discontinuous administration at 12 h intervals allows on and off effects to start within minutes and last 3–5 h. There are no efficacy data for second or third doses during chronic administration.

Calcium Channel Blockers

Calcium antagonists (i.e., CCBs) act chiefly by vasodilation and reduction of the peripheral vascular resistance. CCBs are a heterogeneous group of drugs that can chemically be classified into the DHPs and the non-DHPs, their common pharmacological property being selective inhibition of L-channel opening in vascular smooth muscle and in the myocardium. Distinctions between the DHPs and non-DHPs are reflected in different binding sites on the calcium channel pores and in the greater vascular selectivity of the DHP agents (amlodipine, nifedipine, felodipine). The non-DHPs, by virtue of nodal inhibition, tend to reduce the heart rate (heart rate-lowering agents, verapamil and diltiazem) and explain the antianginal properties.

Nondihydropyridine Verapamil

Among CCBs, verapamil has a large range of approved indications, including all varieties of angina (effort, vasospastic, unstable), supraventricular tachycardias and hypertension. Indirect evidence suggests good safety but with risks of heart block, bradycardia and heart failure. Compared with metoprolol, the antianginal activity was similar [98]. Compared with atenonol in hypertension with CAD, verapamil gave less new diabetes, fewer anginal attacks [103], and less psychological depression [104]. b-Blockade combined with verapamil is not advised (due to risk of heart block); instead, use DHP-b-blockade.

Diltiazem

Diltiazem with its low side-effect profile, has advantages compared with verapamil in the treatment of effort angina [105]. Like verapamil, it acts by peripheral vasodilation, relief of exercise-induced coronary constriction, a modest negative inotropic effect and sinus node inhibition. There are no outcome studies comparing diltiazem and verapamil. As with verapamil, combination with b-blockade, as well as the use in patients with CAD and left ventricular dysfunction, is not advised.

Dihydropyridines

Long-acting nifedipine

This agent is a powerful arterial vasodilator with few serious side-effects. Long-acting nifedipine is especially well-tested in hypertensive anginal patients when added to b-blockade [106]. In ACTION (A Coronary disease Trial Investigating Outcome with Nifedipine), a large placebo-controlled trial long-acting nifedipine in SCAD proved to be safe and reduced the need for coronary angiography and cardiovascular interventions [106]. Contraindications to nifedipine are few (severe aortic stenosis, obstructive cardiomyopathy, or heart failure) and careful combination with b-blockade is usually feasible and desirable. Vasodilatory side effects include headache and ankle edema.

Amlodipine

The very long half-life of amlodipine and its good tolerability make it an effective once-a-day antianginal and antihypertensive agent, setting it apart from agents that are taken either twice or three times daily. Side-effects are few; mainly ankle edema. In patients with CAD and normal blood pressure, amlodipine reduced CV events in a 24-month trial [107]. Exercise-induced ischemia is more effectively reduced by amlodipine than by the b-blocker atenolol while the combination is even better. However, the CCB—b-blocker combination is often underused, even in some studies reporting "optimally treated" stable effort angina.

Others

Felodipine, lacidipine, and lercanidipine share the standard properties of other long-acting DHPs.

Other Anti-ischemic Drugs

Ivabradine

Ivabradine is a specific and selective inhibitor of the I_f ion channel, the principal determinant of the sinoatrial node pacemaker current. Ivabradine reduces the spontaneous ring rate of sinoatrial pacemaker cells and thus slows the heart rate through a mechanism that is not associated with negative inotropic effects. At this time, ivabradine has been approved in Europe for the treatment of chronic stable angina pectoris in adults in sinus rhythm who are unable to tolerate or have contraindications to beta-blocking agents or in patients inadequately controlled with an optimal dose of a beta blocker and whose heart rate is faster than 70 beats/min. Ivabradine has also been approved in Europe for the treatment of chronic heart failure in combination with standard therapy in patients with a heart rate of 75 beats/min or greater.

Ivabradine reduces the peak heart rate during exercise, increases the time to limiting angina when compared with placebo, and is equivalent to atenolol with respect to exercise performance and time to ischemia (ST-segment depression) in patients with stable angina undergoing exercise treadmill testing. Ivabradine has also been shown to reduce the heart rate without any effect on ventilator parameters in patients with obstructive pulmonary disease [108] and to be tolerated in patients with CAD and LV dysfunction [109]. In a randomized trial of 10,917 patients enrolled in the Morbidity—Mortality Evaluation of the If Inhibitor Ivabradine in Patients With Coronary Artery Disease and Left Ventricular Dysfunction (BEAUTIFUL) trial, ivabradine did not reduce the primary endpoint of cardiovascular death, hospitalization for MI, or hospitalization for heart failure. Fewer hospitalizations for MI were observed in the subgroup of patients with a baseline heart rate greater than 70 beats/min who were randomly assigned to ivabradine versus placebo and in patients with a history of limiting angina [110].

Ranolazine

Ranolazine is a piperazine derivative for use in patients with chronic stable angina [111]. Ranolazine is unique among currently approved antianginals in that its anti-ischemic effects are achieved without a clinically meaningful change in heart rate or blood pressure [112]. When studied at high concentrations in in vitro experiments, ranolazine was shown to shift myocardial substrate uptake from fatty acid to glucose and thus was considered to be a potential myocardial metabolic modulator. However, subsequent studies at concentrations of ranolazine consistent with doses tested in clinical trials have indicated that ranolazine exerts favorable effects on ischemia through a reduction in calcium overload in ischemic myocytes via inhibition of the late inward sodium current (I_{Na}) [113,114]. In animal models of ischemia and reperfusion, ranolazine preserves tissue levels of adenosine triphosphate (ATP), improves myocardial contractile function, and reduces the extent of irreversible myocardial injury measured by biomarkers of necrosis and by electron microscopy. Ranolazine has also been suggested to reduce periprocedural myocardial injury in patients undergoing elective PCI [115].

A sustained-release formulation of ranolazine has been studied in three randomized placebo-controlled clinical trials and improved exercise performance and increased the time to ischemia during exercise treadmill testing when used as monotherapy or in combination with the most frequently used doses of atenolol, amlodipine, or diltiazem. Ranolazine also decreases angina frequency and nitroglycerin use when used in conjunction with a beta-blocking agent or calcium channel—blocking agent [111].

When studied in a randomized, blinded, placebo-controlled trial of 6560 patients with non—ST-segment elevation ACS, ranolazine, administered for an average of approximately 1 year, did not add to standard therapy in secondary prevention of major cardiovascular events. However, ranolazine reduced the incidence of recurrent ischemia, in particular, worsening angina, in a significantly more diverse population with established CAD than studied previously with ranolazine. Consistent with previous studies, the reduction in angina and improvement in exercise performance were evident only in patients with a history of chronic angina [116] and was no less in women than in men [117]. Because of its proposed mechanism of action on cardiac myocytes rather than modulation of the heart rate or blood pressure, ranolazine has been studied in patients with angina and ischemia without epicardial CAD. In a pilot study of 20 women with angina and no obstructive CAD but with impaired coronary flow reserve on CMR imaging, ranolazine reduced symptoms with evidence of an improved myocardial perfusion reserve index [118].

Ranolazine should be started at 500 mg twice daily and may be increased to a maximum of 1000 mg twice daily in patients with persistent angina. The most commonly reported adverse effects in clinical studies are nausea, generalized weakness, and constipation. Dizziness has also been reported, as has a small dose-related increase in the corrected QT (QTc) interval, an average of 2–5 ms in the dosage range of 500–1000 mg twice daily.

Nicorandil

Nicorandil is a nicotinamide ester that dilates peripheral and coronary resistance vessels via action on ATP-sensitive potassium channels and possesses a nitrate moiety that promotes systemic venous and coronary vasodilation [119]. As a result of these dual actions, nicorandil reduces preload and afterload and results in an increase in coronary blood flow. In addition to these effects, nicorandil may have cardioprotective actions mediated through the activation of potassium channels. Nicorandil has been associated with ulcerations of the gastrointestinal tract.

Nicorandil has antianginal efficacy similar to that of beta-blocking agents, nitrates, and calcium channel–blocking agents. In a randomized clinical trial (N = 5126), nicorandil reduced the risk for cardiac death, MI, or hospital admission for angina (hazard ratio [HR], 0.83; $P = .014$) in comparison to placebo when added to standard antianginal therapy.

Trimetazidine

Trimetazidine is an anti-ischemic metabolic modulator [120], with similar anti-anginal efficacy to propranolol in doses of 20 mg thrice daily. The heart rate and rate × pressure product at rest and at peak exercise remained unchanged in the trimetazidine group, thus showing a nonmechanical anti-ischemic action [121,122].

Trimetazidine (35 mg twice daily) added to beta-blockade (atenolol) improved effort-induced myocardial ischemia, as reviewed by the European Medicines Agency (EMA) in June 2012 [123], and remains contraindicated in Parkinson's disease and motion disorders (such as tremor (shaking), muscle rigidity and walking disorders and restless leg syndrome). In diabetic persons, trimetazidine improved HbA1c and glycaemia, while increasing forearm glucose uptake [124]. Trimetazidine has not been evaluated in large outcome studies in SCAD patients.

Revascularization

Revascularization Approaches in Stable Coronary Artery Disease

It is widely accepted that the potential benefits of revascularization are proportional to the patient's underlying risk, which makes it essential to quantify the patient's prognosis as accurately as possible (Table 2.4.3). In addition to the patient's risk for major cardiovascular events, sociodemographic factors such as age, physical capacity, ability to adhere to prescribed treatments and lifestyle interventions, overall quality of life, other medical conditions, and patient preferences should be considered. Each of these aspects should be integrated in considering how to best achieve these fundamental two goals of therapy for patients with SCAD. Revascularization approaches are an integral component of an overall management strategy to improve outcomes and are used when needed in addition to optimal medical therapy (OMT). The success of catheter-based or surgical treatment is predicated on the overall success of guideline-directed secondary prevention and lifestyle intervention as a platform for the management of all patients with SCAD. Decisions regarding the best mode of revascularization (catheter based or surgical) should follow a thoughtful assessment of whether and when revascularization is necessary to achieve these goals of therapy and are best made by a multidisciplinary heart team that includes a noninterventional cardiologist, an interventional cardiologist, and a cardiac surgeon. Patients are also critical participants in decision making in terms of their preferences. [125].

Patient Selection for Revascularization

Each of the following considerations may be used to guide decisions regarding the indications for (as well as the approach to) revascularization: (1) the presence and severity of symptoms, (2) physiologic significance of the coronary lesions and other anatomic considerations, (3) extent of myocardial ischemia and the presence of LV dysfunction, and (4) other medical conditions that influence the risks associated with percutaneous or surgical revascularization.

Presence and Severity of Symptoms

A goal of therapy is complete elimination of angina and resumption of full physical function to the extent possible [126]. Mechanical revascularization (catheter based or surgical) should be considered if ischemic symptoms persist after intensification of medical therapy, including stringent risk factor modification, or if unacceptable side effects or the patient's therapeutic preferences limit antianginal therapy.

Significance of Coronary Lesions (and Other Anatomic Considerations)

Seventy percent or greater stenosis of an epicardial coronary artery is considered to be anatomically significant ($\geq$50 for left main coronary stenosis). Multiple additional techniques, such as IVUS and other imaging have been shown to provide enhanced

Table 2.4.3 Risk stratification based on non-invasive testing

High Risk (>3% Annual risk for death or myocardial infarction)

1. Severe resting left ventricular dysfunction (LVEF <35%) not readily explained by noncoronary causes
2. Resting perfusion abnormalities involving >10% of the myocardium without previous known MI
3. High-risk stress findings on the ECG, including
   a. >2-mm ST-segment depression at low workload or persisting into recovery
   b. Exercise-induced ST-segment elevation
   c. Exercise-induced ventricular tachycardia/ventricular fibrillation (VT/VF)
4. Severe stress-induced left ventricle (LV) dysfunction (peak exercise LVEF <45% or drop in LVEF with stress ≥10%)
5. Stress-induced perfusion abnormalities encumbering ≥10% of the myocardium or stress segmental scores indicating multiple vascular territories with abnormalities
6. Stress-induced LV dilation
7. Inducible wall motion abnormality (involving >2 segments or two coronary beds)
8. Wall motion abnormality developing at a low dose of dobutamine (≤10 mg/kg/min) or at a low heart rate (<120 beats/min)
9. Coronary artery calcium (CAC) score >400 Agatston units
10. Multivessel obstructive CAD (>70% stenosis) or left main stenosis (≥50% stenosis) on coronary computed tomography angiography (CCTA)

Intermediate risk (1%–3% annual risk for death or myocardial infarction)

1. Mild to moderate resting LV dysfunction (LVEF of 35%–49%) not readily explained by noncoronary causes
2. Resting perfusion abnormalities involving 5%–9.9% of the myocardium in patients without a history or previous evidence of MI
3. ≥1-mm ST-segment depression occurring with exertional symptoms
4. Stress-induced perfusion abnormalities encumbering 5%–9.9% of the myocardium or stress segmental scores (in multiple segments) indicating one vascular territory with abnormalities but without LV dilation
5. Small wall motion abnormality involving one to two segments and only one coronary bed
6. CAC score of 100–399 Agatston units
7. 1-vessel CAD with ≥70% stenosis or moderate CAD stenosis (50%–69% stenosis) in ≥2 arteries on CCTA

Low Risk (<1% Annual Risk for Death or Myocardial Infarction)

1. Low-risk treadmill score (score ≥5) or no new ST-segment changes or exercise-induced chest pain symptoms when achieving maximal levels of exercise
2. Normal or small myocardial perfusion defect at rest or with stress encumbering ≥5% of the myocardium[a]
3. Normal stress or no change in limited resting wall motion abnormalities during stress
4. CAC score <100 Agatston units
5. No coronary stenosis >50% on CCTA

[a]Although the published data are limited; patients with these findings will probably not be at low risk in the presence of either a high-risk treadmill score or severe resting LV dysfunction (LVEF < 35%).

assessment of the anatomic and functional significance of specific coronary lesions and may complement stress testing to aid in decision-making judgments regarding the potential benefits of revascularization.

Other anatomic features, in addition to lesion severity, also influence the likelihood of success and the approach to revascularization for a given patient [127]. Such features include vessel size, extent of calcification, tortuosity, and relationships to side branches.

Extent of Ischemia and Presence of Left Ventricular Dysfunction

The four major determinants of risk in patients with CAD are the extent of ischemia, the number of vessels diseased, LV function, and the electrical substrate. The extent of ischemia on noninvasive testing is an important predictor of subsequent adverse outcomes and identifies patients in whom revascularization may provide clinical benefit over that of medical therapy beyond the relief of symptoms. The major effect of coronary revascularization is on ischemia, and the magnitude of the benefit versus that of medical therapy is enhanced with LV dysfunction, particularly in the presence of reversibly ischemic jeopardized myocardium. Moreover, the greatest survival benefits of CABG, as well as symptomatic and functional improvements, are evident in patients with impaired LV function (generally defined as an ejection fraction <0.40).

Risks Associated With the Procedure

Patients with SCAD more often than not have other medical conditions, such as renal dysfunction, peripheral atherosclerosis, or pulmonary disease, that may influence the patient's suitability for surgical or percutaneous revascularization. For example, in a patient with peptic ulcer disease and a history of gastrointestinal bleeding, the potential need for long-term dual antiplatelet therapy after a procedure should be considered. Moreover, a patient with three-vessel CAD and impaired LV function who might derive a more durable survival benefit from CABG may be too high risk clinically to undergo surgery and might be a better candidate for multivessel PCI.

In addition, some general principles regarding the choice of treatment in patients with SCAD should be considered:

1. For most patients with chronic angina, revascularization should not constitute the initial management strategy before evidence-based medical therapy (pharmacologic antianginal therapy, disease-modifying treatments, and therapeutic lifestyle intervention) is initiated and optimized.
2. When improvement in survival is not a relevant consideration, the severity of angina or impairment in health status should play a significant role in determining whether revascularization is appropriate to enhance quality of life (i.e., limiting angina while undergoing OMT is a more compelling indication than episodic, exertional angina during minimal medical therapy).

3. The patient's treatment preferences and sociodemographic/clinical circumstances should always be a consideration in guiding which treatment strategy should be used.
4. In certain clinical circumstances it may be difficult to reliably ascertain whether anginal symptoms or anginal equivalents such as exertional dyspnea or fatigue are a direct manifestation of underlying CAD, especially in patients with significant obesity, those who are sedentary, or those who may have coexisting chronic obstructive pulmonary disease. In such settings, symptoms that are either atypical for or nondiagnostic of obstructive CAD may not necessarily improve with revascularization, even when such symptoms coexist with physiologically significant CAD.
5. The decision to proceed with myocardial revascularization in a patient with SCAD should entail a thoughtful, transparent discussion of all potential treatment options, with full disclosure of the anticipated benefits and potential risks associated with PCI or CABG relative to guideline-directed medical therapy. In an elective setting in which urgent/emergency PCI is not being contemplated to reduce the possibility of death or MI, use of a "heart team", as cited earlier, is both prudent and clinically appropriate. Although it is often very common to undertake ad hoc PCI once the patient's coronary anatomy is defined in the catheterization laboratory, it is frequently difficult to have the type of discussion that would involve a complete review of the potential risks and benefits of all treatment options by the "heart team" in this setting. It has been suggested that a "time out", or hitting the "therapeutic pause button", might facilitate a more thorough understanding of what is best for a particular patient, particularly one with extensive multi-vessel CAD. In summary, treatment decisions must be individualized according to the specific clinical features and personal preferences of a given patient (often in collaboration with family members and the patient's referring physician), along with informed discussion about the potential risks and benefits of all three therapeutic options.

FUTURE PERSPECTIVES

Despite the fact that angina was initially described almost two and a half centuries ago, our understanding of the syndrome, its causes, and optimal management continue to evolve. The complex and probably heterogenous causes of myocardial ischemia require continued exploration. We are now confronted by substantial data challenging the paradigm that coronary heart disease requires critical epicardial coronary atherosclerosis or other structural heart disease that results in dramatically increased myocardial O2 demand. Preclinical, translational, and clinical epidemiologic data have all demonstrated abnormalities in coronary artery function that may result in myocardial ischemia in the absence of atherosclerotic obstruction. However, as yet, therapies proposed to address this important syndrome appear to be insufficient. Additional insight into the pathobiology of ischemia in these circumstances may lead to new therapeutic directions. Second, although it is clear that an initial approach of guideline-directed secondary preventive medical therapy and coronary revascularization when necessary is the best approach for most patients with SIHD, there are subsets of patients with indicators of particularly high risk for whom coronary revascularization would seem logical. However, clinical equipoise remains regarding whether such patients, including those with moderate or severe ischemia on noninvasive testing, should routinely undergo coronary revascularization in the absence of symptoms that are refractory to medical therapy. Third, definitive evidence to guide the management of patients with SCAD and other structural heart disease, in particular, concomitant LV dysfunction and mitral valvular disease, remains elusive. Complete revascularization, usually surgical, remains reasonable for patients with multivessel CAD, LV dysfunction, and viable myocardium, particularly when objective evidence of ischemia is present. However, recent studies have challenged this view. Despite our wealth of experience with SCAD, important unanswered questions remain.

REFERENCES

[1] Matthews MB, Julian DG. In: Julian DG, editor. Angina pectoris. 2nd ed. New York: Churchill Livingstone; 1985. p. 1–2.

[2] Cook DG, Shaper AG. Breathlessness, angina pectoris and coronary artery disease. Am J Cardiol 1989;63:921.

[3] Peled N, Abinader EG, Pillar G, et al. Nocturnal ischemic events in patients with obstructive sleep apnea syndrome and ischemic heart disease: effects of continuous positive air pressure treatment. J Am Coll Cardiol 1999;34:1744.

[4] Bogaty P, Kingma Jr JG, Robitaille NM, et al. Attenuation of myocardial ischemia with repeated exercise in subjects with chronic stable angina: relation to myocardial contractility, intensity of exercise and the adenosine triphosphate-sensitive potassium channel. J Am Coll Cardiol 1998;32:1665.

[5] Maseri A. Ischemic heart disease: a rational basis for clinical practice and clinical research. New York: Churchill Livingstone; 1995.

[6] Foreman RD. Mechanisms of cardiac pain. Annu Rev Physiol 1999;61:143.

[7] Rosen SD, Paulesu E, Frith CD, et al. Central nervous pathways mediating angina pectoris. Lancet 1994;344:147.

[8] Cooke RA, Anggiansah A, Chambers JB, et al. A prospective study of oesophageal function in patients with normal coronary angiograms and controls with angina. Gut 1998;42:323.

[9] Chauhan A, Mullins PA, Taylor G, et al. Cardioesophageal reflex: a mechanism for "linked angina" in patients with angiographically proven coronary artery disease. J Am Coll Cardiol 1996;27:1621.

[10] Borjesson M, Albertsson P, Dellborg M, et al. Esophageal dysfunction in syndrome X. Am J Cardiol 1998;82:1187.

[11] Thomas LR, Baden L, Zaleznik DF. Clinical problem-solving. Chest pain with a surprising course. N Engl J Med 1999;341:1134.

[12] Epstein SE, Gerber LH, Borer JS. Chest wall syndrome. A common cause of unexplained cardiac pain. JAMA 1979;241:2793.

[13] Wise CM, Semble EL, Dalton CB. Musculoskeletal chest wall syndromes in patients with noncardiac chest pain: a study of 100 patients. Arch Phys Med Rehabil 1992;73:147.

[14] 27th Bethesda conference: matching the intensity of risk factor management with the hazard for coronary disease events. September 14–15, 1995. J Am Coll Cardiol 1996;27:957.

[15] Winder AF, Jolleys JC, Day LB, et al. Corneal arcus, case finding and definition of individual clinical risk in heterozygous familial hypercholesterolaemia. Clin Genet 1998;54:497.

[16] Chambless LE, Fuchs FD, Linn S, et al. The association of corneal arcus with coronary heart disease and cardiovascular disease mortality in the Lipid Research Clinics Mortality Follow-up Study. Am J Public Health 1990;80:1200.
[17] Ribera M, Pinto X, Argimon JM, et al. Lipid metabolism and apolipoprotein E phenotypes in patients with xanthelasma. Am J Med 1995;99:485.
[18] Klein R, Klein BE, Moss SE, et al. Association of ocular disease and mortality in a diabetic population. Arch Ophthalmol 1999;117:1487.
[19] Gersh BJ, Rihal CS, Rooke TW, et al. Evaluation and management of patients with both peripheral vascular and coronary artery disease. J Am Coll Cardiol 1991;18:203.
[20] Eagle KA, Rihal CS, Foster ED, et al. Long-term survival in patients with coronary artery disease: importance of peripheral vascular disease. J Am Coll Cardiol 1994;23:1091.
[21] Ogren M, Hedblad B, Isacsson SO, et al. Non-invasively detected carotid stenosis and ischaemic heart disease in men with leg arteriosclerosis. Lancet 1993;342:1138.
[22] Gibbons RJ, Chatterjee K, Daley J, et al. Acc/aha/acp-asim guidelines for the management of patients with chronic stable angina: a report of the American college of cardiology/American heart association task force on practice guidelines. J Am Coll Cardiol 1999;33:1999.
[23] Cohn PF, Thompson P, Strauss W, et al. Diastolic heart sounds during static (handgrip) exercise in patients with chest pain. Circulation 1973;47:1217.
[24] Lamas GA, Mitchell GF, Flaker GC, et al. Clinical significance of mitral regurgitation after acute myocardial infarction. Survival and Ventricular Enlargement Investigators. Circulation 1997;96:827.
[25] Sangster JF, Oakley CM. Diastolic murmur of coronary artery stenosis. Br Heart J 1973;35:840.
[26] Vlachopoulos C, Xaplanteris P, Stefanadis C. Mental stress, arterial stiffness, central pressures, and cardiovascular risk. Hypertension 2010;56:e28.
[27] Gullette EC, Blumenthal JA, Babyak M, et al. Effects of mental stress on myocardial ischemia during daily life. JAMA 1997;277:1521.
[28] Muller JE, Mittleman A, Maclure M, et al. Triggering myocardial infarction by sexual activity. Low absolute risk and prevention by regular physical exertion. JAMA 1996;275:1405.
[29] Andrews TC, Fenton T, Toyosaki N, et al. Subsets of ambulatory myocardial ischemia based on heart rate activity. Circadian distribution and response to anti-ischemic medication. Circulation 1993;88:92.
[30] Hillis LD, Braunwald E. Coronary-artery spasm. N Engl J Med 1978;299:695.
[31] Opie LH. The heart: Physiology, from cell to circulation. 3rd ed. Philadelphia: Lippincott-Raven; 1998p 287.
[32] Maseri A, Crea F, Lanza GA. Coronary vasoconstriction: where do we stand in 1999? An important, multifaceted but elusive role (editorial). Cardiologia 1999;44:115.
[33] Epstein SE, Talbot TL. Dynamic coronary tone in precipitation, exacerbation and relief of angina pectoris. Am J Cardiol 1981;48:797.
[34] Baller D, Notohamiprodjo G, Gleichmann U, et al. Improvement in coronary flow reserve determined by positron emission tomography after 6 months of cholesterol-lowering therapy in patients with early stages of coronary atherosclerosis. Circulation 1999;99:2871.
[35] Benhorin J, Banai S, Moriel M, et al. Circadian variations in ischemic threshold and their relation to the occurrence of ischemic episodes. Circulation 1993;87:808.
[36] Gottdiener JS, Krantz DS, Howell RH, et al. Induction of silent myocardial ischemia with mental stress testing: relation to the triggers of ischemia during daily life activities and to ischemic functional severity. J Am Coll Cardiol 1994;24:1645.
[37] Marchant B, Donaldson G, Mridha K, et al. Mechanisms of cold intolerance in patients with angina. J Am Coll Cardiol 1994;23:630.
[38] Kearney MT, Charlesworth A, Cowley AJ, et al. William Heberden revisited: postprandial angina-interval between food and exercise and meal composition are important determinants of time to onset of ischemia and maximal exercise tolerance. J Am Coll Cardiol 1997;29:302.
[39] Baliga RR, Rosen SD, Camici PG, et al. Regional myocardial blood flow redistribution as a cause of postprandial angina pectoris. Circulation 1998;97:1144.
[40] Campeau L. Grading of angina pectoris (letter). Circulation 1976;54:522.

[41] Goldman L, Hashimoto B, Cook EF, et al. Comparative reproducibility and validity of systems for assessing cardiovascular functional class: advantages of a new specific activity scale. Circulation 1981; 64:1227.
[42] Califf RM, Mark DB, Harrell Jr FE, et al. Importance of clinical measures of ischemia in the prognosis of patients with documented coronary artery disease. J Am Coll Cardiol 1988;11:20.
[43] Di Angelantonio E, Sarwar N, Perry P, et al. Major lipids, apolipoproteins, and risk of vascular disease. JAMA 1993;302:2009.
[44] Veeranna V, Zalawadiya SK, Niraj A, et al. Homocysteine and reclassification of cardiovascular disease risk. J Am Coll Cardiol 2011;58:1025.
[45] Zhou YH, Tang JY, Wu MJ, et al. Effect of folic acid supplementation on cardiovascular outcomes: a systematic review and meta-analysis. PLoS ONE 2011;6:e25142.
[46] Ridker PM. High-sensitivity C-reactive protein, vascular imaging, and vulnerable plaque: more evidence to support trials of anti-inflammatory therapy for cardiovascular risk reduction. Circ Cardiovasc Imaging 2011;4:195.
[47] Buckley DI, Fu R, Freeman M, et al. C-reactive protein as a risk factor for coronary heart disease: a systematic review and meta-analyses for the U.S. Preventive Services Task Force. Ann Intern Med 2009;151:483.
[48] Morrow DA. Cardiovascular risk prediction in patients with stable and unstable coronary heart disease. Circulation 2010;121:2681.
[49] Omland T, de Lemos JA, Sabatine MS, et al. A sensitive cardiac troponin T assay in stable coronary artery disease. N Engl J Med 2009;361:2538.
[50] de Lemos JA, Morrow DA, deFilippi CR. Highly sensitive troponin assays and the cardiology community: a love/hate relationship? Clin Chem 2011;57:826.
[51] Sabatine MS, Morrow DA, de Lemos JA, et al. Evaluation of multiple biomarkers of car- diovascular stress for risk prediction and guiding medical therapy in patients with stable coronary disease. Circulation 2012;125:233.
[52] Sharma K, Kohli P, Gulati M. An update on exercise stress testing. Curr Probl Cardiol 2012;37:177.
[53] Bourque JM, Beller GA. Stress myocardial perfusion imaging for assessing prognosis: an update. JACC Cardiovasc Imaging 2011;4:1305.
[54] Gibbons RJ. Noninvasive diagnosis and prognosis assessment in chronic coronary artery disease: stress testing with and without imaging perspective. Circ Cardiovasc Imaging 2008;1:257.
[55] Marcassa C, Bax JJ, Bengel F, et al. Clinical value, cost-effectiveness, and safety of myocardial perfusion scintigraphy: a position statement. Eur Heart J 2008;29:557.
[56] Sicari R, Nihoyannopoulos P, Evangelista A, et al. Stress echocardiography expert consensus statement—executive summary. European Association of Echocardiography (EAE) (a registered branch of the ESC). Eur Heart J 2009;30:278.
[57] Williams MC, Reid JH, McKillop G, et al. Cardiac and coronary CT comprehensive imaging approach in the assessment of coronary heart disease. Heart 2011;97:1198.
[58] Mark DB, Berman DS, Budoff MJ, et al. Accf/acr/aha/nasci/saip/scai/scct 2010 expert consensus document on coronary computed tomographic angiography: a report of the American college of cardiology Foundation task force on expert consensus documents. Circulation 2010;121:2509.
[59] Greenland P, Bonow RO, Brundage BH, et al. ACCF/AHA 2007 clinical expert consensus document on coronary artery calcium scoring by computed tomography in global cardiovascular risk assessment and in evaluation of patients with chest pain: a report of the American college of cardiology foundation clinical expert consensus task force (ACCF/AHA Writing Committee to Update the 2000 Expert Consensus Document on Electron Beam Computed Tomography). Circulation 2007; 115:402.
[60] Bluemke DA, Achenbach S, Budoff M, et al. Noninvasive coronary artery imaging: magnetic resonance angiography and multidetector computed tomography angiography: a scientific statement from the American heart association committee on cardiovascular imaging and intervention of the council on cardiovascular radiology and intervention, and the councils on clinical cardiology and cardiovascular disease in the young. Circulation 2008;118:586.

[61] Blankstein R, DeVore AD. Selecting a noninvasive imaging study after an inconclusive exercise test. Circulation 2010;122:1514.
[62] Di Carli MF, Hachamovitch R. New technology for noninvasive evaluation of coronary artery disease. Circulation 2007;115:1464.
[63] Gaemperli O, Bengel FM, Kaufmann PA. Cardiac hybrid imaging. Eur Heart J 2011;32:2100.
[64] To AC, Desai MY. Role of cardiac magnetic resonance imaging in assessing ischemic and non-ischemic cardiomyopathies. Expert Rev Cardiovasc Ther 2012;10:223.
[65] Claessen BE, Maehara A, Fahy M, et al. Plaque composition by intravascular ultrasound and distal embolization after percutaneous coronary intervention. JACC Cardiovasc Imaging 2012;5:S111.
[66] Chan PH, Alegria-Barrero E, Di Mario C. Tools & techniques: intravascular ultrasound and optical coherence tomography. Euro Interv 2012;7:1343.
[67] Puri R, Worthley MI, Nicholls SJ. Intravascular imaging of vulnerable coronary plaque: current and future concepts. Nat Rev Cardiol 2011;8:131.
[68] Falk E, Wilensky RL. Prediction of coronary events by intravascular imaging. JACC Cardiovasc Imaging 2012;5:S38.
[69] Beltrame JF, Weekes AJ, Morgan C, et al. The prevalence of weekly angina among patients with chronic stable angina in primary care practices: the coronary artery disease in general practice (CADENCE) study. Arch Intern Med 2009;169:1491.
[70] Steg PG, Bhatt DL, Wilson PW, et al. One-year cardiovascular event rates in outpatients with atherothrombosis. JAMA 2007;297:1197.
[71] Arena R, Myers J, Williams MA, et al. Assessment of functional capacity in clinical and research settings: a scientic statement from the American heart association committee on exercise, rehabilitation, and prevention of the council on clinical cardiology and the council on cardiovascular nursing. Circulation 2007;116:329.
[72] Marzilli M, Merz CN, Boden WE, et al. Obstructive coronary atherosclerosis and ischemic heart disease: an elusive link! J Am Coll Cardiol 2012;60:951.
[73] Pepine CJ, Douglas PS. Rethinking stable ischemic heart disease: is this the beginning of a new era? J Am Coll Cardiol 2012;60:957.
[74] Serruys PW, Onuma Y, Garg S, et al. Assessment of the SYNTAX score in the Syntax study. Euro Interv 2009;5:50.
[75] Drozda Jr J, Messer JV, Spertus J, et al. Accf/aha/ama-pcpi 2011 performance measures for adults with coronary artery disease and hypertension: a report of the American college of cardiology foundation/American heart association task force on performance measures and the American medical association−physician consortium for performance improvement. Circulation 2011;124:248.
[76] Smith Jr SC, Benjamin EJ, Bonow RO, et al. AHA/ACCF secondary prevention and risk reduction therapy for patients with coronary and other atherosclerotic vascular disease: 2011 update: a guideline from the American Heart Association and American college of cardiology foundation. Circulation 2011;124:2458.
[77] Kahn R, Robertson RM, Smith R, et al. The impact of prevention on reducing the burden of cardiovascular disease. Circulation 2008;118:576.
[78] Nicholls SJ, Puri R, Anderson T, et al. Effect of Evolocumab on progression of coronary disease in statin-treated patients: the GLAGOV randomized clinical trial. JAMA 2016;316:2373−84.
[79] Ciszewski A, Bilinska ZT, Brydak LB, Kepka C, Kruk M, Romanowska M, Ksiezycka E, Przyluski J, Piotrowski W, Maczynska R, Ruzyllo W. Influenza vaccination in secondary prevention from coronary ischaemic events in coronary artery disease: FLUCAD study. Eur Heart J 2008;29:1350−8.
[80] A randomised, blinded, trial of clopidogrel versus aspirin in patients at risk of ischaemic events (CAPRIE). CAPRIE Steering Committee. Lancet 1996;348:1329−39.
[81] Gurbel PA, Bliden KP, Butler K, Tantry US, Gesheff T, Wei C, Teng R, Antonino MJ, Patil SB, Karunakaran A, Kereiakes DJ, Parris C, Purdy D, Wilson V, Ledley GS, Storey RF. Randomized double-blind assessment of the ONSET and OFFSET of the antiplatelet effects of ticagrelor versus clopidogrel in patients with stable coronary artery disease: the ONSET/OFFSET study. Circulation 2009;120:2577−85.

[82] Jernberg T, Payne CD, Winters KJ, Darstein C, Brandt JT, Jakubowski JA, Naganuma H, Siegbahn A, Wallentin L. Prasugrel achieves greater inhibition of platelet aggregation and a lower rate of non-responders compared with clopidogrel in aspirin-treated patients with stable coronary artery disease. Eur Heart J 2006;27:1166−73.
[83] Wiviott SD, Braunwald E, McCabe CH, Montalescot G, Ruzyllo W, Gottlieb S, Neumann FJ, Ardissino D, De Servi S, Murphy SA, Riesmeyer J, Weerakkody G, Gibson CM, Antman EM. Prasugrel versus clopidogrel in patients with acute coronary syndromes. N Eng J Med 2007;357: 2001−15.
[84] Cannon CP, Harrington RA, James S, Ardissino D, Becker RC, Emanuelsson H, Husted S, Katus H, Keltai M, Khurmi NS, Kontny F, Lewis BS, Steg PG, Storey RF, Wojdyla D, Wallentin L. Comparison of ticagrelor with clopidogrel in patients with a planned invasive strategy for acute coronary syndromes (PLATO): a randomised double-blind study. Lancet 2010;375:283−93.
[85] Patarroyo Aponte MM, Francis GS. Effect of angiotensin-converting enzyme inhibitors and angiotensin receptor antagonists in atherosclerosis prevention. Curr Cardiol Rep 2012;14:433.
[86] Catapano AL, Graham I, De Backer G, et al. ESC/EAS guidelines for the management of dyslipidaemias. Eur Heart J 2016;37(39):2999−3058.
[87] Yusuf S, Wittes J, Friedman L. Overview of results of randomized clinical trials in heart disease. I. Treatments following myocardial infarction. JAMA 1988;260:2088−93.
[88] Bangalore S, Steg G, Deedwania P, Crowley K, Eagle KA, Goto S, Ohman EM, Cannon CP, Smith SC, Zeymer U, Hoffman EB, Messerli FH, Bhatt DL, Investigators RR. beta-Blocker use and clinical outcomes in stable outpatients with and without coronary artery disease. JAMA 2012; 308:1340−9.
[89] Wallace WA, Wellington KL, Chess MA, Liang CS. Comparison of nifedipine gastrointestinal therapeutic system and atenolol on antianginal efficacies and exercise hemodynamic responses in stable angina pectoris. Am J Cardiol 1994;73:23−8.
[90] de Vries RJ, van den Heuvel AF, Lok DJ, Claessens RJ, Bernink PJ, Pasteuning WH, Kingma JH, Dunselman PH. Nifedipine gastrointestinal therapeutic system versus atenolol in stable angina pectoris. The Netherlands Working Group on Cardiovascular Research (WCN). Int J Cardiol 1996; 57:143−50.
[91] Fox KM, Mulcahy D, Findlay I, Ford I, Dargie HJ. The Total Ischaemic Burden European Trial (TIBET). Effects of atenolol, nifedipine SR and their combination on the exercise test and the total ischaemic burden in 608 patients with stable angina. The TIBET Study Group. Eur Heart J 1996;17: 96−103.
[92] van de Ven LL, Vermeulen A, Tans JG, Tans AC, Liem KL, Lageweg NC, Lie KI. Which drug to choose for stable angina pectoris: a comparative study between bisoprolol and nitrates. Int J Cardiol 1995;47:217−23.
[93] Kawanishi DT, Reid CL, Morrison EC, Rahimtoola SH. Response of angina and ischemia to long-term treatment in patients with chronic stable angina: a double- blind randomized individualized dosing trial of nifedipine, propranolol and their combination. J Am Coll Cardiol 1992;19: 409−17.
[94] Meyer TE, Adnams C, Commerford P. Comparison of the efficacy of atenolol and its combination with slow-release nifedipine in chronic stable angina. Cardiovasc Drugs Ther 1993;7:909−13.
[95] Steffensen R, Grande P, Pedersen F, Haunso S. Effects of atenolol and diltiazem on exercise tolerance and ambulatory ischaemia. Int J Cardiol 1993;40:143−53.
[96] Parameshwar J, Keegan J, Mulcahy D, Phadke K, Sparrow J, Sutton GC, Fox KM. Atenolol or nicardipine alone is as efficacious in stable angina as their combination: a double blind randomised trial. Int J Cardiol 1993;40:135−41.
[97] Foale RA. Atenolol versus the fixed combination of atenolol and nifedipine in stable angina pectoris. Eur Heart J 1993;14:1369−74.
[98] Rehnqvist N, Hjemdahl P, Billing E, Bjorkander I, Eriksson SV, Forslund L, Held C, Nasman P, Wallen NH. Effects of metoprolol vs verapamil in patients with stable angina pectoris. The Angina Prognosis Study in Stockholm (APSIS). Eur Heart J 1996;17:76−81.

[99] Hjalmarson A, Goldstein S, Fagerberg B, Wedel H, Waagstein F, Kjekshus J, Wikstrand J, El Allaf D, Vitovec J, Aldershvile J, Halinen M, Dietz R, Neuhaus KL, Janosi A, Thorgeirsson G, Dunselman PH, Gullestad L, Kuch J, Herlitz J, Rickenbacher P, Ball S, Gottlieb S, Deedwania P. Effects of controlled-release metoprolol on total mortality, hospitalizations, and well-being in patients with heart failure: the Metoprolol CR/XL Randomized Intervention Trial in congestive heart failure (MERIT-HF). MERIT-HF Study Group. JAMA 2000;283:1295–302.

[100] The cardiac insufficiency bisoprolol study II (CIBIS-II): a randomised trial. Lancet 1999;353:9–13.

[101] Packer M, Bristow MR, Cohn JN, Colucci WS, Fowler MB, Gilbert EM, Shusterman NH. The effect of carvedilol on morbidity and mortality in patients with chronic heart failure. U.S. Carvedilol Heart Failure Study Group. N Eng J Med 1996;334:1349–55.

[102] Flather MD, Shibata MC, Coats AJ, Van Veldhuisen DJ, Parkhomenko A, Borbola J, Cohen-Solal A, Dumitrascu D, Ferrari R, Lechat P, Soler-Soler J, Tavazzi L, Spinarova L, Toman J, Bohm M, Anker SD, Thompson SG, Poole-Wilson PA. Randomized trial to determine the effect of nebivolol on mortality and cardiovascular hospital admission in elderly patients with heart failure (SENIORS). Eur Heart J 2005;26:215–25.

[103] Pepine CJ, Handberg EM, Cooper-DeHoff RM, Marks RG, Kowey P, Messerli FH, Mancia G, Cangiano JL, Garcia-Barreto D, Keltai M, Erdine S, Bristol HA, Kolb HR, Bakris GL, Cohen JD, Parmley WW. A calcium antagonist vs a non- calcium antagonist hypertension treatment strategy for patients with coronary artery disease. The International Verapamil-Trandolapril Study (INVEST): a randomized controlled trial. JAMA 2003;290:2805–16.

[104] Ried LD, Tueth MJ, Handberg E, Kupfer S, Pepine CJ. A study of antihypertensive drugs and depressive symptoms (SADD-Sx) in patients treated with a calcium antagonist versus an atenolol hypertension treatment strategy in the international verapamil SR-trandolapril study (INVEST). Psychosom Med 2005;67:398–406.

[105] Steffensen R, Grande P, Pedersen F, Haunso S. Effects of atenolol and diltiazem on exercise tolerance and ambulatory ischaemia. Int J Cardiol 1993;40:143–53.

[106] Poole-Wilson PA, Lubsen J, Kirwan BA, van Dalen FJ, Wagener G, Danchin N, Just H, Fox KA, Pocock SJ, Clayton TC, Motro M, Parker JD, Bourassa MG, Dart AM, Hildebrandt P, Hjalmarson A, Kragten JA, Molhoek GP, Otterstad JE, Seabra-Gomes R, Soler-Soler J, Weber S. Effect of long-acting nifedipine on mortality and cardiovascular morbidity in patients with stable angina requiring treatment (ACTION trial): randomised controlled trial. Lancet 2004;364:849–57.

[107] Nissen SE, Tuzcu EM, Libby P, Thompson PD, Ghali M, Garza D, Berman L, Shi H, Buebendorf E, Topol EJ. Effect of antihypertensive agents on cardiovascular events in patients with coronary disease and normal blood pressure: the CAMELOT study: a randomized controlled trial. JAMA 2004;292: 2217–25.

[108] Majewski S, Slomka S, Zielinska-Wyderkiewicz E, et al. Heart rate-lowering efficacy and respiratory safety of ivabradine in patients with obstructive airway disease: a randomized, double-blind, placebo-controlled, crossover study. Am J Cardiovasc Drugs 2012;12:179.

[109] Tendera M, Talajic M, Robertson M, et al. Safety of ivabradine in patients with coronary artery disease and left ventricular systolic dysfunction (from the BEAUTIFUL Holter Sub-study). Am J Cardiol 2011;107:805.

[110] Fox K, Ford I, Steg PG, et al. Relationship between ivabradine treatment and cardiovascular outcomes in patients with stable coronary artery disease and left ventricular systolic dysfunction with limiting angina: a subgroup analysis of the randomized, controlled BEAUTIFUL trial. Eur Heart J 2009;30:2337.

[111] Sossalla S, Maier LS. Role of ranolazine in angina, heart failure, arrhythmias, and diabetes. Pharmacol Ther 2012;133:311.

[112] Stone PH, Chaitman BR, Stocke K, et al. The anti-ischemic mechanism of action of ranolazine in stable ischemic heart disease. J Am Coll Cardiol 2010;56:934.

[113] Boden WE. Ranolazine and its anti-ischemic effects: revisiting an old mechanistic paradigm anew? J Am Coll Cardiol 2010;56:943.

[114] Beyder A, Strege PR, Reyes S, et al. Ranolazine decreases mechanosensitivity of the voltage-gated sodium ion channel Na(v)1.5: a novel mechanism of drug action. Circulation 2012;125:2698.
[115] Pelliccia F, Pasceri V, Marazzi G, et al. A pilot randomized study of ranolazine for reduction of myocardial damage during elective percutaneous coronary intervention. Am Heart J 2012;163:1019.
[116] Wilson SR, Scirica BM, Braunwald E, et al. Efficacy of ranolazine in patients with chronic angina observations from the randomized, double-blind, placebo-controlled MERLIN-TIMI (metabolic efficiency with ranolazine for less ischemia in non−ST-segment elevation acute coronary syndromes) 36 trial. J Am Coll Cardiol 2009;53:1510.
[117] Mega JL, Hochman JS, Scirica BM, et al. Clinical features and outcomes of women with unstable ischemic heart disease: observations from metabolic efficiency with ranolazine for less ischemia in non−ST-elevation acute coronary syndromes−thrombolysis in myocardial infarction 36 (MERLIN-TIMI 36). Circulation 1809;121:2010.
[118] Mehta PK, Goykhman P, Thomson LE, et al. Ranolazine improves angina in women with evidence of myocardial ischemia but no obstructive coronary artery disease. JACC Cardiovasc Imaging 2011;4:514.
[119] Horinaka S. Use of nicorandil in cardiovascular disease and its optimization. Drugs 2011;71:1105.
[120] Tuunanen H, Engblom E, Naum A, Nagren K, Scheinin M, Hesse B, Juhani Airaksinen KE, Nuutila P, Iozzo P, Ukkonen H, Opie LH, Knuuti J. Trimetazidine, a metabolic modulator, has cardiac and extracardiac benefits in idiopathic dilated cardiomyopathy. Circulation 2008;118:1250−8.
[121] Detry JM, Sellier P, Pennaforte S, Cokkinos D, Dargie H, Mathes P. Trimetazidine: a new concept in the treatment of angina. Comparison with propranolol in patients with stable angina. Trimetazidine European Multicenter Study Group. Br J Clin Pharmacol 1994;37:279−88.
[122] El-Kady T, El-Sabban K, Gabaly M, Sabry A, Abdel-Hady S. Effects of trimetazidine on myocardial perfusion and the contractile response of chronically dysfunctional myocardium in ischemic cardiomyopathy: a 24-month study. Am J Cardiovasc Drugs 2005;5:271−8.
[123] Questions and answers on the review of medicines containing trimetazidine (20 mg tablets, 35 mg modified release tablet and 20 mg/ml oral solution); March 9, 2012. http://www.ema.europa.eu/docs/en_GB/document_library/Referrals_document/Trimetazidine_31/WC500129195.pdf.
[124] Fragasso G, Piatti Md PM, Monti L, Palloshi A, Setola E, Puccetti P, Calori G, Lopaschuk GD, Margonato A. Short- and long-term beneficial effects of trimetazidine in patients with diabetes and ischemic cardiomyopathy. Am Heart J 2003;146:e18.
[125] Lamy A, Natarajan M, Yusuf S. Medical treatment, PCI, or CABG for coronary artery disease? BMJ 2011;342:d966.
[126] Fihn SD, Gardin JM, Abrams J, et al. ACCF/AHA/ACP/AATS/PCNA/SCAI/STS guideline for the diagnosis and management of patients with stable ischemic heart disease: a report of the American college of cardiology foundation/American heart association task force on practice guidelines, and the American college of physicians, American association for thoracic surgery, preventive cardiovascular nurses association, Society for cardiovascular angiography and interventions, and Society of thoracic surgeons. Circulation 2012;126:e354.
[127] Levine GN, Bates ER, Blankenship JC, et al. 2011 ACCF/AHA/SCAI guideline for percutaneous coronary intervention: executive summary. A report of the American college of cardiology foundation/American heart association task force on practice guidelines and the society for cardiovascular angiography and interventions. Circulation 2011;124:2574.

CHAPTER 2.5

Acute Coronary Syndromes

Maria Drakopoulou, Konstantinos Toutouzas, Dimitris Tousoulis
1st Cardiology Clinic, 'Hippokration' General Hospital, National and Kapodistrian University of Athens, School of Medicine, Athens, Greece

Contents

INTRODUCTION

An acute coronary syndrome (ACS) is the most ominous manifestation of coronary artery disease [1]. Among other clinical presentations of coronary artery disease, ACS are particularly worrisome as they are prevalent but at the same time portend a worse prognosis [2,3].

ACS represents a life-threatening manifestation of atherosclerosis. Pathological, imaging, and biological observations have demonstrated that atherosclerotic plaque rupture or erosion, with differing degrees of superimposed thrombosis form the basic pathophysiological mechanisms in most conditions of ACS [4–8]. During the last decades, rapid progress has been made in the understanding of ACS pathophysiology and the concept of "vulnerable plaque", the plaque that is considered prone to rupture, has been challenged [8–10]. Indeed, recent studies have provided additional elements for the identification of the vulnerable plaque and/or patient [11–14]. These elements seem to involve in addition to the intrinsic plaque characteristics, factors extrinsic to the plaque

Coronary Artery Disease
ISBN 978-0-12-811908-2, http://dx.doi.org/10.1016/B978-0-12-811908-2.00012-X

that may increase plaque destabilization [15–18]. Overall, plaque destabilization seems to be a biomechanical phenomenon that depends on a complex interplay between structural features, biological processes, and applied exogenous forces and stresses.

This chapter is focused on the pathophysiological and biological processes related to the stepwise progression of atherosclerotic plaques and plaque destabilization leading eventually to ACS. Current methods that can potentially be used to identify high-risk plaques and aid in the management of ACS are also discussed.

DEFINITION

The term "ACS" appeared relatively recently in the medical lexicon. This term was first seen in 1986 in the article entitled: "Flow characteristics of coronary balloon catheters" [19]. In this article, the following sentence was included in the abstract: "Sudden re-occlusion leads frequently to an ACS (acute myocardial infarction [MI], hypotension, arrhythmias) that requires emergency surgery and also leads to permanent myocardial damage of various degrees." Few years later, in a historical article entitled "The Pathogenesis of Coronary Artery Disease and The Acute Coronary Syndromes," ACS was defined as MI, unstable angina (UA) or ischemic sudden death [20]. All three conditions were thought to be a manifestation of the same atherosclerotic coronary artery substrate. This understanding has paved the way for the contemporary use of the term ACS because it unified these "syndromes" through their common pathologic basis. ACS has evolved through the years as a useful term that refers to a spectrum of conditions compatible with acute myocardial ischemia and/or infarction that are usually due to an abrupt reduction in coronary blood flow.

According to current European Society of Cardiology Guidelines (ESC Guidelines), patients with suspected ACS should be differentiated into two groups based on electrocardiogram (ECG) findings [2]:

1. Patients with acute chest pain and persistent (>20 min) ST-segment elevation. This is termed ST-elevation ACS (STE-ACS) and generally reflects an acute total coronary occlusion. Most of these patients will ultimately develop ST-elevation myocardial infarction (STEMI).
2. Patients with acute chest pain but without persistent ST-segment elevation (non-ST elevation acute coronary syndrome, NSTE-ACS). These patients rather have persistent or transient ST-segment depression or T-wave inversion, flat T waves, pseudonormalization of T waves, or no ECG changes at presentation. NSTE-ACS can be further subdivided based on cardiac biomarkers of necrosis. If cardiac biomarkers are elevated and the clinical context is appropriate, the patient is considered to have non-ST elevation myocardial infarction (NSTEMI); otherwise, the patient is deemed to have UA.

The clinical spectrum of NSTE-ACS in current guidelines ranges from patients free of symptoms to patients with ongoing ischemia, electrical, hemodynamic instability or cardiac arrest. When symptomatic, patients with NSTE-ACS may present with (1) rest angina that usually lasts >20 min, (2) new-onset severe angina (class II or III of the

Canadian Cardiovascular Society Classification), (3) recent destabilization of previously stable angina with at least Canadian Cardiovascular Society Class III angina characteristics (crescendo angina), or (4) post-MI angina.

The pathogenesis of myocardial injury is considered in the Third Universal Definition of Myocardial Infarction [2,21]. MI is classified into various types, based on pathological, clinical, and prognostic differences, along with different treatment strategies. MI caused by a primary coronary artery process such as spontaneous plaque rupture is classified as MI type 1 and one related to reduced myocardial oxygen supply and/or increased myocardial oxygen demand (in the absence of a direct coronary artery process) as a MI type 2. An MI that leads to death when no biomarkers are available is classified as type 3. MIs associated with revascularization procedures are classified as type 4 or 5.

PATHOPHYSIOLOGY

The hallmark of ACS is the sudden imbalance between myocardial oxygen consumption (MVO_2) and demand, which is usually the result of an acute coronary event [22,23]. The imbalance may also be caused by other conditions, including excessive myocardial oxygen demand in the setting of a stable flow-limiting lesion, acute coronary insufficiency due to other causes (vasospastic angina, coronary embolism, coronary arteritis), noncoronary causes of myocardial oxygen supply—demand mismatch (hypotension, severe anemia, hypertension, tachycardia, hypertrophic cardiomyopathy, severe aortic stenosis), nonischemic myocardial injury (myocarditis, cardiac contusion, cardiotoxic drugs), and multifactorial causes that are not mutually exclusive (stress cardiomyopathy, pulmonary embolism, severe heart failure, sepsis). In rare cases, ACS may have a nonatherosclerotic etiology such as trauma, dissection, thromboembolism, congenital anomalies, cocaine abuse, or complications of cardiac catheterization.

Stable angina or stable silent ischemia commonly result from increases in myocardial oxygen demand that outstrips the ability of narrowed coronary arteries to increase oxygen delivery. When coming to the pathobiology of ACS, a fairly small fissuring or rupture of a lipid-rich plaque, and occasionally a superficial erosion of a fibrotic plaque, may lead to an acute change in plaque structure and a reduction in coronary blood flow, resulting in exacerbation of angina. Thus, ACS is usually caused by a rupture or eroded atherosclerotic plaque that is then followed by acute thrombosis, with or without concomitant vasoconstriction, causing a sudden and critical reduction in blood flow [24,25]. Transient episodes of thrombotic vessel occlusion at the site of plaque damage may occur, leading to angina at rest. The thrombus that is formed is usually labile and results in temporary vascular occlusion, perhaps lasting only 10—20 min. Major contributors to the reduction of flow at this stage are the release of vasoactive substances by platelets (serotonin, thromboxane A2) and the vasoconstrictive effect of thrombin.

Overall, alterations in perfusion and myocardial oxygen supply probably account for two-thirds of ACS and the remainder may be caused by transient increases in myocardial-oxygen demand. In patients with non-Q-wave MI, the thrombotic occlusion may last up to 1 h. Only ~$^1/_4$ of patients have an infarct-related vessel occluded for more than 1 h, and in these cases the distal myocardial territory is usually supplied by collaterals. In Q-wave MI, larger plaque fissures result in the formation of a fixed and persistent thrombus, which leads to an abrupt cessation of myocardial perfusion for more than 1 h, resulting in transmural necrosis of the involved myocardium. Some cases of sudden coronary death probably involve a rapidly progressive coronary lesion in which plaque disruption—commonly a superficial erosion—and resultant thrombosis lead to ischemic and fatal ventricular arrhythmias in the absence of collateral flow.

PATHOLOGY

Numerical Classification of Histologically Defined Lesion Types

The pathogenesis of atherosclerotic lesions has been seen from microscopic analysis of arteries in different age groups. Based on the histologic classification of the American Heart Association Committee on Vascular Lesions, published in mid-1990s plaque progression can be subdivided into the six distinct categories and different lesion types [26]: type I, intimal thickening; type II, fatty streak; type III, transitional or intermediate lesion; type IV, advanced atheroma with well-defined region of the intima; type V, fibroatheroma or atheroma with overlaid new fibrous connective tissue; and type VI, complicated plaques with surface defects, hematoma or hemorrhage, thrombosis, or a combination of these characteristics.

Morphologic Classification

The numerical classification was replaced by and alternative and simpler classification introduced by Virmani et al. [27–31]. This classification is descriptive and emphasizes the relation between the plaque morphology and clinical presentation (Fig. 2.5.1) [32]. The main lesion types recognized in this classification are: adaptive or diffuse intimal thickening, intimal xanthoma (fatty streak), pathological intimal thickening (PIT), and fibroatheroma. The current updated classification of atherosclerotic lesions based on morphology is depicted in Table 2.5.1 [26,27,33].

Nonatherosclerotic Intimal Lesions

Adaptive or Diffuse Intimal Thickening

Adaptive or diffuse intimal thickening is mostly considered a physiological response to blood flow rather than an atherosclerotic process. These lesions are thought to evolve early in childhood and usually regress over time. Histologically, lesions of intimal thickening contain focal accumulation of smooth muscle cells (SMCs), with proteoglycan-rich extracellular matrix and no signs of inflammation [4,5,26]. Although most of the studies support the idea that intimal proliferation is an adaptive vascular reaction to altered blood

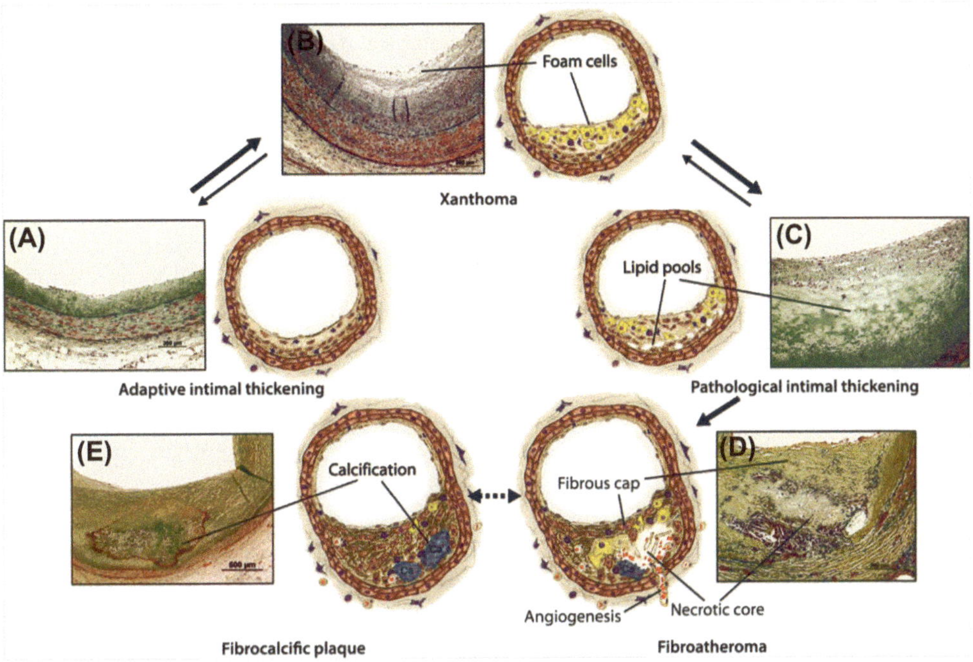

Figure 2.5.1 Lesion types of atherosclerosis and a proposed sequence of their development. (A) Adaptive intimal thickening characterized by smooth muscle cell accumulation within the intima. (B) Intimal xanthoma corresponding to the accumulation of foam cell macrophages within the intima. (C) Pathological intimal thickening denotes the accumulation of extracellular lipid pools in the absence of apparent necrosis. (D) Fibroatheroma indicating the presence of a necrotic core. (E) Calcification of the necrotic core and surrounding tissue, forming fibrocalcific plaque. *(Reproduced with permission from Bentzon JF, Otsuka F, Virmani R, Falk E. Mechanisms of plaque formation and rupture. Circ Res 2014;114:1852–66.)*

flow in coronary arteries, there is some evidence that it may be a precursor of more advanced atherosclerotic lesions. Interestingly, some studies in neonates, adolescents and young adults, have shown that intimal thickening that forms near branch points may enlarge with advancing age and be a precursor to high-risk plaques.

Intimal Xanthomas

Intimal xanthomas (or so-called "fatty streaks") are considered by some as the earliest manifestations of human atherosclerotic disease. "Xanthoma" is a general pathological term that describes focal accumulations of fat-laden macrophages. Histologically, intimal xanthomas are lesions composed of infiltrating macrophage foam cells and, to a lesser extent, lipid-laden SMCs. This type of lesion has been shown to regress in young individuals. As intimal xanthomas do not always convey the mandatory features of more-advanced atherosclerotic plaques they are not considered progression-prone disease.

Table 2.5.1 Updated classification of atherosclerotic lesions based on morphology

Type of lesion	Subtype of lesion	Morphological description
Nonatherosclerotic intimal lesions	Intimal thickening	Natural accumulation of smooth muscle cells (SMCs) in the absence of lipid, macrophage foam cells, and thrombosis.
	Intimal xanthoma	Superficial accumulation of foam cells without a necrotic core, fibrous cap, or thrombosis.
Progressive atherosclerotic lesions	Pathological intimal thickening	Plaque rich in SMCs, with hyaluronan and proteoglycan matrix and focal accumulation of extracellular lipid. Absence of thrombosis.
	Fibroatheroma	During early necrosis: focal macrophage infiltration into areas of lipid pools with an overlying fibrous cap. During late necrosis: loss of matrix and extensive cellular debris with an overlying fibrous cap. With or without calcification. Absence of thrombosis.
	Intraplaque haemorrhage or plaque fissure	Large necrotic core (size $>10\%$ of plaque area) with haemorrhage, and plaque area shows presence of angiogenesis. Necrotic core communicates with the lumen through a fissure. Minimal tear without obvious thrombus.
	Thin-cap fibroatheroma	A thin, fibrous cap (<65 μm) infiltrated by macrophages and lymphocytes, with rare or no SMCs and relatively large underlying necrotic core ($>10\%$ of plaque area). Intraplaque haemorrhage and/or fibrin might be present. Absence of thrombosis.
Lesions with acute thrombi	Plaque rupture	Thin-cap fibroatheroma with cap disruption. Thrombosis is present and might or might not be occlusive. The luminal thrombus communicates with the underlying necrotic core.
	Plaque erosion	Can occur on pathological intimal thickening or on a fibroatheroma. Thrombosis is present and might or might not be occlusive. No communication of the thrombus with the necrotic core
	Calcified nodule	Eruptive (shedding) of calcified nodule with an underlying fibrocalcific plaque with minimal or no necrosis. Thrombosis is usually not occlusive.

Table 2.5.1 Updated classification of atherosclerotic lesions based on morphology—cont'd

Type of lesion	Subtype of lesion	Morphological description
Healed lesions	Healed plaque rupture, erosion, or calcified nodule	Healed lesion composed of SMCs, proteoglycans, and collagen type III with or without underlying disrupted fibrous cap, necrotic core, or nodular calcification. Lesions can contain large areas of calcification with few inflammatory cells and have a small or no necrotic core. The fibrotic or fibrocalcific collagen-rich plaque is associated with significant luminal stenosis. Absence of thrombosis.

Reproduced with permission from Yahagi K, Kolodgie FD, Otsuka F, Finn AV, Davis HR, Joner M, Virmani R. Pathophysiology of native coronary, vein graft, and in-stent atherosclerosis. Nat Rev Cardiol 2016;13:79–98.

Progressive Atherosclerotic Lesions

Pathological Intimal Thickening

PIT is considered to be the earliest progressive lesion and is characterized by the presence of acellular lipid pools and SMCs remnants, within an extracellular matrix (ECM) composed of proteoglycans (biglycan, decorin, and versican) and collagen type III. The defining characteristic of this lesion is the presence of acellular, lipid-rich material in the intima [34]. Small lipid pools are seen beneath the layers of foam cells without gross disruption of the normal architecture of the intima.

The presence of macrophages in PIT may indicate a more advanced stage of atherogenesis as compared with the absence of macrophages. It leads to irreversible disruption of the normal architecture of the intima and leaves behind matrix devoid gruel of lipids and cell debris. When present, macrophages in PIT are typically located away from the area of lipid pools.

Observations in human atheroma specimens, in vitro studies and animal experiments have shown a major contribution of inflammatory activation in the initiation and progression of atherosclerosis and have elucidated the corresponding pathways [35–38]. Indeed, inflammatory activation including both innate and adaptive immunity mechanisms plays a central role in the progression of atherosclerosis [39–43].

At this stage, studies have shown that the initiation of inflammation may be lipid-driven [44,45]. Low-density lipoproteins (LDLs) may cause atherosclerosis by accumulating in the intima where they may be modified by oxidation and aggregation. The modified LDLs, and oxidized lipid derivatives in turn act as chronic stimulators of the innate and adaptive immune response. Both endothelial cells and SMCs are induced to express adhesion molecules (e.g., vascular cell adhesion molecule-1 and intercellular adhesion molecule-1), chemoattractants (e.g., monocyte chemoattractant protein-1), and growth factors (e.g., macrophage colony-stimulating factor and granulocyte-macrophage colony-stimulating factor) that interact with receptors on monocytes and stimulate their homing, migration, and differentiation into macrophages and dendritic cells. In specific, macrophages, recruited as monocytes from the circulating blood, are the predominant cell type of innate immunity in atherosclerosis. The recruitment takes place through adhesion molecules (vascular cell adhesion molecule-1) expressed by vascular endothelial cells. Monocytes enter the intima, and undergo activation to sustain and amplify inflammation locally. In the plaque, these cells express scavenger receptors necessary for uptake of modified lipoproteins and, hence, foam cell formation. Thus, macrophages and dendritic cells serve as deposits of lipids from the insudating lipoproteins and become foam cells at this stage. Although less apparent morphologically, but functionally pivotal, other leukocyte populations reside in plaques including various classes of T lymphocytes, B lymphocytes, mast cells, and dendritic cells.

ECM is also increasingly being recognized as being important in the early steps of inflammation, particularly through its influence on macrophage phenotype [46–48]. Versican has been shown to promote leukocyte adhesion, and interference with the incorporation of versican into ECM blocks monocyte adhesion and dampens inflammatory responses. In the context of vascular disease, versican might be one of the initial amplification steps that contributes to the pathogenesis of inflammation in atherosclerosis, whereby macrophages are recruited into the lipid pools where they eventually undergo necrosis.

Fibroatheroma

The first of the advanced coronary lesions in concordance with the American Heart Association consensus classification is the fibroatheroma [9,33]. As the name implies this lesion comprises of a fibrous cap, defined as a distinct layer of connective tissue completely covering a lipid core. The fibrous cap consists purely of SMCs in a collagenous proteoglycan matrix, with varying degrees of infiltration by macrophages and lymphocytes. The lipid core component of the fibroatheroma contains free cholesterol, cholesterol esters, and cholesterol crystals. The preexisting isolated lipid pools now grow into confluent necrotic cores through the invasion by macrophages. Apoptosis and secondary necrosis of foam cells and SMCs are thought to be an important cause for further necrotic core development, by releasing their lipid contents into the core. Also, lipids may infiltrate the arterial wall directly from the circulating blood. Progressive cholesterol accumulation has deleterious cellular effects, causing further cell apoptosis and promoting inflammation. Lipid particles residing in the core undergo enzymatic modifications, becoming oxidized, and trigger further the release of proinflammatory mediators. Macrophage apoptosis is considered to be of utmost importance in the necrotic core formation, whereas defective phagocytic clearance of apoptotic bodies may also play an important role.

Two unambiguous phases of the necrotic core formation relative to "early" or "late" necrosis have been defined, mainly on the basis of the relative extent of matrix proteoglycans. Early phase necrosis (transition from lipid pool) is associated with infiltration of macrophages into the lipid pool, together with the focal loss of proteoglycans and/or collagen. The necrotic core underlying the "late fibroatheroma" is noticeably deficient in ECM and has more cholesterol clefts, calcification, intraplaque hemorrhage, and surrounding neoangiogenesis, than the earlier phase. At this stage, the expression of hyaluronic and proteoglycan expression is low in the lipid pool and necrotic core, probably owing to active metalloproteinases (MMPs) and/or peptidases of the ADAMTS (A Disintegrin And Metalloproteinase with Thrombospondin Motifs) family.

These morphological attributes provide important mechanistic insight into how necrotic cores initiate and evolve. Late fibroatheromas also have thick fibrous caps composed of collagen type I and III and proteoglycans interspersed with SMCs. Although the precise mechanisms of fibrous cap thinning are not fully understood, ECM degradation facilitated by MMPs and ADAMTS proteases secreted by an overabundance of macrophages are likely to be involved, in addition to possible deficiencies in repair mechanisms.

At this stage, the recruited macrophages express several different polarization phenotypes and exert manifold effects in lesion development. It is now well recognized that at least three subtypes of macrophages can be found in atherosclerotic lesions: M1-like phenotype, the most common form that is activated by cytokines secreted by T cells and attains a pro-inflammatory function, the M2-like phenotype that plays an anti-inflammatory role, promoting thus tissue repair and a novel third phenotype, hemoglobin stimulated macrophage (M(Hb)), induced by intraplaque hemorrhage.

Another major contributor to inflammatory cell recruitment and atherosclerosis progression is the development of vasa vasorum (small network of microvessels for the adventitia, media, and atherosclerotic plaque blood supply) including plaque neovascularization [49–52]. The main stimulus for neovessel formation comes in the form of hypoxia-inducible factors secreted by hypoxic macrophages deep in the lesion's necrotic core. In fact, the anoxemia theory of atherosclerosis postulates that it is the depletion of ATP in macrophages, attributed to the extremely high rates of energy taxing cholesterol uptake, which leads to apoptosis, formation of a necrotic core, and angiogenetic stimuli. The neovessel wall of the vasa vasorum lacks proper structure, in terms of elastic laminae and SMC support, and thus they are leaky, prone to rupture, and provide the plaque with elements crucial for its continued progression, in the form of cholesterol (both lipoprotein and red blood cell membrane derived), cytokines, and leukocytes, straight to the necrotic core, thus establishing a vicious positive feedback loop favoring destabilization and, ultimately, rupture. The presence of an extensive vasa vasorum network can function as a conduit for entry of macrophages and inflammatory factors, facilitating thus, inflammatory cell recruitment and promoting the progression of plaque formation. The relationship between adventitial vasa vasorum and development of atherosclerosis is further supported by experimental and clinical observations [53,54]. Low incidence of atherosclerosis is associated with a significantly lower vasa vasorum density.

In the process of the ongoing inflammatory activation, microRNAs (miRNAs), small non-coding RNA molecules that act as posttranscriptional regulators of gene expression seem to control all major involved cell types [55,56]. Alterations in miRNA signatures have been found in atherosclerotic plaques with macrophage and endothelial-cell-derived miRNAs promoting plaque destabilization (through promotion of inflammation and angiogenesis, respectively). On the contrary, some SMC miRNAs may be modulated to increase cap thickness and modify endothelial shear stress (ESS). Finally, miRNAs seem to be involved in the promotion of calcium deposition. Table 2.5.2 presents the miRNAs that seem to be involved in vulnerable plaque formation and rupture [57].

Thin Fibrous Cap Atheroma

The concept of thin cap fibroatheroma (TCFA) was originally developed from observations of ruptured coronary lesions [28]. A TCFA, conceptually often referred to as a "vulnerable plaque", is the precursor lesion to ACS and is considered to be the plaque that is prone to rupture. Table 2.5.3 summarizes the specific features of vulnerable plaques [25]. The TCFA is characterized by a large "late" necrotic core (usually >25% of the plaque area) encased by a thin intact fibrous cap that is composed predominantly of collagen type I with varying degrees of macrophages and lymphocytes and paucity or absence of SMCs. Hemorrhage

Table 2.5.2 Features of vulnerable plaques

Thrombus
Large necrotic core (>30% of plaque)
Fibrous cap covering necrotic core
Thin (thickness usually <65 μm)
Many macrophages (inflammation)
Few smooth muscle cells (apoptosis)
Expansive remodeling preserving lumen
Neovascularization from vasa vasorum
Plaque hemorrhage
Adventitial/perivascular inflammation
"Spotty" calcification

Reproduced with permission from Falk E, Nakano M, Bentzon JF, Finn AV, Virmani R. Update on acute coronary syndromes: the pathologists' view. Eur Heart J 2013;34:719–28.

Table 2.5.3 MicroRNAs involved in vulnerable plaque formation/rupture

Atherosclerotic plaque vulnerability	miRNAs	Proposed mechanism of action
Plaque inflammation	miRNA-126	VCAM-1 inhibition
	miRNA-155, -222, -424, -503, -9, -17, -20a, and -106a	Regulation of monocyte differentiation into macrophages within the plaque
	miRNA-147, -155, and -342-5p	Activation of plaque macrophages in M1 phenotype, TNFa, and IL-6 upregulation
	miRNA-125a, -146a, -33, and -155	Inhibition of lipid accumulation
	miRNA-15a and -16	Modulation of macrophage apoptosis through bcl-2 receptor targeting
	miRNA-21	Production and secretion of matrix metalloproteinase-9 by human macrophages
	miRNA-34a	VSMC proliferation
	miRNA-210	Tubulogenesis and migration stimulation
	miRNA-146a	Drives peripheral blood mononuclear cells toward a Th1 response
	miRNA-29	miRNA-29 inhibition of elastin expression
	miRNAs-221/222	miRNAs-221/222 promotion of proliferation or cellular death
	miRNA-365	miRNA-365 stimulation of endothelial apoptosis through Bcl2 targeting
	miRNA-100, -127, -145, -133a, and -133b	miRNA-100, -127, -145, -133a, and -133b Highly expressed in symptomatic carotid plaques
Endothelial shear stress	miRNA-143/145	Promotion of VSMCs phenotype switch into the atheroprotective contractile phenotype
	microRNA-126-5p	Limitation of endothelial cell proliferation at low endothelial shear stress sites
	miRNA-92a	Associated with low endothelial shear stress, promotion of Inflammation
Microcalcification	miR-125b	Differentiation of VSMCs into an osteoblast-like phenotype

IL-6, interleukin 6; *TNF*, tumor necrosis factor; *VCAM*, vascular cell adhesion molecule; *VSMC*, vascular smooth muscle cells.

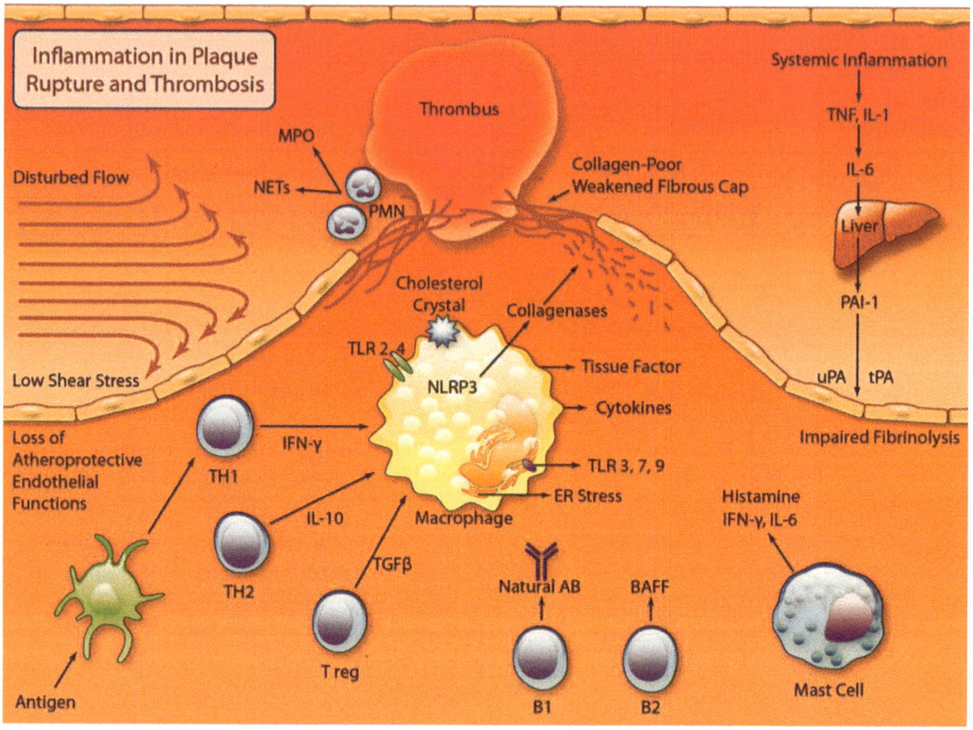

Figure 2.5.2 Inflammation in plaque rupture and thrombosis. Cross-section of the intima of part of an artery affected by atherosclerosis. Altered hydrodynamics (at top left) cause loss of atheroprotective functions of endothelial cells (including vasodilator, anti-inflammatory, profibrinolytic, and anticoagulant properties). Antigens presented on antigen-presenting cells such as dendritic cells (DCs) can activate TH1 lymphocytes to produce interferon-γ (IFN-γ), which activates macrophages (MΦ, yellow). Other subtypes of lymphocytes (shown in blue) include TH2 lymphocytes, which can elaborate the anti-inflammatory cytokine interleukin 10 (IL-10) and regulatory T cells that secrete the anti-inflammatory cytokine transforming growth factor-β (TFG-β). On its surface, the macrophage contains Toll-like receptors (TLRs) 2 and 4, which can bind pathogen-associated molecular patterns and damage-associated molecular patterns (see text). The intracellular TLRs 3, 7, and 9 may also contribute to lipid accumulation and other proatherogenic functions of the macrophage. Macrophages can undergo stress of the endoplasmic reticulum (ER) under atherogenic conditions. Cholesterol crystals found in plaques can activate the NOD-, LRR- and pyrin domain-containing three inflammasome that can generate mature IL-1β from its inactive precursor. The activated macrophage secretes collagenases that can degrade the triple helical interstitial collagen that lends strength to the plaque's fibrous cap. Activated macrophages also express TF, a potent procoagulant, and elaborate proinflammatory cytokines that amplify and sustain the inflammatory process in the plaque. When the plaque ruptures because of a collagen-poor, weakened fibrous cap, blood in the lumen can contact TF in the lipid core, triggering thrombus formation (red). When the thrombus forms, polymorphonuclear leukocytes (PMNs) can accumulate and elaborate myeloperoxidase (MPO), which in turn elaborates the potent pro-oxidant hypochlorous acid. Dying PMNs extrude DNA that can form neutrophil extracellular traps (NETs), which can entrap leukocytes and propagate thrombosis. Other inflammatory cells modulate atherosclerosis. B1 lymphocytes secrete natural antibody that can inhibit plaque inflammation. On the contrary, B2 lymphocytes, in part via B-cell activating factor (BAFF), can promote

and/or calcification is often present and intraplaque vasa vasorum are abundant. The thin fibrous cap is traditionally defined as being <65 μm thick. It seems that macrophages of the plaque can overproduce matrix metalloproteinases (MMPs), specific enzymes with interstitial collagenase activity, causing the thinning of fibrous cap.

The thinning or weakening of the fibrous cap is an important precursor to plaque rupture. In an autopsy study of patients with known coronary artery disease and sudden cardiac death, the mean fibrous cap thickness at the site of rupture was 23 ± 19 μm, with 95% of the ruptured caps measuring <65 μm [27]. TCFAs have been found to be more common among patients presenting with acute plaque rupture than patients with stable plaques and patients that died from noncardiac causes. Although similar to ruptured plaques, TCFAs tend to have a smaller necrotic core, fewer macrophages within the fibrous cap, less plaque burden, and less luminal encroachment than plaques that have ruptured.

Coronary Thrombosis

Acute coronary thrombosis, the event that eventually leads to the development of ACSs may be caused by three distinct morphologic entities: plaque rupture, erosion, and calcified nodule. These processes can occur in the setting of a fibrous cap atheroma or PIT (mostly erosion). The frequency of the coronary lesions with thrombi has been reported to be 55%–65% for ruptures, 30%–35% for erosions, and 2%–7% with calcified nodules in sudden coronary death patients. A single lesion may contain morphological evidence of both rupture and erosion [33]. Fig. 2.5.2 shows in detail the biological processes involved in plaque thrombosis and rupture [58].

Rupture "Plaque rupture" is defined by an area of fibrous cap disruption whereby the overlying thrombus is in continuity with the underlying necrotic core. In ruptured plaques, the underlying necrotic core is usually large in size (>30% of the plaque area) and the disrupted fibrous cap consists mainly of type I collagen and is infiltrated by macrophages and lymphocytes [59]. The SMC content within the fibrous cap at the rupture site may be quite sparse. It is generally a prevailing thought that rupture of the fibrous cap occurs at its weakest point, often near shoulder regions. However, autopsy studies using

inflammation and plaque complication. Mast cells can augment atherogenesis by releasing histamine and the cytokines IFN-γ and IL-6. The consequences of a given plaque rupture depend not only on the solid state of the intimal plaque but also on the fluid phase of blood, as depicted in the top right. Systemic inflammation can give rise to cytokines, culminating in the overproduction of IL-6, the trigger of the hepatic acute phase response. The acute phase reactant fibrinogen (not shown) participates directly in thrombus formation. Another acute phase reactant, plasminogen activator inhibitor-1 (PAI-1), can impair fibrinolysis by inhibiting the endogenous fibrinolytic mediators, urokinase- and tissue-type plasminogen activators (uPA and tPA). *(Reproduced with permission from Libby P, Tabas I, Fredman G, Fisher EA. Inflammation and its resolution as determinants of acute coronary syndromes. Circ Res 2014;114:1867–79.)*

serial sections demonstrate an equal number of ruptures occurring at the mid-portion of the fibrous cap, especially when rupture occurs after exercise. Therefore, it is speculated that different processes may be involved in the final event of plaque rupture, where macrophages in the fibrous cap release MMP to weaken the cap, and high shear stress may also contribute to the disruption of the cap. Microcalcification within a thin fibrous cap (typically 10 μm in diameter) may also facilitate plaque rupture through local increase in stress that leads to interfacial debonding; however, it has been also reported that microcalcification <5 μm in diameter is seemingly less harmful, whereas that >5 μm in diameter may predict plaque rupture [60]. Focal calcification or larger calcified sheets are frequently observed in ruptured plaques, and are mostly located toward the abluminal surface of the necrotic core.

Plaque destabilization followed by rupture of the protective fibrous cap seems is to be governed by several factors. Among them, inflammation decisively controls the metabolism of collagen, a major constituent of the fibrous cap. In fact, recent evidence has shown that the protective fibrous cap displays considerable metabolic activity [61,62]. The balance between synthetic and degradative processes closely controlled by inflammatory mediators regulates the level of collagen in this structure. Indeed, plaques that rupture have thin and friable fibrous caps because of the lack of collagen. Considerable biochemical and experimental data support the role of matrix-degrading proteinases, tightly regulated by inflammatory mediators, as contributors to the dissolution of interstitial collagen that thins and weakens the fibrous cap and hence renders a plaque susceptible to rupture. Death of SMCs in the inflamed atheroma can deplete the plaque of the same cells that synthesize collagen required to form and maintain the fibrous cap.

Once the fibrous cap ruptures and the necrotic core contents are exposed to the flowing blood in the lumen, coagulation cascade is activated in response to the exposure of lipids and tissue factors (TF) which are present in the core. This contact results in thrombus formation. The luminal thrombus is always predominant on platelets (white thrombus) near the actual rupture site and might or might not be obstructive. At sites proximal and distal to the rupture, thrombi usually consist of layers of fibrin (lines of Zahn) interspersed with erythrocytes ("red thrombus") and tend to move to the nearest side branch, thus allowing blood to reflow through the collateral branch.

Erosion "Plaque erosion" is the second most prevalent cause of coronary thrombosis. In this lesion, the luminal thrombus is in direct contact with the intima which is rich in SMCs in a proteoglycan matrix and there is absence of endothelial lining. It is identified when serial sectioning of a thrombosed arterial segment fails to reveal fibrous cap rupture and typically, the endothelium is absent at the erosion site. The erosion lesions tend to be eccentric and are infrequently calcified with intact medial wall [33,63]. The underlying lesion is usually less advanced than in ruptured plaques, and mostly exhibits characteristics of early lesions (PIT in ~16% and early and late fibroatheroma in

50% and 34%, respectively). The area in close proximity to the thrombus is rich in versican, hyaluronan, and type III collagen unlike rupture or stable plaque which are rich in type I collagen and biglycan and decorin. The majority of erosions (56%) lack evidence of calcification, although microcalcification is observed in approximately 40% of lesions; fragmented calcification and sheets of calcification are rare (<2%). Therefore, the most relevant features of plaque erosion include an abundance of SMCs within a proteoglycan matrix and the absence of surface endothelium or a prominent lipid core. Usually, few macrophages and T-lymphocytes are close to the lumen and the plaque is not disrupted. Currently, no distinct morphological features suggestive of erosion-prone plaques have been identified.

It is hypothesized that a selective accumulation of hyaluronan in eroded plaques may promote de-endothelialization and platelet aggregation. Hyaluronan can directly promote the polymerization of fibrin, which may facilitate SMC migration and plaque progression. Although a previous autopsy study in patients with sudden cardiac death demonstrated that the eroded sites at their majority show minimal inflammation characterized by few or absent macrophages and lymphocytes, other evidence support that inflammation may contribute to this mechanism of coronary thrombosis also. Endothelial cells, like SMCs, may undergo apoptosis in response to inflammatory mediators. Loss of endothelial cells can uncover the thrombogenic subendothelial matrix. Endothelial cells can also express proteinases regulated by inflammatory cytokines and oxidized lipoproteins. One of these proteinases, is a key constituent of the subendothelial matrix, and provides an important substrate for adherence of endothelial cells to the intimal surface. Thus, activation of proteases in response to inflammatory stimuli can promote injury to the intima, a possible prelude to local thrombosis resulting from superficial erosion.

Erosions constitute ~40% of cases of thrombotic sudden coronary death [9]. It has been reported that eroded plaques have greater proportion of female, younger age, less % stenosis, less calcification, and less obstructive thrombus as compared with ruptured plaques. Plaque erosions are associated with smoking, especially in premenopausal women.

Calcified Nodule Calcified nodule is the least frequent cause of luminal thrombus. This lesion occurs in heavily calcified arteries and there is little or no underlying necrotic core, a luminal surface that is disrupted by nodules of dense calcium, and overlying thrombus [64]. Although the precise nature of this lesion remains incompletely understood, one hypothesis is that mechanical stress on the heavily calcified coronary arteries might fragment sheets of calcium, resulting in small nodules surrounded by fibrin, which may eventually erupt through the plaque surface. Intraplaque fibrin is also observed in noneruptive nodular calcification which does not have thrombus, potentially arising from surrounding leaky and damaged capillaries. Calcified nodules are more likely seen in older individuals and are preferentially found in patients with tortuous coronary arteries, diabetes mellitus, or chronic renal failure [30]. Eruptive calcified nodules are usually observed in eccentric lesions where protrusion of the nodules into the lumen causes disruption of the overlying endothelium to form a platelet or fibrin-rich thrombus, which is generally nonocclusive.

Lesion Progression

Episodic Rupture and Healing Silent luminal thrombi either from plaque rupture, erosion, or calcified nodule, are eventually repaired to form healed lesions [33]. This process contributes to a substantial increase in plaque burden and luminal narrowing. Silent luminal thrombi organize with granulation tissue, subsequent infiltration by SMCs and deposition of proteoglycans and collagen. As the thrombus propagates proximally and distally and through a progressive healing process, it converts into a fibrous plaque. Progressive healing of silent ruptures is characterized by the initial accumulation of proteoglycans and collagen type III, which is later replaced by collagen type I. Healed ruptures often exhibit multiple layers of necrotic core interspersed by fibrous tissue. The earliest rupture site is always located in the deepest intima, suggestive of previous thrombotic events, which sequentially results in lesion progression. Other healed lesions show no evidence of a preexisting rupture of the fibrous cap, and there is usually no necrotic core. These types of lesions may be the result of healed erosions. The luminal surface in healed lesions will eventually be fully endothelialized, and the underlying plaque consists of SMCs and proteoglycan-rich extracellular matrix. On the basis of autopsy findings for individuals who experienced sudden coronary deaths, healed ruptures are thought to contribute to a considerable increase in plaque burden and also to luminal narrowing. The prevalence of silent plaque rupture, erosion, or calcified nodule, in the clinical setting remains unknown. However, autopsy studies provide evidences of plaque progression from repeated thrombotic events.

Histologic studies have shown that the frequency of healed plaque rupture correlates with the degree of luminal narrowing [27]. Healed plaque rupture was identified in 16% of lesions with <20% diameter stenosis, 19% of lesions with 21%–50% stenosis, and 73% of lesions with >50% stenosis. It has been also reported that 61% of hearts from sudden coronary death patients have healed plaque ruptures and the percentage of stenosis increases with the increasing number of healed plaque ruptures.

Necrotic Core Expansion Necrotic core expansion may occur by entry of blood into the lesion. This can occur through cracks or fissures originating from the luminal surface as an entryway of blood into the lesions. The path is lined by few macrophages and red blood cells and/or fibrin [4,5,51]. The more-dominant mechanism of intraplaque hemorrhage, however, occurs through intraplaque vasa vasorum, which extend into the intima from the adventitia. Intraplaque hemorrhage and specifically red blood cell membranes that contain more free cholesterol than any other part of the body, is the most probable source of free cholesterol seen in advanced necrotic cores [49]. It is unlikely that this free cholesterol is derived from macrophages, given that cholesterol is present in an esterified form in macrophages. It has been shown that intraplaque hemorrhage in lesions from patients with sudden cardiac death was more common in ruptured plaques, TCFAs, and fibroatheromas with late necrotic cores than in early plaques. Moreover, the degree of hemorrhage was proportional to the size of the necrotic core, which also showed greater macrophage infiltration. Intraplaque hemorrhage was also shown to be likely to occur from leaky vasa vasorum. Unlike in the adventitia, the endothelium of neoangiogenic vessels found within the plaque lacks competent junctions, and exhibits membrane blebs, intracytoplasmic vacuoles, and basement membrane detachments. Poorly formed, "leaky" endothelial junctions frequently show an association with inflammation, particularly with macrophages and T lymphocytes, and are highly susceptible to extravasation of red blood cells contributing to intraplaque hemorrhage [49].

Notably, intraplaque hemorrhage is considered to be one of three factors contributing to the sudden increase in lesion size (the others being luminal thrombus and plaque fissure) that can cause the onset of ACS.

Calcification in Advanced Plaques

Progression of coronary atherosclerosis is generally accompanied by an increase in calcification. PIT is known to be the earliest type of lesion exhibiting calcification. At this stage, intimal calcification begins with SMC apoptosis and release of matrix vesicles in the lipid pools and has been defined as "early microcalcification" ($\geq$0.5 μm, and typically <15 μm) [65]. Microcalcification that occurs from macrophage apoptosis or macrophage vesicle releasing in the necrotic core of fibroatheroma is defined as "punctate microcalcification" ($\geq$15 μm). Microcalcification may coalesce into larger masses and involve both the necrotic core and the surrounding extracellular matrix including collagen to form speckled or fragmented calcification. This pattern may also start in the deeper region of the necrotic core close to the media. Sheets or plates of calcium that encompass both the surrounding intimal SMCs and collagen with or without a necrotic core might also be present. These larger fragments of calcium can be visualized by standard imaging techniques, such as radiography or CT. Nodular calcification is the least-common form in the coronary vasculature and is typically observed in tortuous arteries of older individuals or in patients with diabetes or chronic renal failure. The finding of ossification with intervening marrow formation in coronary arteries is rare.

Effects of Plaque Environment and External Stresses

In addition to several intrinsic plaque characteristics that may increase the susceptibility of plaque to rupture, there are factors extrinsic to the plaque that may act as triggers and enhance the likelihood of plaque rupture and subsequent thrombus formation [9,15–17,57,66]. Effects of abnormal exogenous stresses act from the initial stages of

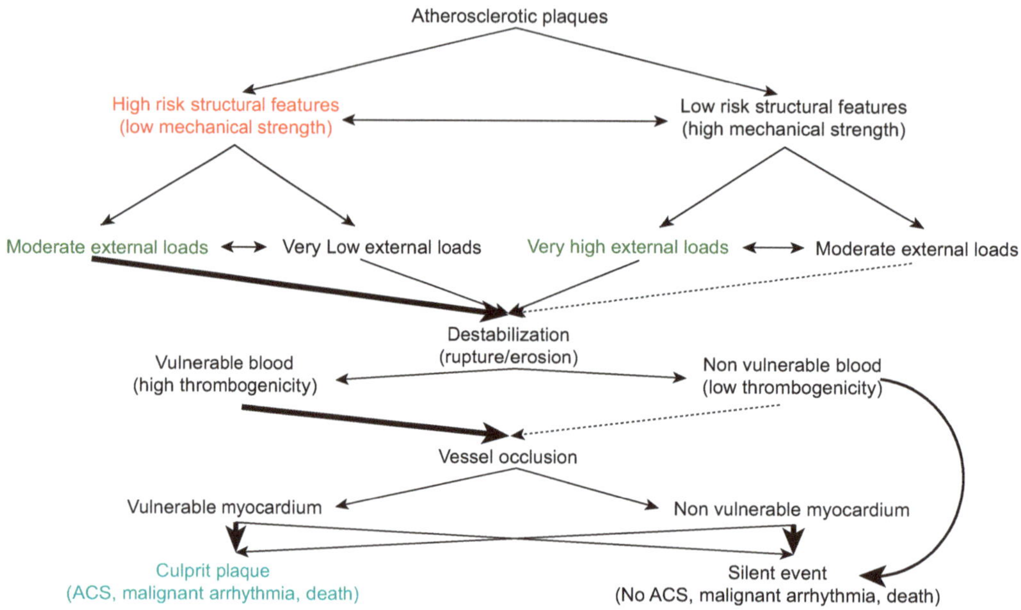

Figure 2.5.3 Projected natural course of atherosclerotic plaques based on their intrinsic and extrinsic features. *(Reproduced with permission from Stefanadis C, Antoniou CK, Tsiachris D, Pietri P. Coronary atherosclerotic vulnerable plaque: current perspectives. J Am Heart Assoc 2017;6:e005543 [Originally published March 17, 2017].)*

atherosclerosis and continue up to the late stages. Normal laminar flow causes relatively stable transverse and shear stresses, with limited oscillatory variations. However, exogenous stress as you may see below, introduces anisotropy of the fibrous cap and rendering it more prone to rupture. Fig. 2.5.3 presents the projected natural course of atherosclerotic plaques based on their intrinsic and extrinsic features [9].

Endothelial Shear Stress

ESS is the tangential force derived by the friction of the flowing blood on the endothelial surface. It seems to play a key role in the pathobiology of atherogenesis, plaque formation, and plaque progression to vulnerability [67,68].

Nonstenotic vulnerable plaques are exposed to low ESS, which perpetuates an ongoing vicious cycle of local inflammation, matrix degradation, and expansive remodeling leading. Recent studies have shown that low ESS is a continue stimulus that perpetuates the atherosclerotic process in a dose-dependent manner leading to plaque vulnerability [15,69]. Low ESS modulates the accumulation of lipids and neovascularization resulting to plaque volume expansion. In the setting of inflammation, low ESS promotes internal elastic lamina fragmentation by increasing mRNA expression and activity of major elastolytic MMPs (MMP 2, 9, 12) and cathepsins (cathepsin K and S) relative to their endogenous inhibitors (TIMPs and cystatin C, respectively) [17]. Inflammatory cells migrate through the fragmented internal elastic lamina into the media where matrix degradation is further promoted followed by expansive remodeling, which is a key pathobiological feature of vulnerable plaque. Low ESS also increases the expression and enzymatic activity of collagenolytic MMPs (MMP 1, 8, 13, and 14), which in turn degrade the collagen in fibrous cap, promoting thus, fibrous cap thinning.

Previous studies have shown that low baseline ESS is associated with inflammation, thin fibrous cap, large lipid core, and expansive remodeling, all of which are key pathobiological fingerprints of vulnerable plaque. Moreover, low ESS is found to be an independent predictor of plaque progression and expansive remodeling with lumen narrowing [13,70].

In contrast to nonstenotic vulnerable plaques, the stenotic vulnerable plaques create a heterogeneous local ESS environment along their length. As blood flow is diverged toward the remaining lumen by the upstream plaque edge, a low-stress area occurs in the upstream shoulders, a high shear stress area at the neck of the plaque and a low/oscillatory ESS in the downstream shoulder. The majority of ruptures occur at the upstream shoulders of the plaque, whereas the downstream regions are more stable and therefore less prone to rupture. Plaque rupture can also occur at the most stenotic part of the plaque likely as a result of local erosion in the setting of very high flow and ESS. However, the pathobiological association of high ESS with plaque erosion is not well investigated.

Wall Stress

Local wall stress is another major local hemodynamic factor that plays a key role in the pathobiology of plaque rupture. Wall stress is proportional to blood pressure and inversely proportional to lumen stenosis. The highest wall stress typically occurs at the upstream shoulders of a stenotic plaque or within a nonstenotic plaque colocalizing with thin fibrous cap, increased macrophage density, and local microcalcifications.

Overall, the dynamic synergism among local hemodynamic factors (ESS and wall stress), plaque composition, and vascular remodeling is a key player in acute plaque disruption.

Vulnerable Plaque Characteristics—Insights From Imaging

Advances in imaging modalities have allowed detailed in vivo correlation of the histopathological features thought to underlie plaque vulnerability. Most of the studies performed over the last year focused either on the detection of or on the identification of vulnerable plaques by invasive imaging modalities. Future studies are anticipated to combine both invasive and noninvasive imaging with the general patient profile. Fig. 2.5.4 presents the history of vulnerable plaque based on histology and imaging studies [11].

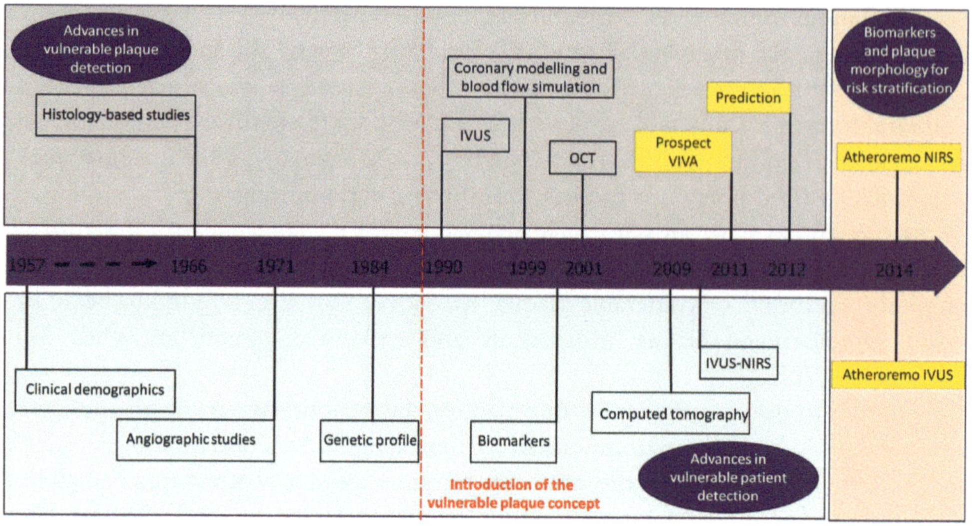

Figure 2.5.4 The history of the vulnerable plaque and vulnerable patient detection. *(Reproduced with permission from Bourantas CV, Garcia-Garcia HM, Torii R, Zhang YJ, Westwood M, Crake T, Serruys PW. Vulnerable plaque detection: an unrealistic quest or a feasible objective with a clinical value? Heart April 2016;102(8):581–9.)*

Invasive Methods in Vulnerable Plaque Imaging

Among invasive imaging modalities, some focus on the structural microanatomy of a plaque [intravascular ultrasound (IVUS), optical coherence tomography (OCT)], some on assessing the metabolic activity (intravascular thermography and elastography), and finally some techniques rely on measuring the chemical composition of a plaque (NIRS) [57,71]. The most promising methods include IVUS, OCT, and NIRS.

Coronary Angiography

Coronary angiography is considered as the "gold standard" for the definition of coronary anatomy. However, coronary angiography fails to give information about arterial wall pathology. Moreover, in the early phases of atherosclerosis the luminal size is not affected by plaque growth due to expansion of the external elastic membrane, a process called "positive remodeling" [72]. As the plaque grows and approaches the lumen, negative remodeling occurs. Positive remodeling is also seen in an acute MI at the site of plaque rupture, whereas negative remodeling and smaller plaque areas may be associated with stable angina. This phenomenon of remodeling, along with the limitation of the method to provide information for the composition of the arterial wall, makes angiography a poor technique for the identification of vulnerable plaques and true atherosclerotic disease burden.

Intravascular Ultrasound

IVUS has a particular role in identifying specific vulnerable plaque characteristics and is considered one of the two invasive coronary atherosclerotic plaque imaging modalities.

This method can visualize the external elastic lamina of the vessel wall, allowing the determination of vessel size, plaque burden and morphology [73,74]. Added modules can enhance tissue characterization, allowing for detection and quantification of different plaque structures. Virtual histology IVUS, iMAP IVUS, and integrated backscatter IVUS are examples of added modules to current IVUS technology.

IVUS and its modular expansions can reliably assess plaque burden, expansive remodeling, as well as presence and relative proportion of necrotic core, calcifications, and neoangiogenesis and neovessel formation (contrast enhanced; penetration depth, 5 mm). At its current form (20/40 MHz transducers), it cannot visualize plaque fibrous caps, especially thinner ones (resolution, >100 lm). Most prospective studies regarding ability of vulnerable plaques to predict events used IVUS-based approaches in their definition of vulnerability [12,13,75,76].

Optical Coherence Tomography

OCT is the second major coronary plaque invasive imaging modality that uses near infrared light (1.3 μm in wavelength). This offers a much greater axial resolution but comes at a price, as it allows a lower penetrance (penetration depth, 1.5 mm) into the vessel wall [77,78].

Given the much smaller wavelength of light in comparison to ultrasound, OCT resolution is ~10 μm, allowing for visualization of fibrous caps, collagen content (polarization-sensitive OCT), macrophages (related to inflammation), neovessels (appearing as microchannels), ruptures, and thrombi. In fact, the excellent resolution of OCT permits measuring the fibrous cap thickness as well as macrophages quantification within the cap, with high agreement with histopathology. A recent study in patients with MI showed that rupture of the plaque usually occurs in sites different than the minimal lumen and at the shoulder of areas with thin fibrous cap (≤90 μm) [79]. Moreover, patients with STEMI had greater plaque disruption and smaller minimal lumen area than patients with NSTEMI.

Several aspects of vulnerability can be visualized by OCT, but not all. For instance, due to the low penetration depth, OCT does not permit proper recognition and estimation of remodeling and necrotic core size. Moreover, other OCT limitations are the high scattering effect of its components, longer image acquisition times (although newer, frequency domain analysis-based processing methods), frequent artifacts, as well as poor discrimination between calcified areas and lipid core, given that both appear as signal poor areas. A newer form of OCT, micro OCT (μ-OCT), can now offer axial and lateral resolution in the order of 1–2 μm by means of advanced frequency domain analysis and use of broad bandwidth light, approaching histology levels. Given that this method may, in the future, offer real-time in vivo images (as well as three-dimensional reconstruction) of the vessel wall at subcellular resolution, it may prove an extremely useful tool for the classification of atherosclerotic plaques and identification of the vulnerable ones, replacing classical histology.

Thermography

Invasive coronary thermography is an approach aiming to detect subtle temperature increases of the vessel wall in areas of heat production, based on the principle that inflammation produces heat. Increased heat production has been shown to accompany inflamed and/or ruptured atherosclerotic plaques [80—89]. Purpose-built catheters include hydrofoil and balloon-based designs, ensuring adequate apposition of the thermistor module on the vessel wall. The latter ensures temporary lumen occlusion, overcoming thus the cooling effect of blood flow, leading to underestimation of local heat production. In patients with ACS heat generation is not a local phenomenon at the culprit lesion but is widespread in the coronary vasculature and can be also detected in nonculprit lesions [85]. A correlation with systemic levels of inflammatory mediators, such as C-reactive protein has also been shown [90]. In addition, increased plaque temperature has been shown to be an independent predictor in patients undergoing percutaneous coronary intervention [91]. Morphological and functional plaque characteristics have been correlated in studies performed with IVUS (positive remodeling) and thermography (increased plaque temperature) [88]. Technical issues have, so far, preclude the use of thermography for plaque vulnerability assessment and prediction of ACS.

Angioscopy

Angioscopy is an invasive imaging modality that evaluates the luminal surface by direct visualization of the interior of blood vessels [92]. Atherosclerotic plaques are seen as well-demarcated white, yellow or orange protrusions into the lumen. Deep yellow and yellow-red lesions denote lipid-rich atheromas with necrotic cores and a thin fibrous cap, respectively, although gray—white color corresponds to fibrous plaques. Patients with UA generally have predominantly yellow lesions, commonly with ruptures and red thrombi. A higher intensity of angioscopic yellow color was associated with thinner fibrous caps by OCT. As an imaging modality, this method may elucidate the morphological interaction between the plaque and thrombus, and assess the effectiveness of percutaneous coronary intervention or antiatherosclerotic drugs. Furthermore, the recently developed molecular angioscopy approach allows to observe LDL-oxide, collagen, and macrophages.

Near-Infrared Spectroscopy

NIRS is a novel intravascular technique that has been developed for the identification of the lipid core within an atherosclerotic lesion [93]. Low sensitivity and high penetration has rendered this method appropriate for assessing the lipid content of plaques, especially in cases of positive remodeling, with large, deep-seated, lipid-laden necrotic cores. The determination of lesion's chemical composition is the unique spectral characteristics of lipid core cholesterol. The results of NIRS are presented as a "chemogram"—a color coded map that indicates the probability of lipid core presence in given locations.

Studies using NIRS have shown that large lipid content, rather than plaque burden, is associated with thin cap fibroatheroma features. Larger lipid core burden has been shown to accurately differentiate between culprit and nonculprit lesion in ST-elevation MI patients and has been associated with higher risk for periprocedural MI.

Because NIRS only provides compositional data, a combination NIRS/IVUS catheter has been designed to provide both plaque characterization as well as chemical composition [94]. An advantage of the NIRS-IVUS system is that it provides both structural (IVUS) and compositional (NIRS) information in a single pullback. A modified form of NIRS, near-infrared autofluorescence, involves active stimulation of lipid components to emit detectable infrared light. Overall, it seems that NIRS is a very promising method for identification of high-risk plaques.

Noninvasive Imaging Systems

Although noninvasive modalities are obviously the less precise method of vulnerable plaque detection, a research in this field is particularly important as these techniques may help to identify high-risk plaques in asymptomatic subjects without the risks associated with an invasive procedure [95].

Computed Tomography Angiography

Contrast enhanced coronary computed tomography angiography (CTA) is typically performed by a computed tomography scanner after the injection of iodinated contrast. The contrast enhances the arterial lumen [96,97]. Advances in technology have allowed for improved image quality with reduced scan times and radiation exposure. This method has significant benefits such as noninvasiveness, the ability to imaging the whole coronary vasculature, and potential for assessing both vessel wall and lumen.

From a clinical standpoint, CTA has emerged as one of the most promising imaging modalities in identifying vulnerable plaque characteristics, approaching histology levels. Thus, events both in the cellular and molecular level can be visualized. CTA can determine the degree of stenosis as well as such atherosclerotic changes as plaque volume, positive vascular remodeling, lipoprotein deposition (noncalcified plaques), and calcification. The presence of positive remodeling, low attenuation and spotty calcification were significantly more frequent in the culprit ACS. Furthermore, in asymptomatic patients, presence of positive remodeling and low attenuation plaque portends a greater risk for development of acute coronary events. Primary prevention studies are needed to evaluate the merit of plaque characterization and to define the population that will benefit the most from such a strategy.

Molecular Imaging

Molecular imaging involves targeting and visualizing specific components of biological processes and has been extensively used in attempts to visualize specific vulnerable plaque components [98]. Any imaging modality can theoretically be used for molecular imaging as long as a proper tracer can be developed. The two main imaging modalities that are used for vulnerable plaque imaging, often in conjunction with CT or magnetic resonance imaging (to provide anatomic background) are positron emission tomography (PET) and single-photon emission computed tomography (SPECT).

Positron Emission Tomography Specific tracers can be used in PET imaging to target the metabolic processes involved in atherosclerosis and plaque disruption [99]. The main pathophysiological processes that can be detected by this noninvasive imaging modality related to vulnerable plaque are inflammation [(18F-fluorodeoxyglucose (^{18}F-FDG)]), microcalcification (18F-sodium fluoride) and hypoxia (18F-fluoromisonidazole) [99,100].

PET-^{18}F-FDG detects active plaque inflammation by assessing glucose utilization as a measure of local metabolic activity. ^{18}F-FDG is a glucose analog that is injected intravenously and is taken up by inflammatory cells that utilize glucose for metabolic activity. In vulnerable plaques, activated macrophages with up to 50-fold higher glycolytic activity than background tissue take up PET-^{18}F-FDG and can be visualized as areas of active inflammation. Although the initial reports concerning the identification of increased plaque inflammation were for carotid arteries, later reports confirmed the clinical value of PET-^{18}F-in detecting vulnerable atherosclerotic coronary plaques.

PET has also been experimentally evaluated for the study of intraplaque hypoxia and assessment of macrophage content. Finally, PET-CT in combination with sodium fluoride as a tracer may reveal sites of active calcification [96]. This is achieved because of preferential binding of the tracer on hydroxyapatite crystals, whose exposed surface is much larger in areas of ongoing crystal formation.

Cardiac PET-CT as a new hybrid imaging technique in the evaluation of patients with coronary artery disease provides complementary information on morphological and functional coronary artery characteristics. Although its role in clinical practice warrants further study, it is clear that it holds great potential for the diagnostic evaluation of these patients.

Single-Photon Emission Computed Tomography Cardiac SPECT imaging has been used for myocardial perfusion and viability for many years and is a well established and the most commonly used imaging modality in these areas. For the identification of specific vulnerable plaque characteristics, Technetium-99m-labeled (99mTc) radiotracers such as 99mTc-sestamibi and 99mTc-tetrofosmin have been used mostly in experimental studies [101].

Thus, although 99mTc SPECT is able to assess atherosclerotic coronary lesions in experimental studies, most of the clinical studies using SPECT for imaging vulnerable plaques focus on the investigation of patients with carotid plaques. Increased uptake of radioisotopes was reported in culprit coronary lesions when compared to the normal arteries, which did not demonstrate tracer accumulation.

Overall, despite promising reports with SPECT imaging, the important application of nuclear cardiac imaging lies in the utilization of PET as SPECT is limited by its poor spatial resolution in imaging coronary arteries and coronary lesions.

Therapeutic Implications

Although the advancement of imaging modalities during the last decades has provided insight into the biological processes involved in atherosclerosis progression, little is known about the relevant targets against which it may be possible to intervene and prevent acute coronary events [9].

Contemporary medical therapy, and in specific statin treatment, has prevented both first and recurrent ACS in broad categories of patients [102–105]. Interestingly, the cholesterol level lowering and event reduction were not associated with significant changes on the degree of stenosis and resulted in only modest reductions in atheroma volume. This observation has led to the hypothesis that lipid lowering with statins may alter qualitative characteristics of atheromas, limiting thus the propensity of plaques to rupture and leading to "plaque stabilization". This effect of statin therapy has been attributed to "pleiotropic effects" [106]. Indeed, inhibition by statins of prenylation of a family of intracellular signaling molecules known as small G proteins, among others, may have direct effects on cells independent of systemic lipid lowering (alter macrophage metabolism, smooth muscle proliferation, endothelial function).

Although a remarkable benefit with statin therapy has been observed, patients appropriately treated with this class of agents are still at considerable risk for ACS. During the last years, novel therapies for lowering LDL cholesterol levels below those achievable with statins alone (inhibition of serum proprotein convertase subtilisin/kexin 9) have provided considerable promise in this regard [105,107]. Therapies that target other aspects of the lipid profile [high-density lipoprotein (HDL)] are still ongoing and results are yet conflicting [102]. Thus, further studies are needed in order to evaluate whether ACS patients would benefit from LDL- or HDL-targeted therapies or a combination of not.

Given the role of inflammation in the pathophysiological aspects of plaque rupture, several studies are assessing the use of specific anti-inflammatory therapies to reduce the risk of a recurrent ACS. Among others, potential targets that are currently under investigation are MMPs, tumor necrosis factor-alpha (TNF-alpha), Interleukin one beta (IL-1-beta), miRNAs, and specialized proresolving mediators (SPMs, a group of structurally related signaling molecules derived from polyunsaturated fatty acids) [41,44,104,108–117]. Current evidence in this field is limited and further studies are needed.

Finally, as thrombosis is considered as the final cause of ACS, apart from antiplatelet and anticoagulant agents, the potential therapeutic possibilities of inhibiting the TF pathway have been explored [59]. Interestingly, Specific inhibition of vascular TF by TF pathway inhibitor was associated with a significant reduction of acute thrombus formation in lipid-rich plaques.

Thus, current ongoing research focused on targeting specific biological processes involved ion plaque progression may lead to the development of new therapeutic strategies in the prevention of ACS.

CONCLUSION

New imaging modalities have advanced the field of vulnerable atherosclerotic plaque research. Many findings determined by histological evaluation of post mortem samples have been verified in vivo and further advanced by invasive and noninvasive imaging. The main features that characterize a vulnerable plaque and predict future coronary events are greater plaque burden, larger lipid and necrotic core, thinner fibrous caps, and inflammatory cells infiltrating the fibrous cap. With further development of noninvasive imaging that minimizes radiation and cost, and invasive imaging systems that minimize risk, it is conceivable that a screening strategy of asymptomatic patients may identify a high-risk population that would benefit from aggressive prophylactic medical or interventional treatment.

REFERENCES

[1] WHO. Cardiovascular Diseases (CVDs) Fact sheet-Reviewed. June 2016.

[2] Roffi M, Patrono C, Collet JP, Mueller C, Valgimigli M, Andreotti F, Bax JJ, Borger MA, Brotons C, Chew DP, Gencer B, Hasenfuss G, Kjeldsen K, Lancellotti P, Landmesser U, Mehilli J, Mukherjee D, Storey RF, Windecker S, Baumgartner H, Gaemperli O, Achenbach S, Agewall S, Badimon L, Baigent C, Bueno H, Bugiardini R, Carerj S, Casselman F, Cuisset T, Erol C, Fitzsimons D, Halle M, Hamm C, Hildick-Smith D, Huber K, Iliodromitis E, James S, Lewis BS, Lip GY, Piepoli MF, Richter D, Rosemann T, Sechtem U, Steg PG, Vrints C, Luis Zamorano J, Management of Acute Coronary Syndromes in Patients Presenting without Persistent STSEotESoC. 2015 ESC Guidelines for the management of acute coronary syndromes in patients presenting without persistent ST-segment elevation: task force for the management of Acute Coronary Syndromes in patients presenting without persistent ST-segment elevation of the European Society of Cardiology (ESC). Eur Heart J 2016;37:267–315.

[3] Task Force on the management of STseamiotESoC, Steg PG, James SK, Atar D, Badano LP, Blomstrom-Lundqvist C, Borger MA, Di Mario C, Dickstein K, Ducrocq G, Fernandez-Aviles F, Gershlick AH, Giannuzzi P, Halvorsen S, Huber K, Juni P, Kastrati A, Knuuti J, Lenzen MJ, Mahaffey KW, Valgimigli M, van't Hof A, Widimsky P, Zahger D. ESC Guidelines for the management of acute myocardial infarction in patients presenting with ST-segment elevation. Eur Heart J 2012;33:2569–619.

[4] Naghavi M, Libby P, Falk E, Casscells SW, Litovsky S, Rumberger J, Badimon JJ, Stefanadis C, Moreno P, Pasterkamp G, Fayad Z, Stone PH, Waxman S, Raggi P, Madjid M, Zarrabi A, Burke A, Yuan C, Fitzgerald PJ, Siscovick DS, de Korte CL, Aikawa M, Airaksinen KE,

Assmann G, Becker CR, Chesebro JH, Farb A, Galis ZS, Jackson C, Jang IK, Koenig W, Lodder RA, March K, Demirovic J, Navab M, Priori SG, Rekhter MD, Bahr R, Grundy SM, Mehran R, Colombo A, Boerwinkle E, Ballantyne C, Insull Jr W, Schwartz RS, Vogel R, Serruys PW, Hansson GK, Faxon DP, Kaul S, Drexler H, Greenland P, Muller JE, Virmani R, Ridker PM, Zipes DP, Shah PK, Willerson JT. From vulnerable plaque to vulnerable patient: a call for new definitions and risk assessment strategies: part II. Circulation 2003;108:1772–8.

[5] Naghavi M, Libby P, Falk E, Casscells SW, Litovsky S, Rumberger J, Badimon JJ, Stefanadis C, Moreno P, Pasterkamp G, Fayad Z, Stone PH, Waxman S, Raggi P, Madjid M, Zarrabi A, Burke A, Yuan C, Fitzgerald PJ, Siscovick DS, de Korte CL, Aikawa M, Juhani Airaksinen KE, Assmann G, Becker CR, Chesebro JH, Farb A, Galis ZS, Jackson C, Jang IK, Koenig W, Lodder RA, March K, Demirovic J, Navab M, Priori SG, Rekhter MD, Bahr R, Grundy SM, Mehran R, Colombo A, Boerwinkle E, Ballantyne C, Insull Jr W, Schwartz RS, Vogel R, Serruys PW, Hansson GK, Faxon DP, Kaul S, Drexler H, Greenland P, Muller JE, Virmani R, Ridker PM, Zipes DP, Shah PK, Willerson JT. From vulnerable plaque to vulnerable patient: a call for new definitions and risk assessment strategies: part I. Circulation 2003;108:1664–72.

[6] Libby P, Pasterkamp G. Requiem for the 'vulnerable plaque'. Eur Heart J 2015;36:2984–7.

[7] Nishimura S, Ehara S, Hasegawa T, Matsumoto K, Yoshikawa J, Shimada K. Cholesterol crystal as a new feature of coronary vulnerable plaques: an optical coherence tomography study. J Cardiol 2017; 69:253–9.

[8] Michail M, Serruys PW, Stettler R, Crake T, Torii R, Tenekecioglu E, Zeng Y, Onuma Y, Mathur A, Bourantas CV. Intravascular multimodality imaging: feasibility and role in the evaluation of coronary plaque pathology. Eur Heart J Cardiovasc Imaging 2017;18.

[9] Stefanadis C, Antoniou CK, Tsiachris D, Pietri P. Coronary atherosclerotic vulnerable plaque: current perspectives. J Am Heart Assoc 2017;6:e005543 [Originally published March 17, 2017].

[10] Pasterkamp G, den Ruijter HM, Libby P. Temporal shifts in clinical presentation and underlying mechanisms of atherosclerotic disease. Nat Rev Cardiol 2017;14:21–9.

[11] Bourantas CV, Garcia-Garcia HM, Torii R, Zhang YJ, Westwood M, Crake T, Serruys PW. Vulnerable plaque detection: an unrealistic quest or a feasible objective with a clinical value? Heart April 2016;102(8):581–9.

[12] Stone GW, Maehara A, Lansky AJ, de Bruyne B, Cristea E, Mintz GS, Mehran R, McPherson J, Farhat N, Marso SP, Parise H, Templin B, White R, Zhang Z, Serruys PW, Investigators P. A prospective natural-history study of coronary atherosclerosis. New Engl J Med 2011;364:226–35.

[13] Stone PH, Saito S, Takahashi S, Makita Y, Nakamura S, Kawasaki T, Takahashi A, Katsuki T, Nakamura S, Namiki A, Hirohata A, Matsumura T, Yamazaki S, Yokoi H, Tanaka S, Otsuji S, Yoshimachi F, Honye J, Harwood D, Reitman M, Coskun AU, Papafaklis MI, Feldman CL, Investigators P. Prediction of progression of coronary artery disease and clinical outcomes using vascular profiling of endothelial shear stress and arterial plaque characteristics: the PREDICTION study. Circulation 2012;126:172–81.

[14] Cheng JM, Garcia-Garcia HM, de Boer SP, Kardys I, Heo JH, Akkerhuis KM, Oemrawsingh RM, van Domburg RT, Ligthart J, Witberg KT, Regar E, Serruys PW, van Geuns RJ, Boersma E. In vivo detection of high-risk coronary plaques by radiofrequency intravascular ultrasound and cardiovascular outcome: results of the ATHEROREMO-IVUS study. Eur Heart J 2014;35:639–47.

[15] Chatzizisis YS, Toutouzas K, Giannopoulos AA, Riga M, Antoniadis AP, Fujinom Y, Mitsouras D, Koutkias VG, Cheimariotis G, Doulaverakis C, Tsampoulatidis I, Chouvarda I, Kompatsiaris I, Nakamura S, Rybicki FJ, Maglaveras N, Tousoulis D, Giannoglou GD. Association of global and local low endothelial shear stress with high-risk plaque using intracoronary 3D optical coherence tomography: introduction of 'shear stress score'. Eur Heart J Cardiovasc Imaging 2016. http://dx.doi.org/10.1093/ehjci/jew134.

[16] Toutouzas K, Chatzizisis YS, Riga M, Giannopoulos A, Antoniadis AP, Tu S, Fujino Y, Mitsouras D, Doulaverakis C, Tsampoulatidis I, Koutkias VG, Bouki K, Li Y, Chouvarda I, Cheimariotis G, Maglaveras N, Kompatsiaris I, Nakamura S, Reiber JH, Rybicki F, Karvounis H, Stefanadis C, Tousoulis D, Giannoglou GD. Accurate and reproducible reconstruction of coronary arteries and endothelial shear stress calculation using 3D OCT: comparative study to 3D IVUS and 3D QCA. Atherosclerosis 2015;240:510–9.

[17] Chatzizisis YS, Antoniadis AP, Wentzel JJ, Giannoglou GD. Vulnerable plaque: the biomechanics of matter. Atherosclerosis 2014;236:351–2.
[18] Ohayon J, Finet G, Le Floc'h S, Cloutier G, Gharib AM, Heroux J, Pettigrew RI. Biomechanics of atherosclerotic coronary plaque: site, stability and in vivo elasticity modeling. Ann Biomed Eng 2014; 42:269–79.
[19] Angelini P, Leachman R, Heibig J. Flow characteristics of coronary balloon catheters. Texas Heart Inst J 1986;13:213–5.
[20] Fuster V, Badimon L, Badimon JJ, Chesebro JH. The pathogenesis of coronary artery disease and the acute coronary syndromes (1). New Engl J Med 1992;326:242–50.
[21] Thygesen K, Alpert JS, Jaffe AS, Simoons ML, Chaitman BR, White HD, Joint ESC/ACCF/AHA/WHF/Task Force for the Universal Definition of Myocardial Infarction, Katus HA, Lindahl B, Morrow DA, Clemmensen PM, Johanson P, Hod H, Underwood R, Bax JJ, Bonow RO, Pinto F, Gibbons RJ, Fox KA, Atar D, Newby LK, Galvani M, Hamm CW, Uretsky BF, Steg PG, Wijns W, Bassand JP, Menasche P, Ravkilde J, Ohman EM, Antman EM, Wallentin LC, Armstrong PW, Simoons ML, Januzzi JL, Nieminen MS, Gheorghiade M, Filippatos G, Luepker RV, Fortmann SP, Rosamond WD, Levy D, Wood D, Smith SC, Hu D, Lopez-Sendon JL, Robertson RM, Weaver D, Tendera M, Bove AA, Parkhomenko AN, Vasilieva EJ, Mendis S. Third universal definition of myocardial infarction. Circulation 2012;126: 2020–35.
[22] Davies MJ. Pathophysiology of acute coronary syndromes. Indian Heart J 2000;52:473–9.
[23] Maseri A, Chierchia S, Davies G. Pathophysiology of coronary occlusion in acute infarction. Circulation 1986;73:233–9.
[24] Fuster V. Lewis A. Conner Memorial Lecture. Mechanisms leading to myocardial infarction: insights from studies of vascular biology. Circulation 1994;90:2126–46.
[25] Falk E, Nakano M, Bentzon JF, Finn AV, Virmani R. Update on acute coronary syndromes: the pathologists' view. Eur Heart J 2013;34:719–28.
[26] Stary HC, Chandler AB, Dinsmore RE, Fuster V, Glagov S, Insull Jr W, Rosenfeld ME, Schwartz CJ, Wagner WD, Wissler RW. A definition of advanced types of atherosclerotic lesions and a histological classification of atherosclerosis. A report from the Committee on vascular lesions of the Council on Arteriosclerosis, American Heart Association. Circulation 1995;92:1355–74.
[27] Virmani R, Kolodgie FD, Burke AP, Farb A, Schwartz SM. Lessons from sudden coronary death: a comprehensive morphological classification scheme for atherosclerotic lesions. Arterioscler Thromb Vasc Biol 2000;20:1262–75.
[28] Virmani R, Burke AP, Kolodgie FD, Farb A. Pathology of the thin-cap fibroatheroma: a type of vulnerable plaque. J Interv Cardiol 2003;16:267–72.
[29] Virmani R, Burke AP, Kolodgie FD, Farb A. Vulnerable plaque: the pathology of unstable coronary lesions. J Interv Cardiol 2002;15:439–46.
[30] Virmani R, Burke AP, Farb A, Kolodgie FD. Pathology of the vulnerable plaque. J Am Coll Cardiol 2006;47:C13–8.
[31] Virmani R, Burke AP, Farb A, Kolodgie FD. Pathology of the unstable plaque. Prog Cardiovasc Dis 2002;44:349–56.
[32] Bentzon JF, Otsuka F, Virmani R, Falk E. Mechanisms of plaque formation and rupture. Circ Res 2014;114:1852–66.
[33] Yahagi K, Kolodgie FD, Otsuka F, Finn AV, Davis HR, Joner M, Virmani R. Pathophysiology of native coronary, vein graft, and in-stent atherosclerosis. Nat Rev Cardiol 2016;13:79–98.
[34] Linton MF, Yancey PG, Davies SS, Jerome WGJ, Linton EF, Vickers KC. In: De Groot LJ, Chrousos G, Dungan K, Feingold KR, Grossman A, Hershman JM, Koch C, Korbonits M, McLachlan R, New M, Purnell J, Rebar R, Singer F, Vinik A, editors. The role of lipids and lipoproteins in atherosclerosis. South Dartmouth (MA): Endotext; 2000.
[35] Schroeder AP, Falk E. Vulnerable and dangerous coronary plaques. Atherosclerosis 1995; 118(Suppl.):S141–9.
[36] Libby P, Sukhova G, Lee RT, Galis ZS. Cytokines regulate vascular functions related to stability of the atherosclerotic plaque. J Cardiovasc Pharmacol 1995;25(Suppl. 2):S9–12.

[37] Kristensen SD, Ravn HB, Falk E. Insights into the pathophysiology of unstable coronary artery disease. Am J Cardiol 1997;80:5E−9E.
[38] Arbustini E, Morbini P, Bello BD, Prati F, Specchia G. From plaque biology to clinical setting. Am Heart J 1999;138:S55−60.
[39] Krams R, Segers D, Mousavi Gourabi B, Maat W, Cheng C, van Pelt C, van Damme LC, de Feyter P, van der Steen T, de Korte CL, Serruys PW. Inflammation and atherosclerosis: mechanisms underlying vulnerable plaque. J Interv Cardiol 2003;16:107−13.
[40] Tousoulis D, Pitsavos C, Stefanadis C. C-reactive protein and lesion morphology. Circulation 2004; 109:e153. Author reply e153.
[41] Tousoulis D, Oikonomou E, Economou EK, Crea F, Kaski JC. Inflammatory cytokines in atherosclerosis: current therapeutic approaches. Eur Heart J 2016;37:1723−32.
[42] Siasos G, Tsigkou V, Oikonomou E, Zaromitidou M, Tsalamandris S, Mourouzis K, Vavuranakis M, Anastasiou M, Vlasis K, Limperi M, Gennimata V, Boletis JN, Papavassiliou AG, Tousoulis D. Circulating biomarkers determining inflammation in atherosclerosis progression. Curr Med Chem 2015; 22:2619−35.
[43] Siasos G, Tousoulis D, Kioufis S, Oikonomou E, Siasou Z, Limperi M, Papavassiliou AG, Stefanadis C. Inflammatory mechanisms in atherosclerosis: the impact of matrix metalloproteinases. Curr Top Med Chem 2012;12:1132−48.
[44] Papageorgiou N, Tousoulis D. Oxidized-LDL immunization for the treatment of atherosclerosis: how far are we? Int J Cardiol 2016;222:93−4.
[45] Vindis C, Escargueil-Blanc I, Uchida K, Elbaz M, Salvayre R, Negre-Salvayre A. Lipid oxidation products and oxidized low-density lipoproteins impair platelet-derived growth factor receptor activity in smooth muscle cells: implication in atherosclerosis. Redox Rep 2007;12:96−100.
[46] Otsuka F, Kramer MC, Woudstra P, Yahagi K, Ladich E, Finn AV, de Winter RJ, Kolodgie FD, Wight TN, Davis HR, Joner M, Virmani R. Natural progression of atherosclerosis from pathologic intimal thickening to late fibroatheroma in human coronary arteries: a pathology study. Atherosclerosis 2015;241:772−82.
[47] Kolodgie FD, Burke AP, Wight TN, Virmani R. The accumulation of specific types of proteoglycans in eroded plaques: a role in coronary thrombosis in the absence of rupture. Curr Opin Lipidol 2004;15:575−82.
[48] Kolodgie FD, Burke AP, Farb A, Weber DK, Kutys R, Wight TN, Virmani R. Differential accumulation of proteoglycans and hyaluronan in culprit lesions: insights into plaque erosion. Arterioscler Thromb Vasc Biol 2002;22:1642−8.
[49] Di Stefano R, Felice F, Balbarini A. Angiogenesis as risk factor for plaque vulnerability. Curr Pharm Des 2009;15:1095−106.
[50] Kolodgie FD, Gold HK, Burke AP, Fowler DR, Kruth HS, Weber DK, Farb A, Guerrero LJ, Hayase M, Kutys R, Narula J, Finn AV, Virmani R. Intraplaque hemorrhage and progression of coronary atheroma. New Engl J Med 2003;349:2316−25.
[51] Carlier S, Kakadiaris IA, Dib N, Vavuranakis M, O'Malley SM, Gul K, Hartley CJ, Metcalfe R, Mehran R, Stefanadis C, Falk E, Stone G, Leon M, Naghavi M. Vasa vasorum imaging: a new window to the clinical detection of vulnerable atherosclerotic plaques. Curr Atheroscler Rep 2005;7: 164−9.
[52] Virmani R, Kolodgie FD, Burke AP, Finn AV, Gold HK, Tulenko TN, Wrenn SP, Narula J. Atherosclerotic plaque progression and vulnerability to rupture: angiogenesis as a source of intraplaque hemorrhage. Arterioscler Thromb Vasc Biol 2005;25:2054−61.
[53] Vavuranakis M, Kakadiaris IA, Papaioannou TG, O'Malley SM, Carlier S, Naghavi M, Stefanadis C. Contrast-enhanced intravascular ultrasound: combining morphology with activity-based assessment of plaque vulnerability. Expert Rev Cardiovasc Ther 2007;5:917−25.
[54] Vavuranakis M, Kalogeras K, Tousoulis D. Vasa vasorum: still an invisible factor? J Am Coll Cardiol 2015;66:2157−8.
[55] Tousoulis D, Androulakis E, Papageorgiou N, Siasos G, Latsios G, Charakida M, Kampoli AM, Oikonomou E, Stefanadis C. Novel biomarkers assessing endothelial dysfunction: role of microRNAs. Curr Top Med Chem 2013;13:1518−26.

[56] Papageorgiou N, Tslamandris S, Giolis A, Tousoulis D. MicroRNAs in cardiovascular disease: perspectives and reality. Cardiol Rev 2016;24:110–8.
[57] Toutouzas K, Benetos G, Karanasos A, Chatzizisis YS, Giannopoulos AA, Tousoulis D. Vulnerable plaque imaging: updates on new pathobiological mechanisms. Eur Heart J 2015;36:3147–54.
[58] Libby P, Tabas I, Fredman G, Fisher EA. Inflammation and its resolution as determinants of acute coronary syndromes. Circ Res 2014;114:1867–79.
[59] Abbate R, Cioni G, Ricci I, Miranda M, Gori AM. Thrombosis and acute coronary syndrome. Thromb Res 2012;129:235–40.
[60] Bobryshev YV, Killingsworth MC, Lord RS, Grabs AJ. Matrix vesicles in the fibrous cap of atherosclerotic plaque: possible contribution to plaque rupture. J Cell Mol Med 2008;12:2073–82.
[61] Teng N, Maghzal GJ, Talib J, Rashid I, Lau AK, Stocker R. The roles of myeloperoxidase in coronary artery disease and its potential implication in plaque rupture. Redox Rep 2017;22:51–73.
[62] Tang D, Yang C, Huang S, Mani V, Zheng J, Woodard PK, Robson P, Teng Z, Dweck M, Fayad ZA. Cap inflammation leads to higher plaque cap strain and lower cap stress: an MRI-PET/CT-based FSI modeling approach. J Biomech 2017;50:121–9.
[63] Eisen A, Giugliano RP, Braunwald E. Updates on acute coronary syndrome: a review. JAMA Cardiol 2016;1:718–30.
[64] Otsuka F, Sakakura K, Yahagi K, Joner M, Virmani R. Has our understanding of calcification in human coronary atherosclerosis progressed? Arterioscler Thromb Vasc Biol 2014;34:724–36.
[65] Alexopoulos N, Raggi P. Calcification in atherosclerosis. Nat Rev Cardiol 2009;6:681–8.
[66] Antoniadis AP, Giannopoulos AA, Wentzel JJ, Joner M, Giannoglou GD, Virmani R, Chatzizisis YS. Impact of local flow haemodynamics on atherosclerosis in coronary artery bifurcations. EuroIntervention 2015;11(Suppl V):V18–22.
[67] Siasos G, Tousoulis D, Siasou Z, Stefanadis C, Papavassiliou AG. Shear stress, protein kinases and atherosclerosis. Curr Med Chem 2007;14:1567–72.
[68] Stone PH, Coskun AU, Yeghiazarians Y, Kinlay S, Popma JJ, Kuntz RE, Feldman CL. Prediction of sites of coronary atherosclerosis progression: in vivo profiling of endothelial shear stress, lumen, and outer vessel wall characteristics to predict vascular behavior. Curr Opin Cardiol 2003;18:458–70.
[69] Tousoulis D, Papageorgiou N, Synetos A, Stefanadis C. Assessing vulnerable plaque: is shear stress enough? Int J Cardiol 2014;172:e135–8.
[70] Andreou I, Antoniadis AP, Shishido K, Papafaklis MI, Koskinas KC, Chatzizisis YS, Coskun AU, Edelman ER, Feldman CL, Stone PH. How do we prevent the vulnerable atherosclerotic plaque from rupturing? Insights from in vivo assessments of plaque, vascular remodeling, and local endothelial shear stress. J Cardiovasc Pharmacol Ther 2015;20:261–75.
[71] Toutouzas K, Stathogiannis K, Synetos A, Karanasos A, Stefanadis C. Vulnerable atherosclerotic plaque: from the basic research laboratory to the clinic. Cardiology 2012;123:248–53.
[72] Batty JA, Subba S, Luke P, Gigi LW, Sinclair H, Kunadian V. Intracoronary imaging in the detection of vulnerable olaques. Curr Cardiol Rep 2016;18:28.
[73] de Korte CL, van der Steen AF, Cepedes EI, Pasterkamp G, Carlier SG, Mastik F, Schoneveld AH, Serruys PW, Bom N. Characterization of plaque components and vulnerability with intravascular ultrasound elastography. Phys Med Biol 2000;45:1465–75.
[74] Garcia-Garcia HM, Mintz GS, Lerman A, Vince DG, Margolis MP, van Es GA, Morel MA, Nair A, Virmani R, Burke AP, Stone GW, Serruys PW. Tissue characterisation using intravascular radiofrequency data analysis: recommendations for acquisition, analysis, interpretation and reporting. EuroIntervention 2009;5:177–89.
[75] Calvert PA, Obaid DR, O'Sullivan M, Shapiro LM, McNab D, Densem CG, Schofield PM, Braganza D, Clarke SC, Ray KK, West NE, Bennett MR. Association between IVUS findings and adverse outcomes in patients with coronary artery disease: the VIVA (VH-IVUS in vulnerable atherosclerosis) study. JACC Cardiovasc Imaging 2011;4:894–901.
[76] Cheng JM, Suoniemi M, Kardys I, Vihervaara T, de Boer SP, Akkerhuis KM, Sysi-Aho M, Ekroos K, Garcia-Garcia HM, Oemrawsingh RM, Regar E, Koenig W, Serruys PW, van Geuns RJ, Boersma E, Laaksonen R. Plasma concentrations of molecular lipid species in relation to coronary plaque characteristics and cardiovascular outcome: results of the ATHEROREMO-IVUS study. Atherosclerosis 2015;243:560–6.

[77] Tearney GJ, Yabushita H, Houser SL, Aretz HT, Jang IK, Schlendorf KH, Kauffman CR, Shishkov M, Halpern EF, Bouma BE. Quantification of macrophage content in atherosclerotic plaques by optical coherence tomography. Circulation 2003;107:113–9.
[78] Chatzizisis YS, Koutkias VG, Toutouzas K, Giannopoulos A, Chouvarda I, Riga M, Antoniadis AP, Cheimariotis G, Doulaverakis C, Tsampoulatidis I, Bouki K, Kompatsiaris I, Stefanadis C, Maglaveras N, Giannoglou GD. Clinical validation of an algorithm for rapid and accurate automated segmentation of intracoronary optical coherence tomography images. Int J Cardiol 2014;172:568–80.
[79] Toutouzas K, Karanasos A, Tsiamis E, Riga M, Drakopoulou M, Synetos A, Papanikolaou A, Tsioufis C, Androulakis A, Stefanadi E, Tousoulis D, Stefanadis C. New insights by optical coherence tomography into the differences and similarities of culprit ruptured plaque morphology in non-ST-elevation myocardial infarction and ST-elevation myocardial infarction. Am Heart J 2011;161:1192–9.
[80] Madjid M, Toutouzas K, Stefanadis C, Willerson JT, Casscells SW. Coronary thermography for detection of vulnerable plaques. J Nucl Cardiol 2007;14:244–9.
[81] Stefanadis C, Diamantopoulos L, Dernellis J, Economou E, Tsiamis E, Toutouzas K, Vlachopoulos C, Toutouzas P. Heat production of atherosclerotic plaques and inflammation assessed by the acute phase proteins in acute coronary syndromes. J Mol Cell Cardiol 2000;32:43–52.
[82] Stefanadis C, Diamantopoulos L, Vlachopoulos C, Tsiamis E, Dernellis J, Toutouzas K, Stefanadi E, Toutouzas P. Thermal heterogeneity within human atherosclerotic coronary arteries detected in vivo: a new method of detection by application of a special thermography catheter. Circulation 1999;99:1965–71.
[83] Stefanadis C, Toutouzas K, Tsiamis E, Mitropoulos I, Tsioufis C, Kallikazaros I, Pitsavos C, Toutouzas P. Thermal heterogeneity in stable human coronary atherosclerotic plaques is underestimated in vivo: the "cooling effect" of blood flow. J Am Coll Cardiol 2003;41:403–8.
[84] Stefanadis C, Toutouzas K, Vavuranakis M, Tsiamis E, Tousoulis D, Panagiotakos DB, Vaina S, Pitsavos C, Toutouzas P. Statin treatment is associated with reduced thermal heterogeneity in human atherosclerotic plaques. Eur Heart J 2002;23:1664–9.
[85] Toutouzas K, Drakopoulou M, Mitropoulos J, Tsiamis E, Vaina S, Vavuranakis M, Markou V, Bosinakou E, Stefanadis C. Elevated plaque temperature in non-culprit de novo atheromatous lesions of patients with acute coronary syndromes. J Am Coll Cardiol 2006;47:301–6.
[86] Toutouzas K, Drakopoulou M, Stefanadi E, Siasos G, Stefanadis C. Intracoronary thermography: does it help us in clinical decision making? J Interv Cardiol 2005;18:485–9.
[87] Toutouzas K, Markou V, Drakopoulou M, Mitropoulos I, Tsiamis E, Stefanadis C. Patients with type two diabetes mellitus: increased local inflammatory activation in culprit atheromatous plaques. Hellenic J Cardiol 2005;46:283–8.
[88] Toutouzas K, Synetos A, Stefanadi E, Vaina S, Markou V, Vavuranakis M, Tsiamis E, Tousoulis D, Stefanadis C. Correlation between morphologic characteristics and local temperature differences in culprit lesions of patients with symptomatic coronary artery disease. J Am Coll Cardiol 2007;49:2264–71.
[89] Toutouzas K, Tsiamis E, Drakopoulou M, Synetos A, Karampelas J, Riga M, Tsioufis C, Tousoulis D, Stefanadi E, Vlasis C, Stefanadis C. Impact of type 2 diabetes mellitus on diffuse inflammatory activation of de novo atheromatous lesions: implications for systemic inflammation. Diabetes Metab 2009;35:299–304.
[90] Stefanadis C, Toutouzas K, Tsiamis E, Vavuranakis M, Tsioufis C, Stefanadi E, Boudoulas H. Relation between local temperature and C-reactive protein levels in patients with coronary artery disease: effects of atorvastatin treatment. Atherosclerosis 2007;192:396–400.
[91] Stefanadis C, Toutouzas K, Tsiamis E, Stratos C, Vavuranakis M, Kallikazaros I, Panagiotakos D, Toutouzas P. Increased local temperature in human coronary atherosclerotic plaques: an independent predictor of clinical outcome in patients undergoing a percutaneous coronary intervention. J Am Coll Cardiol 2001;37:1277–83.
[92] Shibuya M, Fujii K, Hao H, Imanaka T, Fukunaga M, Miki K, Tamaru H, Nakata T, Sawada H, Naito Y, Hirota S, Masuyama T. Ex vivo comparison of angioscopy and histopathology for the evaluation of coronary plaque characteristics. Int J Cardiovasc Imaging 2016;32:863–9.

[93] Puri R, Madder RD, Madden SP, Sum ST, Wolski K, Muller JE, Andrews J, King KL, Kataoka Y, Uno K, Kapadia SR, Tuzcu EM, Nissen SE, Virmani R, Maehara A, Mintz GS, Nicholls SJ. Near-infrared spectroscopy enhances intravascular ultrasound assessment of vulnerable coronary plaque: a combined pathological and in vivo study. Arterioscler Thromb Vasc Biol 2015;35:2423−31.

[94] Kang SJ, Mintz GS, Pu J, Sum ST, Madden SP, Burke AP, Xu K, Goldstein JA, Stone GW, Muller JE, Virmani R, Maehara A. Combined IVUS and NIRS detection of fibroatheromas: histopathological validation in human coronary arteries. JACC Cardiovasc Imaging 2015;8:184−94.

[95] Patel K, Tarkin J, Serruys PW, Tenekecioglu E, Foin N, Zhang YJ, Crake T, Moon J, Mathur A, Bourantas CV. Invasive or non-invasive imaging for detecting high-risk coronary lesions? Expert Rev Cardiovasc Ther 2017;15:165−79.

[96] Rodriguez-Granillo GA, Carrascosa P, Bruining N, Waksman R, Garcia-Garcia HM. Defining the non-vulnerable and vulnerable patients with computed tomography coronary angiography: evaluation of atherosclerotic plaque burden and composition. Eur Heart J Cardiovasc Imaging 2016;17:481−91.

[97] Caixeta A, Genereux P, de Siqueira ME, Baruzzi A, Abizaid A, Stone GW. MSCT identification of vulnerable plaque. JACC Cardiovasc Imaging 2016;9:207−9.

[98] Krishnan S, Otaki Y, Doris M, Slipczuk L, Arnson Y, Rubeaux M, Dey D, Slomka P, Berman DS, Tamarappoo B. Molecular imaging of vulnerable coronary plaque: a pathophysiologic perspective. J Nucl Med 2017;58:359−64.

[99] Chrapko BE, Chrapko M, Nocun A, Stefaniak B, Zubilewicz T, Drop A. Role of 18F-FDG PET/CT in the diagnosis of inflammatory and infectious vascular disease. Nucl Med Rev Cent East Eur 2016;19:28−36.

[100] Chen W, Dilsizian V. Targeted PET/CT imaging of vulnerable atherosclerotic plaques: microcalcification with sodium fluoride and inflammation with fluorodeoxyglucose. Curr Cardiol Rep 2013;15:364.

[101] Annovazzi A, Bonanno E, Arca M, D'Alessandria C, Marcoccia A, Spagnoli LG, Violi F, Scopinaro F, De Toma G, Signore A. 99mTc-interleukin-2 scintigraphy for the in vivo imaging of vulnerable atherosclerotic plaques. Eur J Nucl Med Mol Imaging 2006;33:117−26.

[102] Androulakis E, Zacharia E, Papageorgiou N, Lioudaki E, Bertsias D, Charakida M, Siasos G, Tousoulis D. High-density lipoprotein and low-density lipoprotein therapeutic approaches in acute coronary syndromes. Curr Cardiol Rev 2017. http://dx.doi.org/10.2174/1573403X13666170209145622.

[103] Papageorgiou N, Zacharia E, Briasoulis A, Androulakis E, Tousoulis D. Statins and myocardial infarction: type, dose, and administration time: does it matter? Trends Cardiovasc Med 2016;26:433−41.

[104] Tousoulis D, Psarros C, Demosthenous M, Patel R, Antoniades C, Stefanadis C. Innate and adaptive inflammation as a therapeutic target in vascular disease: the emerging role of statins. J Am Coll Cardiol 2014;63:2491−502.

[105] Papageorgiou N, Tousoulis D, Antoniades C, Giolis A, Briasoulis A, Stefanadis C. The impact of statin administration in acute coronary syndromes. Hellenic J Cardiol 2010;51:250−61.

[106] Akdim F, van Leuven SI, Kastelein JJ, Stroes ES. Pleiotropic effects of statins: stabilization of the vulnerable atherosclerotic plaque? Curr Pharm Des 2007;13:1003−12.

[107] Vlachopoulos C, Terentes-Printzios D, Georgiopoulos G, Skoumas I, Koutagiar I, Ioakeimidis N, Stefanadis C, Tousoulis D. Prediction of cardiovascular events with levels of proprotein convertase subtilisin/kexin type 9: a systematic review and meta-analysis. Atherosclerosis 2016;252:50−60.

[108] Tousoulis D, Simopoulou C, Papageorgiou N, Oikonomou E, Hatzis G, Siasos G, Tsiamis E, Stefanadis C. Endothelial dysfunction in conduit arteries and in microcirculation. Novel therapeutic approaches. Pharmacol Ther 2014;144:253−67.

[109] Achimastos A, Alexandrides T, Alexopoulos D, Athyros V, Bargiota A, Bilianou E, Chrysochoou C, Drogari E, Elisaf M, Ganotakis E, Goudevenos I, Ioannidis I, Kolovou G, Kotsis V, Lekakis I, Liberopoulos E, Melidonis A, Nikolaou V, Ntaios G, Papanas N, Pappas S, Pitsavos C, Rallidis L, Richter D, Skoumas I, Tentolouris N, Tousoulis D, Tselepis A, Tsioufis K, Tziakas D, Tziomalos K, Vardas P, Vlachopoulos C, Vlahakos D. Expert consensus on the rational clinical use of proprotein convertase subtilisin/kexin type 9 (PCSK9) inhibitors. Hormones 2016;15:8−14.

[110] Papageorgiou N, Tousoulis D, Katsargyris A, Charakida M, Androulakis E, Siasos G, Tentolouris C, Stefanadis C. Antioxidant treatment and endothelial dysfunction: is it time for flavonoids? Recent Pat Cardiovasc Drug Discov 2013;8:81−92.
[111] Tousoulis D. Inflammation in atherosclerosis: current therapeutic approaches. Curr Pharm Des 2011; 17:4087−8.
[112] Tousoulis D, Andreou I, Antoniades C, Tentolouris C, Stefanadis C. Role of inflammation and oxidative stress in endothelial progenitor cell function and mobilization: therapeutic implications for cardiovascular diseases. Atherosclerosis 2008;201:236−47.
[113] Tousoulis D, Andreou I, Tsiatas M, Miliou A, Tentolouris C, Siasos G, Papageorgiou N, Papadimitriou CA, Dimopoulos MA, Stefanadis C. Effects of rosuvastatin and allopurinol on circulating endothelial progenitor cells in patients with congestive heart failure: the impact of inflammatory process and oxidative stress. Atherosclerosis 2011;214:151−7.
[114] Tousoulis D, Androulakis E, Papageorgiou N, Briasoulis A, Siasos G, Antoniades C, Stefanadis C. From atherosclerosis to acute coronary syndromes: the role of soluble CD40 ligand. Trends Cardiovasc Med 2010;20:153−64.
[115] Tousoulis D, Antoniades C, Koumallos N, Marinou K, Stefanadi E, Latsios G, Stefanadis C. Novel therapies targeting vascular endothelium. Endothelium 2006;13:411−21.
[116] Tousoulis D, Antoniades C, Koumallos N, Stefanadis C. Pro-inflammatory cytokines in acute coronary syndromes: from bench to bedside. Cytokine Growth Factor Rev 2006;17:225−33.
[117] Tousoulis D, Briasoulis A, Papageorgiou N, Tsioufis C, Tsiamis E, Toutouzas K, Stefanadis C. Oxidative stress and endothelial function: therapeutic interventions. Recent Pat Cardiovasc Drug Discov 2011;6:103−14.

CHAPTER 3

Novel and Experimental Therapies in Advance Atherosclerosis

CHAPTER 3.1

Anti-Inflammatory Treatment

Sotirios Tsalamandris, Evangelos Oikonomou, Georgios Latsios, Dimitris Tousoulis
1st Cardiology Clinic, 'Hippokration' General Hospital, National and Kapodistrian University of Athens, School of Medicine, Athens, Greece

Contents

Coronary Artery Disease
ISBN 978-0-12-811908-2, http://dx.doi.org/10.1016/B978-0-12-811908-2.00013-1

INTRODUCTION

Multiple therapeutic approaches in the cardiovascular field, i.e., statins and antihypertensive agents, have shown their ability to modify immunological and inflammatory responses in parallel to their principal effects as cholesterol-lowering agents or blood pressure reduction. Moreover, novel agents directed against specific targets in the inflammatory cascade are now also being applied in clinical settings. Interest in this novel therapeutic field is expanding rapidly, as demonstrated by the large number of ongoing clinical trials involving this kind of agents. As our knowledge regarding the role of cytokines in atherogenesis expands, newer approaches come to light and many show promising results in the fight against atherosclerosis. This chapter challenges to highlight the actions and effects of anti-inflammatory agents, provides available data on their clinical significance, and deliver further insight regarding their future prospects.

ANGIOTENSIN CONVERTING ENZYME INHIBITORS AND ANGIOTENSIN RECEPTORS BLOCKERS

Angiotensin-converting enzyme inhibitors (ACEIs) and angiotensin receptors blockers (ARBs) are antihypertensive drugs. ACEIs inhibit kinase II, blocking the formation of angiotensin II and preventing its activation of angiotensin I receptors in the adrenal cortex, thereby decreasing aldosterone and its effects on vasculature and reducing vasoconstriction. ACEIs also inhibit the metabolism of bradykinin (BK), which causes nitric oxide (NO)/endothelium-derived relaxing factor mediated vasodilation [1]. Angiotensin II receptor blockers produce direct antagonism of the angiotensin II (AT2) receptors, unlike the ACE inhibitors. They displace the angiotensin II from the AT1 receptor, and produce their blood pressure lowering effects by antagonizing AT1 induced vasoconstriction, aldosterone release, catecholamines release, arginine vasopressin release, water intake, and hypertrophic responses.

ACE in plasma constitutes only a minor part of total ACE in the body. The rest of it is localized on the surface of the endothelium [2]. ACE-I affect endothelium by several different mechanisms. They reduce the formation of angiotensin II contributing to the reversal of endothelial dysfunction via decreased nicotinamide adenine dinucleotide phosphate [NAD(P)H] oxidase expression and diminished O_2- formation by vascular endothelium [3]. On the other hand ACE-I lead to accumulation of BK in the vicinity of endothelial B2 kinin receptors [4]. Hence, ACEI increase the production of vasoprotective, antithrombotic and anti-inflammatory NO and prostaglandin I2 (PGI2) by the endothelium. In the case of chronic administration of ACE-I, the induction of endothelial B1 receptors may also take place [5].

Thus, both, diminished angiotensin II production and accumulation of BK lead to improvement of endothelial cell function and vascular tone and may benefit not only patients with established vascular disease but normotensive patients as well [6]. Importantly, the above is achieved by correction of various functions of the endothelium, not only of e-NO synthase/NO pathway. Indeed, inhibition of endothelial ACE is able to reverse complex phenotype of activated proinflammatory, prothrombotic endothelium toward healthy quiescent phenotype. The correction of endothelial function by ACE-I contributes significantly to their clinical benefits, such as the inhibition of progression of atherothrombosis and a decrease in mortality from myocardial infarction and stroke. They both improve cardiovascular outcomes and rather have a beneficial effect on endothelial function of high-risk individuals with atherosclerosis, as it has been reported in numerous clinical trials.

The HOPE trial examined the effect of ACE inhibitors in 9297 high-risk patients who had evidence of vascular disease or diabetes and one other cardiovascular risk factor without heart failure (HF). These patients were randomly assigned to receive ramipril (10 mg/day) for a 5 years period. The primary study outcome was a composite of myocardial infarction, stroke, or death from cardiovascular causes. The secondary outcomes were defined as death from any cause, the need for revascularization, hospitalization for unstable angina or HF, and complications related to diabetes. Treatment with ramipril reduced the rates of death from cardiovascular causes, myocardial infarction and stroke. In addition, ramipril treatment reduced the rate of revascularization procedures and incidence of HF [7,8].

The EUROPA trial was similar in design to HOPE except that the patients were at lower risk [9]. The 13,655 patients fulfilling the inclusion criteria of the study were given perindopril for a 4 week run-in period and then randomly assigned to perindopril (8 mg/day) or placebo. According to the findings of the study, a significant reduction in the primary end point was observed. Another trial was the Antihypertensive and Lipid-Lowering Treatment to Prevent Heart Attack Trial (ALLHAT) [10,11], a randomized prospective trial of nearly 45,000 patients with hypertension and one additional risk factor for CAD. The ALLHAT was designed to evaluate whether the incidence of adverse coronary and cardiovascular outcomes differed among those randomly assigned to chlorthalidone (12.5 to a maximum of 25 mg/day) compared to one of three other antihypertensive drugs: amlodipine, lisinopril, or doxazosin. At a mean follow-up of almost 5 years, the major finding regarding to diabetes mellitus showed a reduction in the relative risks for combined cardiovascular disease and congestive HF outcomes in diabetes mellitus (DM) type 2 patients. In the QUIET trial, 1750 patients undergoing coronary angiography were followed for 3 years for cardiac end points. No definitive answer for a role for ACE inhibition as an antiatherosclerosis treatment could be demonstrated [12]. In the QUO VADIS Study long-term quinapril treatment significantly reduced clinical ischemic events within 1 year after coronary artery bypass grafting [13].

ACEIs have been found to improve endothelium dependent vasodilation in atherosclerotic coronary vasculature of normotensive patients [14]. In a substudy of the ARAMIS trial, subjects with isolated systolic hypertension were randomized to three progressively increasing telmisartan doses, hydrochrolothiazide or placebo; all three telmisartan doses produced significantly higher reductions in microalbuminuria compared to hydrochlorothiazide or placebo regardless of baseline microalbuminuria including its normal values [15]. Furthermore, ACEIs and ARBs were proved suppress free radical generation, nuclear factor kB activity and C reactive protein in hypertensive subjects [16]. Nevertheless, the Blood Pressure Lowering Treatment Trialists' Collaboration has reported that ACEIs but not ARBs appear to afford some additional cardiovascular protection beyond their ability to reduce BP [17].

In conclusion we understand that well established agents such as ACEI and Ang-II receptor blockers are more than simple blood pressure-lowering drugs. Data suggest that these antihypertensive agents are able to modulate vascular redox state because they actively block angiotensin effects on the vascular wall, thus exerting pro-oxidant, proinflammatory, and proliferative effects on the vasculature and however, is considered to have anti-inflammatory properties.

STATINS

The most widely used and promising agents in cardiovascular disease, hydroxymethylglutaryl-coenzyme A (HMG-CoA) reductase inhibitors, or statins, are renowned for their "pleiotropic" lipid-lowering independent effects. Even though this concept has recently been challenged, data from large meta-analyses confirm that the beneficial effects exerted by statins in vascular disease states are independent of their lipid-lowering properties. Statin's inhibition of the mevalonate pathway and its isoprenoid formation have been proposed to account for many of these nonlipid-lowering effects, because protein isoprenylation is implicated in intracellular signaling events. Asymmetrical dimethyl arginine (ADMA) is believed to be an important mediator of inflammation-induced endothelial dysfunction, leading to atherogenesis. Statins suppress the expression of proinflammatory mediators, resulting in reduced circulating ADMA levels [18,19].

Statins and Their Anti-inflammatory Effects: General Role of Statins in Inflammation

HMG-CoA is the precursor for cholesterol synthesis. HMG-CoA is formed by condensation of acetyl-CoA and acetoacetyl-CoA, catalyzed by HMG-CoA synthase. HMG-CoA reductase catalyzes the production of mevalonate from HMG-CoA, in which the HMG-CoA reductase reaction is the rate-limiting step for cholesterol synthesis. Mevalonate is phosphorylated by two phosphate transfers from ATP, yielding the pyrophosphate derivative. Pyrophosphomevalonate decarboxylase catalyzes ATP-dependent decarboxylation, with dehydration, to yield isopentenyl pyrophosphate (isoprenylation). Isopentenyl pyrophosphate is the first compound in the mevalonate pathway. The pleiotropic effects of statins are mediated by inhibition of protein isoprenylation, altering several signal transduction molecules in the cardiovascular pathways. By inhibiting mevalonate synthesis, statins also prevent the synthesis of other important isoprenoid intermediates of the cholesterol biosynthetic pathway, such as farnesylpyrophosphate and geranylgeranyl pyrophosphate. These isoprenoid intermediates serve as important lipid attachments for a variety of proteins, including the small guanosine triphosphate (GTP) binding proteins Ras, Rho, and Rac. Rho proteins are involved in the expression of proinflammatory cytokines; Rac proteins modulate reactive oxygen species (ROS) generation, whereas Ras proteins are involved in cell proliferation, apoptosis, and cell adhesion [18]. Inhibition of Ras and Rho isoprenylation by statins renders them inactive. It has been demonstrated that coincubation of human vessels with statins and mevalonate reverses the beneficial effects of statins on vascular redox state; such findings highlight the importance of the mevalonate pathway in mediating the pleiotropic effects of statins in humans [20–22].

Effect of Statins on Circulating Proinflammatory Cytokines

Early-stage atherogenesis entails leukocyte/endothelial interaction and accumulation of inflammatory cells. Statins were shown to inhibit adhesion of leukocytes to endothelial cells [23]. Furthermore, HMG-CoA reductase inhibitors regulate the expression of chemokines controlling the migration of leukocytes to subendothelial sites of inflammation, such as monocyte chemoattractant protein (MCP)-1 and interleukin 8 (IL-8). Simvastatin was shown to reduce the expression of proinflammatory cytokines such as IL-6, IL-8, and MCP-1 in peripheral blood mononuclear cells from patients with hypercholesterolemia both in vitro and in vivo [24]. MMPs were also described to be implicated in the process of monocyte/macrophage migration, and it has been revealed that statins diminished their expression. Thus, HMG-CoA reductase inhibitors lower the expression and function of a broad range of MMPs, including interstitial collagenases (MMP-1, MMP-13), gelatinases (MMP-2, MMP-9), and stromelysin (MMP-3).

Effect of Statins on Adhesion Molecules

Leukocyte transendothelial migration is regulated by the cooperative actions of Intercellular adhesion molecule 1 (ICAM-1), vascular cell adhesion molecule 1 (VCAM-1) on both endothelial cells and the leukocytes. Leukocytes cross endothelial cells either through intercellular junctions (paracellular pathway) or through the endothelial cell body (transcellular pathway). Paracellular transendothelial migration involves platelet endothelial cell adhesion molecule (PECAM)-1 and members of the junctional adhesion molecule (JAM) family [25].

The Rho family consists of 22 genes encoding at least 25 proteins in humans, of which the Rho, Rac, and CDc42 proteins have been studied in the most detail. Within the Rho family, there are subgroups of closely related isoforms including Rho (A, B, C) and Rac (1, 2, 3). All Rho family members bind GTP, and most exhibit ATPase activity, cycling between an inactive GDP-bound form and an active GTP-bound form. Rho GTPase function is also regulated by their localization. Rho is required for leukocyte adhesion to endothelial cells, for clustering of ICAM-1 and VCAM-1, and for formation of docking structures around adherent leukocytes. Most Rho family proteins undergo posttranslational modification by prenylation of a cysteine residue located four amino acids from the C terminus, followed by methylation of this cysteine and proteolytic removal of the last three amino acids. The prenyl group helps anchor the protein to membranes [26–28].

Isoprenoids are intermediates in the pathway to cholesterol synthesis, and their production is reduced by statins. As mentioned previously, some of the pleiotropic effects of statins are due to the inhibition of farnesyl-pyrophosphate and geranylgeranyl-pyrophosphate synthesis, the isoprenoids required for prenylation of Rho proteins. Statins thereby modulate Rho family protein subcellular localization, and this affects their stability and activity. Statins also alter Rho expression at the transcriptional level, although the mechanistic basis for this is not clear. Indirect evidence suggests that Rho GTPases regulate PECAM-1 expression on the endothelial cell surface. Statins not only affect the total levels of PECAM-1 in different endothelial cell lines but also modulate PECAM-1 localization. Moreover, C3 transferase treatment mimics the effect of statins on PECAM-1 expression [29,30].

Main Studies

The multicenter JUPITER study [31,32] was designed after the observation that when both low density lipoprotein and C-reactive protein (CRP) were reduced, patients fared better than when only LDL was lowered. After screening some 90,000 prospects, JUPITER enrolled 17,802 patients. The study population was not diagnosed with either CHD or diabetes, and had "normal" LDL levels, but high hsCRP levels. Participants were randomized to either treatment with rosuvastatin daily or placebo. The primary end point composite included a nonfatal MI, nonfatal stroke, hospitalization for unstable angina, an arterial revascularization procedure, or confirmed death from a cardiovascular cause. The secondary end points included the individual components of the primary end point and all-cause mortality. Given prior data that CRP values improved prediction of risk, JUPITER suggested that a seemingly healthy patient population at high risk, previously ineligible for statin therapy, may benefit from rosuvastatin treatment, guided by the CRP level. Attention was focused on the large body of research that points to inflammation as a common denominator in atherothrombosis (i.e., that inflammation matters at a clinical level) [33]. The most significant dual target analysis of the JUPITER study evaluated the effects of rosuvastatin 20 mg versus placebo on the prespecified end points according to on-treatment values of LDL (<70 or ≥70 mg/dL) and CRP (<2 mg/L, or <1 mg/L) [34]. Compared to placebo, those in the treatment group who achieved an LDL <70 mg/dL had a 55% reduction in vascular events per 100-person years. Those who achieved CRP <2 mg/L had a 62% reduction in events. Among those who reached dual reductions, there was a 65% reduction in events compared to a 33% reduction in those who achieved one or neither target. Those who reached an LDL <70 mg/dL and a CRP <1 mg/L enjoyed a 79% reduction in events. Treatment CRP levels predicted the event rate no matter what lipid end point was used, including the ratio of apoB/aopA-1. Therefore, regardless of the lipid profile, the lower the CRP, the better was the prognosis. An editorial concluded that JUPITER provided "key experimental data" that inflammation mediated the benefits of rosuvastatin, but mused at the prospect of comparing the absolute risk reduction of statins with the effects of weight loss and regular exercise, remarking that such a study is unlikely, both for lack of funding and interest [35]. Many of the pleiotropic effects of rosuvastatin are anti-inflammatory, and a catalog of those actions - significant improvement in endothelial function, immune responses, plaque stabilization, vascular remodeling, and oxidative stress—and antithrombotic actions further underpins the observation that targeting CRP is worthwhile and clinically rewarding [36].

The efficacy of statins in cardiovascular diseases, both in primary and secondary prevention setting has been proved in numerous trials [37]. Three trials, the West Of Scotland COronary Prevention Study (WOSCOPS) [38], the Prospective Study Of Pravastatin in the Elderly at Risk (PROSPER) [39], and the ALLHAT-LLT [40], evaluated the use of pravastatin at a dose of 40 mg/day versus no statin therapy in primary prevention. Two of the trials demonstrated reductions in LDL cholesterol concentrations, cardiac mortality and coronary events with pravastatin when compared to placebo—WOSCOPS and PROSPER. Pravastatin is effective also in secondary prevention, as demonstrated by the Cholesterol and Recurrent Events trial [41] and by the Long-term Intervention with Pravastatin in Ischaemic Disease study [42].

There is less extensive data showing cardiovascular benefits with lovastatin and fluvastatin, although it is presumed that they improve outcomes similarly to pravastatin if titrated to equivalent lipid-lowering effects [43—45]. The second generation of statins, comprising atorvastatin and simvastatin have significantly improved efficacy in reducing LDL cholesterol concentrations when compared to the earlier statins. A large body of data supports their use in both primary and secondary prevention [46—51]. Finally, there is a single commercially available drug in the third, high potency generation of statins, namely rosuvastatin.

The efficacy of rosuvastatin was observed in several trials [52,53]. A number of observational studies indicate that statins decrease major cardiovascular outcomes, including mortality, by approximately 30%—40% in treated subjects compared with controls when initiated before or at discharge following an acute coronary syndrome (ACS) [54—60]. Additional meta-analyses of randomized controlled trials demonstrate that early initiation of statin therapy after ACS improves cardiovascular outcomes [61,62]. However, it should be noted that the established benefit of statins has limited clinical trial design due to concerns over the ethics of placebo-controlled randomization in moderate to high-risk populations. These findings stimulated interest in evaluating the direct effects of lipid modifying therapy on atherosclerotic plaque burden using surrogate measures of clinical outcomes, such as carotid intima-media thickness (CIMT) and coronary plaque volume [63—65]. However, CIMT represents a surrogate marker of coronary atherosclerosis. Direct measurement of changes in coronary plaque burden can instead be performed using volumetric intravascular ultrasound (IVUS). IVUS was used to evaluate clinical atheromatous progression in the Reversing Atherosclerosis with Aggressive Lipid

Lowering (REVERSAL) trial. This trial compared treatment with 40 mg pravastatin versus 80 mg atorvastatin in 2163 patients with a mean baseline serum LDL cholesterol of 3.9 mmol/L. Pravastatin reduced serum LDL cholesterol concentrations to 2.8 mmol/L, whereas intensive therapy with atorvastatin reduced LDL cholesterol to 2 mmol/L. In the pravastatin cohort, a 2.7% progression of atheroma volume was observed versus no significant change in the atorvastatin arm [66].

The available data, showing regression of atherosclerosis and a decreased occurrence of cardiovascular events following early initiation of treatment strongly suggests that other effects of statin exist, beyond their ability to lower cholesterol. In particular, in the Atorvastatin for Reduction of MYocardial Damage during Angioplasty—Cell Adhesion Molecules substudy, the protective effect of statins from myocardial injury during percutaneous coronary intervention was related to a reduction in endothelial inflammatory response. These findings may support a strategy of routine use of high dose atorvastatin prior to percutaneous coronary intervention, even if the patient is already using the drug for therapy [67].

A post hoc analysis of the Air Force/Texas Coronary Atherosclerosis Prevention Study showed that patients with LDL cholesterol below the median (3.9 mmol/L) and CRP concentrations above the median (1.60 mg/L) were at increased risk of cardiovascular events. Of interest, lovastatin reduced the risk of a cardiovascular event by 42% compared with placebo [68]. Moreover, the Pravastatin Inflammation/CRP Evaluation trial demonstrated that, in a primary prevention setting, pravastatin reduced median CRP concentrations by 16.9% at 24 weeks, a decrease of 0.2 mg/L. In fact, reductions were noted in the pravastatin group, whereas no change in CRP concentrations was observed in the placebo group. This effect was seen as early as 12 weeks [69]. In the Pravastatin or Atorvastatin Evaluation and Infection Therapy-Thrombolysis in Myocardial Infarction 22 trial, patients with LDL cholesterol concentrations and CRP value below the median had a significant reduction in recurrent myocardial infarction or death from coronary causes [70]. In the REVERSAL trial, CRP concentrations were significantly reduced with pravastatin therapy, versus with atorvastatin therapy [66]. Similar findings were also observed in the Myocardial Ischemia Reduction with Aggressive Cholesterol Lowering study [71]. Taken together, these observations suggest that individuals with normal LDL cholesterol (3.4 mmol/L), but increased CRP concentrations are at increased risk of ACS, and statin therapy could reduce this risk.

Interestingly, a study by Tousoulis et al. examined the effect of combined therapy with atorvastatin and metformin on glucose-induced variations of inflammatory molecules in patients with newly diagnosed diabetes mellitus (DM) type 2. Interestingly, they found that the combination of metformin and atorvastatin partly prevents the glucose-loading induced elevation of glucose levels (at 1 h), suggesting a better response to glucose intake than monotherapy with metformin. In addition, combined treatment with atorvastatin and metformin reduces the postglucose loading levels of tumor necrosis factor alpha (TNF-α) compared to metformin monotherapy and thus showing an attenuation of the inflammatory process [72].

ASPIRIN

Aspirin inhibits cyclooxygenase irreversibly and suppresses the production of prostaglandins and thromboxanes [73]. Therefore, it acts not only as an antiplatelet agent but also as an anti-inflammatory one. In murine models of atherosclerosis, low-dose aspirin improves vascular inflammation and plaque stability [74]. Recent data in ApoE-deficient mice support the hypothesis that aspirin can also reduce fractalkine levels (fractalkine acts both as a chemokine and as an adhesion molecule) and improve the severity of atherosclerotic lesions [75]. Recently, it was also documented in humans that low-dose aspirin treatment reduces chemerin (a peptide chemoattractant for macrophages and an adipokine-regulating adipocyte differentiation and metabolism) secretion by adipocytes through reduction of proinflammatory cytokine secretion by macrophages [76]. Aspirin further protects human endothelial function in the presence of inflammation and decreases the plasma levels of several inflammatory cytokines and markers such as IL-6, CRP, and monocyte colony-stimulating factor (M-CSF) in patients with stable angina [77,78]. Interestingly, the use of aspirin as a primary prevention therapy is associated with a reduction in the development of myocardial infarction, which appears to be directly related to CRP levels, raising the possibility that the anti-inflammatory properties of aspirin may be as important as its antithrombotic effects [79]. Moreover, in primary prevention, aspirin can improve arterial stiffness even after 2 weeks of low dose (160 mg/day) [80]. However, we have to notice that the exact antiatherosclerotic effect of aspirin cannot be determined as reports support the notion that aspirin can reduce hypertriglyceridemia, cholesterol, and free fatty acid production but also increases lipid peroxidation [81,82]. Indeed, the exact benefit of aspirin in chronic CAD has recently been questioned by the results of the CONFIRM study (Coronary CT Angiography Evaluation for Clinical Outcomes study). In this multicenter international registry, 5712 individuals with normal coronary arteries and 4706 individuals with nonobstructive CAD were followed up for a median of 27.2 months, and aspirin use, contrary to statins, was not associated with mortality benefit in patients with nonobstructive CAD irrespective of the status of the plaque [83]. Similarly, data from the prevention of progression of arterial disease and diabetes trial have shown that even in patients with diabetes mellitus and asymptomatic peripheral arterial disease aspirin did not reduce the rate of cardiovascular events [84]. Moreover, in 1884 recipients of low-dose aspirin with chronic kidney disease paired with 1884 nonrecipients, the incidence of atherosclerotic events was significantly higher in the aspirin users [85]. It would be interesting to study the effects of aspirin in groups with elevated versus low levels of circulating inflammatory markers (Table 3.1.1).

Table 3.1.1 Completed and ongoing anti-inflammatory clinical trials in atherosclerosis

Drug/target	Author	Study population	Study size	Results (primary end point)
Canakinumab/IL-1	—	Patients with MI and elevated hsCRP	10,000	Ongoing (time to first occurrence of MACE) [86]
Golimumab/TNF-a	Tam et al.	Ankylosing spondylitis patients	41	Significant increase in the arterial stiffness (PWV) of the placebo group compared with the golimumab treated group [87]
Anakinra/IL-1Ra	Ikonomidis et al.	Rheumatoid arthritis patients with and without CAD	80	Anakinra leads to a greater improvement in endothelial, coronary, and aortic function in RA patients with CAD than In those without [88]
	Abbate et al.	STEMI patients	40	Anakinra for 2 weeks post-STEMI appears to have a neutral effect on recurrent ischemic events, whereas it may prevent new-onset heart failure long term after STEMI.
	—	STEMI patients	99 (recruiting)	Ongoing (CRP levels during the first 14 days period after STEMI)
	Morton et al.	NSTEMI patients	182	IL-1Ra treatment reduces inflammatory markers at 14 days post-NSTEMI [89]
Ustekinumab-Briakinumab/IL-12/23	Tzellos et al.	Psoriasis patients	4653	Patients with psoriasis on anti-IL-12/23 treatment had a significantly higher risk of MACEs compared with placebo [90]

Adalimumab/TNF-a	–	Psoriasis patients	106	Ongoing (change in ascending aorta inflammation as assessed with FDG-PET after 16 weeks of treatment)
	–	RA patients	60	Ongoing (CF-PWV, central aortic pulse pressure and FMD after 24 weeks of treatment)
Tocilizumab/IL-6	–	Rheumatoid arthritis patients treated with tocilizumab versus etanercept	3080	Ongoing (time to occurrence of primary MACE)
	–	Post-NSTEMI patients	120	Ongoing (hsCRP levels during 56 h follow-up)
	–	NSTEMI/STEMI patients with drug or placebo given within 24 h	125	Ongoing (MACE within 30 days)
Inclacumab/P-Selectin blockade	Tenaglia et al.	STEMI patients	544	Reduction in myocardial injury after PCI [91]
Viral Serp-1/Serine Protease inhibition	Lucas et al.	ACS patients	48	Significant reduction in troponin I and CK-MB levels [92]
Etanercept/TNF-α	–	Psoriasis patients starting etanercept and matched controls	50 (recruiting)	Ongoing (change in aortic-coronary inflammation status as assessed with FDG-PET, 3 months after of treatment)
Methotrexate	Choi et al.	RA patients	1240	Methotrexate is associated with lower all-cause and CV mortality risk [93]
	–	Patients with prior MI or angiographically demonstrated multivessel CAD	216	Ongoing (change in arterial inflammation as assessed with FDG-PET at 8 months)
	–	Stable CAD patients with T2DM or MetS	7000	Ongoing (composite of nonfatal MI, nonfatal stroke, and CV death)

Continued

Table 3.1.1 Completed and ongoing anti-inflammatory clinical trials in atherosclerosis—cont'd

Drug/target	Author	Study population	Study size	Results (primary end point)
Methotrexate and colchicine	–	Stable CAD patients with hsCRP > 2 mg/L	120 (recruiting)	Coronary segment endothelial function at 8 weeks
Colchicine	Nidorf et al.	Patients with stable CAD	532	Significant reduction in major cardiac events [94]
	–	Patients referred for PCI	400 (recruiting)	Ongoing (post-PCI IL-6 levels)
	–	Patients with recent (30 days) acute MI	4500	Ongoing (MACE events)
	–	Patients with atherosclerotic vascular disease	106	Ongoing (change in inflammation of the ascending aorta as assessed with FDG-PET at 6 months)
Corticosteroids	Giugliano et al.	Acute MI patients	2646 (meta-analysis)	Corticosteroid therapy in the acute setting after MI shows no significant mortality benefit in randomized control trials [95]

ACS, acute coronary syndrome; *CAD*, coronary artery disease; *CF-PWV*, carotid-femoral pulse wave velocity; *CK*, creatinine kinase; *FDG-PET*, 18-fluorodeoxyglucose positron emission tomography; *FMD*, flow-mediated dilatation; *hsCRP*, high-sensitivity C-reactive protein; *hsTrP*, high-sensitivity troponin T; *IL-1Ra*, interleukin-1 receptor antagonist; *MACE*, major adverse cardiovascular event; *MetS*, metabolic syndrome; *MI*, myocardial infarction; *(N)STEMI*, (non)-ST-segment elevation myocardial infarction; *PCI*, percutaneous coronary intervention; *RA*, rheumatoid arthritis; *T2DM*, type 2 diabetes mellitus; *TNF-a*, tumor necrosis factor-α.

CORTICOSTEROIDS

A few reports that examined intermediate atherosclerotic markers such as arterial stiffness as assessed by pulse wave velocity have shown that the use of corticosteroids can act beneficially in some chronic inflammatory disorders such as polymyalgia rheumatic [96]. However, corticosteroids in the setting of acute myocardial infarction have failed to demonstrate significant mortality benefit, as shown in one meta-analysis of randomized control trials [95]. In addition, long-term use of corticosteroids in rheumatoid arthritis patients is associated with a significantly higher incidence of acute myocardial infarction [97].

METHOTREXATE

Methotrexate (MTX, 4-amino-4-deoxy-N10-methyl pteroylglutamic acid) is an analog of aminopterin (4-aminopteroylglutamic acid), a folic acid antagonist. It prevents the synthesis of nucleotides and serves as a chemotherapeutic agent at high doses, whereas at low-doses, methotrexate is frequently used to target inflammation in systemic inflammatory disease. Various mechanisms of action have been proposed for MTX. The most relevant in the case of inflammatory diseases is the promotion of adenosine release with subsequent suppression of inflammation [98]. The most important mechanism of action of MTX in inflammatory diseases is probably related to the recently described anti-inflammatory activity of adenosine through specific receptors [99]. Another possible mechanism of action of MTX may involve its inhibitory effect on the proliferation of antigen stimulated T cells, which has been reported both in vitro and in patients receiving MTX, or the induction of apoptosis in activated T cells, which may also be mediated by adenosine receptors [100,101]. Furthermore, administration of methotrexate reduces cellular responses to IL-1 while increasing production of anti-inflammatory cytokines such as IL-10 [102]. Finally, methotrexate may also play a role in apoptosis of proinflammatory cells [103].

The effects of methotrexate on cardiovascular outcomes have not been fully explored in patients without rheumatoid arthritis. In a recent meta-analysis and systematic review consisting of case control and cohort studies, administration of low-dose methotrexate in patients with rheumatoid arthritis appeared to have been associated with a significant reduction in cardiovascular events compared with other antirheumatic drugs [104,105].

Interestingly, studies have proved that low-dose methotrexate can decrease the cardiovascular risk of patients with chronic inflammatory disease. In patients with rheumatoid arthritis, methotrexate may provide a substantial survival benefit, largely by reducing cardiovascular mortality [93]. Similar results were also described by van Halm et al. in a case–control study [106]. However, some studies did not confirm the protective role of methotrexate in the incidence of cardiovascular events in rheumatoid arthritis patients [107]. Therefore, further studies have now been designed to test this hypothesis. For example, the Cardiovascular Inflammation Reduction Trial (CIRT) aims to enroll 7000 individual to test the role of methotrexate in the medical management of stable CAD patients with persistent elevation of high-sensitivity CRP [108], whereas the TETHYS trial aims to evaluate its value in reducing infarct size when administered within the first 6 h of admission for ST-elevation myocardial infraction [109]. Furthermore, a recently announced CIRT substudy aims to evaluate the effects of methotrexate in arterial inflammation (as seen on PET-CT scanning) in patients with prior MI or multivessel CAD. It is difficult, however, to definitely conclude that the beneficial effect of methotrexate in cardiovascular incidence is solely due to its anti-inflammatory properties. Studies have recently documented that methotrexate modifies lipid levels increasing both HDL and LDL cholesterol and affects macrophage cholesterol handling opposing foam cell formation [110].

COLCHICINE

Inflammatory processes have been identified as key mediators of the deleterious effects of ischemia/reperfusion in ST-segment elevation myocardial infarction (STEMI) [111,112]. Despite the well-known fact that inflammation plays an important role in coronary artery disease development and progression, there have been few attempts to systematically examine the potential role of anti-inflammatory treatment in this setting, possibly because of a lack in anti-inflammatory agents without the adverse cardiovascular safety profile of corticosteroids and nonsteroidal anti-inflammatory drugs [113—115]. Colchicine is a substance with potent anti-inflammatory properties, suitable for safe use in patients with cardiovascular disease [116]. Colchicine may reduce inflammatory responses associated with atherosclerosis by inhibiting vascular endothelial surface protein synthesis, cytokine production and monocyte migration [117].

In a prospective, double-blinded, placebo-controlled study, performed in primary percutaneous coronary intervention referral hospitals, whereas magnetic resonance imaging (MRI) studies also were performed in patients presenting with STEMI, $\leq$12 h from the onset of chest pain, the investigators tested the hypothesis that a short course of colchicine treatment could lead to reduced infarct size. The results of the study indicate that treatment with colchicine in patients with STEMI undergoing primary percutaneous coronary intervention is associated with smaller infarct size, as defined by both biomarker release and MRI. This effect was accompanied by a substantial treatment-related difference in markers of postmyocardial infarction inflammatory response, namely neutrophil count and CRP, whereas these latter parameters were quite strongly associated with infarct size [118].

Furthermore, in patients with stable coronary artery disease, colchicine has been shown to significantly reduce hsCRP independent of aspirin and atorvastatin use [119], whereas similar results were found in patients with ACS in a different study [120].

In a recent prospective, randomized multicenter study by Nidorf et al., researchers tried to determine whether colchicine can reduce the risk of cardiovascular events in patients with clinically stable coronary disease. After 3 years of follow-up, patients receiving 0.5 mg daily of colchicine had a significant reduction in the rates of major adverse coronary events [94]. A retrospective cross-sectional study concluded that long-term treatment of gout with low-dose colchicine decreased not only CRP levels but also the incidence of myocardial infraction [121].

BLOCKADE OF PROINFLAMMATORY AND DELIVERY OF ANTI-INFLAMMATORY CYTOKINES

Anti-TNF-α Treatment

Several anti-TNF agents (infliximab, etanercept, adalimumab, golimumab, and certolizumab) have been introduced recently against inflammatory and autoimmune diseases [122].

Interestingly, spare preclinical data have shown that anti-TNF treatment can favorably affect atherosclerosis. Recently, in a model that reproduced in vivo interactions between leucocytes and endothelial cells, anti-TNF treatment diminished leucocyte-endothelial interactions induced by inflammatory stimuli [123]. It is currently known that treatment with anti-TNF agents can improve endothelial function and arterial wall properties in patients with autoimmune diseases, such as psoriasis [124]. Moreover, it has been reported that inhibition of TNF-a in rheumatoid arthritis can increase circulating endothelial progenitor cells concurrently with a proportional decrease of disease activity and decrease serum asymmetric dimethyl arginine levels [125].

Clinical studies have revealed that anti-TNF-a therapies in patients with ankylosing spondylitis improve subclinical atherosclerosis and aortic stiffness when compared with placebo controls [126]. A meta-analysis published in 2011 showed that anti-TNF-a treatment in rheumatoid arthritis is associated with a lower risk for all cardiovascular events [127]. Nevertheless, we have to take into consideration that studies with these agents in patients with chronic HF have failed to improve patient outcome [127]. Unfortunately, there are currently no ongoing trials evaluating the exact health or mortality benefit of anti-TNF-a agents in a large cohort of healthy or CAD patients and studies so far have been limited to rheumatological patients.

Anti-interleukin-12/23p40 and Anti-interleukin-17 Agents

Monoclonal antibodies that are targeted against the shared p40 subunit of IL-12 and IL-23 have recently been introduced in the treatment of psoriasis. These cytokines mediate the function of Th17 cells, a distinct subset of T cells that have been identified as a proatherogenic group of leucocytes within the atherosclerotic plaque [128]. Despite these proatherogenic effects, a recent meta-analysis concluded that treatment with such agents (ustekinumab and briakinumab) may even increase the risk for major adverse cardiovascular events when compared with placebo [90]. Brodalumab (an IL-17 receptor A antagonist), ixekizumab, and secukinumab (IL-17A antagonists), also used in psoriasis, are potential antagonists of atherosclerosis, given the proinflammatory nature of IL-17 [129]. IL-17 inhibition attenuates atherosclerosis in animals, but clinical studies in humans are not available to date [130].

Anti-P-selectin Agents

During inflammation, neutrophils, monocytes, or effector T cells migrate to specific tissues in response to infection or injury. The selectins are transmembrane, Ca2+-dependent lectins that mediate leucocyte rolling on vascular surfaces, at their first adhesive step during inflammation and immune surveillance. However, selectins also transduce signals that modulate leucocyte behavior, for example, by enabling integrin-dependent reductions in rolling velocities. The interaction of leucocytes refers not only to endothelial cells but selectins also mediate leucocyte adhesion to activated platelets and to other leucocytes, Thus selectins initiate multicellular adhesive and signaling events during physiological or pathological inflammation and hemostasis [131,132].

L-selectin is expressed on the surfaces of all leucocytes. P-selectin is expressed by megakaryocytes and endothelial cells. It is stored in membranes of α granules of platelets and Weibel–Palade bodies of endothelial cells. TNF-α, IL-1β, or lipopolysaccharide (LPS) increases transcription of mRNA for P-selectin in most mammals, including mice, but not in primates, including humans. P-selectin binds to and acts via its major ligand, protein selected glycoprotein ligand-1 (PSGL-1), expressed by leukocytes [133–135]. E-selectin is constitutively expressed on the surfaces of venular endothelial cells of bone marrow and skin. In most organs, however, endothelial cells must be stimulated with TNF-α, IL-1β, or LPS to synthesize E-selectin. This mechanism of induced expression is conserved in mice and humans [136].

Selectins, and particularly P-selectin, have been shown to contribute to pathological inflammation and thrombosis in many preclinical disease models, including atherosclerosis and ischemia–reperfusion injury [137–139]. Patients with inflammatory and thrombotic disorders typically have higher levels of activated platelets displaying P-selectin, platelet–leucocyte aggregates, and soluble P-selectin in peripheral blood [140,141].

In a study, patients with STEMI were randomized to receive a placebo or a soluble form of PSGL-1 fused to the Fc portion of human IgG1 (a P-selectin antagonist) intravenously as an adjunct to standard thrombolytic therapy but the trial was terminated prematurely due to the lack of efficacy in ST-segment elevation, left ventricular function, clinical outcome, and other endpoints [142]. In a phase II trial [SELECT Acute Coronary Syndrome (SELECT-ACS) trial], patients with non-ST-segment elevation myocardial infarction (NSTEMI) undergoing percutaneous coronary intervention were randomized to receive a placebo or a humanized anti-P-selectin to block P-selectin binding to PSGL-1. Relative to placebo, inclacumab decreased plasma troponin I to levels that approached statistical significance [143]. In the SELECT-Coronary Artery Bypass Surgery trial, inclacumab was administered to patients undergoing saphenous vein grafting but the results have not yet reported.

These initial clinical trials suggest that selectin antagonists may be effective in some cardiovascular disorders, as predicted from the many preclinical studies. Given the inherent risks in extrapolating data from animal models to humans, careful selection of clinical targets and proper trial design will be essential for further progress.

Intereleukin-6 Inhibition

Potential strategies using anti-IL-6 therapies have not been as widely tested as in the case of TNF-a or IL-1. Tocilizumab is a monoclonal antibody that blocks both membrane-bound and circulating IL-6 receptor. Randomized trials in patients with rheumatoid arthritis have shown that tocilizumab increases total, HDL, and LDL cholesterol and triglyceride levels [144]. Moreover, it results in improved endothelial function and decreased aortic stiffness [145]. Currently, several ongoing studies investigate the role of tocilizumabin in NSEMI (Phase 2 trial) and in comparison to etanercept in patients with rheumatoid arthritis and cardiovascular risk factors (Phase 4 trial).

Interleukin-1b Inhibition

Studies in both experimental animal models and patients suggest that the cytokine IL-1 plays an important role in promoting tissue inflammation and remodeling. The IL-1 family of proteins consists of IL-1a and IL-1b, proinflammatory cytokines which act on endothelial and vascular smooth muscle cells (VSMCs). IL-1b plays an important role in up regulating leukocyte adhesion molecules on endothelial cells, proliferation of VSMCs and production of other cytokines which have all been implicated in the development of atherosclerosis [146—148]. IL-1 receptor antagonist (IL-1Ra) is an endogenous, competitive antagonist of the IL-1 receptor, which modulates the biologic actions of IL-1 by prevention of signal transduction [149].

Anakinra is a recombinant form of human IL-1Ra (recombinant nonglycosylated form of human IL-1 receptor antagonist), a competitive antagonist of IL-1 that blocks the actions of IL-1 without any detectable agonist activity [150]. A study in the clinical setting of patients with rheumatoid arthritis but not known CAD was associated with improved coronary flow reserve, left ventricular function, endothelial function and reduced levels of inflammatory biomarkers compared to placebo, after 30 days of treatment with anakinra [151]. However, in a larger study, therapy with anakinra early after non-ST elevation ACS patients inflammatory markers were reduced, but soon after discontinuation of treatment (Day 30) high-sensitive CRP levels had risen again. Moreover, the subsequent cardiac events did not reduced either [89]. These results show the importance of IL-1 as a target in ACS, but also indicate the need for additional studies with anti-IL-1 therapy in ACS to assess duration and safety. The VCU-ART 2 (Virginia Commonwealth University Anakinra Remodeling Trial) study showed that anakinra given for 2 weeks post-STEMI has a neutral effect on recurrent ischaemic events but may lower new HF incidence or mortality at 1 year [152], a finding that will be further evaluated by the more recent VCU-ART-3 trial.

Canakinumab is a high-affinity human anti-interleukin IL-1β monoclonal antibody of the IgG1/κ isotype, which functionally neutralizes the bioactivity of this proinflammatory cytokine. Its specificity is confined to human and marmoset IL-1β [153]. In a large clinical trial (Canakinumab Anti-inflammatory Thrombosis Outcomes Study) the administration of Canakinumab evaluate whether IL-1β inhibition as compared to placebo can reduce rates of recurrent myocardial infarction, stroke, and cardiovascular death among stable coronary artery disease patients who remain at high vascular risk due to persistent elevations of hsCRP (2 mg/L) despite contemporary secondary prevention strategies. In this study, subjects with well-controlled diabetes mellitus and high cardiovascular risk, significantly reduces IL-6 and fibrinogen as markers of inflammation, without major effect on low-density lipoprotein cholesterol or high-density lipoprotein cholesterol [86,154].

LEUKOTRIENES

Leukotrienes are arachidonic acid (AA) derived lipid mediators of inflammation. The initial step in the formation of LTs is catalyzed by 5-lipoxygenase (5-LO) in conjunction with its five lipoxygenase activating protein (FLAP) [155,156]. Inflammation plays an important role in atherosclerosis process. In particular, many enzymes associated with the 5-LO pathway are abundantly expressed in arterial walls of patients afflicted with various lesion stages of atherosclerosis of the aorta and of coronary arteries [157].

In a randomized, placebo-controlled trial, Hakonarson et al. demonstrated that an inhibitor of FLAP (DG-031) in patients with specific at-risk variants of two genes in the leukotriene pathway, DG-031 led to significant and dose-dependent suppression of hsCRP that is associated with increased risk of AMI events [158]. Even Allayee et al. randomized patients with moderate-to-severe asthma to receive either montelukast, theophylline, or placebo for 6 months. Asthmatic patients receiving montelukast and, to some extent, low-dose theophylline decreased certain inflammatory and lipid CVD risk factors [159].

In ACS, a double blind trial with placebo involving 191 patients by Tardif et al. demonstrated that a potent 5-LO inhibitor (VIA-2291) reduces leukotriene production. Specifically, a significant reduction of urine leukotriene LTE4 was obtained in all dose groups with the absence of serious adverse events. In this study, a subset of patients underwent a 64-slice coronary CT examination at baseline continued on study medication for a total of 24 weeks and underwent a repeat scan. A significant reduction in noncalcified plaque volume at 24 weeks versus placebo was observed in VIA-2291 treated groups in patients in whom this end point was analyzable [160].

In a registry-based nation-wide Swedish cohort of almost 7 million subjects followed for 3.5 years, montelukast use was associated with a borderline significant protective effect on recurrent stroke, whereas in the same study a stratified analysis revealed that montelukast use significantly reduced the risk of recurrent myocardial infarction in males providing a first indication for beneficial effects of the CysLT1 receptor antagonist montelukast in the secondary prevention of stroke and myocardial infarction [161].

SERPINS

Serine protease inhibitors (serpins) are evolutionary old, structurally conserved molecules which encompass nearly all branches of life. Serpins were termed after their capability to inhibit serine proteases, but mounting evidence suggests that they may achieve a greater deal of functions, ranging from embryological growth to synaptic plasticity, development of both myeloid and lymphoid immune cells, and modulation of apoptosis. Serpins are mainly extracellular molecules, although some of them (namely, ovserpins, or clade B serpins) mostly act inside the cells, being either ubiquitously or tissue-specifically expressed [162,163]. In mammals, serpins exhibit both antithrombotic and prothrombotic effects, whereas in viruses' serpins demonstrate anti-inflammatory and antithrombotic properties making them an ideal therapeutic target for cardiovascular disease.

Serp-1 is a secreted glycoprotein of the serpin superfamily that inhibits tissue and urokinase-type plasminogen activators, plasmin, and factor Xa in vitro. It exhibits potent anti-inflammatory and antiatherogenic activity in vivo. Originally identified as a virulence factor that enabled poxviruses to reduce the host's inflammatory response to infection, purified recombinant Serp-1 has been shown to inhibit the early (24–72 h) activation and infiltration of monocytes/macrophages and T lymphocytes in a wide range of animal models of arterial angioplasty injury, stent implant, arthritis, and transplant [164,165]. A double-blind pilot trial included 48 ACS patients undergoing percutaneous coronary intervention randomly assigned to Serp-1 or to placebo. Serp-1 was given by intravenous bolus immediately before intervention and 24 and 48 h later. The study found a significant reduction in myocardial damage biomarkers supports further assessment of Serp-1 in ACS patients undergoing stent deployment [166].

ANTIDEPRESSANTS

Depression has been associated with increased inflammatory process. Although antidepressive medication has anti-inflammatory effect in major depression, a study evaluated the impact of antidepressive medication on the expression of proinflammatory cytokines and acute phase response proteins, in patients with HF and major depression. Results demonstrated that treatment of patients with HF and major depression with tricyclic antidepressants and/or serotonin/norepinephrine reuptake inhibitors, is associated with lower levels of TNF-alpha and CRP, suggesting that the type of antidepressive treatment may have a significant effect on the underlying inflammatory process [167].

DIET

Mediterranean Diet

Results from the ATTICA study have shown the advantages of the traditional Mediterranean diet in cardiovascular risk. Metabolic disorders such as diabetes, hyper-lipidemia and abdominal obesity are traditional risk factors in the development of cardiovascular disease [168,169]. Recently, the concept of metabolic syndrome (MetS)—a clustering of coexisting risk factors—has received much attention [170]. It is a common condition, present in approximately 20%–30% of the global adult population and has become a leading health problem [171]. The main components of MetS are insulin resistance, central adiposity, dyslipidemia, arterial hypertension and endothelial dysfunction. Over the years, various additional abnormalities such as chronic inflammation and thrombotic tendency have been added to the entity of the syndrome, all further promoting atherosclerosis [172]. Ridker et al. found that each component of the metabolic syndrome is significantly associated with higher levels of CRP [173]. Moreover, a recent systematic review and meta-analysis found that the metabolic syndrome was associated with a twofold increase in cardiovascular events implying that patients with the metabolic syndrome were at higher risk for CVD compared to those without the syndrome [174].

As inflammatory processes may be a common ground for the development of atherosclerosis that lead to plaque rapture and thrombosis, dietary patterns rich in antioxidant components have been proposed for the prevention and treatment of CVD. The traditional Mediterranean diet, which supplies an abundance of fruit, green vegetables, legumes, nuts and whole grains, associated with moderate consumption of olive oil and alcohol (especially with meals) and low consumption of red meat and butter has shown anti-inflammatory effects [175,176]. Specific nutrients such as n−3 fatty acids found in fish [177], polyphenols [178,179] from extra virgin oil and antioxidant components including β-carotene, vitamin E, vitamin C, and flavonoids [180] from fruit and vegetables have consistently been shown to be associated with lower levels of inflammation. As a result, the Mediterranean diet due to its attribution to the lower generation of the proinflammatory milieu may also exert protective effects regarding the development of the MetS. Thus, the aim of this work was to evaluate whether adherence to an anti-inflammatory diet was associated with 10-year CVD incidence in the ATTICA study, and whether this was affected by the presence of MetS among apparently healthy adults at entry examination.

It found that level of adherence to the anti-inflammatory diet was borderline associated with the 10-year CVD incidence. The latter association remained even after adjusting for a wide range of potential confounders including lifestyle factors, clinical history features and biochemical markers. This protective effect was indicative in the overall sample and significant among MetS free participants. Thus, it could be hypothesized that atherosclerosis is a low-grade inflammatory disease and because the generation of the inflammatory milieu is a common ground for both CVD and MetS [181], increased consumption of antioxidants (fruit, vegetables, fiber) can moderate the above mentioned relationship by mitigating this low grade systemic inflammation [169]. However, the anti-inflammatory diet was not protective against CVD outcomes among MetS participants (i.e., high-CVD risk individuals). A possible explanation could be that atherosclerotic mechanisms, initiated from earlier stages of life because the participants already had MetS, may have influenced the effect of anti-inflammatory diet on CVD risk [182—184]. In any case, the reported findings have a considerable importance for public health because they revealed that higher adherence to an anti-inflammatory diet was associated with reduced 10-year risk for CVD, whereas the presence of MetS seems to impede the preventive role of this diet, a fact that needs special attention by public health strategies.

In conclusion, the anti-inflammatory diet may diminish the risk of CVD in healthy individuals, but may be impossible to reverse the results of several years of exposure to oxidative stress among participants with established MetS. Thus, the presented findings suggest that favoring anti-inflammatory diet at early stages of life could be found to act preventively as regards to the development of CVD and challenge us to evaluate potential anti-inflammatory and antioxidant therapies that could be applied on public health action plans.

Oils

ICAM-1, VCAM-1, and TNF-α, are implicated in atherogenesis. In addition, several types of oil as part of different types of diet are associated with the initiation of atherosclerosis and consequently with the risk of cardiovascular disease. However, the exact role of oil consumption on vascular inflammation was unclear. In a parallel study, investigators assessed the acute effects of extra virgin olive oil, soy oil, corn oil and cod liver oil on circulating soluble(s) forms of adhesion molecules and TNF-α. They concluded that acute consumption of all types of oil decreased significantly ICAM-1 levels. In addition, olive oil, soy oil and cod liver oil decreased significantly TNF-α levels and moreover, the absolute change in TNF-α levels was correlated with the absolute change in ICAM-1 levels. These findings indicate that acute consumption of specific types of oil is associated with specific significant anti-inflammatory effects [185].

In a similar study investigators evaluated the effect of acute consumption of maize oil, cod liver oil, soya oil and extra virgin olive oil on endothelial function, oxidative stress and inflammation. However, all types of oils had no effect on serum levels of sVCAM-1 suggesting that they don't affect the inflammatory process through this pathway [186].

Omega-3 Polyunsaturated Fatty Acids

As mentioned earlier, metabolic syndrome and its components, dyslipidemia, glucose intolerance, diabetes mellitus, obesity, arterial hypertension are characterized by chronic inflammation and an imbalance between proinflammatory and anti-inflammatory cytokines [187–189]. Omega-3 polyunsaturated fatty acids (PUFAs) compete with AA as substrates for the formation of proinflammatory mediators, such as leukotrienes, prostaglandins, and cytokines. In addition to competitive inhibition of the n−6 fatty acid pathway, n−3 PUFAs may also inhibit production of inflammatory and fibrotic mediators including C reactive protein, interleukins, TNF alpha, matrix metalloproteinases 2 and 9, and tissue inhibitors of metalloproteinase [190–192]. In a cohort study of subjects with MetS, Tousoulis et al. documented a reduction in IL-6 levels which importantly was accompanied by an improvement in endothelial function and arterial stiffness. These results provided further inside into how omega-3 PUFAs administration can alter arterial wall properties and improve cardiovascular outcome of subjects with MetS.

Furthermore, a similar study, evaluated the effects of Ω-3 PUFAs supplementation on smoking-induced impairment of arterial function and inflammation. It was found that Ω-3 PUFAs improved endothelial function and the elastic properties of the arterial tree in healthy smokers, with a parallel anti-inflammatory effect [193].

ANTIOXIDANTS

Several antioxidant vitamins have been examined for the treatment of the cardiovascular disease such as vitamin C, vitamin E, and beta carotene [194,195]. Antioxidants act through different mechanisms to prevent oxidant-induced cell damages acting either directly or indirectly. They can reduce the generation of ROS, scavenge ROS, or interfere with ROS-induced alterations. Modulating mitochondrial activity is an important possibility to control ROS production. Hence, the use of peroxisome proliferator-activated receptor alpha (PPARα) agonist to reduce fatty acid oxidation and of trace elements such as selenium as antioxidant and other antioxidants such as vitamins E and C, contribute to the prevention of diabetes-induced cardiovascular dysfunction [196,197].

Our studies on subjects with diabetes mellitus and coronary artery disease showed beneficial effects of vitamin C on specific measured markers [198,199]. Despite the fact that there was no impact on inflammatory process of patients with DM type 2 and CAD, vitamin C significantly increased forearm vasodilatory response to reactive hyperemia (endothelial function, a well-known index of atherosclerosis) [19,198]. Similarly, short term treatment with high doses of vitamin C not only improved reactive hyperemia, but also decreased plasma levels of tissue plasminogen activator (tPA) and von Willebrand factor implicating an impact of antioxidants on thrombosis/fibrinolysis system, which is known to play an important role in coronary atherosclerosis [199].

Numerous randomized trials have focused on the effects of antioxidant vitamins in different populations. A randomized trial in 4495 subjects with one or more risk factors for CAD found that daily supplementation with 300 mg of Vitamin E had no effect on primary endpoint of cardiovascular death, nonfatal myocardial infarction, or nonfatal stroke [200]. Furthermore, the role of vitamin C in smokers is under investigation. Smoking is associated with endothelial dysfunction and increased levels of proinflammatory cytokines such as interleukines 1b and 6, TNF-a, and adhesion molecules [201]. We have also shown that combined administration of high doses of vitamins C (2 g/day) and E (800IU/day) has nonspecific antiinflammatpory effect, because it decreases serum levels of TNF-α, serum levels of soluble forms of vascular cell adhesion molecule (sVCAM-1) and intercellular adhesion molecule (sICAM-1), as well as the expression of proinflammatory cytokines such as interleukines 1b and 6 [202]. In similar type of case-controlled studies, we have also shown that combined administration of vitamin C (2 g/day) and vitamin E (800IU/day) significantly increased endothelial function having also important antithrombotic effect in chronic smokers [203]. It seems that by protecting vascular endothelium from the deleterious effect of reactive oxygen species, we prevent the release of a number of prothrombotic molecules from vascular endothelium, such as von Willebrand factor or plasminogen activator inhibitor-1 (PAI-1) [204]. Despite the simultaneous decrease of antithrombotic molecules released from the injured endothelium (such as tPA), it seems that protecting endothelial cells from oxidative injury, significantly decreases the PAI-1/tPA ratio in smokers.

As far as secondary prevention of CAD is concerned, all but one study of Vitamin E has found little or no benefit [200,205]. The HOPE trial evaluated vitamin E (400 IU/day) in 9541 patients over the age of 55 years [205]. After a mean follow-up of 4.5 years, there was no effect of Vitamin E on the primary endpoint (cardiovascular death, myocardial infarction, or stroke) or any of its individual components or any of the secondary endpoints. The HPS randomized 20,536 patients who had a history of CAD, vascular disease, or diabetes mellitus to 5 years of antioxidant vitamin supplementation (vitamin E 600 mg, vitamin c 250 mg, and beta carotene 20 mg) or placebo [206]. The results of the study showed that supplementation had no effect on cardiovascular mortality and nonfatal myocardial infarction. Vitamin E was found to preserve endothelial vasomotor function in patients with coronary atherosclerosis. It has been demonstrated in the past that patients with severe atherosclerosis have an imbalance of oxidant/antioxidant status [207]. Vitamin E supplementation corrects the oxidant/antioxidant balance in plasma, but it doesn't seem to have any direct effect inside the atherosclerotic plaques [207].

Recent studies suggested that administration of vitamin E may prevent the beneficial effects of statins on the progression of coronary artery stenosis or increase the morbidity and mortality in CAD postmenopausal women [208]. In addition, we have shown that the beneficial effect of atorvastatin on endothelial function and inflammatory process was negatively affected by the coadministration of vitamin E [209]. Similarly, other studies reported that combined administration of vitamins E/C and atorvastatin, may be associated with no effect on coronary artery calcification or the progression of coronary atherosclerosis [210].

TARGETING POSTISCHEMIC REPERFUSION INJURY

Re-establishment of coronary perfusion after an acute ischemic event causes the so-called reperfusion injury, which is attributed to oxidative stress, accumulation of reactive oxygen species, and ultimately diminished mitochondrial ability to produce energy in the form of ATP [211]. Bendavia is an intravenously administered mitochondrial targeting peptide that has been shown to reduce myocardial infarct size and attenuate coronary no-reflow in experimental models when given before reperfusion [212]. Therefore, the EMBRACE STEMI study enrolled 300 patients with first anterior STEMI to test whether Bendavia is superior to placebo for the reduction of myocardial infarction size [213]. The first preliminary reports of this study have recently been announced and have shown that Bendavia infusion is safe although it did not achieve a significant reduction in the infarct size. However, future trials are planning to test if higher doses can be beneficial in patients with systolic HF.

EXPERIMENTAL APPROACHES IN ANIMAL MODELS

Future directions concerning the targeted modification of the inflammatory cascade can be derived from experimental models. However, the extrapolation of animal outcomes to human patients is not straightforward as significant differences exist in the pathophysiology of atherosclerosis between different species [214].

Delivery of Anti-Inflammatory Cytokines

Delivery of anti-inflammatory cytokines for therapeutic purposes has been attempted in experimental animal models with promising results. It is now possible to locally deliver anti-inflammatory cytokines to specific atherosclerosis sites via adenoviruses. Such is the case of IL-19 which is an anti-inflammatory cytokine that promotes the Th2 anti-inflammatory cell phenotype [215]. Furthermore, a recently developed method exploits stealth liposomes that are coupled with IL-10 and act to specifically and reliably transfer the atheroprotective cytokine to the desired site in atherosclerotic plaques [216]. It is obvious that these methods may have unknown risks, and studies on safety and efficacy are warranted before they can be applied in humans.

Targeting Intracellular Signaling

Certain studies have succeeded in limiting atherosclerosis by manipulating intracellular signaling in animal models. For example, an oligonucleotide that acted as a decoy binding site for NF-kB succeeded in preventing intimal hyperplasia following balloon angioplasty [217]. An adenovirus-expressed suppressor of cytokine signaling 3 has also been successful in limiting inflammation though not in a model of atherosclerosis [218], and similar efforts have been done trying to block Janus kinase 3 (JAK3) signaling [219]. However, given the ubiquitous nature of these molecules (especially NF-kB), it is important to guarantee that treatment does not interfere with normal processes in order to minimize adverse effects.

Promoting the Actions of T-Regulatory Cells

Administration of anti-CD3 antibodies in mice with progressed atherosclerosis managed to suppress atherogenesis and was linked to upregulated transforming growth factor (TGF)-b and Foxp3 expression in lymphoid organs [220,221]. In addition, oral calcitriol administration to hyperlipidemic mice blocked atherosclerosis possibly by inducing the expansion of T-regulatory cells [222], whereas local delivery of IL-2 suppressed atherogenesis in already existing lesions through the same mechanism [223].

CONCLUSION

Atherosclerosis is considered an inflammatory disease, and the significant role of cytokines in the initiation and maintenance of a proinflammatory state is well established. In recent years, novel therapeutic approaches capable of modulating cytokine production or their actions have been tested or are under investigation. Indeed, broad-based immunomodulatory therapies or specific molecules against inflammatory cytokines are now under investigation for the treatment of high-risk atherosclerotic subjects generating new hopes for the future. However, so far, there are no definitive data on the benefit of targeted anti-inflammatory therapy in atherosclerosis and therefore further studies are needed. Hopefully, in the next few years, we are going to have new data that will provide answers and guidance for our next steps in the fight against atherosclerosis and coronary heart disease.

REFERENCES

[1] Tousoulis D, et al. Novel therapies targeting vascular endothelium. Endothelium 2006;13(6): 411–21.

[2] Zhuo JL, et al. Perindopril chronically inhibits angiotensin-converting enzyme in both the endothelium and adventitia of the internal mammary artery in patients with ischemic heart disease. Circulation 1997;96(1):174–82.

[3] Nickenig G, Harrison DG. The AT(1)-type angiotensin receptor in oxidative stress and atherogenesis: part I: oxidative stress and atherogenesis. Circulation 2002;105(3):393–6.

[4] Faggiotto A, Paoletti R. State-of-the-Art lecture. Statins and blockers of the renin-angiotensin system: vascular protection beyond their primary mode of action. Hypertension 1999;34(4 Pt 2): 987–96.

[5] Marin-Castano ME, et al. Induction of functional bradykinin b(1)-receptors in normotensive rats and mice under chronic angiotensin-converting enzyme inhibitor treatment. Circulation 2002;105(5): 627–32.

[6] Mancini GB, et al. Angiotensin-converting enzyme inhibition with quinapril improves endothelial vasomotor dysfunction in patients with coronary artery disease. The TREND (Trial on Reversing ENdothelial Dysfunction) Study. Circulation 1996;94(3):258–65.

[7] Yusuf S, et al. Effects of an angiotensin-converting-enzyme inhibitor, ramipril, on cardiovascular events in high-risk patients. The Heart Outcomes Prevention Evaluation Study Investigators. N Engl J Med 2000;342(3):145–53.

[8] Dagenais GR, et al. Effects of ramipril on coronary events in high-risk persons: results of the Heart Outcomes Prevention Evaluation Study. Circulation 2001;104(5):522–6.

[9] Fox KM. Efficacy of perindopril in reduction of cardiovascular events among patients with stable coronary artery disease: randomised, double-blind, placebo-controlled, multicentre trial (the EUROPA study). Lancet 2003;362(9386):782–8.

[10] Major outcomes in high-risk hypertensive patients randomized to angiotensin-converting enzyme inhibitor or calcium channel blocker vs diuretic: the Antihypertensive and Lipid-Lowering Treatment to Prevent Heart Attack Trial (ALLHAT). JAMA 2002;288(23):2981–97.

[11] Einhorn PT, et al. ALLHAT: still providing correct answers after 7 years. Curr Opin Cardiol 2010; 25(4):355–65.

[12] Cashin-Hemphill L, et al. Angiotensin-converting enzyme inhibition as antiatherosclerotic therapy: no answer yet. QUIET Investigators. QUinapril Ischemic Event Trial. Am J Cardiol 1999;83(1): 43–7.

[13] Oosterga M, et al. Effects of quinapril on clinical outcome after coronary artery bypass grafting (The QUO VADIS Study). QUinapril on Vascular Ace and Determinants of Ischemia. Am J Cardiol 2001; 87(5):542–6.

[14] Tousoulis D, et al. Novel therapeutic strategies targeting vascular endothelium in essential hypertension. Expert Opin Investig Drugs 2010;19(11):1395–412.

[15] Vogt L, et al. The angiotensin II receptor antagonist telmisartan reduces urinary albumin excretion in patients with isolated systolic hypertension: results of a randomized, double-blind, placebo-controlled trial. J Hypertens 2005;23(11):2055–61.

[16] Volpe M, et al. Angiotensin-II receptor blockers: benefits beyond blood pressure reduction? J Hum Hypertens 2005;19(5):331–9.

[17] Turnbull F, et al. Blood pressure-dependent and independent effects of agents that inhibit the renin-angiotensin system. J Hypertens 2007;25(5):951–8.

[18] Wolfrum S, Jensen KS, Liao JK. Endothelium-dependent effects of statins. Arterioscler Thromb Vasc Biol 2003;23(5):729–36.

[19] Tousoulis D, et al. Effects of atorvastatin and vitamin C on forearm hyperaemic blood flow, asymmentrical dimethylarginine levels and the inflammatory process in patients with type 2 diabetes mellitus. Heart 2007;93(2):244–6.

[20] Goldstein JL, Brown MS. Regulation of the mevalonate pathway. Nature 1990;343(6257):425–30.

[21] Antoniades C, et al. Rapid, direct effects of statin treatment on arterial redox state and nitric oxide bioavailability in human atherosclerosis via tetrahydrobiopterin-mediated endothelial nitric oxide synthase coupling. Circulation 2011;124(3):335−45.
[22] Tousoulis D, et al. Innate and adaptive inflammation as a therapeutic target in vascular disease: the emerging role of statins. J Am Coll Cardiol 2014;63(23):2491−502.
[23] Kimura M, et al. Effects of fluvastatin on leukocyte-endothelial cell adhesion in hypercholesterolemic rats. Arterioscler Thromb Vasc Biol 1997;17(8):1521−6.
[24] Rezaie-Majd A, et al. Simvastatin reduces expression of cytokines interleukin-6, interleukin-8, and monocyte chemoattractant protein-1 in circulating monocytes from hypercholesterolemic patients. Arterioscler Thromb Vasc Biol 2002;22(7):1194−9.
[25] Wang S, et al. PECAM-1, alpha6 integrins and neutrophil elastase cooperate in mediating neutrophil transmigration. J Cell Sci 2005;118(Pt 9):2067−76.
[26] Wennerberg K, Der CJ. Rho-family GTPases: it's not only Rac and Rho (and I like it). J Cell Sci 2004;117(Pt 8):1301−12.
[27] Carman CV, et al. Endothelial cells proactively form microvilli-like membrane projections upon intercellular adhesion molecule 1 engagement of leukocyte LFA-1. J Immunol 2003;171(11): 6135−44.
[28] Wojciak-Stothard B, Williams L, Ridley AJ. Monocyte adhesion and spreading on human endothelial cells is dependent on Rho-regulated receptor clustering. J Cell Biol 1999;145(6):1293−307.
[29] Holstein SA, Wohlford-Lenane CL, Hohl RJ. Consequences of mevalonate depletion. Differential transcriptional, translational, and post-translational up-regulation of Ras, Rap1a, RhoA, and RhoB. J Biol Chem 2002;277(12):10678−82.
[30] Wei H, et al. Statin-inhibited endothelial permeability could be associated with its effect on PECAM-1 in endothelial cells. FEBS Lett 2005;579(5):1272−8.
[31] Ridker PM. Rosuvastatin in the primary prevention of cardiovascular disease among patients with low levels of low-density lipoprotein cholesterol and elevated high-sensitivity C-reactive protein: rationale and design of the JUPITER trial. Circulation 2003;108(19):2292−7.
[32] Ridker PM, et al. Baseline characteristics of participants in the JUPITER trial, a randomized placebo-controlled primary prevention trial of statin therapy among individuals with low low-density lipoprotein cholesterol and elevated high-sensitivity C-reactive protein. Am J Cardiol 2007; 100(11):1659−64.
[33] Libby P, et al. Inflammation in atherosclerosis: transition from theory to practice. Circ J 2010;74(2): 213−20.
[34] Ridker PM, et al. Reduction in C-reactive protein and LDL cholesterol and cardiovascular event rates after initiation of rosuvastatin: a prospective study of the JUPITER trial. Lancet 2009; 373(9670):1175−82.
[35] Despres JP. Bringing JUPITER down to earth. Lancet 2009;373(9670):1147−8.
[36] Kostapanos MS, Milionis HJ, Elisaf MS. An overview of the extra-lipid effects of rosuvastatin. J Cardiovasc Pharmacol Ther 2008;13(3):157−74.
[37] Biasucci LM, Biasillo G, Stefanelli A. Inflammatory markers, cholesterol and statins: pathophysiological role and clinical importance. Clin Chem Lab Med 2010;48(12):1685−91.
[38] Sattar N, et al. Metabolic syndrome with and without C-reactive protein as a predictor of coronary heart disease and diabetes in the West of Scotland Coronary Prevention Study. Circulation 2003; 108(4):414−9.
[39] Shepherd J, et al. Pravastatin in elderly individuals at risk of vascular disease (PROSPER): a randomised controlled trial. Lancet 2002;360(9346):1623−30.
[40] Major outcomes in moderately hypercholesterolemic, hypertensive patients randomized to pravastatin vs usual care: the Antihypertensive and Lipid-Lowering Treatment to Prevent Heart Attack Trial (ALLHAT-LLT). JAMA 2002;288(23):2998−3007.
[41] Sacks FM, et al. The effect of pravastatin on coronary events after myocardial infarction in patients with average cholesterol levels. Cholesterol and Recurrent Events Trial investigators. N Engl J Med 1996;335(14):1001−9.

[42] Prevention of cardiovascular events and death with pravastatin in patients with coronary heart disease and a broad range of initial cholesterol levels. The Long-Term Intervention with Pravastatin in Ischaemic Disease (LIPID) Study Group. N Engl J Med 1998;339(19):1349–57.
[43] Serruys PW, et al. Fluvastatin for prevention of cardiac events following successful first percutaneous coronary intervention: a randomized controlled trial. JAMA 2002;287(24):3215–22.
[44] Holdaas H, et al. Effect of fluvastatin on cardiac outcomes in renal transplant recipients: a multicentre, randomised, placebo-controlled trial. Lancet 2003;361(9374):2024–31.
[45] Downs JR, et al. Primary prevention of acute coronary events with lovastatin in men and women with average cholesterol levels: results of AFCAPS/TexCAPS. Air Force/Texas Coronary Atherosclerosis Prevention Study. JAMA 1998;279(20):1615–22.
[46] Randomised trial of cholesterol lowering in 4444 patients with coronary heart disease: the Scandinavian Simvastatin Survival Study (4S). Lancet 1994;344(8934):1383–9.
[47] MRC/BHF Heart Protection Study of cholesterol lowering with simvastatin in 20,536 high-risk individuals: a randomised placebo-controlled trial. Lancet 2002;360(9326):7–22.
[48] Sever PS, et al. Prevention of coronary and stroke events with atorvastatin in hypertensive patients who have average or lower-than-average cholesterol concentrations, in the Anglo-Scandinavian Cardiac Outcomes Trial—Lipid Lowering Arm (ASCOT-LLA): a multicentre randomised controlled trial. Lancet 2003;361(9364):1149–58.
[49] Colhoun HM, et al. Primary prevention of cardiovascular disease with atorvastatin in type 2 diabetes in the Collaborative Atorvastatin Diabetes Study (CARDS): multicentre randomised placebo-controlled trial. Lancet 2004;364(9435):685–96.
[50] LaRosa JC, et al. Safety and efficacy of atorvastatin-induced very low-density lipoprotein cholesterol levels in patients with coronary heart disease (a post hoc analysis of the treating to new targets [TNT] study). Am J Cardiol 2007;100(5):747–52.
[51] Pedersen TR, et al. High-dose atorvastatin vs usual-dose simvastatin for secondary prevention after myocardial infarction: the IDEAL study: a randomized controlled trial. JAMA 2005;294(19):2437–45.
[52] Jones PH, et al. Comparison of the efficacy and safety of rosuvastatin versus atorvastatin, simvastatin, and pravastatin across doses (STELLAR* Trial). Am J Cardiol 2003;92(2):152–60.
[53] Olsson AG, et al. Effects of rosuvastatin and atorvastatin compared over 52 weeks of treatment in patients with hypercholesterolemia. Am Heart J 2002;144(6):1044–51.
[54] Bybee KA, et al. Effect of concomitant or very early statin administration on in-hospital mortality and reinfarction in patients with acute myocardial infarction. Am J Cardiol 2001;87(6):771–4. a7.
[55] Stenestrand U, Wallentin L. Early statin treatment following acute myocardial infarction and 1-year survival. JAMA 2001;285(4):430–6.
[56] Walter DH, et al. Benefits of immediate initiation of statin therapy following successful coronary stent implantation in patients with stable and unstable angina pectoris and Q-wave acute myocardial infarction. Am J Cardiol 2002;89(1):1–6.
[57] Ferrieres J, et al. Effect of early initiation of statins on survival in patients with acute myocardial infarction (the USIC 2000 Registry). Am J Cardiol 2005;95(4):486–9.
[58] Fonarow GC, et al. Effect of statin use within the first 24 hours of admission for acute myocardial infarction on early morbidity and mortality. Am J Cardiol 2005;96(5):611–6.
[59] Lenderink T, et al. Patients using statin treatment within 24 h after admission for ST-elevation acute coronary syndromes had lower mortality than non-users: a report from the first Euro Heart Survey on acute coronary syndromes. Eur Heart J 2006;27(15):1799–804.
[60] Nagashima M, et al. Effect of early statin treatment at standard doses on long-term clinical outcomes in patients with acute myocardial infarction (the Heart Institute of Japan, Department of Cardiology Statin Evaluation Program). Am J Cardiol 2007;99(11):1523–8.
[61] Hulten E, et al. The effect of early, intensive statin therapy on acute coronary syndrome: a meta-analysis of randomized controlled trials. Arch Intern Med 2006;166(17):1814–21.
[62] Bavry AA, et al. Long-term benefit of statin therapy initiated during hospitalization for an acute coronary syndrome: a systematic review of randomized trials. Am J Cardiovasc Drugs 2007;7(2):135–41.

[63] Smilde TJ, et al. Effect of aggressive versus conventional lipid lowering on atherosclerosis progression in familial hypercholesterolaemia (ASAP): a prospective, randomised, double-blind trial. Lancet 2001;357(9256):577–81.
[64] Taylor AJ, et al. ARBITER: arterial biology for the investigation of the treatment effects of reducing cholesterol: a randomized trial comparing the effects of atorvastatin and pravastatin on carotid intima medial thickness. Circulation 2002;106(16):2055–60.
[65] Crouse 3rd JR, et al. Effect of rosuvastatin on progression of carotid intima-media thickness in low-risk individuals with subclinical atherosclerosis: the METEOR Trial. JAMA 2007;297(12):1344–53.
[66] Nissen SE, et al. Effect of intensive compared with moderate lipid-lowering therapy on progression of coronary atherosclerosis: a randomized controlled trial. JAMA 2004;291(9):1071–80.
[67] Patti G, et al. Protection from procedural myocardial injury by atorvastatin is associated with lower levels of adhesion molecules after percutaneous coronary intervention: results from the ARMYDA-CAMs (Atorvastatin for Reduction of MYocardial Damage during Angioplasty-Cell Adhesion Molecules) substudy. J Am Coll Cardiol 2006;48(8):1560–6.
[68] Ridker PM, et al. Measurement of C-reactive protein for the targeting of statin therapy in the primary prevention of acute coronary events. N Engl J Med 2001;344(26):1959–65.
[69] Albert MA, et al. Effect of statin therapy on C-reactive protein levels: the pravastatin inflammation/CRP evaluation (PRINCE): a randomized trial and cohort study. JAMA 2001;286(1):64–70.
[70] Cannon CP, et al. Intensive versus moderate lipid lowering with statins after acute coronary syndromes. N Engl J Med 2004;350(15):1495–504.
[71] Schwartz GG, et al. Effects of atorvastatin on early recurrent ischemic events in acute coronary syndromes: the MIRACL study: a randomized controlled trial. JAMA 2001;285(13):1711–8.
[72] Tousoulis D, et al. Combined effects of atorvastatin and metformin on glucose-induced variations of inflammatory process in patients with diabetes mellitus. Int J Cardiol 2011;149(1):46–9.
[73] Toth L, Muszbek L, Komaromi I. Mechanism of the irreversible inhibition of human cyclooxygenase-1 by aspirin as predicted by QM/MM calculations. J Mol Graph Model 2013;40: 99–109.
[74] Cyrus T, et al. Effect of low-dose aspirin on vascular inflammation, plaque stability, and atherogenesis in low-density lipoprotein receptor-deficient mice. Circulation 2002;106(10):1282–7.
[75] Liu H, et al. Aspirin inhibits fractalkine expression in atherosclerotic plaques and reduces atherosclerosis in ApoE gene knockout mice. Cardiovasc Drugs Ther 2010;24(1):17–24.
[76] Herova M, et al. Low dose aspirin is associated with plasma chemerin levels and may reduce adipose tissue inflammation. Atherosclerosis 2014;235(2):256–62.
[77] Kharbanda RK, et al. Prevention of inflammation-induced endothelial dysfunction: a novel vasculo-protective action of aspirin. Circulation 2002;105(22):2600–4.
[78] Ikonomidis I, et al. Increased proinflammatory cytokines in patients with chronic stable angina and their reduction by aspirin. Circulation 1999;100(8):793–8.
[79] Ridker PM, et al. Inflammation, aspirin, and the risk of cardiovascular disease in apparently healthy men. N Engl J Med 1997;336(14):973–9.
[80] Pietri P, et al. Beneficial effects of low-dose aspirin on aortic stiffness in hypertensive patients. Vasc Med 2014;19(6):452–7.
[81] van Diepen JA, et al. Aspirin reduces hypertriglyceridemia by lowering VLDL-triglyceride production in mice fed a high-fat diet. Am J Physiol Endocrinol Metab 2011;301(6):E1099–107.
[82] He Z, et al. Aspirin regulates hepatocellular lipid metabolism by activating AMPK signaling pathway. J Toxicol Sci 2015;40(1):127–36.
[83] Chow BJ, et al. Prognostic and therapeutic implications of statin and aspirin therapy in individuals with nonobstructive coronary artery disease: results from the CONFIRM (COronary CT Angiography EvaluatioN for Clinical Outcomes: an InteRnational Multicenter registry) registry. Arterioscler Thromb Vasc Biol 2015;35(4):981–9.
[84] Belch J, et al. The prevention of progression of arterial disease and diabetes (POPADAD) trial: factorial randomised placebo controlled trial of aspirin and antioxidants in patients with diabetes and asymptomatic peripheral arterial disease. Bmj 2008;337:a1840.

[85] Kim AJ, et al. Low-dose aspirin for prevention of cardiovascular disease in patients with chronic kidney disease. PLoS One 2014;9(8):e104179.
[86] Ridker PM. Closing the loop on inflammation and atherothrombosis: why perform the CIRT and CANTOS trials? Trans Am Clin Climatol Assoc 2013;124:174−90.
[87] Tam LS, Kitas GD, Gonzalez-Gay MA. Can suppression of inflammation by anti-TNF prevent progression of subclinical atherosclerosis in inflammatory arthritis? Rheumatol Oxf 2014;53(6): 1108−19.
[88] Ikonomidis I, et al. Increased benefit of interleukin-1 inhibition on vascular function, myocardial deformation, and twisting in patients with coronary artery disease and coexisting rheumatoid arthritis. Circ Cardiovasc Imaging 2014;7(4):619−28.
[89] Morton AC, et al. The effect of interleukin-1 receptor antagonist therapy on markers of inflammation in non-ST elevation acute coronary syndromes: the MRC-ILA Heart Study. Eur Heart J 2015; 36(6):377−84.
[90] Tzellos T, Kyrgidis A, Zouboulis CC. Re-evaluation of the risk for major adverse cardiovascular events in patients treated with anti-IL-12/23 biological agents for chronic plaque psoriasis: a meta-analysis of randomized controlled trials. J Eur Acad Dermatol Venereol 2013;27(5):622−7.
[91] Tenaglia AN, et al. Levels of expression of P-selectin, E-selectin, and intercellular adhesion molecule-1 in coronary atherectomy specimens from patients with stable and unstable angina pectoris. Am J Cardiol 1997;79(6):742−7.
[92] Lucas A, et al. Transplant vasculopathy: viral anti-inflammatory serpin regulation of atherogenesis. J Heart Lung Transpl 2000;19(11):1029−38.
[93] Choi HK, et al. Methotrexate and mortality in patients with rheumatoid arthritis: a prospective study. Lancet 2002;359(9313):1173−7.
[94] Nidorf SM, et al. Low-dose colchicine for secondary prevention of cardiovascular disease. J Am Coll Cardiol 2013;61(4):404−10.
[95] Giugliano GR, et al. Meta-analysis of corticosteroid treatment in acute myocardial infarction. Am J Cardiol 2003;91(9):1055−9.
[96] Schillaci G, et al. Aortic stiffness is increased in polymyalgia rheumatica and improves after steroid treatment. Ann Rheum Dis 2012;71(7):1151−6.
[97] Avina-Zubieta JA, et al. Immediate and past cumulative effects of oral glucocorticoids on the risk of acute myocardial infarction in rheumatoid arthritis: a population-based study. Rheumatol Oxf 2013; 52(1):68−75.
[98] Tian H, Cronstein BN. Understanding the mechanisms of action of methotrexate: implications for the treatment of rheumatoid arthritis. Bull NYU Hosp Jt Dis 2007;65(3):168−73.
[99] Cronstein B. How does methotrexate suppress inflammation? Clin Exp Rheumatol 2010;28(5 Suppl 61):S21−3.
[100] Genestier L, et al. Immunosuppressive properties of methotrexate: apoptosis and clonal deletion of activated peripheral T cells. J Clin Invest 1998;102(2):322−8.
[101] Mirabet M, et al. Expression of A2B adenosine receptors in human lymphocytes: their role in T cell activation. J Cell Sci 1999;112(Pt 4):491−502.
[102] Bulgarelli A, et al. Treatment with methotrexate inhibits atherogenesis in cholesterol-fed rabbits. J Cardiovasc Pharmacol 2012;59(4):308−14.
[103] Wessels JA, Huizinga TW, Guchelaar HJ. Recent insights in the pharmacological actions of methotrexate in the treatment of rheumatoid arthritis. Rheumatol Oxf 2008;47(3):249−55.
[104] Micha R, et al. Systematic review and meta-analysis of methotrexate use and risk of cardiovascular disease. Am J Cardiol 2011;108(9):1362−70.
[105] Westlake SL, et al. The effect of methotrexate on cardiovascular disease in patients with rheumatoid arthritis: a systematic literature review. Rheumatol Oxf 2010;49(2):295−307.
[106] van Halm VP, et al. Disease-modifying antirheumatic drugs are associated with a reduced risk for cardiovascular disease in patients with rheumatoid arthritis: a case control study. Arthritis Res Ther 2006;8(5):R151.
[107] Greenberg JD, et al. Tumour necrosis factor antagonist use and associated risk reduction of cardiovascular events among patients with rheumatoid arthritis. Ann Rheum Dis 2011;70(4):576−82.

[108] Ridker PM. Testing the inflammatory hypothesis of atherothrombosis: scientific rationale for the cardiovascular inflammation reduction trial (CIRT). J Thromb Haemost 2009;7(Suppl 1):332–9.
[109] Moreira DM, et al. Rationale and design of the TETHYS trial: the effects of methotrexate therapy on myocardial infarction with ST-segment elevation. Cardiology 2013;126(3):167–70.
[110] Ronda N, et al. Newly identified antiatherosclerotic activity of methotrexate and adalimumab: complementary effects on lipoprotein function and macrophage cholesterol metabolism. Arthritis Rheumatol 2015;67(5):1155–64.
[111] Christia P, Frangogiannis NG. Targeting inflammatory pathways in myocardial infarction. Eur J Clin Invest 2013;43(9):986–95.
[112] Turer AT, Hill JA. Pathogenesis of myocardial ischemia-reperfusion injury and rationale for therapy. Am J Cardiol 2010;106(3):360–8.
[113] Fitzgerald GA. Coxibs and cardiovascular disease. N Engl J Med 2004;351(17):1709–11.
[114] Nussinovitch U, et al. Glucocorticoids and the cardiovascular system: state of the art. Curr Pharm Des 2010;16(32):3574–85.
[115] Olsen AM, et al. Cause-specific cardiovascular risk associated with nonsteroidal anti-inflammatory drugs among myocardial infarction patients—a nationwide study. PLoS One 2013;8(1):e54309.
[116] Deftereos S, et al. Colchicine and the heart: pushing the envelope. J Am Coll Cardiol 2013;62(20): 1817–25.
[117] Wesley 2nd RB, et al. Extracellular matrix modulates macrophage functions characteristic to atheroma: collagen type I enhances acquisition of resident macrophage traits by human peripheral blood monocytes in vitro. Arterioscler Thromb Vasc Biol 1998;18(3):432–40.
[118] Deftereos S, et al. Anti-inflammatory treatment with Colchicine in acute myocardial infarction: a pilot study. Circulation 2015;132(15):1395–403.
[119] Nidorf M, Thompson PL. Effect of colchicine (0.5 mg twice daily) on high-sensitivity C-reactive protein independent of aspirin and atorvastatin in patients with stable coronary artery disease. Am J Cardiol 2007;99(6):805–7.
[120] Raju NC, et al. Effect of colchicine compared with placebo on high sensitivity C-reactive protein in patients with acute coronary syndrome or acute stroke: a pilot randomized controlled trial. J Thromb Thrombolysis 2012;33(1):88–94.
[121] Crittenden DB, et al. Colchicine use is associated with decreased prevalence of myocardial infarction in patients with gout. J Rheumatol 2012;39(7):1458–64.
[122] Geiler J, Buch M, McDermott MF. Anti-TNF treatment in rheumatoid arthritis. Curr Pharm Des 2011;17(29):3141–54.
[123] Rios-Navarro C, et al. Differential effects of anti-TNF-alpha and anti-IL-12/23 agents on human leukocyte-endothelial cell interactions. Eur J Pharmacol 2015;765:355–65.
[124] Avgerinou G, et al. Anti-tumor necrosis factor alpha treatment with adalimumab improves significantly endothelial function and decreases inflammatory process in patients with chronic psoriasis. Int J Cardiol 2011;151(3):382–3.
[125] Spinelli FR, et al. Effect of therapeutic inhibition of TNF on circulating endothelial progenitor cells in patients with rheumatoid arthritis. Mediat Inflamm 2013;2013:537539.
[126] Tam LS, et al. The effects of golimumab on subclinical atherosclerosis and arterial stiffness in ankylosing spondylitis-a randomized, placebo-controlled pilot trial. Rheumatol Oxf 2014;53(6): 1065–74.
[127] Barnabe C, Martin BJ, Ghali WA. Systematic review and meta-analysis: anti-tumor necrosis factor alpha therapy and cardiovascular events in rheumatoid arthritis. Arthritis Care Res 2011;63(4): 522–9.
[128] Gao Q, et al. A critical function of Th17 proinflammatory cells in the development of atherosclerotic plaque in mice. J Immunol 2010;185(10):5820–7.
[129] Coimbra S, Figueiredo A, Santos-Silva A. Brodalumab: an evidence-based review of its potential in the treatment of moderate-to-severe psoriasis. Core Evid 2014;9:89–97.
[130] Erbel C, et al. Inhibition of IL-17A attenuates atherosclerotic lesion development in apoE-deficient mice. J Immunol 2009;183(12):8167–75.
[131] McEver RP, Zhu C. Rolling cell adhesion. Annu Rev Cell Dev Biol 2010;26:363–96.

[132] Ley K, et al. Getting to the site of inflammation: the leukocyte adhesion cascade updated. Nat Rev Immunol 2007;7(9):678–89.
[133] Yao L, et al. Divergent inducible expression of P-selectin and E-selectin in mice and primates. Blood 1999;94(11):3820–8.
[134] Liu Z, et al. Differential regulation of human and murine P-selectin expression and function in vivo. J Exp Med 2010;207(13):2975–87.
[135] Frenette PS, et al. P-Selectin glycoprotein ligand 1 (PSGL-1) is expressed on platelets and can mediate platelet-endothelial interactions in vivo. J Exp Med 2000;191(8):1413–22.
[136] Vestweber D, Blanks JE. Mechanisms that regulate the function of the selectins and their ligands. Physiol Rev 1999;79(1):181–213.
[137] Dong ZM, et al. The combined role of P- and E-selectins in atherosclerosis. J Clin Invest 1998; 102(1):145–52.
[138] An G, et al. P-selectin glycoprotein ligand-1 is highly expressed on Ly-6Chi monocytes and a major determinant for Ly-6Chi monocyte recruitment to sites of atherosclerosis in mice. Circulation 2008; 117(25):3227–37.
[139] Thiagarajan RR, Winn RK, Harlan JM. The role of leukocyte and endothelial adhesion molecules in ischemia-reperfusion injury. Thromb Haemost 1997;78(1):310–4.
[140] Blann AD, Nadar SK, Lip GY. The adhesion molecule P-selectin and cardiovascular disease. Eur Heart J 2003;24(24):2166–79.
[141] Totani L, Evangelista V. Platelet-leukocyte interactions in cardiovascular disease and beyond. Arterioscler Thromb Vasc Biol 2010;30(12):2357–61.
[142] Mertens P, et al. Recombinant P-selectin glycoprotein ligand-immunoglobulin, a P-selectin antagonist, as an adjunct to thrombolysis in acute myocardial infarction. The P-Selectin Antagonist Limiting Myonecrosis (PSALM) trial. Am Heart J 2006;152(1):125.e1–8.
[143] Tardif JC, et al. Effects of the P-selectin antagonist inclacumab on myocardial damage after percutaneous coronary intervention for non-ST-segment elevation myocardial infarction: results of the SELECT-ACS trial. J Am Coll Cardiol 2013;61(20):2048–55.
[144] Kawashiri SY, et al. Effects of the anti-interleukin-6 receptor antibody, tocilizumab, on serum lipid levels in patients with rheumatoid arthritis. Rheumatol Int 2011;31(4):451–6.
[145] Protogerou AD, et al. A pilot study of endothelial dysfunction and aortic stiffness after interleukin-6 receptor inhibition in rheumatoid arthritis. Atherosclerosis 2011;219(2):734–6.
[146] Dinarello CA. Immunological and inflammatory functions of the interleukin-1 family. Annu Rev Immunol 2009;27:519–50.
[147] Tamaru M, et al. Interleukin-1beta induces tissue- and cell type-specific expression of adhesion molecules in vivo. Arterioscler Thromb Vasc Biol 1998;18(8):1292–303.
[148] Bevilacqua MP, et al. Interleukin-1 activation of vascular endothelium. Effects on procoagulant activity and leukocyte adhesion. Am J Pathol 1985;121(3):394–403.
[149] Arend WP, et al. Interleukin-1 receptor antagonist: role in biology. Annu Rev Immunol 1998;16: 27–55.
[150] Fleischmann RM, et al. Anakinra, a recombinant human interleukin-1 receptor antagonist (r-metHuIL-1ra), in patients with rheumatoid arthritis: a large, international, multicenter, placebo-controlled trial. Arthritis Rheum 2003;48(4):927–34.
[151] Ikonomidis I, et al. Inhibition of interleukin-1 by anakinra improves vascular and left ventricular function in patients with rheumatoid arthritis. Circulation 2008;117(20):2662–9.
[152] Abbate A, et al. Comparative safety of interleukin-1 blockade with anakinra in patients with ST-segment elevation acute myocardial infarction (from the VCU-ART and VCU-ART2 pilot studies). Am J Cardiol 2015;115(3):288–92.
[153] Chakraborty A, et al. Pharmacokinetic and pharmacodynamic properties of canakinumab, a human anti-interleukin-1beta monoclonal antibody. Clin Pharmacokinet 2012;51(6):e1–18.
[154] Ridker PM, et al. Effects of interleukin-1beta inhibition with canakinumab on hemoglobin A1c, lipids, C-reactive protein, interleukin-6, and fibrinogen: a phase IIb randomized, placebo-controlled trial. Circulation 2012;126(23):2739–48.
[155] Funk CD. Leukotriene inflammatory mediators meet their match. Sci Transl Med 2011;3(66):66ps3.

[156] Sampson AP. FLAP inhibitors for the treatment of inflammatory diseases. Curr Opin Investig Drugs 2009;10(11):1163–72.
[157] Riccioni G, et al. Leukotrienes in atherosclerosis: new target insights and future therapy perspectives. Mediat Inflamm 2009;2009:737282.
[158] Hakonarson H, et al. Effects of a 5-lipoxygenase-activating protein inhibitor on biomarkers associated with risk of myocardial infarction: a randomized trial. JAMA 2005;293(18):2245–56.
[159] Allayee H, et al. The effect of montelukast and low-dose theophylline on cardiovascular disease risk factors in asthmatics. Chest 2007;132(3):868–74.
[160] Tardif JC, et al. Treatment with 5-lipoxygenase inhibitor VIA-2291 (Atreleuton) in patients with recent acute coronary syndrome. Circ Cardiovasc Imaging 2010;3(3):298–307.
[161] Ingelsson E, Yin L, Back M. Nationwide cohort study of the leukotriene receptor antagonist montelukast and incident or recurrent cardiovascular disease. J Allergy Clin Immunol 2012;129(3). 702.e2–707.e2.
[162] Gettins PG. Serpin structure, mechanism, and function. Chem Rev 2002;102(12):4751–804.
[163] Huntington JA. Serpin structure, function and dysfunction. J Thromb Haemost 2011;9(Suppl. 1): 26–34.
[164] Nash P, et al. Inhibitory specificity of the anti-inflammatory myxoma virus serpin, SERP-1. J Biol Chem 1998;273(33):20982–91.
[165] Viswanathan K, et al. Myxoma viral serpin, Serp-1, inhibits human monocyte adhesion through regulation of actin-binding protein filamin B. J Leukoc Biol 2009;85(3):418–26.
[166] Tardif JC, et al. A randomized controlled, phase 2 trial of the viral serpin Serp-1 in patients with acute coronary syndromes undergoing percutaneous coronary intervention. Circ Cardiovasc Interv 2010; 3(6):543–8.
[167] Tousoulis D, et al. Antidepressive treatment as a modulator of inflammatory process in patients with heart failure: effects on proinflammatory cytokines and acute phase protein levels. Int J Cardiol 2009; 134(2):238–43.
[168] Kastorini CM, et al. Metabolic syndrome, adherence to the Mediterranean diet and 10-year cardiovascular disease incidence: the ATTICA study. Atherosclerosis 2016;246:87–93.
[169] Chrysohoou C, et al. Adherence to the Mediterranean diet attenuates inflammation and coagulation process in healthy adults: the ATTICA study. J Am Coll Cardiol 2004;44(1):152–8.
[170] Esposito K, et al. Metabolic syndrome and risk of cancer: a systematic review and meta-analysis. Diabetes Care 2012;35(11):2402–11.
[171] Grundy SM. Metabolic syndrome pandemic. Arterioscler Thromb Vasc Biol 2008;28(4):629–36.
[172] Papageorgiou N, Tousoulis D. Is HDL a prognostic biomarker for coronary atherosclerosis? Int J Cardiol 2014;174(3):465–7.
[173] Ridker PM, et al. C-reactive protein, the metabolic syndrome, and risk of incident cardiovascular events: an 8-year follow-up of 14 719 initially healthy American women. Circulation 2003; 107(3):391–7.
[174] Mottillo S, et al. The metabolic syndrome and cardiovascular risk a systematic review and meta-analysis. J Am Coll Cardiol 2010;56(14):1113–32.
[175] Perk J, et al. European guidelines on cardiovascular disease prevention in clinical practice (version 2012): the fifth joint task force of the European society of cardiology and other societies on cardiovascular disease prevention in clinical practice (constituted by representatives of nine societies and by invited experts). Int J Behav Med 2012;19(4):403–88.
[176] Kastorini CM, et al. The effect of Mediterranean diet on metabolic syndrome and its components: a meta-analysis of 50 studies and 534,906 individuals. J Am Coll Cardiol 2011;57(11):1299–313.
[177] Tangney CC, et al. Adherence to a Mediterranean-type dietary pattern and cognitive decline in a community population. Am J Clin Nutr 2011;93(3):601–7.
[178] Giugliano D, Ceriello A, Esposito K. The effects of diet on inflammation: emphasis on the metabolic syndrome. J Am Coll Cardiol 2006;48(4):677–85.
[179] Calabriso N, et al. Extra virgin olive oil rich in polyphenols modulates VEGF-induced angiogenic responses by preventing NADPH oxidase activity and expression. J Nutr Biochem 2016;28: 19–29.

[180] van Herpen-Broekmans WM, et al. Serum carotenoids and vitamins in relation to markers of endothelial function and inflammation. Eur J Epidemiol 2004;19(10):915–21.

[181] Synetos A, et al. Metabolic syndrome predicts plaque rupture in patients with acute myocardial infarction. An optical coherence study. Int J Cardiol 2016;209:139–41.

[182] Czernichow S, et al. Effects of long-term antioxidant supplementation and association of serum antioxidant concentrations with risk of metabolic syndrome in adults. Am J Clin Nutr 2009;90(2):329–35.

[183] Hercberg S, et al. The SU.VI.MAX Study: a randomized, placebo-controlled trial of the health effects of antioxidant vitamins and minerals. Arch Intern Med 2004;164(21):2335–42.

[184] Widmer RJ, et al. The Mediterranean diet, its components, and cardiovascular disease. Am J Med 2015;128(3):229–38.

[185] Papageorgiou N, et al. Divergent anti-inflammatory effects of different oil acute consumption on healthy individuals. Eur J Clin Nutr 2011;65(4):514–9.

[186] Tousoulis D, et al. Acute effects of different types of oil consumption on endothelial function, oxidative stress status and vascular inflammation in healthy volunteers. Br J Nutr 2010;103(1):43–9.

[187] Siasos G, et al. Inflammatory markers in hyperlipidemia: from experimental models to clinical practice. Curr Pharm Des 2011;17(37):4132–46.

[188] Charakida M, et al. Inflammatory and thrombotic processes are associated with vascular dysfunction in children with familial hypercholesterolemia. Atherosclerosis 2009;204(2):532–7.

[189] Stefanadi E, et al. Inflammatory markers in essential hypertension: potential clinical implications. Curr Vasc Pharmacol 2010;8(4):509–16.

[190] He K, et al. Associations of dietary long-chain n-3 polyunsaturated fatty acids and fish with biomarkers of inflammation and endothelial activation (from the Multi-Ethnic Study of Atherosclerosis [MESA]). Am J Cardiol 2009;103(9):1238–43.

[191] Poudyal H, et al. Omega-3 fatty acids and metabolic syndrome: effects and emerging mechanisms of action. Prog Lipid Res 2011;50(4):372–87.

[192] Bloomer RJ, et al. Effect of eicosapentaenoic and docosahexaenoic acid on resting and exercise-induced inflammatory and oxidative stress biomarkers: a randomized, placebo controlled, cross-over study. Lipids Health Dis 2009;8:36.

[193] Siasos G, et al. Effects of omega-3 fatty acids on endothelial function, arterial wall properties, inflammatory and fibrinolytic status in smokers: a cross over study. Int J Cardiol 2013;166(2):340–6.

[194] Reaven PD, et al. Effect of dietary antioxidant combinations in humans. Protection of LDL by vitamin E but not by beta-carotene. Arterioscler Thromb 1993;13(4):590–600.

[195] Jialal I, Fuller CJ, Huet BA. The effect of alpha-tocopherol supplementation on LDL oxidation. A dose-response study. Arterioscler Thromb Vasc Biol 1995;15(2):190–8.

[196] Turan B. Role of antioxidants in redox regulation of diabetic cardiovascular complications. Curr Pharm Biotechnol 2010;11(8):819–36.

[197] Vassort G, Turan B. Protective role of antioxidants in diabetes-induced cardiac dysfunction. Cardiovasc Toxicol 2010;10(2):73–86.

[198] Antoniades C, et al. Vascular endothelium and inflammatory process, in patients with combined Type 2 diabetes mellitus and coronary atherosclerosis: the effects of vitamin C. Diabet Med 2004;21(6):552–8.

[199] Tousoulis D, et al. Vitamin C affects thrombosis/fibrinolysis system and reactive hyperemia in patients with type 2 diabetes and coronary artery disease. Diabetes Care 2003;26(10):2749–53.

[200] Kushi LH, et al. Dietary antioxidant vitamins and death from coronary heart disease in postmenopausal women. N Engl J Med 1996;334(18):1156–62.

[201] Furumoto T, et al. Association of cardiovascular risk factors and endothelial dysfunction in Japanese hypertensive patients: implications for early atherosclerosis. Hypertens Res 2002;25(3):475–80.

[202] Tousoulis D, et al. Effects of combined administration of vitamins C and E on reactive hyperemia and inflammatory process in chronic smokers. Atherosclerosis 2003;170(2):261–7.

[203] Antoniades C, et al. Effects of antioxidant vitamins C and E on endothelial function and thrombosis/fibrinolysis system in smokers. Thromb Haemost 2003;89(6):990–5.

[204] Antoniades C, et al. Oxidative stress, antioxidant vitamins, and atherosclerosis. From basic research to clinical practice. Herz 2003;28(7):628–38.

[205] Yusuf S, et al. Vitamin E supplementation and cardiovascular events in high-risk patients. The heart outcomes prevention evaluation study investigators. N Engl J Med 2000;342(3):154–60.

[206] MRC/BHF Heart Protection Study of antioxidant vitamin supplementation in 20,536 high-risk individuals: a randomised placebo-controlled trial. Lancet 2002;360(9326):23–33.

[207] Micheletta F, et al. Vitamin E supplementation in patients with carotid atherosclerosis: reversal of altered oxidative stress status in plasma but not in plaque. Arterioscler Thromb Vasc Biol 2004; 24(1):136–40.

[208] Brown BG, et al. Simvastatin and niacin, antioxidant vitamins, or the combination for the prevention of coronary disease. N Engl J Med 2001;345(22):1583–92.

[209] Tousoulis D, et al. Effects of combined administration of low dose atorvastatin and vitamin E on inflammatory markers and endothelial function in patients with heart failure. Eur J Heart Fail 2005;7(7):1126–32.

[210] Arad Y, et al. Treatment of asymptomatic adults with elevated coronary calcium scores with atorvastatin, vitamin C, and vitamin E: the St. Francis Heart Study randomized clinical trial. J Am Coll Cardiol 2005;46(1):166–72.

[211] Brown DA, Sabbah HN, Shaikh SR. Mitochondrial inner membrane lipids and proteins as targets for decreasing cardiac ischemia/reperfusion injury. Pharmacol Ther 2013;140(3):258–66.

[212] Cho J, et al. Potent mitochondria-targeted peptides reduce myocardial infarction in rats. Coron Artery Dis 2007;18(3):215–20.

[213] Chakrabarti AK, et al. Rationale and design of the EMBRACE STEMI study: a phase 2a, randomized, double-blind, placebo-controlled trial to evaluate the safety, tolerability and efficacy of intravenous Bendavia on reperfusion injury in patients treated with standard therapy including primary percutaneous coronary intervention and stenting for ST-segment elevation myocardial infarction. Am Heart J 2013;165(4):509–14. e7.

[214] Ramji DP, Davies TS. Cytokines in atherosclerosis: key players in all stages of disease and promising therapeutic targets. Cytokine Growth Factor Rev 2015;26(6):673–85.

[215] Tian Y, et al. Expression and suppressive effects of interleukin-19 on vascular smooth muscle cell pathophysiology and development of intimal hyperplasia. Am J Pathol 2008;173(3):901–9.

[216] Almer G, et al. Interleukin-10: an anti-inflammatory marker to target atherosclerotic lesions via PEGylated liposomes. Mol Pharm 2013;10(1):175–86.

[217] Yoshimura S, et al. Inhibition of intimal hyperplasia after balloon injury in rat carotid artery model using cis-element 'decoy' of nuclear factor-kappaB binding site as a novel molecular strategy. Gene Ther 2001;8(21):1635–42.

[218] Shouda T, et al. Induction of the cytokine signal regulator SOCS3/CIS3 as a therapeutic strategy for treating inflammatory arthritis. J Clin Invest 2001;108(12):1781–8.

[219] Changelian PS, et al. Prevention of organ allograft rejection by a specific Janus kinase 3 inhibitor. Science 2003;302(5646):875–8.

[220] Sasaki N, et al. Oral anti-CD3 antibody treatment induces regulatory T cells and inhibits the development of atherosclerosis in mice. Circulation 2009;120(20):1996–2005.

[221] Steffens S, et al. Short-term treatment with anti-CD3 antibody reduces the development and progression of atherosclerosis in mice. Circulation 2006;114(18):1977–84.

[222] Takeda M, et al. Oral administration of an active form of vitamin D3 (calcitriol) decreases atherosclerosis in mice by inducing regulatory T cells and immature dendritic cells with tolerogenic functions. Arterioscler Thromb Vasc Biol 2010;30(12):2495–503.

[223] Dietrich T, et al. Local delivery of IL-2 reduces atherosclerosis via expansion of regulatory T cells. Atherosclerosis 2012;220(2):329–36.

CHAPTER 3.2

Anti-Oxidant Treatment

Sotirios Tsalamandris, Evangelos Oikonomou, Aggelos Papamikroulis, Dimitris Tousoulis
1st Cardiology Clinic, 'Hippokration' General Hospital, National and Kapodistrian University of Athens, School of Medicine, Athens, Greece

Contents

INTRODUCTION

Nowadays, there is a large body of experimental evidence in animals and epidemiologic data in humans, obtained from studies that examined plasma levels of antioxidants, supporting the oxidation hypothesis. Unfortunately, limited data exist concerning the impact of antioxidant regimens.

Antioxidants are divided into three major categories of molecules: (1) antioxidant proteins that bind metal ions, reducing the hydroxyl radical production (e.g., transferrin for iron (Fe) ion, ceruloplasmin and albumin for copper); (2) intracellular antioxidant enzymes such as superoxide dismutase, catalase, and glutathione peroxidase (whose

Coronary Artery Disease
ISBN 978-0-12-811908-2, http://dx.doi.org/10.1016/B978-0-12-811908-2.00014-3

enzymatic activity is selenium dependent). The above enzymes are responsible for the trapping of free radicals produced in situ, as by-products of cellular metabolism; (3) extracellular antioxidant substances such as α-tocopherol, β-carotene, vitamin C, and others.

Beyond these natural antioxidants, there is also a large number of pharmaceutical antioxidant agents used in clinical practice. The best-studied antioxidants used for the treatment of cardiovascular disease are vitamin E and vitamin C. However, the antioxidant capacity of these "antioxidants" cannot easily be determined because the existent markers of oxidative stress such as malonyldialdehyde, lipid hydroperoxides, or F2-isoprostanes are rather unreliable [1].

STUDIES OF ANTIOXIDANTS AS PRIMARY OR SECONDARY PREVENTION MEASURES IN CORONARY ARTERY DISEASE

Observational Studies on Antioxidants and Risk of Coronary Artery Disease (Summarized in Table 3.2.1)

Large-scale population-based studies were designed to investigate the association between decreased plasma antioxidant levels and risk for atherosclerosis. In these early studies, research was focused principally on the antioxidants β-carotene, ascorbic acid, and α-tocopherol and how their intake was related to atherogenesis and major adverse events. This effort aims to provide information whether dietary antioxidants could explain the different cardiovascular risk observed between patient populations, as well as to provide ample evidence to introduce these compounds as dietary supplements in order to decrease the risk of coronary artery disease (CAD).

Optimism and enthusiasm for trials on antioxidant supplementation came from studies like the Health Professionals Follow-up Study [2], the Nurses' Health Study [3], the Eastern Finland Study [4], the Atherosclerosis Risk in Communities (ARIC) trial [5], the Rotterdam study [6], the Iowa Women's Health Study [7], and the National Health and Nutritional Examination Study (NHANES III) [8,9].

Findings from the NHANES III, Eastern Finland Study, ARIC trial, and Rotterdam study suggested that ascorbic acid was associated with a decreased risk for CAD [4—6,9]. On the other hand, the Health Professionals Follow-up Study, Nurses' Health Study, and Iowa Women's Study all reported that vitamin C had no protective role in the prevention of CAD [10].

Despite the great excitement derived from these epidemiological studies concerning the protective profit of antioxidants, when considered together, it becomes definite that they have a number of serious limitations, which may have influenced the results. The metric by which antioxidant intake was assessed, the lack of measures of plasma antioxidant levels and other potential confounders such as healthy lifestyle behaviors that correlate with antioxidant intake may have played a role in the positive outcomes [11].

Table 3.2.1 Observational studies on antioxidants and coronary artery disease risk

Study	Study type	Vitamin type	Number of patients	Age	Population characteristics	Duration (years)	Outcome	RR
Health Professionals Follow-up Study [2]	Prospective cohort study	B-carotene	39,910	40–75	Male professionals free of CAD, diabetes or hypercholesterolemia	4	β-carotene decreases cardiovascular events Greater risk reduction for former or never smokers	0.71
Nurses Health Study [3]	Prospective cohort study	Carotenoids	73,286	30–55	Healthy female nurses	12	Inverse association between dietary β-carotenoid levels and incident CAD	0.74
Eastern Finland Study [4]	Prospective cohort study	Vitamin C	1,605	42–60	Men without either symptomatic CAD or ischemia on exercise testing at entry	6	Vitamin C deficiency is a risk factor for CAD	3.5
ARIC trial [5]	Prospective cohort study	Dietary and supplemental antioxidants (Vitamin C, α-tocopherol and provitamin A carotenoids)	6318 females and 4989 males	45–64	Dietary criteria	4	Limited support for the hypothesis that dietary vitamin C and alpha-tocopherol may protect against atherosclerotic disease, especially in individuals > 55 years old.	
Rotterdam study [6]	Prospective cohort study	Dietary beta-carotene, vitamin C, and vitamin E	4,367	55–94	No previous cardiovascular disease	4	Vitamin C intake was significantly inversely associated with peripheral arterial disease in women In men, vitamin E intake was inversely associated with peripheral arterial disease	0.64 and 0.67

Continued

Table 3.2.1 Observational studies on antioxidants and coronary artery disease risk—cont'd

Study	Study type	Vitamin type	Number of patients	Age	Population characteristics	Duration (years)	Outcome	RR
Iowa Women's Health Study [7]	Prospective cohort study	Dietary antioxidant vitamins	34,486	55–69	Postmenopausal women with no cardiovascular disease	7	In postmenopausal women the intake of vitamin E from food is inversely associated with the risk of death from CAD. Intakes of vitamins A and C was not associated with lower CAD mortality	0.42 for the highest quartile of vitamin E consumption
National Health and Nutritional Examination Study (NHANES III) [8]	Prospective cohort study	Serum carotenoids	13,293	35–90	Adults enrolled in the Third National Health and Nutrition Examination Survey, 1988–1994, with measured serum carotenoids and mortality follow-up through 2006	18	Higher α and β-carotene levels were associated with lower rates of cardiovascular mortality	1.38

ARIC, Atherosclerosis Risk in Communities; *CAD*, coronary artery disease; *RR*, relative risk.
In this column we present the relative risks where available.

Early Clinical Trials on Antioxidants and Cardiovascular Disease Prevention (Summarized in Table 3.2.2)

The encouraging findings from the aforementioned observational studies set up the basis to support a number of randomized primary and secondary prevention clinical trials.

The primary prevention Alpha-Tocopherol, Beta-Carotene Cancer Prevention (ATBC) study examined cardiovascular disease as a secondary outcome in a total of 29,133 male smokers randomized to α-tocopherol, β-carotene, a combination of the two, or placebo. At 8 years of follow-up, neither antioxidant was found to decrease significantly cardiovascular disease [12–14]. Similar results were observed in the Beta-Carotene and Retinol Efficacy Trial, which was early terminated after finding that β-carotene and retinol (vitamin A) had no considerable or statistically important effect on incident cardiovascular disease [15,16]. Negative findings were also reported in the Vitamin E Atherosclerosis Prevention Study [17].

Vitamin C has been studied in combinations with vitamin E, beta carotene, and other vitamins and minerals in large-scale clinical trials that reported no significant cardiovascular effects. Vitamin C alone has only been evaluated among 8171 women at high risk for cardiovascular disease [18] and 13,887 men at low-intermediate risk for cardiovascular disease [19], and no effect on cardiovascular disease was found with the use of 500 mg/day of vitamin C. Therefore, the clinical utility of vitamin C alone in preventing cardiovascular disease among those at low initial risk of cardiovascular disease remains uncertain.

Despite the negative or neutral results reported in the majority of prevention studies, the Women's Health Study on 40,000 women had favorable results [20]. Furthermore, the Cambridge Heart Antioxidant Study [21], a secondary prevention trial, observed a reduced cardiovascular risk in patients with established disease owing to the use of antioxidants. An interesting study that was carried out to determine whether vitamin C infusion before elective percutaneous coronary intervention is able to affect the incidence of periprocedural myocardial injury revealed that a 3 g vitamin C infusion within 6 h before intervention can reduce periprocedural myocardial injury [22].

To elucidate the complex associations between antioxidant and cardiovascular risk, several meta-analyses have been conducted. In a pooled analysis of nine cohorts, vitamin C supplement use exceeding 700 mg/day was significantly associated with a 25% reduction in coronary heart disease risk [23]. However, in one analysis that included data from 188,209 individuals who participated in 15 placebo-controlled trials, antioxidants failed to favorably influence the cardiovascular events (relative risk [RR] = 1.00, 95% confidence interval [CI] 0.96–01.03). This meta-analysis, however, had a few limitations, such as the use of pooled and nonpatient-level data and could not account for potential differences as a result of supplement dose and follow-up times [24]. Another meta-analysis identified 50 randomized controlled trials including 294,478 in total

Table 3.2.2 Prospective clinical trials on antioxidants and cardiovascular disease prevention

Study	Prevention	Vitamin type	Vitamin dose	Number of patients	Age	Population characteristics	Duration (years)	Outcome	RR
Alpha-Tocopherol—Beta-Carotene Cancer Prevention Study [12]	Primary	Vitamin E	50 mg	22,269	50–69	Male smokers	4.7	Decreases angina pectoris events	0.91 (0.83–0.99)
Alpha-Tocopherol—Beta-Carotene Cancer Prevention Study [136]	Secondary	Vitamin E	50 mg	1862	50–69	Male smokers who had a history of MI	5.3	38% reduction in the occurrence of MI	0.62 (0.41–0.96)
Alpha-Tocopherol—Beta-Carotene Cancer Prevention Study [12]	Primary	b-Carotene	20 mg	27,271	50–69	Male smokers with no history of MI	6.1	No effect on CAD	1.03 (0.91–1.16)
Alpha-Tocopherol—Beta-Carotene Cancer Prevention Study [12]	Primary	Vitamin E+ b-Carotene	50 mg of Vitamin E and 20 mg of b-carotene	27,271	50–69	Male smokers with no history of MI	6.1	No effect on CAD	0.97 (0.86–1.09)
Beta-Carotene and Retinol Efficacy Trial [15]	Primary	b-carotene + retinol	30 mg b-carotene, 25,000 IU retinol	18,314	45–69	Heavy smokers and asbestos workers	4	Borderline increases CAD mortality	1.26 (0.99–1.61)
Cambridge Heart Antioxidant Study (CHAOS) [21]	Secondary	Vitamin E	400 or 800 IU	2002	62	Subjects with CAD	1.4	Decreases nonfatal acute MI	0.23 (0.11–0.47)
Heart Outcomes Prevention Evaluation trial [137]	Primary	Vitamin E	400 IU	9541	>55	High risk for CAD subjects	4.5	No effect on MI, stroke	1.05 (0.95–1.16)
Heart Protection Study [138]	Secondary	Vitamin C + Vitamin E + b-carotene	600 mg vitamin E, 250 mg vitamin C, 20 mg b-carotene	20,536	40–80	Subjects at high CAD risk	5	No effect on CAD mortality	1.05 (0.95–1.15)
Women's Health Study [20]	Primary	α-tocopherol	600 IU of natural-source vitamin-E on alternate days	39,876	≥45	Healthy women	10.1	Decrease in cardiovascular mortality in healthy women, but no effect on the incidence of myocardial infarction	0.76
SU.VI.MAX [139]	Primary	Vitamin C+ Vitamin E+ b-carotene + zinc + selenium	120 mg + 300 mg + 6 mg + 20 mg + 100 μg	1162	53	Healthy Subjects	7.2	No effect on carotid atherosclerosis and arterial stiffness	-

Women's Antioxidant Cardiovascular Study [18]	Primary	b-carotene	50 mg on alternate days	8171	Postme-nopausal women >40	Subjects with CAD history/risk	9,4	No effect MI, Stroke, CAD	1.02 (0.92–1.13)
Women's Antioxidant Cardiovascular Study [18]	Secondary	Vitamin C	500 mg	8171	Postme-nopausal women >40	Subjects with CAD history/risk	9.4	No effect on MI, Stroke, CAD	1.02 (0.92–1.13)
Women's Antioxidant Cardiovascular Study [18]	Primary	Vitamin E	600 IU	8171	Postme-nopausal women >40	Subjects with CAD history/risk	9.4	No effect MI, Stroke, CAD	0.94 (0.85–1.04)
Physicians' Health Study [140]	Primary	b-Carotene	50 mg on alternate days	22,071	40–84	Healthy male physicians	12	No effect on MI	0.96 (0.84–1.09)
Physicians' Health Study II [19]	Primary	Vitamin E	400 IU	13,887	>50	Males at low/medium risk	10	No effect on cardiovascular mortality	1.07 (0.90–1.28)
Physicians' Health Study II [19]	Primary	Vitamin C	500 mg/day	13,887	>50	Males at low/medium risk	10	No effect on cardiovascular mortality	1.02(0.85–1.21)
Physicians' Health Study II [19]	Secondary	Vitamin E	400 IU	754	>50	Males with cardiovascular disease	10	No effect on cardiovascular mortality	0.82 (0.62–1.09)
Physicians' Health Study II [19]	Secondary	Vitamin C	500 mg/day	754	>50	Males with cardiovascular disease	10	No effect on cardiovascular mortality	0.96 (0.72–1.27)
Skin Cancer Prevention Study [141]	Primary	b-Carotene	50 mg	1805	<85	Subjects with skin cancer patients	8.2	No effect on CAD mortality	1.16 (0.82–1.64)
Gruppo Italiano per lo Studio della Sopravvivenza nell'Infarto Miocardico-Prevention Trial [142]	Secondary	Vitamin E	300 mg	11,324	No limits	Post-MI adults subjects	3.5	No effect on MI, stroke	0.98 (0.87–1.10)

Continued

Table 3.2.2 Prospective clinical trials on antioxidants and cardiovascular disease prevention—cont'd

Study	Prevention	Vitamin type	Vitamin dose	Number of patients	Age	Population characteristics	Duration (years)	Outcome	RR
Secondary Prevention with Antioxidants of Cardiovascular Disease in End-stage Renal Disease [143]	Secondary	Vitamin E	800 IU	196	40–75	Hemodialysis patients	2	Decreases CVD	0.46 (0.27–0.78)
Microalbuminuria Cardiovascular Renal Outcomes/ HOPE [144]	Secondary	Vitamin E	400 IU	3654	65	Diabetic subjects	4.5	No effect on MI, stroke	1.03 (0.88–1.21)
Antioxidant Supplementation in Atherosclerosis Prevention Study [145]	Secondary	Vitamin C + Vitamin E	182 IU a-tocopherol, 500 mg vitamin C	520	45–69	Hypercholesterolemic subjects	3	Slow the progression of IMT thickening in men but not in women	-
Primary Prevention Project [146]	Primary	Vitamin E	300 mg	4495	Mean age 64 years	Subjects at CAD risk	3.6	No effect on MI, stroke	1.07 (0.74–1.56)
Probucol and multivitamins in the prevention of restenosis [147]	Secondary	Probucol or Vitamin C + vitamin E + b-carotene or both	500 mg, 500 mg, 700 IU, 30,000 IU	317	Mean age 58.5 year	Subjects before and after coronary angioplasty	6 months	No beneficial effect of antioxidant	-
HOPE-TOO [148]	Secondary	Vitamin E	400 IU	2007	Mean age 66 years	Subjects with vascular disease or diabetes	4	No effect on cardiovascular mortality	1.04(0.96–1.14)
Ujeyl A et al. [149]	Primary	Vitamins C and E	500-1000 mg 400-800 IU Twice/day	40	–	Adults 2 years postheart transplantation	10	Delay of allograft vasculopathy with antioxidants	0.19(0.05–0.76)

Devaraj S et al. [150]	Secondary	Vitamin E	1200 IU/day	90	52–68	Patients with CAD	2	α-Tocopherol supplementation in patients with CAD was safe and significantly reduced plasma biomarkers of oxidative stress and inflammation but had no significant effect on carotid intima-medial thickness during 2 years of follow-up	0.5
Wanq ZJ et al. [22]	Primary	Vitamin C	3 g	532	–	Patients scheduled for elective PCI for de novo lesions in native coronary arteries	1	Vitamin C infusion before PCI was associated with a lower incidence of periprocedural MI	-
Women's Angiographic Vitamin and Estrogen (WAVE) Trial [151]	Secondary	Vitamin C + vitamin E + estrogen	0.625 equine estrogen/day, 400 IU vitamin E and 500 mg vitamin C bid	423	Postme-nopausal	Postmenopausal women with CAD	5	No effect on MI, stroke	

CAD, coronary artery disease; *IMT*, intima-media thickness; *MI*, myocardial infarction; *PCI*, percutaneous coronary intervention; *RR*, relative risk.
In the RR column we present the relative risk with the 95% confidence intervals where available.

(156,663 in the treatment or intervention group and 137,815 in the control or referent group) to evaluate the effect of antioxidants on cardiovascular disease. Using a fixed effects model, the primary analysis found that vitamins or antioxidant supplements did not reduce the risk for major cardiovascular events (RR=1.00; 95% CI 0.98–1.02), leading to the conclusion that antioxidant vitamins or supplements had no beneficial effect in the primary or secondary prevention of cardiovascular disease [25].

Other Antioxidants and Related Clinical Trials (Summarized in Table 3.2.3)

Coenzyme Q10 (CoQ10)

Coenzyme Q10 (2,3-dimethoxyl-5 meth-6-decaprenyl benzoquinone) is a key component of the electron transport chain. It can be supplemented orally and is derived from mevalonic acid and phenylalanine [26]. The mechanism of CoQ10 action is not likely to be secondary to a superoxide scavenger effect given CoQ10's hydrophobic properties. CoQ10 increases the ability of Complexes I and II to transfer electron down the mitochondrial electron transport chain and therefore can reduce mitochondrial superoxide production [27]. CoQ10 may also exert an antioxidant effect by reducing lipid peroxidation at the level of the plasma membrane [28]. Lee BJ et al. have shown that higher plasma levels of CoQ10 were significantly associated with reduced risk of cardiovascular disease, indicating its potential cardioprotective role [29].

A few studies have demonstrated favorable results regarding the correlation between CoQ10 supplementation, leading to higher serum levels of CoQ10, and decrease in inflammatory molecules [30–32]. These findings advocate the role of CoQ10 as an antioxidant.

On the other hand, there is very limited information to determine whether CoQ10 would prevent cardiovascular disease; however, a meta-analysis of 12 trials, including 362 patients in total, found that CoQ10 supplements decreased blood pressure with a drop in systolic blood pressure of 11–17 mmHg and diastolic blood pressure of 8–10 mmHg compared with placebo. Despite these optimistic findings, due to the unreliability of some of the studies included, no recommendations could be made [33]. Another meta-analysis including 194 patients found that CoQ10 improved endothelial function, as assessed by flow-mediated dilatation [34].

Although these early studies are encouraging, this antioxidant remains to be studied in adequately powered large-scale randomized trials.

Resveratrol (RES)

Resveratrol (3,5,4′-trihydroxystilbene) is a polyphenolic compound found in the skin of many berries and grapes. Its reported benefits are wide ranging and include antioxidant, anticancer, cardioprotective, and antiaging effects, and it is thought to be primarily responsible for the beneficial effects of red wine [35]. Specifically in regard to cardioprotection,

Table 3.2.3 Other antioxidants and relevant clinical trials

Study	Prevention	Antioxidant type	Antioxidant dose	Number of patients	Age	Population characteristics	Duration	Outcome
Lee BJ et al. [30]	Secondary	CoQ10	300 mg/day	51	55.4–83.2	Patients with CAD	12 weeks	Significantly lower levels of inflammatory markers (TNF-a and IL-6)
Lee BJ et al. [31]	Secondary	CoQ10	60 mg/day or 150 mg/day	51	69.8–84.6	Patients with CAD	12 weeks	CoQ10 supplementation at a dosage of 150 mg decreases C-reactive protein, IL-6, homocysteine, malondialdehyde and superoxidase dismutase
Tiano L et al. [32]	Secondary	CoQ10	300 mg/day	38	Mean age 55 ± 4	CAD patients	1 month	Improvement in the endothelium-dependent vasodilation, in the endothelium extracellular superoxide dismutase activity
Tome-Carneiro et al. [48]	Secondary	Resveratrol	8 mg/day	35	47–75	Hypertensive male patients with type II diabetes mellitus	1 year	Treatment decreases inflammatory cytokines, such as C–C motif chemokine ligand 3, tumor necrosis factor-α, IL-6, -8-1β and C-X-C motif chem0kine ligand 2
Tome-Carneiro et al. [49]	Secondary	Resveratrol	8 mg/day	75	48–72	Stable-CAD patients	1	Resveratrol increases the anti-inflammatory serum adiponectin and decreases the thrombogenic plasminogen activator inhibitor type 1
Maqyar K. et al. [50]	Secondary	Resveratrol	10 mg/day	40	–	Postinfarction Caucasian patients	3 months	Improves left ventricle diastolic function, endothelial function, reduces LDL-cholesterol levels

Continued

Table 3.2.3 Other antioxidants and relevant clinical trials—cont'd

Study	Prevention	Antioxidant type	Antioxidant dose	Number of patients	Age	Population characteristics	Duration	Outcome
Potsdam arm of the European prospective Investigation into Cancer and nutrition study [71]	Primary	Chocolate	Mean intake ~6 g/day	19,357	35–65	Healthy adults	8 years	Chocolate consumption appears to lower cardiovascular disease risk, in part through reducing blood pressure. The inverse association may be stronger for stroke than for myocardial infarction.
Zomer E. et al. [72]	Primary	Chocolate	100 g/day	2013	–	People with hypertension who met the criteria for metabolic syndrome, with no history of cardiovascular disease and not receiving antihypertensive therapy	10 years	The blood pressure and cholesterol lowering effects of dark chocolate consumption are beneficial in the prevention of cardiovascular events in a population with metabolic syndrome.
Inami S. et al. [77]	Primary	Polyphenol 70S	500 mg/day of catechin, which is equivalent to 6 cups of green tea per day	40	28.6–49.7	Healthy adult volunteers	4 weeks	reduces circulating Ox-LDL
Widlansky ME et al. [78]	Secondary	Epigallocatechin gallate (EGCG)	300 mg/day	42	48–68	Clinically stable CAD patients	2 weeks	Epigallocatechin gallate acutely improves endothelial function
Siani A et al. [85]	Primary	L-Arginine	–	6	39±4	Healthy adults	1 week	Reduces blood pressure
Siasos G et al. [152]	Primary	L-Arginine	21 g/day	12	–	Healthy smokers	2 days	Improves endothelial function
Kelly BS et al. [87]	Primary and secondary	L-Arginine	9 g/day for 9 days following by 18 g/day for 9 more days	16	–	6 normotensive volunteers and 10 kidney transplant patients	18 days	Restores NO-mediated vasodilators response and reduction in blood pressure

Yin WH et al. [89]	Secondary	L-Arginine and vitamin C	10 g/day L-arginine or 500 mg/day vitamin C, as active control	31	–	Stable CAD patients	6 weeks	Consumption of either L-arginine or vitamin C significantly increases brachial artery flow-mediated dilatation
Lee BJ et al. [96]	Secondary	Acetyl-L-carnitine	1000 mg/day	47	61.3–83.8	CAD patients	12 weeks	increases activity of antioxidant enzymes
Selenium Therapy in Coronary Artery disease Patients (SETCAP) Study [104]	Secondary	Sodium selenite	200 or 500 μg/day	465	61–70	CAD patients	12 weeks	Dose-dependent increase in glutathione peroxide-1 (GPx-1) protein expression and activity
Mohammadpour AH et al. [153]	secondary	Pentoxifylline	400 mg × 3/day	40	44.43 –63.15	CAD patients	2 months	Decreases serum levels of Intercellular adhesion molecule 1 and vascular cell adhesion molecule 1
Fernandes JL et al. [111]	Secondary	Pentoxifylline	400 mg/day	64	18–80	Patients with Acute Coronary Syndromes (ACS)	6 months	Significant reduction of CRP and TNF-alpha
Sobeninin IA et al. [120]	Secondary	Garlic powder	150 mg × 2/day	51	40–55	CAD patients	12 months	Decreases LDL cholesterol levels and cardiovascular risk
Wu YZ et al. [121]	Secondary	Ginkgo biloba extract	87.5 mg/day	80	45–74	CAD patients	2 weeks	Increases left anterior descending coronary artery blood flow
Enseleit F et al. [123]	secondary	Pycnogenol	200 mg/day	23	≥18	CAD patients	8 weeks	Improves flow-mediated dilatation and decreases 15-F_{2t}-isoprostane

CAD, coronary artery disease; *CRP*, C-reactive protein; *IL*, interleukin; *LDL*, low density lipoprotein; *ox-LDL*, oxidized low density lipoprotein; *TNF-alpha*, tumor necrosis factor alpha.

the drug has been shown to be antiatherogenic [35], reduces low density lipoprotein (LDL) while increases high-density lipoprotein (HDL) [36–38], reduces oxidative stress by upregulating antioxidant enzymes such as superoxide dismutase, glutathione reductase, and catalase [39,40], improves glucose homeostasis [38,41], prevents platelet aggregation [42–44] and monocyte recruitment [45], promotes coronary vasorelaxation [46,47]. As resveratrol has so many actions it is an intriguing preventive and therapeutic option for the treatment of CAD, which is an extremely complex disease.

Early studies concerning the supplementation of diet with resveratrol exhibit promising results [48–50]. More specifically, dietary intake of resveratrol decreases inflammatory cytokines and LDL and improves left ventricle diastolic and endothelial function.

Flavonoids

Flavonoids are constitutes from several polyphenols. They are found in multiple fruits, vegetables, and beverages such as red wine, tea, cocoa, and dark chocolate [51]. Flavonoids have been tested as antioxidant therapy for cardiovascular disease because epidemiological data suggest improved cardiovascular outcomes in individuals with high intake of dietary flavonoid [51–53]. Moreover, in vitro and cellular studies suggest the antioxidant efficacy of these substances [54–57]. Nevertheless, flavonoids have limited bioavailability after oral consumption that suggests that beyond their scavenging activity they have an additional role in cell-to-cell interactions that may be the key mechanism mediating beneficial cardiovascular effects [57]. This hypothesis is further confirmed by evidence showing that inhibition of nicotinamide adenine dinucleotide phosphate (NADPH) oxidase activity occurs through molecular pathways that restrain angiotensin converting enzyme [58,59], increase endothelial nitric oxide synthase (eNOS)-specific NO production through the estrogen receptor [60], and alter COX-2 expression [60]. Recently, we have shown that concord grape juice, a rich source of flavonoids, ameliorates endothelial function and improves vascular elasticity of the arterial circuit in healthy smokers and attenuates acute smoking-induced impairment of arterial wall properties [61]. However, studies investigating the antihypertensive effects of flavonoids are inconclusive. In several small studies of short duration, dark chocolate therapy has demonstrated blood pressure lowering effects in hypertensive patients [62–65]. Nevertheless, studies in normotensive and prehypertensive individuals have demonstrated no benefit [66,67]. The specific flavonoids and combination of flavonoids that exert the largest beneficial effects remain unknown.

Dark Chocolate/Cocoa

Dark chocolate/cocoa has been consumed for at least 2500 years thanks to its good taste, and for a long time it has been regarded as a medicine [68]. It is naturally polyphenol-rich food. The favorable cardiovascular effects of cocoa have been shown by several randomized studies. These effects include amelioration of systemic hypertension, enhancement of endothelial function, improvement of arterial stiffness, glycemic control, and platelet adhesiveness [69,70], which are mediated by polyphenols [71,72]. Hence, direct dietary supplementation with flavonol instead of whole chocolate consumption deserves further study.

Tea and Catechins

Lower cardiovascular risk following tea consumption has been observed for many years, but the molecular pathways remain unidentified [73]. Tea, especially black and green tea, contains catechins, which are monomeric flavonols that include epicatechin, epigallocatechin, epigallocatechingallate, and epicatechingallate. In vitro studies have confirmed that polyphenolic compounds and tea catechins can effectively scavenge reactive oxygen species (ROS), by enhancing production and/or bioavailability of endothelium-derived nitric oxide (NO), or have indirect antioxidant activity by effecting transcription factors and enzyme activities [74—76]. Inami et al. administered 500 mg of Polyphenon 70S, extracted from green tea leaves with hot water. Its compounds include epigallocatechingallate (35.6%), epicatechingallate (10.2%), gallocatechingallate (3.8%), catechingallate (0.6%), epigallocatechin (15.0%), and epicatechin (8.7%). In the catechin group, each individual received Polyphenon 70S after a meal once a day per os capsules containing 500 mg of catechin. 500 mg of catechin is equal to the dose containing six or seven cups of green tea [77]. Widlansky et al. administered 300 mg of epigallocatechin gallate (EGCG) [78]. Both studies showed favorable results regarding cardiovascular risk, such as reduction of oxidized LDL and improvement in endothelial function, respectively. In a meta-analysis of 10 cohort studies and 7 case—control studies, Peters et al. demonstrated a decrease by 11% in the incidence rate of myocardial infarction with an increase in tea consumption by three cups per day [51].

L-Arginine

L-Arginine is an amino acid as well as the main substrate of eNOS in a reaction whose end product is NO and depends on tetrahydrobiopterin [79]. Potential dietary sources include milk products, beef, wheat germ, nuts, and soybeans. Reduction in intracellular levels of L-arginine has been observed in hypertensive subjects [80,81]. Although L-arginine deficiency itself does not appears to lead to uncoupling of eNOS [82], it may lead to lower bioavailability of NO, probably contributing to hypertension. Interestingly, short-term treatment with L-arginine improves endothelial function and vascular elastic properties of healthy smokers [61,83], supposing that L-arginine supplementation might limit hypertension by reestablishing physiological NO bioavailability. This concept is supported by studies demonstrating the antihypertensive effect of L-arginine supplementation in salt-sensitive rats [84], healthy human subjects [85], hypertensive diabetics [86] patients with chronic kidney disease [87], and diabetic patients in combination with N-acetylcysteine, a precursor of glutathione [86].

L-arginine's antihypertensive response may be mediated in part by its suppressive effects on angiotensin II and endothelin-1, and its potentiating effects on insulin [88]. Findings from another study suggest that oral L-arginine supplement improved endothelial function and reduced LDL oxidation in stable CAD patients [89].

Nevertheless, potential toxic side effects of homocysteine due to L-arginine supplementation have raised new concerns [90]. The majority of L-arginine is converted to creatine, leading to increased homocysteine levels [91]. Homocysteine can increase oxidative stress [92]. According to a recent study, this mechanism correlates with L-arginine metabolism in humans [93], suggesting a potential pathway for neutralizing the eNOS-related antioxidant effects of L-arginine.

Acetyl-L-Carnitine and α-Lipoic Acid

Acetyl-L-carnitine is a key compound in the transport of fatty acids into mitochondria for beta-oxidation. The antioxidant mechanism of acetyl-L-carnitine supplementation appears to be secondary to reductions in mitochondrial ROS production in synergy with concomitant α-lipoic acid therapy [94]. The exact intramitochondrial mechanisms of acetyl-L-carnitine's effects are not clear and prior work in older rats demonstrates acetyl-L-carnitine potential to be prooxidative when used alone [95]. The effect of acetyl-L-carnitine was investigated by Lee et al. [96], who conclude that CAD patients might benefit from using acetyl-L-carnitine supplements to increase their antioxidation capacity, because acetyl-L-carnitine supplementation at a dose of 1000 mg/day was found to be associated with a significant reduction in oxidative stress and an increase in antioxidant enzymes activities (catalase, superoxide dismutase).

α-Lipoic acid is a dithiol compound synthesized from octanoic acid in mitochondria. The in vivo and in vitro effects of α-lipoic aid have been thoroughly reviewed elsewhere [97,98]. α-lipoic acid has moderate oral bioavailability [99]. Although α-lipoic acid is a potent in vitro antioxidant, the limited plasma concentrations achievable with supplementation and rapid clearance of α-lipoic acid suggest that free radical scavenging and antioxidant recycling are less likely to be the primary in vivo activity of α-lipoic acid. Participation in mitochondrial-associated metabolic pathways, in cell signaling that may improve coupling of eNOS, and in anti-inflammatory responses are among the potential beneficial effects of α-lipoic acid supplementation [98,100].

Human data with respect to the antihypertensive effects of these compounds are limited to two small studies that have shown some promising results. Consistent with animal data, combined acetyl-L-carnitine and α-lipoic acid therapy reduced systolic blood pressure in CAD patients with hypertension and/or metabolic syndrome at the time of enrollment. Also consistent with prior cell culture and animal work [101], type 2 diabetic subjects supplemented with 2 g/day of acetyl-L-carnitine showed significantly lowered blood pressure and improved insulin sensitivity [102].

Selenium

Selenium is a nonmetal trace element indispensable for the formation of selenoproteins, such as the antioxidant enzymes glutathione peroxidase and thioredoxin reductase. Dietary sources of selenium include nuts, particularly Brazilian nuts, cereals, meat, tuna, eggs, and mushrooms. Clinical trials on selenium supplementation for the primary prevention of cardiovascular disease have been conducted. Twelve randomized controlled trials with a total of 19,715 patients who were followed up for 3 months or longer were included in a meta-analysis. According to this analysis, selenium supplementation had no effect on all-cause or cardiovascular mortality and no effect on any major adverse cardiovascular event. Taking into deep account all the available data, it is clear that there is not sufficient data to justify the dietary supplementation with selenium for primary prevention of cardiovascular disease [103]. As far as the secondary prevention of CAD is concerned Schnabel et al. found that sodium selenite supplementation increases glutathione peroxidase-1 activity in endothelial cells and in CAD patients [104]. Future studies though have to demonstrate whether long-term CAD outcome can be improved.

Pentoxifylline

Pentoxifylline is a phosphodiesterase inhibitor with immune-modulating properties and ability to inhibit T helper 1 lymphocytes activity. In animal models, pharmacological treatment with pentoxifylline has resulted in reduced atherosclerosis and modification in plaque morphology [105—107]. In humans, only few studies have focused specifically in modulating this particular immune response. Pentoxifylline has been shown to increase interleukin 10 (IL-10), reduce IL-6, and inhibit C-reactive protein (CRP) expression and activity in peripheral blood cells [108,109] and has been extensively used in patients with peripheral vascular disease [110]. Fernandes et al. demonstrated a beneficial effect of pentoxifylline in the immunomodulation of the inflammatory activity in patients with CAD with a significant decrease in proinflammatory markers and increase in anti-inflammatory cytokines toward a more stable immune phenotype [111].

Garlic (Allium sativum *L. fam. Alliaceae)*

Garlic (*Allium sativum* L. fam. Alliaceae) is one of the most studied herbal remedies. Traditionally, it was considered as first-line treatment of infections, rheumatic disorders, heart problems, diabetes mellitus, and many other pathologies.

Several experiments have confirmed its beneficial effects on hypertension, bacterial infections, and hyperglycemia [112]. Clinically, garlic has been evaluated for a number of conditions, including hypertension, hypercholesterolemia, diabetes, and for the prevention of arteriosclerosis and cancer [112]. Several studies have indicated that garlic and its constituents inhibit key enzymes involved in cholesterol and fatty acid synthesis [113—116]. However, the clinical evidence is far from compelling. Garlic contains a number of different biologically active compounds and sulfur-containing amino acids, such as allicin, ajoene, S-allylcysteine, S-methylcysteine, diallyl disulfide, and sulfoxides, that may be responsible for its antiatherosclerotic activity [115,117]. In recent years, anti-atherosclerotic and cardiovascular protective effects of garlic have been extensively evaluated. Evidence from numerous studies demonstrate that garlic-based preparations can enhance fibrinolytic activity, inhibit platelet aggregation, and reduce blood pressure [118] and normalize plasma lipids by inhibiting human squalene monooxygenase and 3-hydroxy-3-methylglutaryl-coenzyme A (HMG-CoA) reductase, as well as other key enzymes involved in cholesterol biosynthesis [119]. Sobenin et al. demonstrated that a 12-month treatment with garlic tablets results in a significant reduction of 1.5 and 1.3 times in cardiovascular risk in men and women, respectively. Cardiovascular risk reduction was mainly mediated by a decrease in LDL cholesterol (32.9 mg/dL in men and 27.3 mg/dL in women). Interestingly, the most outstanding effects were observed in men. On the other hand, the insufficient sample size of the female group may have played a critical role in the result [120].

Ginkgo Biloba Extract

Ginkgo biloba extract is a specific and complex product prepared from the dried leaves of Ginkgo biloba that has been used therapeutically in traditional Chinese medicine for 5000 years. The pharmacologically active constituents, flavonol glycosides and the terpene lactones, have been identified. The terpene lactones comprise ginkgo lides A, B, C and biloba lides. The extract scavenges excess free radicals and pretreatment with ginkgo biloba extract reduces damage by free radicals in patients undergoing coronary bypass surgery. Release of NO and cytokines, such as prostacyclines was shown to be stimulated. Recently, Wu et al. suggested that Ginkgo biloba extract treatment in CAD patients led to an increase of left anterior descending coronary artery blood flow, as it was assessed by significant improvement in maximal diastolic peak velocity, maximal systolic peak velocity, and diastolic time velocity integral, which might be correlated to the restoration of the delicate equilibrium between NO and endothelin 1 [121].

*Pycnogenol (*Pinus pinaster *ssp.* atlantica*)*

Pycnogenol (*Pinus pinaster* ssp. *atlantica*) is a bark extract of the French maritime pine tree that has been used in traditional medicine. Pycnogenol exerts anti-inflammatory effects, inhibits platelet aggregation LDL peroxidation, and lipid peroxidation [122]. Enseleit et al. reported that pycnogenol improved the endothelial function of CAD patients by reducing oxidative stress [123].

Intervention via Mediterranean Diet

Based on positive findings from observational cohort studies of higher dietary intake or supplementation of antioxidants [124–127], the American Heart Association recommended the consumption of a balanced diet with emphasis on antioxidant-rich fruits, vegetables, and whole grains [128]. The findings from clinical trials on antioxidant supplements have led to the resurgence of the interest on studying dietary antioxidants that are consumed through fruit, vegetable, and nut intake. The beneficial antioxidant effects of diet have been examined extensively in the PREvention con DIetaMEDiterranea study, a multicenter randomized controlled clinical trial that evaluated the effects of a Mediterranean diet on cardiovascular outcomes in individuals without cardiovascular disease. Participants were subgrouped in a randomized way to a Mediterranean diet containing nuts or extra-virgin olive oil, or a control low-fat diet. After a median period of 4.8 years, there was a 28%–30% reduction in major adverse cardiovascular events in individuals randomized to the supplemented Mediterranean diet [129]. These findings were attributed, in part, to a reduction in oxidant stress, as markers of oxidant stress (i.e., oxidized low-density lipoproteins and malondialdehyde) were decreased in individuals assigned to the Mediterranean diet [130]. Although it is not clear what dietary antioxidants are responsible for the cardiovascular risk reduction, it is evident that dietary intervention with the Mediterranean diet offered the best possible outcomes of all the antioxidants studied.

FAILURE OF ANTIOXIDANTS? WHY?

Pathophysiologically, oxidative stress has been suggested to contribute to the central mechanism by which all clinical risk factors (hypertension, hyperlipidemia, smoking, and/or diabetes mellitus) led to endothelial dysfunction. Understandably, given this view, the reduction of oxidative would be anticipated to have a potentially favorable impact on CAD. Despite the abundance of laboratory and observational evidence indicating that antioxidants should prevent CAD, the results from large-scale, randomized, placebo-controlled clinical trials have been rather controversial. There are many potential and speculative explanations for this.

First, an obvious reason could be that the wrong antioxidant or combination of antioxidant has been studied. Perhaps the vitamins used in the antioxidant trials were selected because they were available and not ideal or equivalent to the natural source. The synthetic versions of the antioxidants tested in clinical intervention trials may not completely mimic the natural forms of antioxidants and as a result the cardiovascular risk reduction seen in dietary studies may be associated with other ingredients in the foods consumed [131]. For example, potential additive or antagonistic effects of other antioxidants administered concomitantly with vitamin E have been reported. In particular, vitamin C is thought to be important in the regeneration of the original form of vitamin E, acting as a co-antioxidant. Another issue associated with the use of vitamin supplementation in clinical trials is their metabolism. Vitamin E is a fat-soluble vitamin and its absorption occurs in the small intestine along with other fat soluble nutrients. Consequently, the bioavailability of vitamin E is impacted by the consumption of dietary fats and plant sterols, which share similar digestion and absorption pathways. In contrast, a-tocopherol also acts as a pro-oxidant molecule by its reaction with lipid peroxyl radicals and conversion into an oxidized product, a-tocopheroxyl free radical. High levels of this vitamin may displace other fat soluble antioxidants (i.e., g-tocopherol), disrupting the natural balance of antioxidant system. The decreased absorption of natural g-tocopherol caused by α-tocopherol supplements can result in a significant reduction of plasma g-tocopherol levels. γ-Tocopherol can provide greater antioxidant protection against specific oxidants. Vitamin E may also inhibit human cytosolic glutathione S-transferases, which are involved in the detoxification of drugs and endogenous toxins.

Second, the dosage of these vitamins varies among different studies. Not all antioxidants are functionally similar and those included in clinical trials may have different threshold effects for retracting cellular response related to oxidative stress and the pathogenesis of atherosclerosis. The optimal dose of antioxidants remains controversial; 800 units of vitamin E and 500 mg of vitamin C have been suggested to be the effective threshold doses. As it has also been proposed with other agents, there is individual variability in metabolism and in response to vitamin supplementation as well, meaning that same doses of vitamins to everybody will not lead to the same antioxidant effect.

Moreover, although there are several markers of oxidative stress (F2-isoprostanes, thiobarbituric acid reactive substances, oxidized LDL, and free oxygen radicals), so far, none of the candidate markers or methods has been widely accepted as the gold standard one to use in clinical trials. Interestingly, many studies, in which positive results of antioxidant supplementation were observed, reported a significant increase in plasma antioxidant levels, whereas other studies did not include a metric system to determine whether the dose of antioxidant given actually reduced oxidant stress and did not examine dose—response [132].

Another potential explanation may be the inadequate duration of the antioxidant therapy. When considering adequate duration therapy, it is important to keep in mind that in animal studies antioxidants were proved to be effective through the prevention of the earliest forms of atherosclerotic lesion formation [133]. Therefore, even clinical trials lasting years may be too short to show any benefit. A trial several decades long may be required to see some impact of vitamin E supplementation. When primary or secondary prevention in adults is studied, we are looking at the effects of 40 or 50 years of oxidative stress on the endothelium, and it might be unrealistic to expect that we can turn that around with 1 or 2 years of antioxidant therapy. There may be an impact, but follow-up has not been long enough to prove it.

A final possible explanation for the incapacity of clinical trials to highlight the profits of antioxidants would be that the wrong patients may have been studied. The majority of randomized clinical trials enrolled patients with preexisting cardiovascular event or at risk of cardiovascular disease. Although the selection of high-risk population in order to conduct a randomized trial reduces the required sample size due to the higher event rate, these patients may be less amenable to a positive and protective effect with antioxidant supplementation, particularly when treated for a short period of time, as mentioned earlier. Some patients who would exert no benefit from supplemental antioxidants may have been included in the studies. This is mostly important for patients being treated with statins, as studies on statins plus antioxidant vitamins have found that there was no additional benefit from the vitamin supplements [134]. In contrast, instead of treating everyone with antioxidant supplements, as has been done in previous trials, perhaps only patients who are deficient in antioxidants or who have a proven increase in levels of oxidative stress should be treated. This may explain why trials in select populations, such as renal failure or postcardiac transplantation populations, have been more likely to demonstrate a benefit from antioxidant therapies [21,135].

CONCLUSION

A healthy endothelium is vital for the preservation and advancement of vascular homeostasis. In contrast, oxidative stress has a key role in the progression of endothelial dysfunction as well as in atherosclerotic progression. Despite the supportive observational epidemiologic and animal data, most clinical trials have failed to support the putative action of antioxidant treatment in preventing atherosclerotic disease and its complications. Well-designed new clinical trials are necessary to establish the exact role of oxidative stress in the pathogenesis of atherosclerosis and the precise impact of antioxidant therapy on primary and secondary prevention of CAD. Using the best antioxidants at the right dose for the optimal duration and in the right population will provide the most trustworthy and indisputable results.

REFERENCES

[1] Antoniades C, et al. Oxidative stress, antioxidant vitamins, and atherosclerosis. From basic research to clinical practice. Herz 2003;28(7):628–38.
[2] Rimm EB, et al. Vitamin E consumption and the risk of coronary heart disease in men. N Engl J Med 1993;328(20):1450–6.
[3] Osganian SK, et al. Dietary carotenoids and risk of coronary artery disease in women. Am J Clin Nutr 2003;77(6):1390–9.
[4] Nyyssonen K, et al. Vitamin C deficiency and risk of myocardial infarction: prospective population study of men from eastern Finland. Bmj 1997;314(7081):634–8.
[5] Kritchevsky SB, et al. Dietary antioxidants and carotid artery wall thickness. The ARIC Study. Atherosclerosis risk in communities study. Circulation 1995;92(8):2142–50.
[6] Klipstein-Grobusch K, et al. Dietary antioxidants and peripheral arterial disease: the Rotterdam study. Am J Epidemiol 2001;154(2):145–9.
[7] Kushi LH, et al. Dietary antioxidant vitamins and death from coronary heart disease in postmenopausal women. N Engl J Med 1996;334(18):1156–62.

[8] Shardell MD, et al. Low-serum carotenoid concentrations and carotenoid interactions predict mortality in US adults: the Third National Health and Nutrition Examination Survey. Nutr Res 2011;31(3):178–89.
[9] Simon JA, Hudes ES. Serum ascorbic acid and cardiovascular disease prevalence in U.S. adults: the Third National Health and Nutrition Examination Survey (NHANES III). Ann Epidemiol 1999;9(6):358–65.
[10] Osganian SK, et al. Vitamin C and risk of coronary heart disease in women. J Am Coll Cardiol 2003;42(2):246–52.
[11] Leopold JA. Antioxidants and coronary artery disease: from pathophysiology to preventive therapy. Coron Artery Dis 2015;26(2):176–83.
[12] Virtamo J, et al. Effect of vitamin E and beta carotene on the incidence of primary nonfatal myocardial infarction and fatal coronary heart disease. Arch Intern Med 1998;158(6):668–75.
[13] The Alpha-Tocopherol, Beta Carotene Cancer Prevention Study Group. The effect of vitamin E and beta carotene on the incidence of lung cancer and other cancers in male smokers. N Engl J Med 1994;330(15):1029–35.
[14] Rapola JM, et al. Effect of vitamin E and beta carotene on the incidence of angina pectoris. A randomized, double-blind, controlled trial. JAMA 1996;275(9):693–8.
[15] Omenn GS, et al. Effects of a combination of beta carotene and vitamin A on lung cancer and cardiovascular disease. N Engl J Med 1996;334(18):1150–5.
[16] Redlich CA, et al. Effect of long-term beta-carotene and vitamin A on serum cholesterol and triglyceride levels among participants in the Carotene and Retinol Efficacy Trial (CARET). Atherosclerosis 1999;145(2):425–32.
[17] Hodis HN, et al. Alpha-tocopherol supplementation in healthy individuals reduces low-density lipoprotein oxidation but not atherosclerosis: the Vitamin E Atherosclerosis Prevention Study (VEAPS). Circulation 2002;106(12):1453–9.
[18] Cook NR, et al. A randomized factorial trial of vitamins C and E and beta carotene in the secondary prevention of cardiovascular events in women: results from the Women's Antioxidant Cardiovascular Study. Arch Intern Med 2007;167(15):1610–8.
[19] Sesso HD, et al. Vitamins E and C in the prevention of cardiovascular disease in men: the Physicians' Health Study II randomized controlled trial. JAMA 2008;300(18):2123–33.
[20] Lee IM, et al. Vitamin E in the primary prevention of cardiovascular disease and cancer: the Women's Health Study: a randomized controlled trial. JAMA 2005;294(1):56–65.
[21] Stephens NG, et al. Randomised controlled trial of vitamin E in patients with coronary disease: Cambridge Heart Antioxidant Study (CHAOS). Lancet 1996;347(9004):781–6.
[22] Wang ZJ, et al. The effect of intravenous vitamin C infusion on periprocedural myocardial injury for patients undergoing elective percutaneous coronary intervention. Can J Cardiol 2014;30(1):96–101.
[23] Knekt P, et al. Antioxidant vitamins and coronary heart disease risk: a pooled analysis of 9 cohorts. Am J Clin Nutr 2004;80(6):1508–20.
[24] Ye Y, Li J, Yuan Z. Effect of antioxidant vitamin supplementation on cardiovascular outcomes: a meta-analysis of randomized controlled trials. PLoS One 2013;8(2):e56803.
[25] Myung SK, et al. Efficacy of vitamin and antioxidant supplements in prevention of cardiovascular disease: systematic review and meta-analysis of randomised controlled trials. Bmj 2013;346:f10.
[26] Overvad K, et al. Coenzyme Q10 in health and disease. Eur J Clin Nutr 1999;53(10):764–70.
[27] McCarty MF. Coenzyme Q versus hypertension: does CoQ decrease endothelial superoxide generation? Med Hypotheses 1999;53(4):300–4.
[28] Houston MC. Nutraceuticals, vitamins, antioxidants, and minerals in the prevention and treatment of hypertension. Prog Cardiovasc Dis 2005;47(6):396–449.
[29] Lee BJ, et al. The relationship between coenzyme Q10, oxidative stress, and antioxidant enzymes activities and coronary artery disease. Sci World J 2012;2012:792756.
[30] Lee BJ, et al. Effects of coenzyme Q10 supplementation (300 mg/day) on antioxidation and anti-inflammation in coronary artery disease patients during statins therapy: a randomized, placebo-controlled trial. Nutr J 2013;12(1):142.

[31] Lee BJ, et al. Effects of coenzyme Q10 supplementation on inflammatory markers (high-sensitivity C-reactive protein, interleukin-6, and homocysteine) in patients with coronary artery disease. Nutrition 2012;28(7-8):767–72.
[32] Tiano L, et al. Effect of coenzyme Q10 administration on endothelial function and extracellular superoxide dismutase in patients with ischaemic heart disease: a double-blind, randomized controlled study. Eur Heart J 2007;28(18):2249–55.
[33] Rosenfeldt FL, et al. Coenzyme Q10 in the treatment of hypertension: a meta-analysis of the clinical trials. J Hum Hypertens 2007;21(4):297–306.
[34] Gao L, et al. Effects of coenzyme Q10 on vascular endothelial function in humans: a meta-analysis of randomized controlled trials. Atherosclerosis 2012;221(2):311–6.
[35] Chu LM, et al. Resveratrol in the prevention and treatment of coronary artery disease. Curr Atheroscler Rep 2011;13(6):439–46.
[36] Do GM, et al. Long-term effects of resveratrol supplementation on suppression of atherogenic lesion formation and cholesterol synthesis in apo E-deficient mice. Biochem Biophys Res Commun 2008; 374(1):55–9.
[37] Rocha KK, et al. Resveratrol toxicity: effects on risk factors for atherosclerosis and hepatic oxidative stress in standard and high-fat diets. Food Chem Toxicol 2009;47(6):1362–7.
[38] Robich MP, et al. Resveratrol improves myocardial perfusion in a swine model of hypercholesterolemia and chronic myocardial ischemia. Circulation 2010;122(11 Suppl.):S142–9.
[39] Berrougui H, et al. A new insight into resveratrol as an atheroprotective compound: inhibition of lipid peroxidation and enhancement of cholesterol efflux. Atherosclerosis 2009;207(2):420–7.
[40] Li Y, Cao Z, Zhu H. Upregulation of endogenous antioxidants and phase 2 enzymes by the red wine polyphenol, resveratrol in cultured aortic smooth muscle cells leads to cytoprotection against oxidative and electrophilic stress. Pharmacol Res 2006;53(1):6–15.
[41] Fullerton MD, Steinberg GR. SIRT1 takes a backseat to AMPK in the regulation of insulin sensitivity by resveratrol. Diabetes 2010;59(3):551–3.
[42] Olas B, et al. Effect of resveratrol, a natural polyphenolic compound, on platelet activation induced by endotoxin or thrombin. Thromb Res 2002;107(3-4):141–5.
[43] Yang Y, et al. Inhibitory effects of resveratrol on platelet activation induced by thromboxane a(2) receptor agonist in human platelets. Am J Chin Med 2011;39(1):145–59.
[44] Lin KH, et al. Mechanisms of resveratrol-induced platelet apoptosis. Cardiovasc Res 2009;83(3): 575–85.
[45] Deng YH, et al. Inhibition of TNF-alpha-mediated endothelial cell-monocyte cell adhesion and adhesion molecules expression by the resveratrol derivative, trans-3,5,4′-trimethoxystilbene. Phytother Res 2011;25(3):451–7.
[46] Rivera L, et al. Long-term resveratrol administration reduces metabolic disturbances and lowers blood pressure in obese Zucker rats. Biochem Pharmacol 2009;77(6):1053–63.
[47] Behbahani J, et al. Resveratrol and small artery compliance and remodeling in the spontaneously hypertensive rat. Am J Hypertens 2010;23(12):1273–8.
[48] Tome-Carneiro J, et al. One-year supplementation with a grape extract containing resveratrol modulates inflammatory-related microRNAs and cytokines expression in peripheral blood mononuclear cells of type 2 diabetes and hypertensive patients with coronary artery disease. Pharmacol Res 2013; 72:69–82.
[49] Tome-Carneiro J, et al. Grape resveratrol increases serum adiponectin and downregulates inflammatory genes in peripheral blood mononuclear cells: a triple-blind, placebo-controlled, one-year clinical trial in patients with stable coronary artery disease. Cardiovasc Drugs Ther 2013;27(1):37–48.
[50] Magyar K, et al. Cardioprotection by resveratrol: a human clinical trial in patients with stable coronary artery disease. Clin Hemorheol Microcirc 2012;50(3):179–87.
[51] Peters U, Poole C, Arab L. Does tea affect cardiovascular disease? A meta-analysis. Am J Epidemiol 2001;154(6):495–503.
[52] Bazzano LA, et al. Fruit and vegetable intake and risk of cardiovascular disease in US adults: the first national health and nutrition examination survey epidemiologic follow-up study. Am J Clin Nutr 2002;76(1):93–9.

[53] Siasos G, et al. Flavonoids in atherosclerosis: an overview of their mechanisms of action. Curr Med Chem 2013;20(21):2641–60.
[54] Aviram M, Fuhrman B. Wine flavonoids protect against LDL oxidation and atherosclerosis. Ann N Y Acad Sci 2002;957:146–61.
[55] Lotito SB, Fraga CG. (+)-Catechin prevents human plasma oxidation. Free Radic Biol Med 1998; 24(3):435–41.
[56] Stein JH, et al. Purple grape juice improves endothelial function and reduces the susceptibility of LDL cholesterol to oxidation in patients with coronary artery disease. Circulation 1999;100(10):1050–5.
[57] Lotito SB, Frei B. Consumption of flavonoid-rich foods and increased plasma antioxidant capacity in humans: cause, consequence, or epiphenomenon? Free Radic Biol Med 2006;41(12):1727–46.
[58] Aviram M, Dornfeld L. Pomegranate juice consumption inhibits serum angiotensin converting enzyme activity and reduces systolic blood pressure. Atherosclerosis 2001;158(1):195–8.
[59] Aviram M, et al. Pomegranate juice consumption for 3 years by patients with carotid artery stenosis reduces common carotid intima-media thickness, blood pressure and LDL oxidation. Clin Nutr 2004;23(3):423–33.
[60] Anter E, et al. Activation of endothelial nitric-oxide synthase by the p38 MAPK in response to black tea polyphenols. J Biol Chem 2004;279(45):46637–43.
[61] Siasos G, et al. Favorable effects of concord grape juice on endothelial function and arterial stiffness in healthy smokers. Am J Hypertens 2014;27(1):38–45.
[62] Taubert D, et al. Chocolate and blood pressure in elderly individuals with isolated systolic hypertension. JAMA 2003;290(8):1029–30.
[63] Grassi D, et al. Short-term administration of dark chocolate is followed by a significant increase in insulin sensitivity and a decrease in blood pressure in healthy persons. Am J Clin Nutr 2005;81(3): 611–4.
[64] Taubert D, et al. Effects of low habitual cocoa intake on blood pressure and bioactive nitric oxide: a randomized controlled trial. JAMA 2007;298(1):49–60.
[65] Grassi D, et al. Blood pressure is reduced and insulin sensitivity increased in glucose-intolerant, hypertensive subjects after 15 days of consuming high-polyphenol dark chocolate. J Nutr 2008;138(9): 1671–6.
[66] Ried K, Frank OR, Stocks NP. Dark chocolate or tomato extract for prehypertension: a randomised controlled trial. BMC Complement Altern Med 2009;9:22.
[67] Zilkens RR, et al. Red wine and beer elevate blood pressure in normotensive men. Hypertension 2005;45(5):874–9.
[68] Vlachopoulos C, Alexopoulos N, Stefanadis C. Effect of dark chocolate on arterial function in healthy individuals: cocoa instead of ambrosia? Curr Hypertens Rep 2006;8(3):205–11.
[69] Sudano I, et al. Cocoa, blood pressure, and vascular function. Curr Hypertens Rep 2012;14(4): 279–84.
[70] Latif R. Chocolate/cocoa and human health: a review. Neth J Med 2013;71(2):63–8.
[71] Buijsse B, et al. Chocolate consumption in relation to blood pressure and risk of cardiovascular disease in German adults. Eur Heart J 2010;31(13):1616–23.
[72] Zomer E, et al. The effectiveness and cost effectiveness of dark chocolate consumption as prevention therapy in people at high risk of cardiovascular disease: best case scenario analysis using a Markov model. Bmj 2012;344:e3657.
[73] Vita JA. Tea consumption and cardiovascular disease: effects on endothelial function. J Nutr 2003; 133(10):3293s–7s.
[74] Higdon JV, Frei B. Tea catechins and polyphenols: health effects, metabolism, and antioxidant functions. Crit Rev Food Sci Nutr 2003;43(1):89–143.
[75] Karim M, McCormick K, Kappagoda CT. Effects of cocoa extracts on endothelium-dependent relaxation. J Nutr 2000;130(8S Suppl.):2105s–8s.
[76] Lorenz M, et al. A constituent of green tea, epigallocatechin-3-gallate, activates endothelial nitric oxide synthase by a phosphatidylinositol-3-OH-kinase-, cAMP-dependent protein kinase-, and Akt-dependent pathway and leads to endothelial-dependent vasorelaxation. J Biol Chem 2004; 279(7):6190–5.

[77] Inami S, et al. Tea catechin consumption reduces circulating oxidized low-density lipoprotein. Int Heart J 2007;48(6):725−32.

[78] Widlansky ME, et al. Acute EGCG supplementation reverses endothelial dysfunction in patients with coronary artery disease. J Am Coll Nutr 2007;26(2):95−102.

[79] Tiefenbacher CP. Tetrahydrobiopterin: a critical cofactor for eNOS and a strategy in the treatment of endothelial dysfunction? Am J Physiol Heart Circ Physiol 2001;280(6):H2484−8.

[80] Schlaich MP, et al. Impaired L-arginine transport and endothelial function in hypertensive and genetically predisposed normotensive subjects. Circulation 2004;110(24):3680−6.

[81] Wang D, et al. Asymmetric dimethylarginine, oxidative stress, and vascular nitric oxide synthase in essential hypertension. Am J Physiol Regul Integr Comp Physiol 2009;296(2):R195−200.

[82] Bevers LM, et al. Tetrahydrobiopterin, but not L-arginine, decreases NO synthase uncoupling in cells expressing high levels of endothelial NO synthase. Hypertension 2006;47(1):87−94.

[83] Siasos G, et al. Short-term treatment with L-arginine prevents the smoking-induced impairment of endothelial function and vascular elastic properties in young individuals. Int J Cardiol 2008;126(3): 394−9.

[84] Chen PY, Sanders PW. L-arginine abrogates salt-sensitive hypertension in Dahl/Rapp rats. J Clin Invest 1991;88(5):1559−67.

[85] Siani A, et al. Blood pressure and metabolic changes during dietary L-arginine supplementation in humans. Am J Hypertens 2000;13(5 Pt 1):547−51.

[86] Martina V, et al. Long-term N-acetylcysteine and L-arginine administration reduces endothelial activation and systolic blood pressure in hypertensive patients with type 2 diabetes. Diabetes Care 2008; 31(5):940−4.

[87] Kelly BS, et al. Oral arginine improves blood pressure in renal transplant and hemodialysis patients. JPEN J Parenter Enteral Nutr 2001;25(4):194−202.

[88] Gokce N. L-arginine and hypertension. J Nutr 2004;134(10 Suppl.):2807S−11S. Discussion 2818S-2819S.

[89] Yin WH, et al. L-arginine improves endothelial function and reduces LDL oxidation in patients with stable coronary artery disease. Clin Nutr 2005;24(6):988−97.

[90] Loscalzo J. Adverse effects of supplemental L-arginine in atherosclerosis: consequences of methylation stress in a complex catabolism? Arterioscler Thromb Vasc Biol 2003;23(1):3−5.

[91] Persky AM, Brazeau GA. Clinical pharmacology of the dietary supplement creatine monohydrate. Pharmacol Rev 2001;53(2):161−76.

[92] Tyagi N, et al. Mechanisms of homocysteine-induced oxidative stress. Am J Physiol Heart Circ Physiol 2005;289(6):H2649−56.

[93] Jahangir E, et al. The effect of L-arginine and creatine on vascular function and homocysteine metabolism. Vasc Med 2009;14(3):239−48.

[94] Hagen TM, et al. Mitochondrial decay in the aging rat heart: evidence for improvement by dietary supplementation with acetyl-L-carnitine and/or lipoic acid. Ann N Y Acad Sci 2002;959:491−507.

[95] Hagen TM, et al. Acetyl-L-carnitine fed to old rats partially restores mitochondrial function and ambulatory activity. Proc Natl Acad Sci USA 1998;95(16):9562−6.

[96] Lee BJ, et al. Effects of L-carnitine supplementation on oxidative stress and antioxidant enzymes activities in patients with coronary artery disease: a randomized, placebo-controlled trial. Nutr J 2014;13:79.

[97] Shay KP, et al. Alpha-lipoic acid as a dietary supplement: molecular mechanisms and therapeutic potential. Biochim Biophys Acta 2009;1790(10):1149−60.

[98] Smith AR, et al. Lipoic acid as a potential therapy for chronic diseases associated with oxidative stress. Curr Med Chem 2004;11(9):1135−46.

[99] Teichert J, et al. Investigations on the pharmacokinetics of alpha-lipoic acid in healthy volunteers. Int J Clin Pharmacol Ther 1998;36(12):625−8.

[100] Zhang WJ, Frei B. Alpha-lipoic acid inhibits TNF-alpha-induced NF-kappaB activation and adhesion molecule expression in human aortic endothelial cells. Faseb J 2001;15(13):2423−32.

[101] Addabbo F, Montagnani M, Goligorsky MS. Mitochondria and reactive oxygen species. Hypertension 2009;53(6):885−92.

[102] Ruggenenti P, et al. Ameliorating hypertension and insulin resistance in subjects at increased cardiovascular risk: effects of acetyl-L-carnitine therapy. Hypertension 2009;54(3):567–74.
[103] Rees K, et al. Selenium supplementation for the primary prevention of cardiovascular disease. Cochrane Database Syst Rev 2013;(1):Cd009671.
[104] Schnabel R, et al. Selenium supplementation improves antioxidant capacity in vitro and in vivo in patients with coronary artery disease the SElenium Therapy in Coronary Artery disease Patients (SETCAP) Study. Am Heart J 2008;156(6). 1201.e1–11.
[105] Laurat E, et al. In vivo downregulation of T helper cell 1 immune responses reduces atherogenesis in apolipoprotein E-knockout mice. Circulation 2001;104(2):197–202.
[106] Lutgens E, et al. Both early and delayed anti-CD40L antibody treatment induces a stable plaque phenotype. Proc Natl Acad Sci USA 2000;97(13):7464–9.
[107] Zhao L, et al. Selective interleukin-12 synthesis defect in 12/15-lipoxygenase-deficient macrophages associated with reduced atherosclerosis in a mouse model of familial hypercholesterolemia. J Biol Chem 2002;277(38):35350–6.
[108] D'Hellencourt CL, et al. Differential regulation of TNF alpha, IL-1 beta, IL-6, IL-8, TNF beta, and IL-10 by pentoxifylline. Int J Immunopharmacol 1996;18(12):739–48.
[109] Ramani M, et al. Interleukin-10 and pentoxifylline inhibit C-reactive protein-induced tissue factor gene expression in peripheral human blood monocytes. FEBS Lett 1994;356(1):86–8.
[110] Ward A, Clissold SP. Pentoxifylline. A review of its pharmacodynamic and pharmacokinetic properties, and its therapeutic efficacy. Drugs 1987;34(1):50–97.
[111] Fernandes JL, et al. Pentoxifylline reduces pro-inflammatory and increases anti-inflammatory activity in patients with coronary artery disease—a randomized placebo-controlled study. Atherosclerosis 2008;196(1):434–42.
[112] Aviello G, et al. Garlic: empiricism or science? Nat Prod Commun 2009;4(12):1785–96.
[113] Gebhardt R. Multiple inhibitory effects of garlic extracts on cholesterol biosynthesis in hepatocytes. Lipids 1993;28(7):613–9.
[114] Liu L, Yeh YY. Water-soluble organosulfur compounds of garlic inhibit fatty acid and triglyceride syntheses in cultured rat hepatocytes. Lipids 2001;36(4):395–400.
[115] Yeh YY, Liu L. Cholesterol-lowering effect of garlic extracts and organosulfur compounds: human and animal studies. J Nutr 2001;131(3s):989s–93s.
[116] Yeh YY, Yeh SM. Garlic reduces plasma lipids by inhibiting hepatic cholesterol and triacylglycerol synthesis. Lipids 1994;29(3):189–93.
[117] Block E. The chemistry of garlic and onions. Sci Am 1985;252(3):114–9.
[118] Agarwal KC. Therapeutic actions of garlic constituents. Med Res Rev 1996;16(1):111–24.
[119] Gupta N, Porter TD. Garlic and garlic-derived compounds inhibit human squalene monooxygenase. J Nutr 2001;131(6):1662–7.
[120] Sobenin IA, et al. The effects of time-released garlic powder tablets on multifunctional cardiovascular risk in patients with coronary artery disease. Lipids Health Dis 2010;9:119.
[121] Wu YZ, et al. Ginkgo biloba extract improves coronary artery circulation in patients with coronary artery disease: contribution of plasma nitric oxide and endothelin-1. Phytother Res 2008;22(6):734–9.
[122] Fitzpatrick DF, Bing B, Rohdewald P. Endothelium-dependent vascular effects of pycnogenol. J Cardiovasc Pharmacol 1998;32(4):509–15.
[123] Enseleit F, et al. Effects of pycnogenol on endothelial function in patients with stable coronary artery disease: a double-blind, randomized, placebo-controlled, cross-over study. Eur Heart J 2012;33(13):1589–97.
[124] Ames BN, Shigenaga MK, Hagen TM. Oxidants, antioxidants, and the degenerative diseases of aging. Proc Natl Acad Sci USA 1993;90(17):7915–22.
[125] Jha P, et al. The antioxidant vitamins and cardiovascular disease. A critical review of epidemiologic and clinical trial data. Ann Intern Med 1995;123(11):860–72.
[126] Gaziano JM, et al. A prospective study of consumption of carotenoids in fruits and vegetables and decreased cardiovascular mortality in the elderly. Ann Epidemiol 1995;5(4):255–60.

[127] Losonczy KG, Harris TB, Havlik RJ. Vitamin E and vitamin C supplement use and risk of all-cause and coronary heart disease mortality in older persons: the Established Populations for Epidemiologic Studies of the Elderly. Am J Clin Nutr 1996;64(2):190–6.
[128] Tribble DL. AHA Science Advisory. Antioxidant consumption and risk of coronary heart disease: emphasis on vitamin C, vitamin E, and beta-carotene: a statement for healthcare professionals from the American Heart Association. Circulation 1999;99(4):591–5.
[129] Estruch R, et al. Primary prevention of cardiovascular disease with a Mediterranean diet. N Engl J Med 2013;368(14):1279–90.
[130] Fito M, et al. Effect of a traditional Mediterranean diet on lipoprotein oxidation: a randomized controlled trial. Arch Intern Med 2007;167(11):1195–203.
[131] Steinhubl SR. Why have antioxidants failed in clinical trials? Am J Cardiol 2008;101(10a):14d–9d.
[132] Roberts 2nd LJ, et al. The relationship between dose of vitamin E and suppression of oxidative stress in humans. Free Radic Biol Med 2007;43(10):1388–93.
[133] Kaliora AC, Dedoussis GV, Schmidt H. Dietary antioxidants in preventing atherogenesis. Atherosclerosis 2006;187(1):1–17.
[134] Arad Y, et al. Treatment of asymptomatic adults with elevated coronary calcium scores with atorvastatin, vitamin C, and vitamin E: the St. Francis Heart Study randomized clinical trial. J Am Coll Cardiol 2005;46(1):166–72.
[135] Fang JC, et al. Effect of vitamins C and E on progression of transplant-associated arteriosclerosis: a randomised trial. Lancet 2002;359(9312):1108–13.
[136] Rapola JM, et al. Randomised trial of alpha-tocopherol and beta-carotene supplements on incidence of major coronary events in men with previous myocardial infarction. Lancet 1997;349(9067): 1715–20.
[137] Yusuf S, et al. Vitamin E supplementation and cardiovascular events in high-risk patients. The heart outcomes prevention evaluation study Investigators. N Engl J Med 2000;342(3):154–60.
[138] MRC/BHF Heart Protection Study of antioxidant vitamin supplementation in 20,536 high-risk individuals: a randomised placebo-controlled trial. Lancet 2002;360(9326):23–33.
[139] Zureik M, et al. Effects of long-term daily low-dose supplementation with antioxidant vitamins and minerals on structure and function of large arteries. Arterioscler Thromb Vasc Biol 2004;24(8): 1485–91.
[140] Hennekens CH, et al. Lack of effect of long-term supplementation with beta carotene on the incidence of malignant neoplasms and cardiovascular disease. N Engl J Med 1996;334(18):1145–9.
[141] Greenberg ER, et al. Mortality associated with low plasma concentration of beta carotene and the effect of oral supplementation. JAMA 1996;275(9):699–703.
[142] Dietary supplementation with n-3 polyunsaturated fatty acids and vitamin E after myocardial infarction: results of the GISSI-Prevenzione trial. Gruppo Italiano per lo Studio della Sopravvivenza nell'Infarto miocardico. Lancet 1999;354(9177):447–55.
[143] Boaz M, et al. Secondary prevention with antioxidants of cardiovascular disease in endstage renal disease (SPACE): randomised placebo-controlled trial. Lancet 2000;356(9237):1213–8.
[144] Lonn E, et al. Effects of vitamin E on cardiovascular and microvascular outcomes in high-risk patients with diabetes: results of the HOPE study and MICRO-HOPE substudy. Diabetes Care 2002;25(11): 1919–27.
[145] Salonen JT, et al. Antioxidant Supplementation in Atherosclerosis Prevention (ASAP) study: a randomized trial of the effect of vitamins E and C on 3-year progression of carotid atherosclerosis. J Intern Med 2000;248(5):377–86.
[146] de Gaetano G. Low-dose aspirin and vitamin E in people at cardiovascular risk: a randomised trial in general practice. Collaborative Group of the Primary Prevention Project. Lancet 2001;357(9250): 89–95.
[147] Tardif JC, et al. Probucol and multivitamins in the prevention of restenosis after coronary angioplasty. Multivitamins and Probucol Study Group. N Engl J Med 1997;337(6):365–72.
[148] Lonn E, et al. Effects of long-term vitamin E supplementation on cardiovascular events and cancer: a randomized controlled trial. JAMA 2005;293(11):1338–47.

[149] Ujeyl A, et al. Early vitamin C and E supplementation and cardiac allograft vasculopathy: 10-year follow-up from a randomized, controlled study. J Heart Lung Transpl 2011;30(10):1200–1.
[150] Devaraj S, et al. Effect of high-dose alpha-tocopherol supplementation on biomarkers of oxidative stress and inflammation and carotid atherosclerosis in patients with coronary artery disease. Am J Clin Nutr 2007;86(5):1392–8.
[151] Waters DD, et al. Effects of hormone replacement therapy and antioxidant vitamin supplements on coronary atherosclerosis in postmenopausal women: a randomized controlled trial. JAMA 2002; 288(19):2432–40.
[152] Siasos G, et al. The impact of oral L-arginine supplementation on acute smoking-induced endothelial injury and arterial performance. Am J Hypertens 2009;22(6):586–92.
[153] Mohammadpour AH, et al. Pentoxifylline decreases serum level of adhesion molecules in atherosclerosis patients. Iran Biomed J 2014;18(1):23–7.

CHAPTER 3.3

Gene Therapy

Debbie Falconer[1], Nikolaos Papageorgiou[2,3], Charalambos Antoniades[4], Dimitris Tousoulis[2]

[1]University College London Hospital, London, United Kingdom; [2]1st Cardiology Clinic, 'Hippokration' General Hospital, National and Kapodistrian University of Athens, School of Medicine, Athens, Greece; [3]St Bartholomew's Hospital, London, United Kingdom; [4]University of Oxford, Oxford, United Kingdom

Contents

INTRODUCTION

Atherosclerosis is a progressive pathophysiological process that results in narrowing of arteries around the body. This subsequently decreases blood supply downstream, which can manifest in a variety of clinical presentations including ischemic heart disease, peripheral vascular disease, and stroke [1]. The underlying pathology is a chronic process that occurs at points of disturbed blood flow, such as branch points [2].

Approaches to managing atherogenesis are centered around minimizing the risk factors that contribute to plaque progression. This can be through lifestyle alterations or pharmacological means. When these techniques are unlikely to succeed, procedures are available to restore blood supply such as primary coronary intervention (PCI) or coronary artery bypass grafting (CABG). Gene therapy is the process where genetic material is transfected to a new host where it will alter the phenotype of the cell. Certainly, it is not in use currently in clinical practice, but animal studies have shown promise that this technique may act as an adjunct to current practice.

Coronary Artery Disease
ISBN 978-0-12-811908-2, http://dx.doi.org/10.1016/B978-0-12-811908-2.00015-5

ATHEROSCLEROTIC PROCESS

The development of atherosclerotic plaques is a complex interaction between genetics and environment. It is characterized by endothelial dysfunction [3], vascular inflammation [4], and oxidative stress [5], resulting in the formation of a plaque below the endothelium.

Low-density lipoprotein (LDL) is retained below the endothelium and subjected to oxidation by free radicals. Resident macrophages consume the oxidized LDL to become foam cells, which produce growth factors that attract smooth muscle cells (SMCs) [6]. The cells proliferate, resulting in atherosclerotic plaques. Stable plaques are those with a thick fibrous cap made of strong connective tissue. Over time, further inflammatory cells accumulate within the cap, weakening the fibrous material and making it prone to rupture [7]. Rupture can lead to complete occlusion of the vessel through thrombus formation or downstream occlusion from an embolism. As plaque formation involves an array of steps, so too are there a large number of genes that can be targeted [8].

GENE THERAPY

Gene therapy involves modification of gene expression to alter the properties of transfected cells. Vectors carrying genes can be delivered systemically, potentially initiating change throughout the vascular endothelium, or locally, e.g., within a stent deployed during PCI, thus limiting side effects to one site [3].

There is no perfect way to deliver these genes to the area of desired effect. Broadly, there are viral and nonviral vectors (Table 3.3.1 and Fig. 3.3.1). As a rule, nonviral vectors are generally less effectual, as they interact less efficiently with host cells compared with viral vectors [9].

Viral vectors have a greater transduction rate [9,10], yet they are immunogenic. The inflammatory response potentiates platelet aggregation, and thus can result in thrombus formation at the area of treatment [11]. Viral vectors include adenovirus, adeno-associated virus, lentivirus, retrovirus, and herpes simplex virus. Adenoviruses are used most commonly to deliver genes in atherosclerosis as they grow high titers and express genes in nonmitotic cells—ideal for targeting the endothelium [12]. Transduction efficiency in in vivo trials has been quoted at 50% [13,14]. However, there exist several disadvantages, notably, their high immunogenicity and formation of neutralizing antibodies [15]. Others too, are flawed. Retroviruses integrate their

Table 3.3.1 Viral vectors: pros and cons

Viral vectors	Pros	Cons
Adenovirus	High in vivo expression	Highly immunogenic
Adeno-associated virus	Low immunogenicity	Small vector capacity
Retrovirus	Low immunogenicity	Low in vivo expression
Lentivirus	Long-term transgene expression	Low in vivo expression

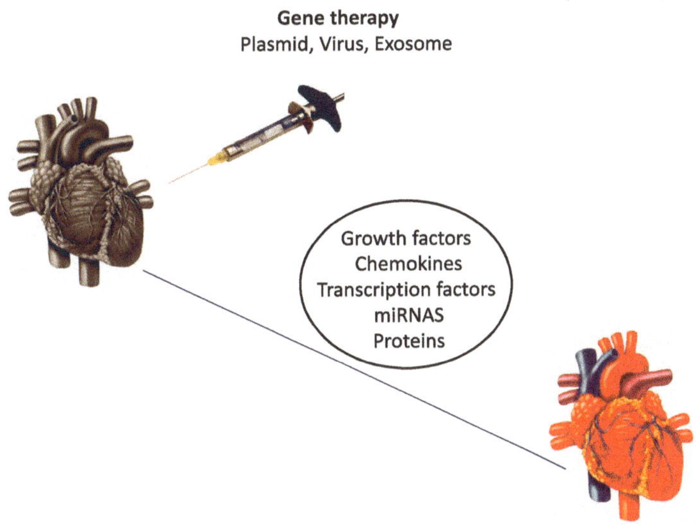

Figure 3.3.1 Gene therapy to improve heart function.

DNA permanently [16], but only in mitotic cells, and retro- and lentiviruses can insert a maximum of 8 kilo bases (kb) of material compared to adenoviruses (30 kb) [17]. Retroviruses, although an efficient way to integrate DNA into host cells, have resulted in T-cell lymphoproliferative disease in four patients with X-linked severe combined immunodeficiency (SCID) [18].

Nonviral vectors include naked plasmid DNA and chemical transmitters, such as liposomes. They do not induce an inflammatory response and can carry genetic material of vast quantities, yet are limited by poor uptake into cells and subsequently poor entry into the nucleus to enable transcription [19].

SYSTEMIC VERSUS LOCALIZED THERAPY

Systemic delivery of genes allows atheromatous plaques to be targeted throughout the body, but risks unwanted transgene expression and side effects. Some of the processes targeted, such as inflammation, are not uniquely seen in atherosclerosis. Therefore, altering these genes may change many other pathophysiological pathways. Gene therapy at local points of interest may circumvent this problem [20].

Bypass grafting involves side stepping the occluded vessel, which can be in the coronary (CABG) or peripheral vasculature, with a venous graft from another part of the body. CABG is a generally successful procedure, but the grafts are subject to accelerated atherosclerosis, leading ultimately to luminal narrowing and graft failure [21,22].

PCI involves reinstating blood supply to cardiac muscle, this time by unblocking remaining occluded vessels using a stent. These stents, like the grafts, are prone to restenosis. Drug-eluting stents are already in use, coated with drugs such as tacrolimus, which reduce in-stent restenosis [23]. As both grafts and stents are outside the body for some time and are deployed at a local site of plaque formation, these are very interesting prospects for transferring gene therapy to clinical practice.

GENE THERAPY: POTENTIAL TARGETS

Endothelial Nitric Oxide Synthase

Nitric oxide (NO) is key in endothelial hemostasis. It is produced in endothelial cells by the conversion of L-arginine to L-citrulline by the endothelial isoform of the enzyme nitric oxide synthase (NOS) [24]. NO is linked with atherosclerosis through its variety of roles; it regulates vascular tone and inhibits atherosclerotic processes such as SMC proliferation [25]. Multiple animal studies have demonstrated that overexpression of NOS results in decreased plaque progression, the use of endothelial NOS (eNOS) showing particular promise. Hayashi et al. [26] induced atherosclerosis in rabbits through balloon injury and high cholesterol diet. Systemic eNOS delivery using an adenoviral vector resulted in regression of atherosclerosis. Interestingly, there was no regression when the rabbits were transfected with inducible NOS (iNOS). A different group used adenoviral vectors to transfer eNOS to rat arteries, which had been subjected to balloon-induced injury. Neointimal hyperplasia was reduced 2 weeks postinjury in the transfected groups compared with controls. This was attributed, at least in part, to a reduction seen in SMC proliferation [27].

However, this is not as straightforward as it may seem. eNOS relies on a number of cofactors to function appropriately. If it becomes uncoupled from its cofactors, superoxide will be generated instead of NO, producing unwanted effects [28]. Interestingly, overexpression of eNOS in the apoE-deficient mouse actually increased atherosclerotic plaque formation. The vessels of these mouse models produced large quantities of superoxide, which may explain these findings [29]. 5,6,7,8-Tetrahydrobiopterin (BH4) is a cofactor of eNOS [30], the production of which relies on the enzyme GTP cyclohydrolase I (GTP-CHI) as the rate-limiting step. Levels of GTP vary within the population due to genetic variation [31]. It has been demonstrated that overexpression of GTP-CHI increased BH4 synthesis and reduced atherosclerosis in apoE-deficient mice [32,33].

Local delivery of NOS using veins grafts and stents has also proven to be effective in animal models. Matsumoto et al. [34] showed gene transfer of eNOS inhibited intimal hyperplasia of canine vein graft as a result of increased local production of NO. Ohta et al. [35] transfected venous grafts in cholesterol-fed rabbits using the hemagglutinating virus of Japan (hvJ), showing significantly smaller neointimal formation in the transfected group. However, the group noted that this delivery method resulted in SMC loss of the vascular wall, potentially reducing gene expression. This is an important factor to consider when delivering genes locally. Adenoviral gene delivery of eNOS to carotid arteries of goats [36] and of iNOS to carotid arteries of pigs [37] using jugular venous grafts, resulted in reduced neointimal hyperplasia in both studies.

In-stent restenosis was significantly decreased in the carotid arteries of rats that received stents coated with iNOS, compared with controls [38]. iNOS has also been delivered locally using balloon-based catheter systems. Lipoplexes [39] and viruses [40] have proved successful vectors in transferring iNOS, resulting in reduced in-stent restenosis in porcine coronary arteries. In short, overexpression of NOS has been shown to reduce plaque formation, graft neointima formation and in-stent restenosis in several animal models.

Antioxidant Enzymes

Oxidation of LDLs permits their uptake into macrophages [5]. Gene therapy can target either delivery of antioxidative enzymes to the vascular wall or those enzymes associated with high-density lipoproteins (HDL).

Antioxidative enzymes include heme oxygenase-1 (HO-1), superoxide dismutase (SOD), and catalase. HO-1 cleaves heme to biliverdin [41], the products of which are antioxidant in nature [42]. SOD1, 2, and 3 release hydrogen peroxide from superoxide [43], then hydrogen peroxide is catalyzed to water and oxygen by catalase [44]. These enzymes are all potential targets in gene therapy. Low levels of HO-1 are linked with atherogenesis [45]. SOD and catalase levels vary during the different stages of atherosclerosis [46].

Systemic adenoviral delivery of HO-1 was studied in apo-E-deficient mice. There were beneficial effects in both young (14 weeks old) and older (20 weeks) mice receiving the recombinant virus [47]. Significantly reduced intimal thickening in venous grafting was also found in rats treated with HO-1-containing adenovirus [48]. The beneficial effects have been attributed not only to antioxidation [49], but also to increased eNOS expression [50].

Adenoviral delivery of SOD3 to arteries of rats with cuff-induced injury resulted in reduced SMC proliferation and augmented iNOS expression. The observed effect was reduced neointimal thickness [51]. Another group investigated the effects of local catheter-based delivery of SOD while simultaneously stenting the aortas of rabbits. The treated group showed reduced neointimal thickness at 28 days, but no significant difference at 90 days [52].

However, not all studies are so clear-cut. Manganese SOD has been shown to reverse endothelial dysfunction in the absence, but not the presence of plaques [53]. Copper/zinc-SOD gene transfer in LDL receptor-deficient mice [54] and SOD3 gene transfer in apoE-deficient mice [55] did not demonstrate any significant decrease in plaque formation. One possible explanation is the difference in gene delivery method-the successful studies used local therapies [4].

HDLs are felt to be protective in atherogenesis, attributed to the antioxidant effect of its structural proteins apoA-I and -II [56,57] and of its related enzymes lecithin-cholesterol transferase (LCAT) [58], platelet-activating factor acetyl hydrolase [59] (PAF-AH), and paraoxonase [60] (PON). HDL-related genes are expressed at the site of cholesterol synthesis, the liver. Viruses interact easily with hepatocytes, meaning the genes can be delivered efficiently to this site [61]. Gene studies have looked at overexpression of apoA-I and HDL-related enzymes, observing this suppressed the development of atherosclerosis in a variety of animal models.

Using apo-E deficient mice, viral vector delivery of apoA-I reduced progression of both native atherosclerosis and allograft vasculopathy [62]. Another group found the same effect, even over a longer follow up of 20 weeks [63]. Mice models with LDL deficiency showed reduced atherogenesis with systemic delivery of apoA-I [64].

Overexpression of PON1 [65], PON2 [66], and PON3 [67] in ApoE mice models resulted in significantly reduced plaque progression. PON-1 transfection also inhibited atherosclerosis in rabbits fed high-cholesterol diet with balloon-induced injury [68] and in a mouse model with metabolic syndrome [69]. LCAT gene transfer protected against plaque formation in rabbits fed high-cholesterol diets [70] and a model of metabolic syndrome [58].

Anti-inflammatory Enzymes and Proteins

Inflammation is another key process in plaque formation, arising from the activities of many genes [4]. Some studies have focused on preventing inflammatory cell migration. Genes of interest include nuclear factor kappa B (NF-kappa B), MCP-1, peroxisome proliferator-activated receptors (PPARs), protein 35K, and interleukin-10 (IL-10). The transcription factor NF-kappa B controls the expression of proinflammatory genes [71], whereas MCP-1 is a proinflammatory chemokine that promotes monocyte infiltration [72]. In contrast, the PPAR family (PPARalpha, gamma, and delta) is a group of nuclear receptors that possess anti-inflammatory characteristics [73]. IL-10 is also anti-inflammatory in nature, produced by Th2-type T cells, B cells, monocytes, and macrophages [74].

As NF-kappa B and MCP-1 are both proinflammatory mediators, it makes sense to antagonize their action. Intramuscular anti-MCP1 markedly reduced restenotic changes in balloon-induced injury in rats and monkeys [75]. Antagonizing MCP-1 also reduced plaque formation in apoE mice [76], vein graft-thickening in mice models [77], and in-stent restenosis in hypercholesteremic rabbits and monkeys [78]. Blockade of MCP-1 by transfecting an N-terminal deletion mutant of the MCP-1 gene not only limited progression of plaques, but also reduced destabilization of established plaques in apoE-knockout mice [79].

35K is a protein that inactivates most chemokines with two cysteines near their amino terminus [80]. 35K transfer reduced macrophage accumulation and proliferation in jugular vein grafts of rabbits, resulting in reduced neointimal thickening [81]. Recombinant adenoviral transfer of 35K to apoE mice also reduced atherosclerosis in carotid-caval vein grafts [82].

As the PPAR family and IL-10 are anti-inflammatory, overexpression should protect against plaque formation. Adenoviruses containing a mutant form of PPARgamma were delivered to the carotid arteries of rats after balloon-induced injury. The PPARgamma mutant strongly inhibited vascular SMC proliferation and migration, increased apoptosis, and reduced neointimal formation [83]. Plaque area and macrophage infiltration were reduced in apoE-deficient mice after transfer of IL-10 using an adenoviral vector [84] and hvJ-liposomal vector [85]. LDL-deficient mice have been transfected with IL-10 through intravenous injection [86] and bone marrow transfer [87], resulting in reduced inflammatory cell infiltration and reduced plaque formation.

After macrophages are attracted to the plaque, they uptake modified LDL to become foam cells [88]. The LDL is transported across the cell membrane via a scavenger receptor A(SRA). Mice with deleted SRA have reduced atherosclerotic lesions; a 60% reduction in plaque size was found in the aorta [89]. CD36 is a class B scavenger receptor. CD36-deficient bone marrow was transplanted into apoE-null mice, resulting in a profound reduction in plaque size. Macrophages expressing CD36 were then reintroduced and the lesions increased in size 2.1-fold [90].

Not only can the receptors be antagonized, but their actions can be promoted to increase uptake of LDL in desirous areas. Lectin-like oxidized LDL receptor-1 (LOX-1) is a major receptor for oxidized LDL in the endothelium. In ApoE-deficient mice, LOX-1 was transfected via adenovirus. The increased expression in the liver enhanced LDL uptake from the blood, preventing plaque formation [91].

Cell Cycle Regulators

SMC proliferation is an event that leads to plaque progression and is regulated by multiple genes. Those genes that regulate the cell cycle, and therefore proliferation of the cells, have been the target of a number of animal models.

Cyclin-dependent kinase-2 is required for transition from G1 to S phase in the cell cycle. Suzuki et al. [92] demonstrated that overexpression of antisense oligonucleotides to Cdk-2 reduced neointima formation in coronary artery grafts of mouse models. The gene p53 has a number of role in the cell cycle; it can induce apoptosis as well as arrest the cycle at G1/S phase [93]. Bai et al. [94] demonstrated that overexpression of p53 significantly reduced neointima formation in venous grafts to carotid arteries in rabbits. The p53 gene itself was transferred using adenoviral vectors in porcine grafts [95], where it acted to upregulate SMC apoptosis, thereby inhibiting neointimal proliferation. Transfer of the retinoblastoma [96] and Nogo-B genes [97] (in human saphenous explant model and porcine vein graft, respectively) has been shown to reduce SMC proliferation.

The Project of Ex-vivo Vein Graft Engineering via Transfection (PREVENT) trial (see Table 3.3.2) looked at the safety and biological efficacy of intraoperative gene therapy in patients undergoing CABG. E2 factor (E2F) is a key transcription factor involved in neointimal formation, and therefore graft failure. "Decoy" antisense oligonucleotides were transfected to venous grafts ex vivo. In the small, preliminary study of 41 patients graft restenosis was reduced in these patients [98]. However, the larger follow-up study of 3014 patients over showed no significant benefit over placebo [99]. It is unknown if there is a delayed benefit over a longer period of time.

Antimatrix Metalloproteinases

Matrix metalloproteinases (MMP) are important regulators in SMC migration and neointima formation [100]. Their role in plaque prevention has been investigated in preventing graft disease.

George et al. transfected tissue inhibitor of metalloproteinases (TIMP)-1 in human saphenous vein cultures. There was decreased neointimal thickening and SMC migration, but no effects on SMC proliferation or cell viability [101] The same group found that TIMP-2 transfer resulted in reduced SMC migration and neointimal formation in vein cultures [102]. In vivo studies were later performed in porcine saphenous grafts to the carotid artery. Here, it was found that TIMP-3 (but not TIMP-2) transfer resulted in reduced neointima formation. The in vivo study demonstrated that TIMP-3 not only inhibited migration of SMCs, but also resulted in SMC apoptosis [103].

Eefting et al. created two hybrid proteins, binding TIMP-1 to protease inhibitors. The first protein, consisting of TIMP-1 and amino-terminal fragment of urokinase-type PA (ATF) [104], the second a combination of ATF and bovine pancreas trypsin inhibitor [105]. The hybrid proteins were transfected to hypercholesterolemic mouse models, resulting in reduced vein-graft thickening in both cases.

Anti-microRNA

MicroRNAs (miRNAs) are noncoding RNA molecules that can regulate gene expression at the posttranscriptional level. miRNAs are involved in all steps of atherogenesis, including endothelial damage and dysfunction, monocyte-wall invasion and activation, lipoprotein formation and platelet, and vascular SMC function. MicroRNA does not modulate only a single gene, but many target genes in differing ways. Therefore, delivery of a specific miRNA may downregulate a crucial pathway in atherogenesis, but it could also potentially affect other pathways in unpredictable ways [106].

Table 3.3.2 Evidence from clinical trials

Gene	Trial	Number	Delivery method	Result	*P* value
E2F "decoy" [99]	PREVENT	41	Ex vivo oligonucleotide transfection to vein grafts	Therapy is safe and feasible; fewer graft occlusions seen in E2F group.	<.0001
E2F "decoy" [100]	PREVENT	3014		No significant beneficial effect seen in graft failure of E2F group	.66
VEGF [116]		5	Plasmid injection to myocardium	Therapy is safe	a
VEGF [120]		13		Reduced are of ischemia seen on electrochemical mapping	.0009[a]
VEGF [119]		7		Reduced reversible ischemia seen on SPECT	.016
VEGF [122]	NORTHERN	93		No benefit in VEGF therapy in any end point studied	(multiple end-points)
VEGF [121]	REVASC	67	Adenoviral injection to myocardium	Improved exercise time compared with controls	.026
VEGF [117]		10		Treatment is safe and feasible	a
VEGF [123]	NOVA	17		No improvement in exercise capacity or myocardial perfusion	
VEGF [118,119]		21		Therapy is safe	
HGF [124]		18		Therapy is safe, myocardial perfusion is improved	<.05
VEGF [126,127]	KAT	103	Plasmid/adenoviral injection to coronary arteries	Therapy is safe, myocardial perfusion is improved	<.05
FGF [128]	AGENT	79		Improved exercise tolerance in subgroup with exercise tolerance <10 min	.01
HGF [129]		18		Therapy is safe	a

AGENT, Angiogenic GENe Therapy; *E2F*, E2 factor; *FGF*, fibroblast growth factor; *HGF*, hepatocyte growth factor; *KAT*, Kuopio Angiogenesis Trial; *NORTHERN*, NOGA angiogenesis Revascularization Therapy: assessment by RadioNuclide imaging; *NOVA*, NOGA Delivery of VEGF for Angina; *PREVENT*, Project of Ex-vivo Vein Graft Engineering via Transfection; *REVASC, Randomized Evaluation of VEGF for Angiogenesis; VEGF*, vascular endothelial growth factor.

[a]No placebo.

Gene delivery of mRNA-145 has been investigated by Lovren et al. Lentivirus was used to transfer microRNA-145 to vascular SMCs of apoE knockout mice. Treated mice has smaller plaques in their aortic sinuses, ascending aortas, and brachiocephalic arteries [107]. Rayner et al. showed that microRNA-33 is key in regulating cholesterol metabolism [108] and that delivering anti-microRNA-33 results in regression of atherosclerosis in mouse models. The treated group had fewer macrophages, fewer LDL particles and more stable plaques [109]. In African green monkeys, the same group demonstrated a reduction in very LDLs with antisense oligonucleotides [110] This finding was actually not seen in mouse models.

Antisense oligonucleotides have also been used to study the effects of miR-122, the most abundant microRNA found in the liver. In mouse models, miRNA-122 inhibition resulted in decreased plasma cholesterol levels and a significant improvement in liver steatosis [111]. miRNA-122 was antagonized in nonhuman primates using locked-nucleic acid oligonucleotides. There was a long-lasting decrease in plasma cholesterol [112].

Antineogenesis

Cytokines such as vascular endothelial growth factor (VEGF), fibroblast growth factor (FGF), and hepatocyte growth factor (HGF) are angiogenic in nature. This, in theory, could reestablish blood supply, improving perfusion and reducing ischemia downstream [113]. Stents eluting plasmids carrying the VEGF-2 gene were implanted in the iliac arteries of normocholesterolemic and hypercholesterolemic rabbits. There was greater reendothelialization in the treated group after 10 days, and a significantly larger luminal cross-section after 3 months in the treated rabbits compared with placebo. Moreover, there was a 2.4-fold greater NO production in the treated group [114].

VEGF, HGF, and FGF have been delivered to human myocardium in early clinical trials via myocardial and intracoronary injection (Table 3.2.2).

Intramyocardial injection of plasmids and adenoviral vectors carrying VEGF has been carried out and shown to be a safe and practical therapy. Phase 1 trials highlighted the safety of the therapy [115—118]. Sarkar et al. studied a small group of end-stage angina patients, in whom further revascularization was unfeasible. There was reduced reversible ischemia on single photon emission computerized tomography (SPECT) and a significant improvement in symptoms using the Canadian Cardiovascular Society class [119]. Another small study of 13 patients showed reduced area of myocardial ischemia on electrochemical mapping after transfection with naked plasmid DNA coding for VEGF [120]. In a larger study of 67 patients with severe angina, the group treated with intramyocardial delivery of VEGF showed improved exercise duration and angina symptoms compared with the control group that received maximal medical therapy [121]. Other studies have painted more of a mixed picture, however. The NORTHERN

(NOGA angiogenesis Revascularization Therapy: assessment by RadioNuclide imaging) trial of 93 patients, who received plasmid-containing VEGF or placebo showed no difference in myocardial perfusion measure by SPECT at 3 or 6 months [122]. There was also no difference in exercise treadmill time between the two groups. Injections of adenoviral vectors of VEGF also failed to show any improvement in exercise capacity or myocardial perfusion in the NOVA (NOGA Delivery of VEGF for Angina) trial of 17 patients [123]. Yuan et al. [124] showed that injection of HGF was safe and that myocardial perfusion was increased in a dose-dependent manner. This was a small trial of 18 patients, so larger, randomized trials are needed before clear conclusions are drawn.

Drug-coated balloons can deliver a short burst of a therapy to a localized target [125]. In the KAT (Kuopio Angiogenesis Trial) [126], patients already undergoing coronary artery stenting also received catheter-delivered VEGF adenovirus, VEGF plasmid liposome, or placebo (Ringer's lactate). There was no difference in restenosis rate between the groups after 6 months, and there was improved myocardial perfusion in the group that received the adenoviral vector. Long-term follow-up of these patients over 8 years demonstrated no difference in mortality between these groups [127]. Catheter-based delivery of FGF by adenovirus showed an improvement in patients who received the therapy compared with controls. A significant improvement in exercise time was seen in patients with a baseline exercise tolerance of less than 10 min [128]. Local delivery of HGF was found to be safe [129], though no trials have proven the efficacy yet.

CONCLUSIONS

Both local and systemic gene delivery have shown promising results in animal models. The large number of potential targets is encouraging, but makes it difficult to know where efforts should be focused in progressing therapies from animal models to clinical practice. The early human clinical trials have shown gene transfer can be safe, but the larger phase 2 trials produced mixed results.

One problem is that potential targets, such as inflammation or cell cycle regulation, are not unique to atherosclerosis. There must be a balance between altering the gene in question, while not interfering with normal cell function around the body [130] This highlights the importance of locally delivered and disease-specific therapy.

As with all animal models, the results do not always predict outcomes in humans. Most of the studies discussed use smaller animals, and very few use nonhuman primates. Before moving to clinical trials, porcine models offer very valuable experience in testing intravascular delivery methods and imaging possibilities [130].

Gene therapy is an exciting prospect, but more knowledge is required about vector use and downstream effects of altering genes before it will be used in clinical practice.

REFERENCES

[1] Weber C, Noels H. Atherosclerosis: current pathogenesis and therapeutic options. Nat Med 2011;17: 1410–22.
[2] Moore KJ, Tabas I. Macrophages in the pathogenesis of atherosclerosis. Cell 2011;145:341–55.
[3] Lewandowski AJ, Davis EF, Lazdam M, Leeson P. From gene to epigene-based therapies targeting the vascular endothelium. Curr Vasc Pharmacol 2012;10:125–37.
[4] Van-Assche T, Huygelen V, Crabtree MJ, Antoniades C. Gene therapy targeting inflammation in atherosclerosis. Curr Pharm Des 2011;17:4210–23.
[5] Hulsmans M, Holvoet P. The vicious circle between oxidative stress and inflammation in atherosclerosis. J Cell Mol Med 2010;14:70–8.
[6] Campbell JH, Campbell GR. Cell biology of atherosclerosis. J Hypertens 1994;12:S129–32.
[7] Drakopoulou M, Toutouzas K, Michelongona A, Tousoulis D, Stefanadis C. Vulnerable plaque and inflammation: potential clinical strategies. Curr Pharm Des 2011;17:4190–209.
[8] Van-Assche T, Huygelen V, Crabtree MJ, Antoniades C. Gene delivery strategies targeting stable atheromatous plaque. Curr Pharm Des 2013;19:1626–37.
[9] Theoharis S, Manunta M, Tan PH. Gene delivery to vascular endothelium using chemical vectors: implications for cardiovascular gene therapy. Expert Opin Biol Ther 2007;7:627–43.
[10] Rutanen J, Rissanen TT, Kivelä A, Vajanto I, Ylä-Herttuala S. Clinical applications of vascular gene therapy. Curr Cardiol Rep 2001;3:29–36.
[11] Nayak S, Herzog RW. Progress and prospects: immune responses to viral vectors. Gene Ther 2010; 17:295–304.
[12] Wickham TJ, et al. Targeted adenovirus gene transfer to endothelial and smooth muscle cells by using bispecific antibodies. J Virol 1996;70:6831–8.
[13] Feldman LJ, Steg G. Optimal techniques for arterial gene transfer. Cardiovasc Res 1997;35:391–404.
[14] Tan PH, et al. Effect of vectors on human endothelial cell signal transduction: implications for cardiovascular gene therapy. Arterioscler Thromb Vasc Biol 2006;26:462–7.
[15] Mac Gabhann F, Annex BH, Popel AS. Gene therapy from the perspective of systems biology. Curr Opin Mol Ther 2010;12:570–7.
[16] Rainov NG, Ren H. Clinical trials with retrovirus mediated gene therapy—what have we learned? J Neurooncol 2003;65:227–36.
[17] Hall K, Blair Zajdel ME, Blair GE. Unity and diversity in the human adenoviruses: exploiting alternative entry pathways for gene therapy. Biochem J 2010;431:321–36.
[18] Nair V. Retrovirus-induced oncogenesis and safety of retroviral vectors. Curr Opin Mol Ther 2008; 10:431–8.
[19] Budker V, Gurevich V, Hagstrom JE, Bortzov F, Wolff JA. pH-sensitive, cationic liposomes: a new synthetic virus-like vector. Nat Biotechnol 1996;14:760–4.
[20] Francis SC, et al. Genetic targeting for cardiovascular therapeutics: are we near the summit or just beginning the climb? Physiol Genomics 2001;7:79–94.
[21] Shukla N, Jeremy JY. Pathophysiology of saphenous vein graft failure: a brief overview of interventions. Curr Opin Pharmacol 2012;12:114–20.
[22] Margaritis M, Channon KM, Antoniades C. Statins and vein graft failure in coronary bypass surgery. Curr Opin Pharmacol 2012;12:172–80.
[23] Jukema JW, Verschuren JJW, Ahmed TAN, Quax PHA. Restenosis after PCI. Part 1: pathophysiology and risk factors. Nat Rev Cardiol 2012;9:53–62.
[24] Anderson TJ. Nitric oxide, atherosclerosis and the clinical relevance of endothelial dysfunction. Heart Fail Rev 2003;8:71–86.
[25] Palmer RM, Ferrige AG, Moncada S. Nitric oxide release accounts for the biological activity of endothelium-derived relaxing factor. Nature 1987;327:524–6.
[26] Hayashi T, et al. Gene transfer of endothelial NO synthase, but not eNOS plus inducible NOS, regressed atherosclerosis in rabbits. Cardiovasc Res 2004;61:339–51.
[27] Janssens S, et al. Human endothelial nitric oxide synthase gene transfer inhibits vascular smooth muscle cell proliferation and neointima formation after balloon injury in rats. Circulation 1998;97:1274–81.

[28] Vásquez-Vivar J, et al. Superoxide generation by endothelial nitric oxide synthase: the influence of cofactors. Proc Natl Acad Sci USA 1998;95:9220–5.
[29] Ozaki M, et al. Overexpression of endothelial nitric oxide synthase accelerates atherosclerotic lesion formation in apoE-deficient mice. J Clin Invest 2002;110:331–40.
[30] Crabtree MJ, et al. Quantitative regulation of intracellular endothelial nitric-oxide synthase (eNOS) coupling by both tetrahydrobiopterin-eNOS stoichiometry and biopterin redox status: insights from cells with tet-regulated GTP cyclohydrolase I expression. J Biol Chem 2009;284:1136–44.
[31] Thöny B, Auerbach G, Blau N. Tetrahydrobiopterin biosynthesis, regeneration and functions. Biochem J 2000;347(Pt 1):1–16.
[32] Alp NJ, McAteer MA, Khoo J, Choudhury RP, Channon KM. Increased endothelial tetrahydrobiopterin synthesis by targeted transgenic GTP-cyclohydrolase I overexpression reduces endothelial dysfunction and atherosclerosis in ApoE-knockout mice. Arterioscler Thromb Vasc Biol 2004;24:445–50.
[33] Ali ZA, et al. CCR2-mediated antiinflammatory effects of endothelial tetrahydrobiopterin inhibit vascular injury-induced accelerated atherosclerosis. Circulation 2008;118:S71–7.
[34] Matsumoto T, et al. Hemagglutinating virus of Japan-liposome-mediated gene transfer of endothelial cell nitric oxide synthase inhibits intimal hyperplasia of canine vein grafts under conditions of poor runoff. J Vasc Surg 1998;27:135–44.
[35] Ohta S, et al. Intraluminal gene transfer of endothelial cell-nitric oxide synthase suppresses intimal hyperplasia of vein grafts in cholesterol-fed rabbit: a limited biological effect as a result of the loss of medial smooth muscle cells. Surgery 2002;131:644–53.
[36] Wu S, Wang X, Guo L, Zi J. Adenovirus mediated endothelial nitric oxide synthase gene transfer prevents restenosis of vein grafts. ASAIO J 2004;1992(50):272–7.
[37] Kibbe MR, et al. Adenovirus-mediated gene transfer of human inducible nitric oxide synthase in porcine vein grafts inhibits intimal hyperplasia. J Vasc Surg 2001;34:156–65.
[38] Fishbein I, et al. Local delivery of gene vectors from bare-metal stents by use of a biodegradable synthetic complex inhibits in-stent restenosis in rat carotid arteries. Circulation 2008;117:2096–103.
[39] Muhs A, et al. Preclinical evaluation of inducible nitric oxide synthase lipoplex gene therapy for inhibition of stent-induced vascular neointimal lesion formation. Hum Gene Ther 2003;14:375–83.
[40] Wang K, et al. Local adenoviral-mediated inducible nitric oxide synthase gene transfer inhibits neointimal formation in the porcine coronary stented model. Mol Ther 2003;7:597–603.
[41] Willis D, Moore AR, Frederick R, Willoughby DA. Heme oxygenase: a novel target for the modulation of the inflammatory response. Nat Med 1996;2:87–90.
[42] Platt JL, Nath KA. Heme oxygenase: protective gene or Trojan horse. Nat Med 1998;4:1364–5.
[43] Johnson F, Giulivi C. Superoxide dismutases and their impact upon human health. Mol Asp Med 2005;26:340–52.
[44] Nicholls P. Classical catalase: ancient and modern. Arch Biochem Biophys 2012;525:95–101.
[45] Chen Y-H, et al. Microsatellite polymorphism in promoter of heme oxygenase-1 gene is associated with susceptibility to coronary artery disease in type 2 diabetic patients. Hum Genet 2002;111:1–8.
[46] Gupta S, Sodhi S, Mahajan V. Correlation of antioxidants with lipid peroxidation and lipid profile in patients suffering from coronary artery disease. Expert Opin Ther Targets 2009;13:889–94.
[47] Juan S-H, et al. Adenovirus-mediated heme oxygenase-1 gene transfer inhibits the development of atherosclerosis in apolipoprotein e–deficient mice. Circulation 2001;104:1519–25.
[48] Bouche D, et al. Inhibition of graft arteriosclerosis development in rat aortas following heme oxygenase-1 gene transfer. Transpl Immunol 2002;9:235–8.
[49] Datla SR, et al. Induction of heme oxygenase-1 in vivo suppresses NADPH oxidase derived oxidative stress. Hypertension 2007;1979(50):636–42.
[50] Ahmad M, et al. Heme oxygenase-1 gene expression increases vascular relaxation and decreases inducible nitric oxide synthase in diabetic rats. Cell Mol Biol 2005;51:371–6.
[51] Ozumi K, et al. Extracellular superoxide dismutase overexpression reduces cuff-induced arterial neointimal formation. Atherosclerosis 2005;181:55–62.

[52] Bräsen JH, et al. Extracellular superoxide dismutase accelerates endothelial recovery and inhibits in-stent restenosis in stented atherosclerotic Watanabe heritable hyperlipidemic rabbit aorta. J Am Coll Cardiol 2007;50:2249–53.
[53] Zanetti M, et al. Gene transfer of manganese superoxide dismutase reverses vascular dysfunction in the absence but not in the presence of atherosclerotic plaque. Hum Gene Ther 2001;12:1407–16.
[54] Yang H, et al. Retardation of atherosclerosis by overexpression of catalase or both Cu/Zn-superoxide dismutase and catalase in mice lacking apolipoprotein E. Circ Res 2004;95:1075–81.
[55] Laukkanen MO, et al. Gene transfer of extracellular superoxide dismutase to atherosclerotic mice. Antioxid Redox Signal 2001;3:397–402.
[56] Garner B, Waldeck AR, Witting PK, Rye KA, Stocker R. Oxidation of high density lipoproteins. II. Evidence for direct reduction of lipid hydroperoxides by methionine residues of apolipoproteins AI and AII. J Biol Chem 1998;273:6088–95.
[57] Garner B, et al. Oxidation of high density lipoproteins. I. Formation of methionine sulfoxide in apolipoproteins AI and AII is an early event that accompanies lipid peroxidation and can be enhanced by alpha-tocopherol. J Biol Chem 1998;273:6080–7.
[58] Mertens A, et al. Increased low-density lipoprotein oxidation and impaired high-density lipoprotein antioxidant defense are associated with increased macrophage homing and atherosclerosis in dyslipidemic obese mice: LCAT gene transfer decreases atherosclerosis. Circulation 2003;107: 1640–6.
[59] Quarck R, et al. Adenovirus-mediated gene transfer of human platelet-activating factor-acetylhydrolase prevents injury-induced neointima formation and reduces spontaneous atherosclerosis in apolipoprotein E-deficient mice. Circulation 2001;103:2495–500.
[60] Aviram M, et al. Paraoxonase inhibits high-density lipoprotein oxidation and preserves its functions. A possible peroxidative role for paraoxonase. J Clin Invest 1998;101:1581–90.
[61] Smith RM, Wu GY. Hepatocyte-directed gene delivery by receptor-mediated endocytosis. Semin Liver Dis 1999;19:83–92.
[62] Feng Y, et al. Wild-type apo A-I and apo A-I(Milano) gene transfer reduce native and transplant arteriosclerosis to a similar extent. J Mol Med 2009;87:287–97.
[63] Pastore L, et al. Helper-dependent adenoviral vector-mediated long-term expression of human apolipoprotein A-I reduces atherosclerosis in apo E-deficient mice. Gene 2004;327:153–60.
[64] Belalcazar LM, et al. Long-term stable expression of human apolipoprotein A-I mediated by helper-dependent adenovirus gene transfer inhibits atherosclerosis progression and remodels atherosclerotic plaques in a mouse model of familial hypercholesterolemia. Circulation 2003; 107:2726–32.
[65] Zhang C, et al. Studies on protective effects of human paraoxonases 1 and 3 on atherosclerosis in apolipoprotein E knockout mice. Gene Ther 2010;17:626–33.
[66] Ng CJ, Hama SY, Bourquard N, Navab M, Reddy ST. Adenovirus mediated expression of human paraoxonase 2 protects against the development of atherosclerosis in apolipoprotein E-deficient mice. Mol Genet Metab 2006;89:368–73.
[67] Ng CJ, et al. Adenovirus-mediated expression of human paraoxonase 3 protects against the progression of atherosclerosis in apolipoprotein E-deficient mice. Arterioscler Thromb Vasc Biol 2007;27: 1368–74.
[68] Miyoshi M, et al. Gene delivery of paraoxonase-1 inhibits neointimal hyperplasia after arterial balloon-injury in rabbits fed a high-fat diet. Hypertens Res 2007;30:85–91.
[69] Mackness B, Quarck R, Verreth W, Mackness M, Holvoet P. Human paraoxonase-1 overexpression inhibits atherosclerosis in a mouse model of metabolic syndrome. Arterioscler Thromb Vasc Biol 2006;26:1545–50.
[70] Hoeg JM, et al. Overexpression of lecithin:cholesterol acyltransferase in transgenic rabbits prevents diet-induced atherosclerosis. Proc Natl Acad Sci USA 1996;93:11448–53.
[71] Monaco C, Paleolog E. Nuclear factor kappaB: a potential therapeutic target in atherosclerosis and thrombosis. Cardiovasc Res 2004;61:671–82.
[72] Kitamoto S, Egashira K. Anti-monocyte chemoattractant protein-1 gene therapy for cardiovascular diseases. Expert Rev Cardiovasc Ther 2003;1:393–400.

[73] Ricote M, Li AC, Willson TM, Kelly CJ, Glass CK. The peroxisome proliferator-activated receptor-gamma is a negative regulator of macrophage activation. Nature 1998;391:79–82.
[74] Wakkach A, Cottrez F, Groux H. Can interleukin-10 be used as a true immunoregulatory cytokine? Eur Cytokine Netw 2000;11:153–60.
[75] Usui M, et al. Anti-monocyte chemoattractant protein-1 gene therapy inhibits restenotic changes (neointimal hyperplasia) after balloon injury in rats and monkeys. FASEB J 2002;16:1838–40.
[76] Ni W, et al. New anti-monocyte chemoattractant protein-1 gene therapy attenuates atherosclerosis in apolipoprotein E-knockout mice. Circulation 2001;103:2096–101.
[77] Schepers A, et al. Anti-MCP-1 gene therapy inhibits vascular smooth muscle cells proliferation and attenuates vein graft thickening both in vitro and in vivo. Arterioscler Thromb Vasc Biol 2006;26: 2063–9.
[78] Egashira K, et al. Local delivery of anti-monocyte chemoattractant protein-1 by gene-eluting stents attenuates in-stent stenosis in rabbits and monkeys. Arterioscler Thromb Vasc Biol 2007;27: 2563–8.
[79] Inoue S, et al. Anti-monocyte chemoattractant protein-1 gene therapy limits progression and destabilization of established atherosclerosis in apolipoprotein e–knockout mice. Circulation 2002; 106:2700–6.
[80] Burns JM, Dairaghi DJ, Deitz M, Tsang M, Schall TJ. Comprehensive mapping of poxvirus vCCI chemokine-binding protein. Expanded range of ligand interactions and unusual dissociation kinetics. J Biol Chem 2002;277:2785–9.
[81] Puhakka HL, et al. Effects of vaccinia virus anti-inflammatory protein 35K and TIMP-1 gene transfers on vein graft stenosis in rabbits. In Vivo 2005;19:515–21.
[82] Ali ZA, et al. Gene transfer of a broad spectrum CC-chemokine inhibitor reduces vein graft atherosclerosis in apolipoprotein E-knockout mice. Circulation 2005;112:I235–41.
[83] Lim S, et al. Effect of a peroxisome proliferator-activated receptor gamma sumoylation mutant on neointimal formation after balloon injury in rats. Atherosclerosis 2009;206:411–7.
[84] Yoshioka T, et al. Adeno-associated virus vector-mediated interleukin-10 gene transfer inhibits atherosclerosis in apolipoprotein E-deficient mice. Gene Ther 2004;11:1772–9.
[85] Namiki M, et al. Intramuscular gene transfer of interleukin-10 cDNA reduces atherosclerosis in apolipoprotein E-knockout mice. Atherosclerosis 2004;172:21–9.
[86] Liu Y, et al. Inhibition of atherogenesis in LDLR knockout mice by systemic delivery of adeno-associated virus type 2-hIL-10. Atherosclerosis 2006;188:19–27.
[87] Han X, Kitamoto S, Wang H, Boisvert WA. Interleukin-10 overexpression in macrophages suppresses atherosclerosis in hyperlipidemic mice. FASEB J 2010;24:2869–80.
[88] Brown MS, Goldstein JL. Lipoprotein metabolism in the macrophage: implications for cholesterol deposition in atherosclerosis. Annu Rev Biochem 1983;52:223–61.
[89] Babaev VR, et al. Reduced atherosclerotic lesions in mice deficient for total or macrophage-specific expression of scavenger receptor-A. Arterioscler Thromb Vasc Biol 2000;20:2593–9.
[90] Febbraio M, Guy E, Silverstein RL. Stem cell transplantation reveals that absence of macrophage CD36 is protective against atherosclerosis. Arterioscler Thromb Vasc Biol 2004;24:2333–8.
[91] Ishigaki Y, et al. Impact of plasma oxidized low-density lipoprotein removal on atherosclerosis. Circulation 2008;118:75–83.
[92] Suzuki J, et al. Prevention of graft coronary arteriosclerosis by antisense cdk2 kinase oligonucleotide. Nat Med 1997;3:900–3.
[93] Bieging KT, Mello SS, Attardi LD. Unravelling mechanisms of p53-mediated tumour suppression. Nat Rev Cancer 2014;14:359–70.
[94] Bai H, et al. Inhibition of intimal hyperplasia after vein grafting by in vivo transfer of human senescent cell-derived inhibitor-1 gene. Gene Ther 1998;5:761–9.
[95] Wan S, George SJ, Nicklin SA, Yim APC, Baker AH. Overexpression of p53 increases lumen size and blocks neointima formation in porcine interposition vein grafts. Mol Ther 2004;9: 689–98.
[96] Lamfers ML, et al. Adenoviral delivery of a constitutively active retinoblastoma mutant inhibits neo-intima formation in a human explant model for vein graft disease. Vasc Pharmacol 2002;39:293–301.

[97] Kritz AB, et al. In vivo modulation of Nogo-B attenuates neointima formation. Mol Ther 2008;16: 1798–804.
[98] Mann MJ, et al. Ex-vivo gene therapy of human vascular bypass grafts with E2F decoy: the PREVENT single-centre, randomised, controlled trial. Lancet 1999;354:1493–8.
[99] Alexander JH, et al. Efficacy and safety of edifoligide, an E2F transcription factor decoy, for prevention of vein graft failure following coronary artery bypass graft surgery: PREVENT IV: a randomized controlled trial. JAMA 2005;294:2446–54.
[100] Birkedal-Hansen H, et al. Matrix metalloproteinases: a review. Crit Rev Oral Biol Med 1993;4: 197–250.
[101] George SJ, Johnson JL, Angelini GD, Newby AC, Baker AH. Adenovirus-mediated gene transfer of the human TIMP-1 gene inhibits smooth muscle cell migration and neointimal formation in human saphenous vein. Hum Gene Ther 1998;9:867–77.
[102] George SJ, Baker AH, Angelini GD, Newby AC. Gene transfer of tissue inhibitor of metalloproteinase-2 inhibits metalloproteinase activity and neointima formation in human saphenous veins. Gene Ther 1998;5:1552–60.
[103] George SJ, Lloyd CT, Angelini GD, Newby AC, Baker AH. Inhibition of late vein graft neointima formation in human and porcine models by adenovirus-mediated overexpression of tissue inhibitor of metalloproteinase-3. Circulation 2000;101:296–304.
[104] Eefting D, et al. In vivo suppression of vein graft disease by nonviral, electroporation-mediated, gene transfer of tissue inhibitor of metalloproteinase-1 linked to the amino terminal fragment of urokinase (TIMP-1.ATF), a cell-surface directed matrix metalloproteinase inhibitor. J Vasc Surg 2010;51: 429–37.
[105] Eefting D, et al. A novel urokinase receptor-targeted inhibitor for plasmin and matrix metalloproteinases suppresses vein graft disease. Cardiovasc Res 2010;88:367–75.
[106] Economou EK, et al. The role of microRNAs in coronary artery disease: from pathophysiology to diagnosis and treatment. Atherosclerosis 2015;241:624–33.
[107] Lovren F, et al. MicroRNA-145 targeted therapy reduces atherosclerosis. Circulation 2012;126: S81–90.
[108] Rayner KJ, et al. MiR-33 contributes to the regulation of cholesterol homeostasis. Science 2010;328: 1570–3.
[109] Rayner KJ, et al. Antagonism of miR-33 in mice promotes reverse cholesterol transport and regression of atherosclerosis. J Clin Invest 2011;121:2921–31.
[110] Rayner KJ, et al. Inhibition of miR-33a/b in non-human primates raises plasma HDL and lowers VLDL triglycerides. Nature 2011;478:404–7.
[111] Esau C, et al. miR-122 regulation of lipid metabolism revealed by in vivo antisense targeting. Cell Metab 2006;3:87–98.
[112] Elmén J, et al. LNA-mediated microRNA silencing in non-human primates. Nature 2008;452: 896–9.
[113] Atluri P, Woo YJ. Pro-angiogenic cytokines as cardiovascular therapeutics: assessing the potential. BioDrugs 2008;22:209–22.
[114] Walter DH, et al. Local gene transfer of phVEGF-2 plasmid by gene-eluting stents: an alternative strategy for inhibition of restenosis. Circulation 2004;110:36–45.
[115] Losordo DW, et al. Gene therapy for myocardial angiogenesis: initial clinical results with direct myocardial injection of phVEGF165 as sole therapy for myocardial ischemia. Circulation 1998;98: 2800–4.
[116] Fuchs S, et al. A randomized, double-blind, placebo-controlled, multicenter, pilot study of the safety and feasibility of catheter-based intramyocardial injection of AdVEGF121 in patients with refractory advanced coronary artery disease. Catheter Cardiovasc Interv 2006;68:372–8.
[117] Rosengart TK, et al. Angiogenesis gene therapy: phase I assessment of direct intramyocardial administration of an adenovirus vector expressing VEGF121 cDNA to individuals with clinically significant severe coronary artery disease. Circulation 1999;100:468–74.

[118] Rosengart TK, et al. Six-month assessment of a phase I trial of angiogenic gene therapy for the treatment of coronary artery disease using direct intramyocardial administration of an adenovirus vector expressing the VEGF121 cDNA. Ann Surg 1999;230:466–70. Discussion 470–472.
[119] Sarkar N, et al. Effects of intramyocardial injection of phVEGF-A165 as sole therapy in patients with refractory coronary artery disease—12-month follow-up: angiogenic gene therapy. J Intern Med 2001;250:373–81.
[120] Vale PR, et al. Left ventricular electromechanical mapping to assess efficacy of phVEGF(165) gene transfer for therapeutic angiogenesis in chronic myocardial ischemia. Circulation 2000;102: 965–74.
[121] Stewart DJ, et al. Angiogenic gene therapy in patients with nonrevascularizable ischemic heart disease: a phase 2 randomized, controlled trial of AdVEGF(121) (AdVEGF121) versus maximum medical treatment. Gene Ther 2006;13:1503–11.
[122] Stewart DJ, et al. VEGF gene therapy fails to improve perfusion of ischemic myocardium in patients with advanced coronary disease: results of the NORTHERN trial. Mol Ther 2009;17:1109–15.
[123] Kastrup J, et al. A randomised, double-blind, placebo-controlled, multicentre study of the safety and efficacy of BIOBYPASS (AdGVVEGF121.10NH) gene therapy in patients with refractory advanced coronary artery disease: the NOVA trial. EuroIntervention 2011;6:813–8.
[124] Yuan B, et al. Short-term safety and curative effect of recombinant adenovirus carrying hepatocyte growth factor gene on ischemic cardiac disease. In Vivo 2008;22:629–32.
[125] Gertz ZM, Wilensky RL. Local drug delivery for treatment of coronary and peripheral artery disease. Cardiovasc Ther 2011;29:e54–66.
[126] Hedman M, et al. Safety and feasibility of catheter-based local intracoronary vascular endothelial growth factor gene transfer in the prevention of postangioplasty and in-stent restenosis and in the treatment of chronic myocardial ischemia: phase II results of the Kuopio Angiogenesis Trial (KAT). Circulation 2003;107:2677–83.
[127] Hedman M, et al. Eight-year safety follow-up of coronary artery disease patients after local intracoronary VEGF gene transfer. Gene Ther 2009;16:629–34.
[128] Grines CL, et al. Angiogenic gene therapy (AGENT) trial in patients with stable angina pectoris. Circulation 2002;105:1291–7.
[129] Yang Z-J, et al. Phase I clinical trial on intracoronary administration of Ad-hHGF treating severe coronary artery disease. Mol Biol Rep 2009;36:1323–9.
[130] Kivelä AM, Huusko J, Ylä-Herttuala S. Prospect and progress of gene therapy in treating atherosclerosis. Expert Opin Biol Ther 2015;15:1699–712.

CHAPTER 3.4

Stem-Cell Therapy

Georgia Vogiatzi[1], Alexandros Briasoulis[2], Sotirios Tsalamandris[1], Dimitris Tousoulis[1]

[1] 1st Cardiology Clinic, 'Hippokration' General Hospital, National and Kapodistrian University of Athens, School of Medicine, Athens, Greece; [2] Mayo Clinic, Rochester, MN, United States

Contents

INTRODUCTION

Despite continuous therapeutic advancements, atherosclerotic heart disease poses a major health burden [1] and current therapies (including percutaneous intervention or surgical revascularization, optimized medical therapy, devices, and lifestyle modification) do not completely reverse the pathophysiology. New approaches in advance atherosclerosis in coronary arteries based on stem cell therapy may offer novel avenues to reduce the prevalence coronary disease worldwide [2].

Until recently, the heart was considered as a solid organ with limited regenerative capacity and hypertrophy of cardiomyocytes was considered the unique adaptive mechanism in myocardial injury [3]. Careful experimental activity confuted this traditional view showing a continuous cellular turnover of cardiomyocytes [3–5]. At the same time, the capacity of extracardiac stem cells to differentiate after myocardial infarct and improve functional recovery gave birth to numerous preclinical and clinical research efforts studying the rational of cell-based cardiac therapy. Application of stem cell-based treatment in clinical cardiology could improve standards of care for many patients [6].

Coronary Artery Disease
ISBN 978-0-12-811908-2, http://dx.doi.org/10.1016/B978-0-12-811908-2.00016-7

POTENTIAL SOURCES FOR STEM CELL THERAPY (FIG. 3.4.1)

Embryonic Stem Cells

Embryonic stem cells (ECSs) constitute the most primitive group of all stem cells as they have the broadest developmental potential. As indicated in experimental studies, these cells expressed cardiac-specific proteins and evaluated their ability to engraft when injected into damaged myocardium and improve myocardial performance by forming stable grafts over several weeks after implantation [7,8]. It has been shown that human-induced pluripotent stem cells can differentiate into functional cardiomyocytes and these cells may be suitable for therapeutic application [9] while preclinical studies have proved that they have the ability to differentiate into myocytes with the structural and functional properties of cardiomyocytes [10]. These studies show that functionally intact cardiomyocytes can be generated from ESCs in humans in vitro. However, the use of ECSs is associated with ethical controversies about their malignant potential. Legal problems called into doubt the realistic use of ESCs in humans.

Adult Bone Marrow Stem Cells

Adult bone marrow stem cells (BMCs) are capable of differentiation along several lineages such as hematopoietic, mesenchymal, stromal, and further stem cells. These unspecialized cells are found in within a tissue or organ and have the ability to develop

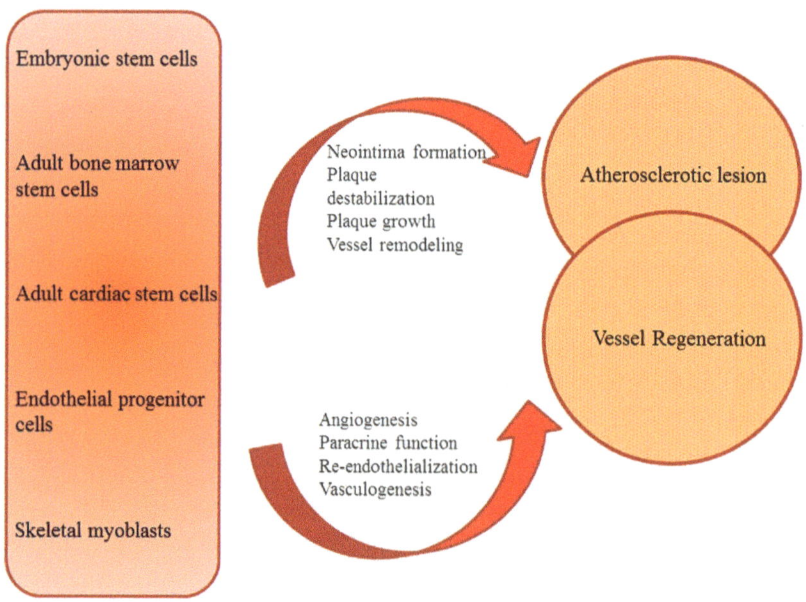

Figure 3.4.1 Graphical summary of stem cells' functions in advance atherosclerosis.

into specialized cells and repair the tissue in which they are found. In human heart, these stem cells are intrinsically programmed to generate cardiac tissue [11]. They are characterized by different cell surface markers. The most studied are CD34 (cluster of differentiation 34), c-Kit [(CD117, tyrosine kinase receptor of stem cell factor receptor (SCF-R)], stem cell antigen-1 (Sca-1), lineage marker (lin), CD133, CD105, CD34, CD31 (PECAM-1, platelet and endothelial cells adhesion molecule 1), and CD45 (LCA, leukocyte common antigen) [12]. CD34 is expressed on primitive stem cells and differentiated progenitor cells, and is absent on mature hematopoietic cells, whereas the presence or absence of lin, c-kit, and sca-1 can further distinguish cell populations. The latter two are markers of primitive stem cells, whereas lin^+ cells are committed to a particular lineage. Many of these BMCs are found within a tissue or organ and have the ability to develop into specialized cells and repair the tissue in which they are found [13]. BMCs have been used in several animal models of myocardial infarction.

Adult Cardiac Stem Cells

Cardiac stem cells (CSCs) were first isolated in the rat heart by Beltrami et al. as cells positive for the surface receptor c-kit [3,14]. In addition to c-kit, other specific markers define CSCs also and lead to creation of the above cells types: (1) $Sca\text{-}1^+$, (2) $Isl1^+$, and (3) CSCs derived from cardiospheres [15], which have properties as the ability to divide without limitation and symmetrically in order to generate two daughter stem cells to expand the stem cell compartment of the heart [16]. These tissue resident CSCs are self-renewing and expandable in cell culture, multipotent, and can give rise to cardiomyocytes. Their multipotentiality allows them to differentiate along the three main cardiac lineages: myocytes, endothelial cells, and smooth muscle cells [14]. After their injection in the ischemic heart, the formation of the above-mentioned cell types contributes to the regeneration of myocardium and the improvement of its contractility [17]. CSCs are considered an attractive population for cell therapy despite the ethical questions concerning their organ-specific delivery and the premature isolation and delivery techniques.

Endothelial Progenitor Cells

Endothelial progenitor cells (EPCs) have been extensively studied for almost 20 years now. They are a subset of the total mononuclear cell population that possesses an enhanced potential for differentiation within the endothelial-cell lineage [18]. Typically, EPCs exist in both bone marrow and systemic circulation and are selected via the expression of $CD34^+$ kinase insert domain receptor markers. They have been shown to predict independently the progression of coronary artery disease and death from cardiovascular causes and might, therefore, be a useful tool for risk stratification [19]. EPCs participate in vascular growth and regeneration, and their potential use for treatment of myocardial injury or disease has been evaluated in experimental studies [18]. In a murine model of myocardial infarction, EPCs contributed to neoangiogenesis by incorporation into the endothelium of small vessels in the infarct border zone [20]. Further evidence for the ability of EPCs to regenerate the myocardium arises from reports that they can differentiate into cardiomyocytes when cocultured with cardiomyocytes [21]. However, these observations must be interpreted with caution and further clinical investigations are needed.

Skeletal Myoblasts

In contrast with cardiac muscle, skeletal muscle has the capacity for self-repair because of a resident population of skeletal myoblasts (SMs), avoiding thereby ethical considerations or immunological problems [22]. SMs, once activated, fuse to form new muscle fibers [23]. Preclinical data from a variety of experimental studies have demonstrated the capacity for myoblasts to engraft and form striated myotubules, after their intramural implantation or arterial delivery. Subsequently, they enhance the function of the damaged myocardium after transplantation into infarcted myocardium [24,25]. Lately, preliminary human studies focusing on patients with ischemic heart disease indicate successful myoblast transplantation into the scar of the infarcted myocardium [26,27]. However, skeletal myoblasts fail to build classic cell-to-cell connections with cardiomyocytes [28] and they have arrhythmiogenic potential, which has been reported in several trials [26].

EXPERIMENTAL STUDIES/BASIC RESEARCH

To date, several experimental studies (Table 3.4.1) have been performed in the settings of advance atherosclerosis, using transplanted cells in an attempt to repair damaged myocardium [29]. Previous studies in the animal models have suggested that BMCs or EPCs are mainly used because of their easy availability and safety, although the capacity of BMCs to transdifferentiate into cardiac muscle for myocardiogenesis is either lacking or minimal [30]. ESCs transplantation in animal models (mice, rats) has revealed an improvement in cardiac structure and function after their differentiation into cardiomyocytes with better results when compared with the bone marrow-derived cells [31–33]. Data from experimental studies in first Department of Cardiology of the University of Athens have revealed that the direct administration of EPCs derived from BM, such as their indirect mobilization with granulocyte-colony stimulating factor (G-CSF) administration reduced the levels of proinflammatory factors and improved the stability of the atheromatic plaque [34,35]. Similarly, administration of mesenchymal stem cells onto scarred myocardium repaired cardiac performance to nearly normal level [36,37]. Moreover, autologous transplantation of CSCs with the main advantage of the elimination of the potential for a potent immunological response against the allograft in the host haw turn CSCs into a convenient source for autologous stem-cell therapy [38,39]. Initial studies of CSCs demonstrated that they were located in the necrotic area and not the spared myocardium, even though the cells were injected at the border between the ischemic and healthy myocardium [3]. The injured myocardium provides a milieu that

Table 3.4.1 Most relevant experimental studies using stem cell therapy

Study	Animal model	Stem cell type	Outcome	References
Min et al.	Rat	EPCs	• Differentiation into CMs • Improved cardiac function	[31]
Singla et al.	Mouse	EPCs	• Differentiation into CMs, VSMC, and ECs • Improved cardiac structure and function	[32]
Kolossov et al.	Mouse	BMCs/EPCs	• Differentiation into CMs after ESCs injection • NS effect after BMCs injection	[33]
Mangi et al.	Rat	MSCs	• Reduce intramyocardial inflammation and hypertrophy of CMs • Regeneration of 80%–90% of lost myocardial volume	[36]
Miyahara et al.	Rat	MSCs	• Formation of new vessels and few CMs • Triggered angiogenesis through paracrine pathways • Improved cardiac function	[37]
Ellison et al.	Mouse	CSCs	• Homing in damaged myocardium • Regeneration of lost CMs • Improved cardiac function	[40]
Bailey et al.	Mouse	C-kit$^+$ CSCs	• Improved myocardial remodeling • Reduction in the cell death rate • Induction of endogenous repair mechanisms • Differentiation in CMs	[42]
Wang et al.	Mouse	Sca-1$^+$ CSCs	• Significant angiogenesis in peri-infarction regions, • Minimal differentiation into ECs • Low rate of transdifferentiation into CMs • Attenuation of adverse structural and functional abnormalities through paracrine effects	[43]

BMCs, bone marrow-derived stem cells; *CMs*, cardiomyocytes; *CSCs*, cardiac stem cells; *EPCs*, endothelial progenitor cells; *ESCs*, endothelial cells; *MSCs*, mesenchymal stem cells; *NS*, no significant; *VSMCs*, vascular smooth muscle cells.

supports the homing, nesting, survival, and differentiation of CSCs [40]. However, CSCs are significantly more difficult to harvest and isolate than bone marrow-derived mononuclear cells [39]. Furthermore, the number of cells that can be obtained from an individual is quite small; however, these cells can be expanded in culture [38]. Thus, it is feasible that cells isolated from very small fragments of human myocardium and expanded many-fold in vitro could reach numbers appropriate for in vivo transplantation while retaining their differentiation potential [41]. Among the different CSCs subtypes, c-kit$^+$ CSCs have been more frequently utilized in experimental studies because they are cell populations with well-defined characteristics and more accurate extraction and expansion methods. Moreover, compared to the other adult stem cells in the heart, they appear to be in the most primitive stage of differentiation. Experimental models of myocardial infarction have demonstrated a reduction in the cell death rate, induction of endogenous repair mechanisms, and the contribution of new cells to the injured myocardium, resulting in satisfactory improvement of myocardial remodeling [42]. Finally, when cells from the Sca-1$^+$ CSCs subtype were transplanted in a mouse cardiac-infarction model, significant angiogenesis was observed in the peri-infarction regions. However, the absolute number of newly differentiated cardiomyocytes was low and most likely insufficient to make substantial direct contributions to left ventricular structure and function [43].

The potential mechanisms of stem cell therapy in cardiac repair are represented in a schematic way in Fig. 3.4.2.

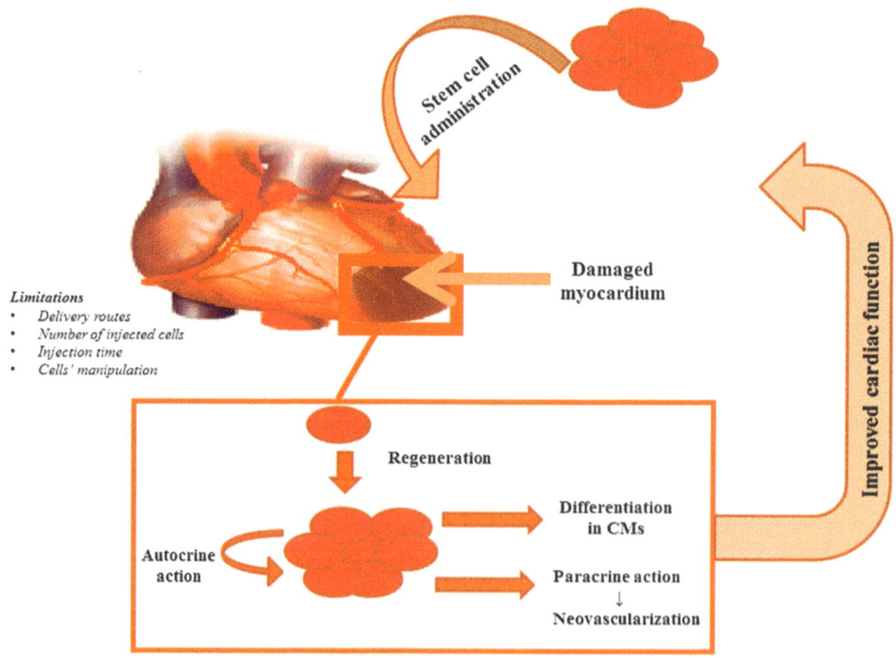

Figure 3.4.2 Schematic representation of the potentials mechanisms of stem cells involved in heart repair. *CMs*, cardiomyocytes.

CLINICAL STUDIES

The capacity of stem cells to differentiate and the large body of preclinical evidence motivated the translation of the stem cell therapeutic approach into human clinical studies and made cell replacement therapy a promising addition to heal the damaged heart [44,45]. Transplantation of ESCs, BMCs, SMs, and EPCs was employed to find the optimal cell type for cardiac repair [46–48]. The results from large clinical trials albeit the good safety profiles have demonstrated divergent findings in cardiac function in patients with myocardial infarction [49–55]. The most important are summarized in Table 3.4.2. Different methods have been employed to deliver stem cells to the damaged myocardium. The most frequent are intracoronary and intramyocardial approaches with diverse advantages and disadvantages for each one (Table 3.4.3). Most of the studies have used intracoronary delivery as preferred administration route [49]. Cells are delivered over the lumen of an inflated over the wire balloon catheter placed in the reopened coronary artery after percutaneous coronary intervention [56]. The direct delivery to the infarcted myocardium and the relatively low invasiveness are the main advantages. Direct intramyocardial injection of cells has also been used, especially in models of chronic ischaemic cardiomyopathy, and facilitates cell delivery despite the potential arrhythmiogenic effects, and the need for concomitant open heart surgery [57]. Most of the studies have resulted in beneficial outcomes for the cardiac function. The largest experience in the field of stem cell transplantation is the placebo controlled, double-blind, multicenter REPAIR-AMI (Reinfusion of Enriched Progenitor cells And Infarct Remodeling in Acute Myocardial Infarction) trial which enrolled 204 patients [56]. Stem cell transplantation resulted in an increase of left ventricular ejection fraction (LVEF) of 2.5% after 4 months, whereas at 1 year a significant reduction in the prespecified combined clinical end point of death, recurrent myocardial infarction, and revascularization was observed. Interestingly, patients with lower baseline EF experienced more pronounced beneficial effects [56]. In the Doppler substudy of the REPAIR-AMI, microvascular function of the infarct-related artery was restored after intracoronary transplantation of BMCs in patients with reperfused acute myocardial infarction [58]. Another randomized controlled trial with 60 patients who had sustained an ST-elevation myocardial infarction (STEMI) with subsequent percutaneous reperfusion, the bone marrow transfer to enhance ST-elevation infarct regeneration trial, had similar outcomes with an improvement in global LVEF in 6 months as assessed by cardiac magnetic resonance imaging (MRI) [59]. However, these promising results were not sustained in a longer follow-up of 18 months, where no significant differences in contractile performance could be found between the two groups, raising concerns that BM transplantation may not translate into permanent myocardial repair [60]. In the Autologous Stem cell Transplantation in Acute Myocardial Infarction trial with 100 patients, treatment with BMCs did not alter either left ventricular function or infarct size or scar tissue remodeling [61]. Similarly, LEUVEN-AMI was also a randomized, double-blind, placebo-controlled study in 67 patients. Compared

Table 3.4.2 Most important randomized controlled stem cell therapy trials

Study	Stem cell type	Patients	Imaging modality	Outcome	References
Chen et al.	MSCs intracoronary	69 (34/35)	SPECT Echo CEM	- LVEF improvement - Retained result in follow-up (3 months) - No deaths or malignant arrhythmias	[50]
Li et al.	PBSCs intracoronary	70 (35/35)	Echo	- LVEF improvement in 6 months - No deaths	[51]
Choi et al.	PBSCs intracoronary	73 (10/62)	Echo MRI	- NS improvement in LVEF - Safe, feasible	[52]
Gyöngyösi et al. (MYSTAR study)	BMCs intramyocardial and intracoronary	60	Echo	- Moderate improvement in myocardial infarct size and LVEF - Safe, feasible	[53]
Meluzín et al.	BMCs intracoronary	60 (40/20)	Echo SPECT	- Short-term LVEF improvement, partially lost during the 12-month follow-up - Safe	[54]
Krause et al.	BMCs intramyo-cardially	20	Echo CA	- LVEF and electromechanical parameters improvement after 12-months - Safe	[55]
REPAIR-AMI	BMCs intracoronary	101/103	Echo	- LVEF improvement - Sustained results after 12 months	[56]
BOOST trial	BMCs intracoronary	60 (30/30)	MRI	- LVEF improvement after 6 months - Reduced infarct size - NS in scar tissue remodeling - Not sustained results after 18 months	[59]
ASTAMI	BMCs intracoronary	97 (47/50)	SPECT Echo MRI	- NS improvement in LVEF or infarct size	[61]

LEUVEN-AMI	BMCs intracoronary	67 (33/34)	MRI	- LVEF improvement - Reduced infarct size - Not sustained results after 4 months	[62]
SCIPIO	CSCs	33 (20/13)	MRI	- LVEF improvement after 12 months - Reduced infarct size - Improvement in NYHA class - Improvement in QoL - Safe, feasible	[63]
CADUCEUS	CDCs	25 (17/8)	MRI	- Scar mass reduction - Increased in regional contractility at 6-month - NS improvement in LVEF - Safe,feasible	[64]

ASTAMI Study, autologous stem cell transplantation in acute myocardial infarction study; *BMCs*, bone marrow-derived cells; *BOOST Study*, bone marrow transfer to enhance ST-elevation infarct regeneration study; *CA*, coronary angiography; *CADUCEUS Study*, cardiosphere-derived autologous stem cell store verse ventricular dysfunction study; *CDCs*, cardiosphere-derived stem cells; *CEM*, cardiac electromechanical mapping; *CSCs*, cardiac stem cells; *LVEF*, left ventricular ejection fraction; *MRI*, magnetic resonance imaging; *MSCs*, mesenchymal stem cells; *MYSTAR Study*, Myocardial stem cell administration after acute myocardial infarction study; *NS*, no significant; *NYHA classification*, New York Heart Association functional classification; *PBSCs*, peripheral blood stem cells; *QoL*, quality of life; *REPAIR-AMI Study*, reinfusion of enriched progenitor cells and infarct remodeling in acute myocardial infarction study; *SCIPIO Study*, stem cell infusion in patients with ischemic cardiomyopathy; *SPECT*, single positron emission computer tomography; *TDI*, tissue doppler imaging.

Table 3.4.3 Characteristics of the two frequent administration routes of stem cell therapy

Characteristics	Intracoronary delivery	Intramyocardial delivery
Ease to perform	++	+
Unlimited amount of injectable cells	+	−
Unlimited size of transplanted cells	−	+
Intramyocardial accumulation of the implanted cells	±	+
Quick washout	+	−
Temporary occlusion of the artery	+	−
Myocardial perforation	−	+

with placebo, BMCs transfer was associated with a significant reduction in infarct size and a better recovery of regional systolic function; however, there was no meaningful difference in LVEF after 4 months [62]. Similar results were also obtained with the use of CSCs as a source of cell therapy. The Stem Cell Infusion in Patients with Ischemic cardiomyopathy study, the first study of CSC treatment in humans, received approval to conduct a phase 1 clinical trial using CSCs [63]. The LVEF was improved by 8% at 4 months after stem cell administration, in contrast to no change in the control patients. In addition, the increase in LVEF was even higher 1 year after the administration of CSCs, suggesting that the CSCs continued to improve LVEF overtime. Furthermore, this improvement was coupled with a concomitant decrease in infarct size and an increase in the left ventricle viable mass, implying the robust regeneration of myocardial tissue. Importantly, the functional benefits of CSCs were inversely related to the baseline functional status of the myocardial region. That is, the lower the baseline function, the greater the improvement afforded by CSCs infusion. This study also demonstrated the feasibility of isolating and expanding CSCs obtained from cardiac tissue during coronary artery bypass grafting surgery [63]. The subtypes of CSCs, cardiosphere-derived CSCs (CDCs), were studied in two different randomized trials, the Cardiosphere-derived autologous stem cell store versus ventricular dysfunction and the ALLSTAR (Allogeneic heart stem cells to achieve myocardial regeneration), using autologous and allogenic CDCs, respectively [64,65]. The first one revealed the feasibility and safety of this cell therapy with no significant improvement in LVEF, whereas the results of the ongoing ALLSTAR study are expected. More recently, a phase I/II randomized, double-blind, placebo-controlled clinical trial was proposed by a European study group initiative [Safety and Efficacy Evaluation of Intracoronary Infusion of Allogeneic Human Cardiac Stem Cells in Patients With AMI (CAREMI)] to investigate the therapeutic role of allogeneic CSCs in patients with congestive heart failure due to myocardial infarction. Despite the favorable results observed in some individuals treated with stem cells, certain clinical expectations were not achieved because the levels of improvement in cardiac tissue structure or functionality were less than those previously observed in experimental trials [66]. Beyond these landmark studies, there are more than 20 smaller trials in the field of stem cell transplantation currently published. Even in those with a positive outcome, the small increase in LVEF and its clinical significance remain an important issue. Nevertheless, it should be mentioned that many of the interventions with an evidence-based lifesaving effect in myocardial infarct, including therapy with b-blockers and angiotensin converting enzyme inhibitors, also provide only moderate yet clinically meaningful increases in LVEF. Albeit several challenges, such as optimization of the therapeutic protocols and technological improvement of stem cell delivery devices, it is believed that the current ongoing studies will accelerate the further development and acceptance of stem cell therapy and the optimal selection of stem cell type.

LIMITATIONS BEFORE CLINICAL APPLICABILITY

Despite the magnitude of benefit demonstrated in animal studies, it remains unclear why these trials have failed to translate preclinical expectations in humans. Clinical trials have shown only a modest or negligible improvement of cardiac function [39]. Various explanations for this discrepancy have been suggested. First, the rather small number and the selection of patients, and their preparation have to be considered. Second, factors related to the cell administration, such as the injection of an insufficient number of cells, the biodistribution of injected cells, the use of wrong cell type, the extent of cellular differentiation of these cells within the target area, the use of an inappropriate delivery method, or possible inappropriate timing relative to the age of the infarct [67]. The use of proper cell type is another serious matter. Despite all efforts, the specific cell types conferring therapeutic effects have not been yet defined [68,69]. Intense debate is focused on definition of specific markers of endothelial and other vascular progenitor cells. Moreover, differences between animal models and humans, including the absence of comorbidities in the animal model, should be considered [39]. Furthermore, many of the clinical studies suffer from methodological limitations; lack of randomization and blinding, or use of methods other than MRI, the gold standard for serial cardiac functional and structural evaluation, while it is difficult to compare individual studies as they differ in methodology [70]. Last, results of chronic postinfraction studies are less supportive for stem cell therapy than acute myocardial infarction studies. Compared to chronic disease, several trials in acute phase of myocardial infarction found that cell treatment significantly improved left ventricular function [71]. The effect, however, was transient. So far only REPAIR-AMI study reported improved left ventricular contractility at 1 year after cell therapy [72].

FUTURE PERSPECTIVES

The traditional concept of heart muscle as a postmitotic organ has been abandoned and the contribution of stem cells to atherosclerosis progression and vascular remodeling after injury in humans is established. Damaged myocardium can regenerate from the stem cells that enter the cell cycle, contributing to changes in the myocardial mass. Early animal studies reported very promising results for stem cell therapy, with robust remasculinization. However, despite intensive research, the principal mechanisms have not yet been fully elucidated. Moreover, there is an ongoing debate in the field of basic stem cell research as to which cell type is more suitable. This lack of knowledge translates into the currently available clinical trials, which were only partially able to reproduce the groundbreaking findings from animal studies. There is a need to put even more efforts in order to clarify the principal regenerative properties of different stem cells and to develop promising approaches for their clinical application. It is important to emphasize in specific patient-related factors that may have an impact on cell therapy outcomes, such as severity of atherosclerosis or cardiovascular risk factors. Although cell therapy holds therapeutic promise for clinical applicability, further clinical trials powered to detect differences in mortality and to identify the most beneficial approaches are warranted in order conclude whether stem cell transplantation will eventually join the standard therapeutic armamentarium and improve both the mortality rate and quality of life.

HIGHLIGHTS

- It has been proven that the myocardium has regenerative capacity and the contribution of stem cells to atherosclerosis progression and vascular remodeling after injury in humans is established.
- Potential sources for stem cell therapy in damaged myocardium are embryonic stem cells, adult BMCs, adult CSCs, EPCs, and SMs.
- Results from numerous studies in basic research have signalized the significant potential of stem cell-based therapy in advance atherosclerotic lesions.
- Clinical trials have shown good safety profile but only transient results in cardiac function, and further basic research is necessary in order to clarify the principal cellular mechanisms and develop promising approaches with clinical applicability.

REFERENCES

[1] Murray CJ, Lopez AD. Global mortality, disability, and the contribution of risk factors: Global Burden of Disease Study. Lancet 1997;349(9063):1436–42.
[2] Korbling M, Estrov Z. Adult stem cells for tissue repair – a new therapeutic concept? N Engl J Med 2003;349(6):570–82.
[3] Beltrami AP, et al. Adult cardiac stem cells are multipotent and support myocardial regeneration. Cell 2003;114(6):763–76.
[4] Bearzi C, et al. Human cardiac stem cells. Proc Natl Acad Sci USA 2007;104(35):14068–73.
[5] Barile L, et al. Endogenous cardiac stem cells. Prog Cardiovasc Dis 2007;50(1):31–48.
[6] Dotsenko O. Stem/Progenitor cells, atherosclerosis and cardiovascular regeneration. Open Cardiovasc Med J 2010;4:97–104.

[7] Klug MG, et al. Genetically selected cardiomyocytes from differentiating embronic stem cells form stable intracardiac grafts. J Clin Invest 1996;98(1):216–24.
[8] Caspi O, et al. Transplantation of human embryonic stem cell-derived cardiomyocytes improves myocardial performance in infarcted rat hearts. J Am Coll Cardiol 2007;50(19):1884–93.
[9] Zhang J, et al. Functional cardiomyocytes derived from human induced pluripotent stem cells. Circ Res 2009;104(4):e30–41.
[10] Kehat I, et al. Human embryonic stem cells can differentiate into myocytes with structural and functional properties of cardiomyocytes. J Clin Invest 2001;108(3):407–14.
[11] Barile L, et al. Cardiac stem cells: isolation, expansion and experimental use for myocardial regeneration. Nat Clin Pract Cardiovasc Med 2007;(4 Suppl. 1):S9–s14.
[12] Smith RR, et al. Stem cells in the heart: what's the buzz all about?—Part 1: preclinical considerations. Heart Rhythm 2008;5(5):749–57.
[13] Tosh D, Slack JM. How cells change their phenotype. Nat Rev Mol Cell Biol 2002;3(3):187–94.
[14] Di Felice V, et al. Cardiac stem cell research: an elephant in the room? Anat Rec Hob 2009;292(3): 449–54.
[15] Leite CF, et al. Multipotent stem cells of the heart-do they have therapeutic promise? Front Physiol 2015;6:123.
[16] Urbanek K, et al. Myocardial regeneration by activation of multipotent cardiac stem cells in ischemic heart failure. Proc Natl Acad Sci USA 2005;102(24):8692–7.
[17] Dawn B, et al. Cardiac stem cells delivered intravascularly traverse the vessel barrier, regenerate infarcted myocardium, and improve cardiac function. Proc Natl Acad Sci U. S. a 2005;102(10): 3766–71.
[18] Sekiguchi H, Ii M, Losordo DW. The relative potency and safety of endothelial progenitor cells and unselected mononuclear cells for recovery from myocardial infarction and ischemia. J Cell Physiol 2009;219(2):235–42.
[19] Werner N, et al. Circulating endothelial progenitor cells and cardiovascular outcomes. N Engl J Med 2005;353(10):999–1007.
[20] Asahara T, et al. Bone marrow origin of endothelial progenitor cells responsible for postnatal vasculogenesis in physiological and pathological neovascularization. Circ Res 1999;85(3):221–8.
[21] Badorff C, et al. Transdifferentiation of blood-derived human adult endothelial progenitor cells into functionally active cardiomyocytes. Circulation 2003;107(7):1024–32.
[22] Bischoff R. Regeneration of single skeletal muscle fibers in vitro. Anat Rec 1975;182(2):215–35.
[23] Jain M, et al. Cell therapy attenuates deleterious ventricular remodeling and improves cardiac performance after myocardial infarction. Circulation 2001;103(14):1920–7.
[24] Pouzet B, et al. Factors affecting functional outcome after autologous skeletal myoblast transplantation. Ann Thorac Surg 2001;71(3):844–50. discussion 850–1.
[25] Pagani FD, et al. Autologous skeletal myoblasts transplanted to ischemia-damaged myocardium in humans. Histological analysis of cell survival and differentiation. J Am Coll Cardiol 2003;41(5):879–88.
[26] Menasche P, et al. Autologous skeletal myoblast transplantation for severe postinfarction left ventricular dysfunction. J Am Coll Cardiol 2003;41(7):1078–83.
[27] Herreros J, et al. Autologous intramyocardial injection of cultured skeletal muscle-derived stem cells in patients with non-acute myocardial infarction. Eur Heart J 2003;24(22):2012–20.
[28] Menasche P, et al. Myoblast transplantation for heart failure. Lancet 2001;357(9252):279–80.
[29] Tomita S, et al. Autologous transplantation of bone marrow cells improves damaged heart function. Circulation 1999;100(19 Suppl.):II247–56.
[30] Li RK, et al. In vivo survival and function of transplanted rat cardiomyocytes. Circ Res 1996;78(2): 283–8.
[31] Min JY, et al. Long-term improvement of cardiac function in rats after infarction by transplantation of embryonic stem cells. J Thorac Cardiovasc Surg 2003;125(2):361–9.
[32] Singla DK, et al. Transplantation of embryonic stem cells into the infarcted mouse heart: formation of multiple cell types. J Mol Cell Cardiol 2006;40(1):195–200.
[33] Kolossov E, et al. Engraftment of engineered ES cell-derived cardiomyocytes but not BM cells restores contractile function to the infarcted myocardium. J Exp Med 2006;203(10):2315–27.

[34] Tousoulis D, et al. Effects of intramuscular administration of hematopoietic progenitor and co-administration of erythropoietin on blood perfusion, angiogenic factors and vascularization in a murine model of limb ischemia. Int J Cardiol 2013;169(3):229–31.
[35] Tousoulis D, et al. Effects of direct infusion of bone marrow-derived progenitor cells and indirect mobilization of hematopoietic progenitor cells on atherosclerotic plaque and inflammatory process in atherosclerosis. Int J Cardiol 2013;168(5):4769–74.
[36] Mangi AA, et al. Mesenchymal stem cells modified with Akt prevent remodeling and restore performance of infarcted hearts. Nat Med 2003;9(9):1195–201.
[37] Miyahara Y, et al. Monolayered mesenchymal stem cells repair scarred myocardium after myocardial infarction. Nat Med 2006;12(4):459–65.
[38] Gaetani R, et al. Differentiation of human adult cardiac stem cells exposed to extremely low-frequency electromagnetic fields. Cardiovasc Res 2009;82(3):411–20.
[39] Lovell MJ, Mathur A. Republished review: cardiac stem cell therapy: progress from the bench to bedside. Postgrad Med J 2011;87(1030):558–64.
[40] Ellison GM, et al. Adult c-kit(pos) cardiac stem cells are necessary and sufficient for functional cardiac regeneration and repair. Cell 2013;154(4):827–42.
[41] Messina E, et al. Isolation and expansion of adult cardiac stem cells from human and murine heart. Circ Res 2004;95(9):911–21.
[42] Bailey B, et al. Cardiac stem cell genetic engineering using the alphaMHC promoter. Regen Med 2009;4(6):823–33.
[43] Wang X, et al. The role of the sca-1+/CD31- cardiac progenitor cell population in postinfarction left ventricular remodeling. Stem Cells 2006;24(7):1779–88.
[44] Orlic D, et al. Bone marrow cells regenerate infarcted myocardium. Nature 2001;410(6829):701–5.
[45] Leri A, Kajstura J, Anversa P. Role of cardiac stem cells in cardiac pathophysiology: a paradigm shift in human myocardial biology. Circ Res 2011;109(8):941–61.
[46] Taylor DA, et al. Regenerating functional myocardium: improved performance after skeletal myoblast transplantation. Nat Med 1998;4(8):929–33.
[47] Power C, Rasko JE. Promises and challenges of stem cell research for regenerative medicine. Ann Intern Med 2011;155(10):706–13. w217.
[48] Strauer BE, Steinhoff G. 10 years of intracoronary and intramyocardial bone marrow stem cell therapy of the heart: from the methodological origin to clinical practice. J Am Coll Cardiol 2011;58(11):1095–104.
[49] Strauer BE, et al. Intracoronary, human autologous stem cell transplantation for myocardial regeneration following myocardial infarction. Dtsch Med Wochenschr 2001;126(34–35):932–8.
[50] Chen SL, et al. Improvement of cardiac function after transplantation of autologous bone marrow mesenchymal stem cells in patients with acute myocardial infarction. Chin Med J Engl 2004;117(10):1443–8.
[51] Li ZQ, et al. The clinical study of autologous peripheral blood stem cell transplantation by intracoronary infusion in patients with acute myocardial infarction (AMI). Int J Cardiol 2007;115(1):52–6.
[52] Choi JH, et al. Lack of additional benefit of intracoronary transplantation of autologous peripheral blood stem cell in patients with acute myocardial infarction. Circ J 2007;71(4):486–94.
[53] Gyongyosi M, et al. Combined delivery approach of bone marrow mononuclear stem cells early and late after myocardial infarction: the MYSTAR prospective, randomized study. Nat Clin Pract Cardiovasc Med 2009;6(1):70–81.
[54] Meluzin J, et al. Three-, 6-, and 12-month results of autologous transplantation of mononuclear bone marrow cells in patients with acute myocardial infarction. Int J Cardiol 2008;128(2):185–92.
[55] Krause K, et al. Percutaneous intramyocardial stem cell injection in patients with acute myocardial infarction: first-in-man study. Heart 2009;95(14):1145–52.
[56] Schachinger V, et al. Intracoronary bone marrow-derived progenitor cells in acute myocardial infarction. N Engl J Med 2006;355(12):1210–21.
[57] Dib N, et al. Safety and feasibility of autologous myoblast transplantation in patients with ischemic cardiomyopathy: four-year follow-up. Circulation 2005;112(12):1748–55.

[58] Erbs S, et al. Restoration of microvascular function in the infarct-related artery by intracoronary transplantation of bone marrow progenitor cells in patients with acute myocardial infarction: the Doppler Substudy of the Reinfusion of Enriched Progenitor Cells and Infarct Remodeling in Acute Myocardial Infarction (REPAIR-AMI) trial. Circulation 2007;116(4):366–74.
[59] Wollert KC, et al. Intracoronary autologous bone-marrow cell transfer after myocardial infarction: the BOOST randomised controlled clinical trial. Lancet 2004;364(9429):141–8.
[60] Meyer GP, et al. Intracoronary bone marrow cell transfer after myocardial infarction: eighteen months' follow-up data from the randomized, controlled BOOST (BOne marrOw transfer to enhance ST-elevation infarct regeneration) trial. Circulation 2006;113(10):1287–94.
[61] Lunde K, et al. Intracoronary injection of mononuclear bone marrow cells in acute myocardial infarction. N Engl J Med 2006;355(12):1199–209.
[62] Janssens S, et al. Autologous bone marrow-derived stem-cell transfer in patients with ST-segment elevation myocardial infarction: double-blind, randomised controlled trial. Lancet 2006;367(9505): 113–21.
[63] Chugh AR, et al. Administration of cardiac stem cells in patients with ischemic cardiomyopathy: the SCIPIO trial: surgical aspects and interim analysis of myocardial function and viability by magnetic resonance. Circulation 2012;126(11 Suppl. 1):S54–64.
[64] Makkar RR, et al. Intracoronary cardiosphere-derived cells for heart regeneration after myocardial infarction (CADUCEUS): a prospective, randomised phase 1 trial. Lancet 2012;379(9819):895–904.
[65] Chakravarty T, et al. ALLogeneic Heart STem Cells to Achieve Myocardial Regeneration (ALLSTAR) trial: Rationale & Design. Cell Transpl 2016;26(2):205–14.
[66] Crisostomo V, et al. Allogeneic cardiac stem cell administration for acute myocardial infarction. Expert Rev Cardiovasc Ther 2015;13(3):285–99.
[67] Hoshino K, et al. In vivo tracking in cardiac stem cell-based therapy. Prog Cardiovasc Dis 2007;49(6): 414–20.
[68] Yoder MC. Defining human endothelial progenitor cells. J Thromb Haemost 2009;(7 Suppl. 1): 49–52.
[69] Sugiyama S, et al. Characterization of smooth muscle-like cells in circulating human peripheral blood. Atherosclerosis 2006;187(2):351–62.
[70] Dimmeler S, Burchfield J, Zeiher AM. Cell-based therapy of myocardial infarction. Arterioscler Thromb Vasc Biol 2008;28(2):208–16.
[71] Martin-Rendon E, et al. Autologous bone marrow stem cells to treat acute myocardial infarction: a systematic review. Eur Heart J 2008;29(15):1807–18.
[72] Schachinger V, et al. Improved clinical outcome after intracoronary administration of bone-marrow-derived progenitor cells in acute myocardial infarction: final 1-year results of the REPAIR-AMI trial. Eur Heart J 2006;27(23):2775–83.

3.5 Diagnostic Strategies-Novel Imaging Modalities-Multimodalities and Hybrid Techniques

CHAPTER 3.5.1

Non-invasive Imaging Techniques in Coronary Artery Disease

Constantina Aggeli[1], Sofia Mavrogeni[2], Dimitris Tousoulis[1]
[1]1st Cardiology Clinic, 'Hippokration' General Hospital, National and Kapodistrian University of Athens, School of Medicine, Athens, Greece; [2]Onassis Cardiac Surgery Center, Athens, Greece

Contents

Although mortality from coronary heart disease (CHD) has fallen substantially in many countries, CHD remains the leading cause of death in adults in developed countries and is a rapidly increasing cause of death for adults in developing countries.

On the other hand, there is a wide armamentarium of cardiovascular imaging techniques in the evaluation of patients with and without CHD. The transthoracic

Coronary Artery Disease
ISBN 978-0-12-811908-2, http://dx.doi.org/10.1016/B978-0-12-811908-2.00017-9

echocardiography is the oldest and the most widely used technique in the evaluation of patients with cardiovascular diseases.

One major issue is the appropriate application of cardiovascular imaging modalities in clinical practice. Screening of the general population is not cost effective. As a result it is necessary to identify "at-risk" individuals. Understanding the theory of conditional probability (Bayes' theorem), as applied to medical decision making, is critical in understanding the utility of testing for the diagnosis of CHD in asymptomatic individuals. Knowledge of the pretest probability of disease, along with the sensitivity and specificity of the tests performed, is necessary to determine the predictive value of the testing performed. Risk of CHD can be estimated in any individual via risk scores, and this can be further refined with functional and nonfunctional investigations. There is general agreement regarding the need for investigation of symptomatic patients suspected of CHD and subsequent instigation of therapies. Screening asymptomatic individuals, however, is controversial but potentially allows early detection and more accurate risk estimation. Estimating risk an individual is important not only for implementation of effective management strategies but also for reassurance and psychosocial security. Some asymptomatic patients who have an occupation associated with public safety (e.g., airline pilots, bus drivers, truck drivers) or a high-risk occupation or avocation (e.g., scuba divers) are required by regulatory agencies or governing bodies to have routine testing as part of a periodic health assessment.

Transthoracic echocardiography being a versatile, cost-effective, and nonionizing technique has become an extension of clinical examination. Furthermore, it offers significant diagnostic as well as prognostic data for patients with known or suspected coronary artery disease (CAD). Ultrasound imaging currently does not provide detailed demonstration of the whole coronary vessel anatomy. The proximal left main coronary artery and right coronary artery (RCA) are often depicted particularly in children. It is worth noting that both invasive coronary angiography and lately computed tomographic (CT) coronary angiography remain the imaging modalities of choice for accurate evaluation of the anatomy of coronary artery.

However, echocardiography offers functional assessment of segmental and global left ventricular (LV) systolic function both at rest and using stressors to produce myocardial ischemia. Recording two-dimensional (2D) echocardiographic views, the physician was expected to mentally reconstruct the complex structure of the heart, resulting in geometrical assumptions, which in turn could underestimate the validity of clinical findings. It is very important to point out that three-dimensional echocardiographic (3DE) imaging represents a major innovation in cardiovascular imaging. The usefulness of 3DE has been summarized mainly in the evaluation of cardiac chamber volume and mass, avoiding geometric assumptions; the assessment of regional LV wall motion; presentation of realistic views of heart valves; and volumetric assessment of regurgitant lesions and shunts with three-dimensional (3D) color Doppler imaging.

Transthoracic echocardiography should be the tool evaluating patients with acute or chronic CAD. Regarding the patient with acute precordial pain, transthoracic echocardiographic study should give an accurate and quick view excluding the presence of acute aortic dissection even using the palm—echo machines. At the intensive care unit, the patient with acute myocardial infarction (AMI) should be evaluated with transthoracic echocardiography before and after the primary percutaneous coronary intervention (PCI) or the thrombolytic therapy.

However, the infarcted area appears as slightly/severe hypokinetic or akinetic area leading to slight or severe depression of the LV function. Concerning the severely depressed LV due to extended anterior wall myocardial infarction, the presence of low vortices or thrombus at the apex is very crucial (Fig. 3.5.1.1). Furthermore, using transthoracic echo data on the anatomy and function of the right ventricle (RV) is very useful. Concerning hemodynamic data, the transthoracic study provides valuable information in place of in addition to central line (subclavian) recordings.

It is widely known that coronary anatomy varies from patient to patient. Evaluating the myocardial dysfunction it is very crucial to offer information matching to coronary distribution. Segmental wall motion abnormalities seen by echocardiography correspond closely with the coronary artery blood supply to the myocardium. The left anterior

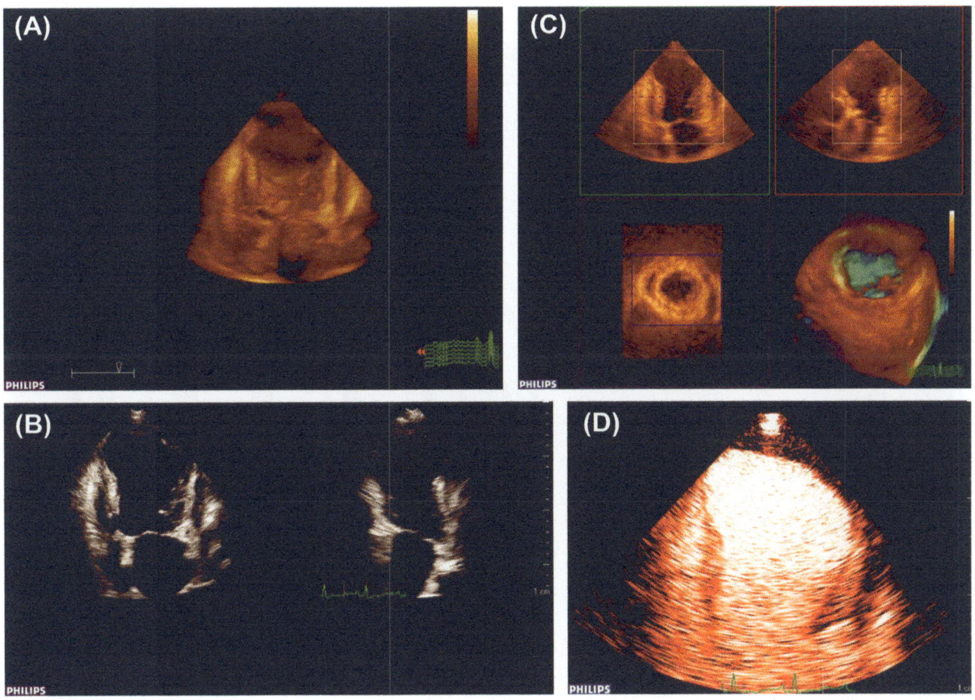

Figure 3.5.1.1 (A—D): Four-chamber view of a patient with extended anterior myocardial infarction using two- and three-dimensional images. The dilation of the apex and the middle portions of the left ventricle, the absence of the myocardial perfusion at these segments, and the presence of thin apical thrombi are shown.

descending artery supplies the anterior portion of the interventricular septum via septal perforating branches and the anterior wall via diagonal branches. The posterior descending artery (PDA) supplies the inferior aspect of the ventricular septum and the inferior free wall. The lateral wall is supplied by obtuse marginal branches of the circumflex artery. The posterior LV wall may be supplied by extension branches from the RCA or by obtuse marginal branches of the circumflex arteries. There is marked individual variability in the blood supply to the LV apex. In some cases, the left anterior descending (LAD) artery extends around the apex to supply the apical segment of the inferior wall. In other cases, the PDA extends around the apex to supply the apical segment of the anterior wall. More commonly, the blood supply to the apex arises from both the LAD and the posterior descending coronary arteries.

According to the American and European Guidelines on Echo, the LV divides into 17 segments to proceed to the CAD. The Wall Motion Score Index (WMSI) is a quantitative index of segmental myocardial systolic function. The scoring system is widely known and includes each of the 17 segments evaluated as normal, 0; hypokinetic, 1; severely hypokinetic, 2; akinetic, 3; and dyskinetic, 4. The sum of all the numbers divided by 17 is the WMSI (Fig. 3.5.1.2).

The presence of acute pericarditis and mechanical complications of AMI are rare in the era of primary PCI. The mechanical complications of AMI are the following: rupture of papillary muscles leading to acute mitral regurgitation (MR), intraventricular rupture, the presence of pseudoaneurysm, or rupture of the free wall. Transthoracic

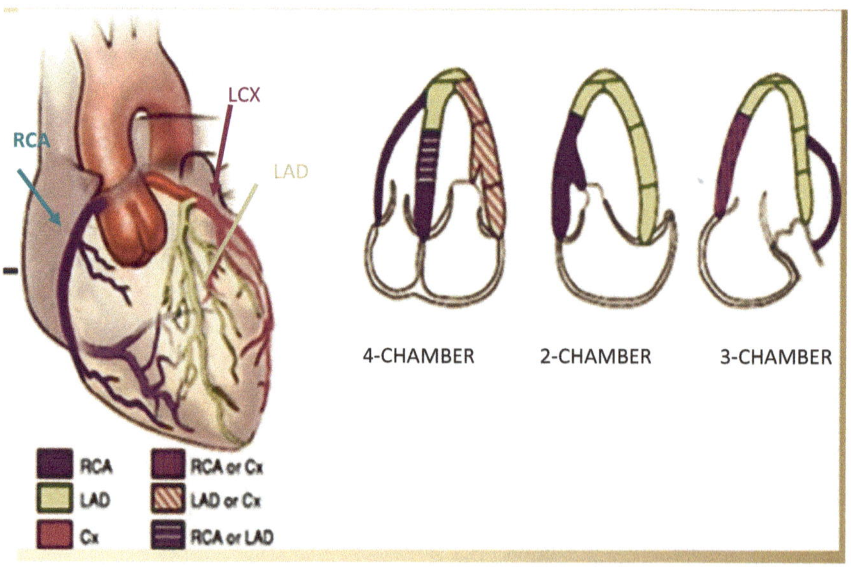

Figure 3.5.1.2 Coronary flow distribution. *Cx*, circumflex; *LAD*, left anterior descending; *RCA*, right coronary artery.

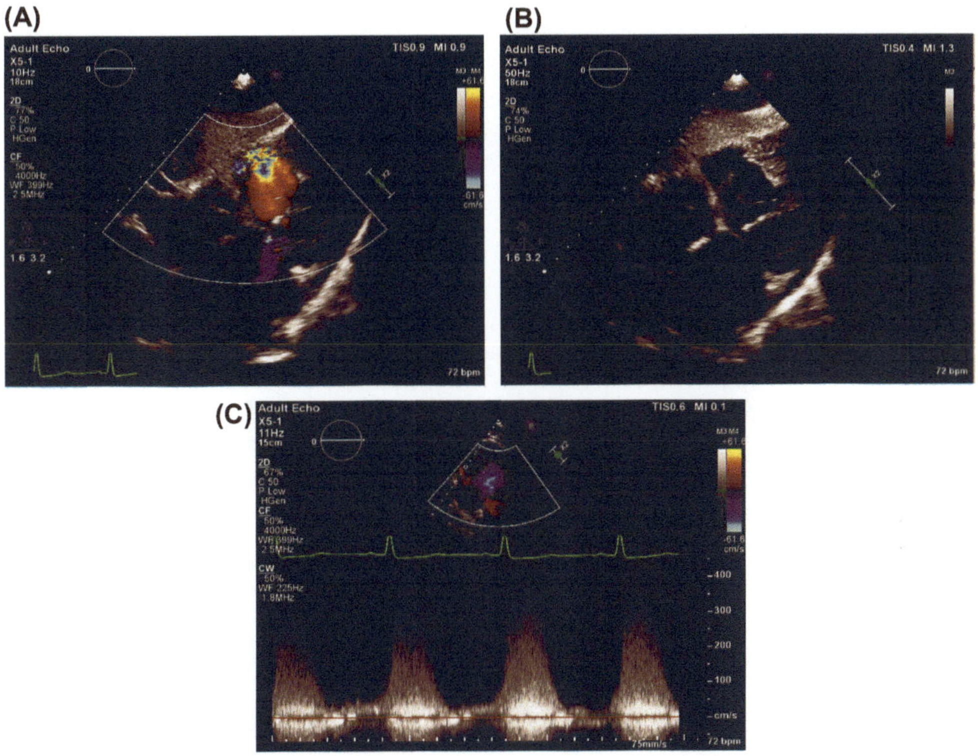

Figure 3.5.1.3 Substernal view depicts the rupture of the intraventricular septum (IVS) with (A) and without color (B). The high velocities derived by continuous Doppler are consisted with the IVS rupture (C).

echocardiography is a unique technique for the evaluation of these severely ill patients. Figs. 3.5.1.3 and 3.5.1.4 demonstrate these complications.

Ischemic MR is a very challenging issue. Two-dimensional echocardiography and/or 3DE should be used in to accurately evaluate the severity of MR (Fig. 3.5.1.5). Most commonly, the cause of the murmur is MR due to papillary muscle dysfunction, abnormal wall motion of the segment underlying a papillary muscle, or papillary muscle rupture. The presence and severity of MR are evaluated using Doppler techniques, and the cause is inferred from 2D or 3D imaging. Ischemic MR typically results from failure of adequate coaptation of the anterior leaflet against a tethered posterior leaflet, which results in a posteriorly directed regurgitant jet. Partial or complete rupture of the papillary muscle is a catastrophic complication that can be recognized as a flail leaflet with an attached mass (the papillary muscle head) that prolapses into the LA in systole.

Dobutamine stress echo (DSE) is a well-established technique for the diagnostic assessment of CAD. The advent of contrast agents has significantly boosted its diagnostic accuracy, as it allows better delineation of endocardial borders, thus improving the

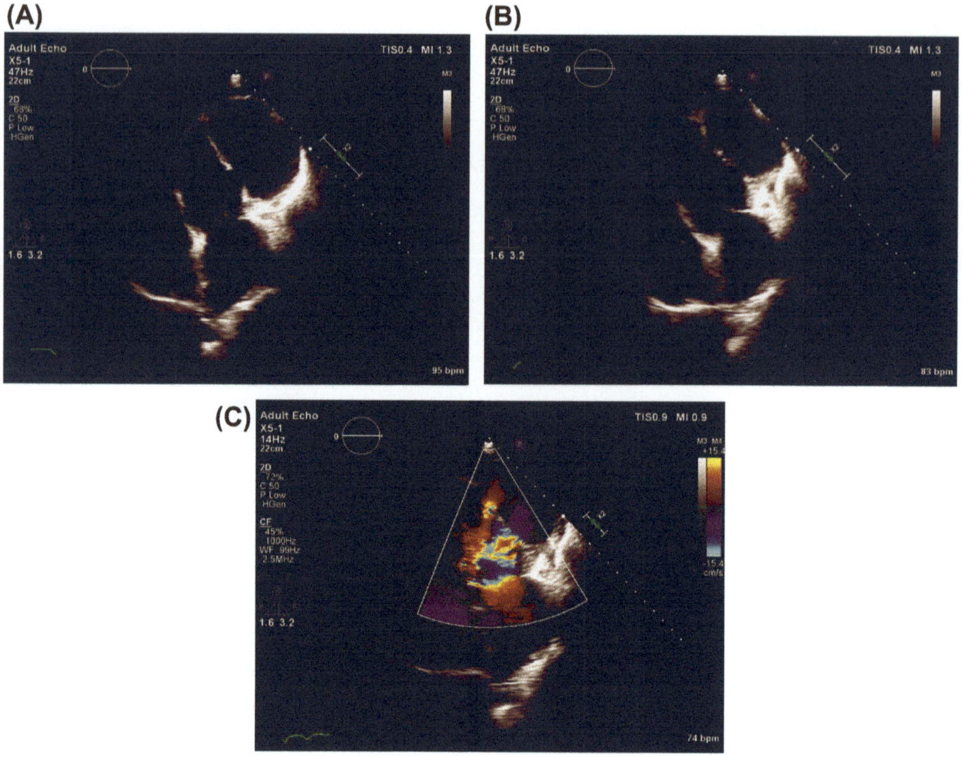

Figure 3.5.1.4 Four-chamber view of patient with a nonrevascularized myocardial infarction of the posterolateral wall. The pseudoaneurysm was depicted at the lateral wall before (A and B) and after (C) using color Doppler.

accuracy of image interpretation. DSE has the unique ability to assess wall motion and perfusion, simultaneously increasing the diagnostic accuracy of the technique (sensitivity 88%, specificity 80%). Moreover, with the aging of population, the prevalence of CAD has significantly increased, rendering stress echocardiography a frequently utilized examination for the assessment of patients with known or suspected CAD.

Therefore, the American Heart Association/American College of Cardiology (AHA/ACC) was motivated to publish the appropriateness criteria as means to avoid unnecessary examinations and consequently large costs, to promote more effective implementation of cardiac imaging in clinical practice.

In addition to the diagnostic value, DSE offers excellent prognostic data for the group of patients with suspected or known CAD. Data stratification according to age illustrates that dobutamine stress contrast echo and appropriateness indication exhibits its most prognostic significance in the middle-aged group. Stress-induced echocardiographic

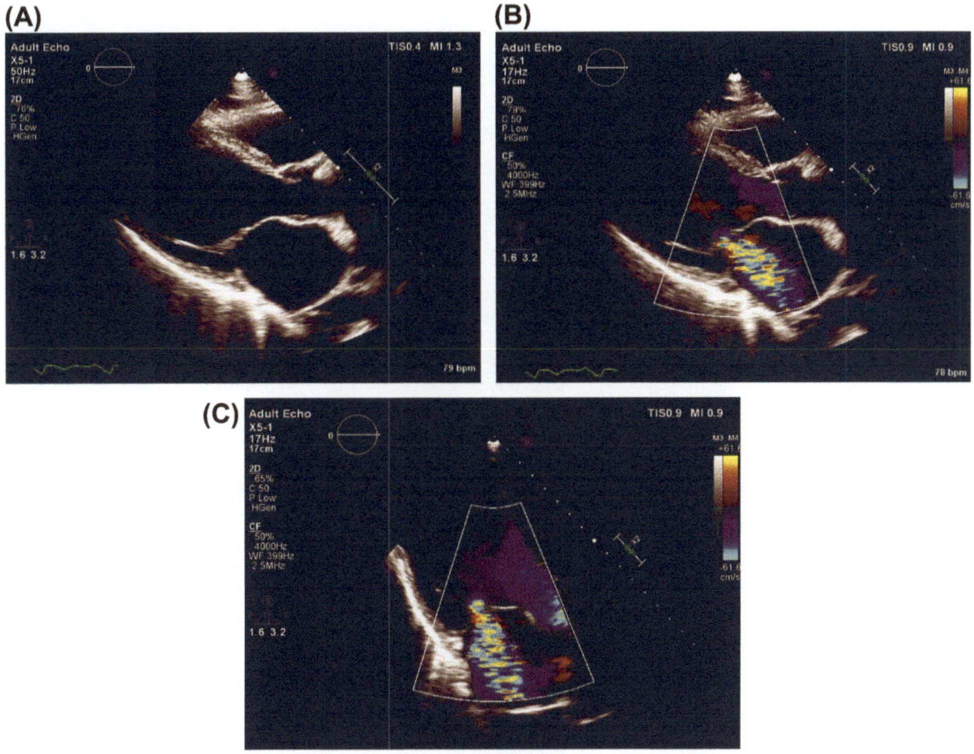

Figure 3.5.1.5 Long axis (A and B) and three-chamber (C) views of a patient with posterolateral infarction leading to severe ischemic mitral valve regurgitation.

abnormalities were not independently associated with increased events in the youngest group. A positive stress echo examination probably was not a major predictor of all-cause mortality in younger patients. This is anticipated since younger patients tend to sustain acute coronary syndromes from vulnerable atheromatous plaques that do not cause major narrowing of epicardial arteries. On the other hand, older patients tend to have more diffuse multivessel disease with major lumen narrowing and various comorbidities, which could affect event rates in a different fashion.

It is worth noting that the diagnostic and prognostic value of each stress test varies according to the age, sex, and presence of risk factors or other comorbidities. Ischemic response based on positive wall motion abnormalities and/or myocardial perfusion defects has high positive prognostic value mainly in middle-aged men and women, in patients with diabetics and hypertension, or in patients with chronic kidney disease (Figs. 3.5.1.6 and 3.5.1.7). Newer echocardiographic techniques increase the diagnostic accuracy and offer useful data to delineate the prognostic value of stress contrast echo (Fig. 3.5.1.8).

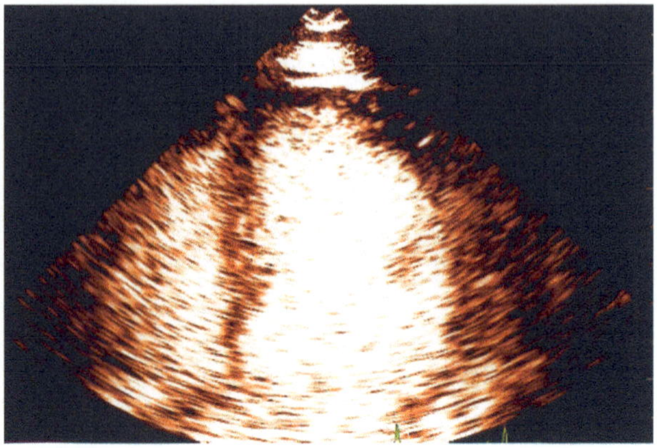

Figure 3.5.1.6 Four-chamber view of the left ventricle depicts the abnormal perfusion of the apex at rest consisting with severe left anterior descending disease.

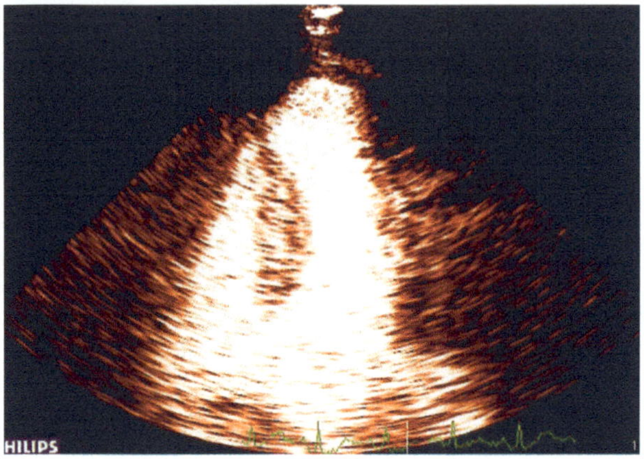

Figure 3.5.1.7 Four-chamber view of patient with ischemic response at the apex and mid-anterolateral wall during dobutamine stress contrast echo. Coronary angiography demonstrated severe mid-left anterior descending lesion.

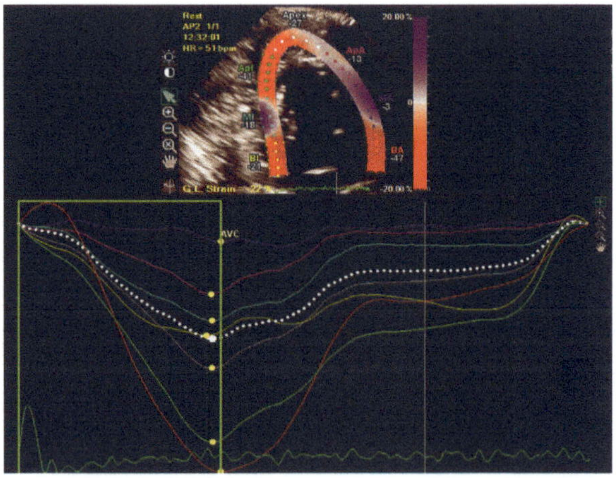

Figure 3.5.1.8 Speckle tracking imaging during of the dobutamine stress echo demonstrating low myocardial deformation at the lateral wall, consistent with myocardial ischemia.

MYOCARDIAL VIABILITY

The presence of myocardial viability in clinical decision making is crucial. Distinguishing irreversible myocardial infarction from viable myocardium that might benefit from reperfusion therapy is important in clinical decision making. Akinesis or hypokinesis at rest may be due to myocardial stunning or hibernation rather than due to irreversible myocardial damage.

The combination of echo parameters increases the accuracy in distinguishing the viable myocardium. The following data are consistent with myocardial viability: myocardial thickening >6 mm, biphasic response to dobutamine infusion, and presence of myocardial perfusion, and speckle tracking of −13% serves as a cutoff point to distinguish viable from nonviable myocardium.

During DSE, the viable myocardium shows an improvement in wall thickening in response to low-dose (2.5–5 to 10 μg/kg/min) dobutamine. At higher dobutamine doses, myocardium supplied by a patent coronary artery shows further enhancement of myocardial thickening, whereas ischemia with hypokinesis or akinesis is seen when significant coronary stenosis is present. This biphasic response—wall motion improvement at low doses and worsening at high doses of dobutamine—identifies areas of myocardium likely to benefit from percutaneous or surgical reperfusion, even late after myocardial infarction with a high sensitivity of 80% and specificity of 78%.

ROLE OF SINGLE-PHOTON EMISSION COMPUTED TOMOGRAPHY IN THE DIAGNOSIS OF CORONARY ARTERY DISEASE

Myocardial perfusion single-photon emission computed tomography (SPECT) is one of the most important and commonly performed noninvasive cardiac imaging tests. SPECT plays a key role in the diagnosis of cardiovascular disease, establishing prognosis, assessing the effectiveness of therapy, and evaluating myocardial viability. The success of this imaging modality has, in large part, been due to advanced technology that continues to improve and expand the field. This includes progressive improvement of SPECT technologies, new radiopharmaceuticals, and new software. Thallium and technetium are the commonly used radioisotopes (Fig. 3.5.1.9).

Stress Protocols

Different imaging protocols can be used. These can be combined with different modalities of stress testing including exercise and mainly vasodilator protocols. The physiologic principle in CAD diagnosis is a challenge to the coronary flow reserve leading to differential uptake of isotope based on regional heterogeneity of flow in stenosed and nonstenosed territories.

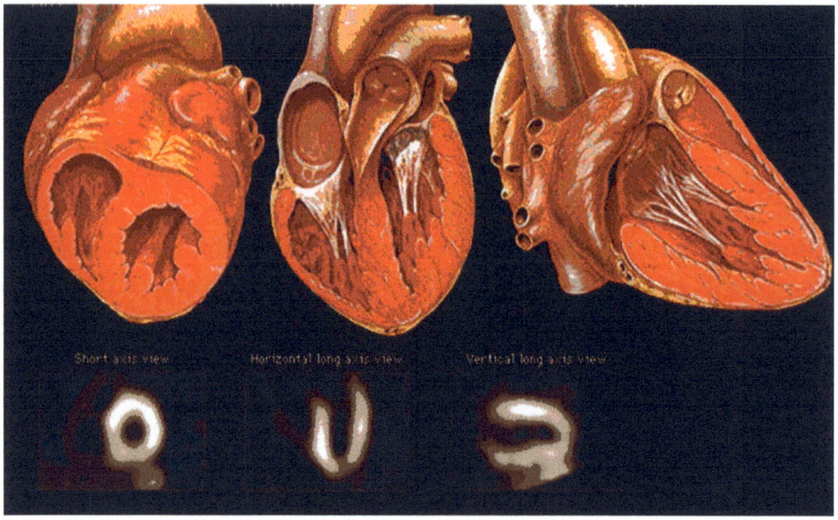

Figure 3.5.1.9 Three standard orthogonal slices displayed for myocardial perfusion imaging (MPI). From left to right: short axis, horizontal long axis, and vertical long axis. Most uptake takes place in left ventricular myocardium due to its thicker wall, therefore MPI generally contains left ventricular information.

The adenosine A2a receptor primarily mediates coronary vasodilation and is the main target of nonselective pharmacologic vasodilators (adenosine, dipyridamole). However, stimulation of A1 receptors by these agents causes atrioventricular (AV) block, and A2b/A3 receptors cause bronchoconstriction. More recently, selective A2a receptor agonists such as regadenoson have become available for pharmacologic SPECT.

In case of contraindication to exercise stress testing, pharmacological stress must be considered. Thalium-201 (201 Tl-chloride) and technetium-99m (99mTC) sestamibi and tetrofosmin are the currently used radiopharmaceuticals. There are several artifacts and interpretation pitfalls that can potentially compromise SPECT study. It is essential for both the technologist and the interpreting physician to be aware of these potential sources of error, take appropriate steps to limit them beforehand, wherever possible correct them if they do occur, and when they cannot be eliminated, recognize their potential impact on the interpretation of the study.

The major indications for SPECT are diagnosis of CAD, risk stratification in patients with known CAD, as well as assessment of therapy and intervention and myocardial viability before percutaneous intervention or bypass surgery. Therefore, close collaboration between the referring and interpreting physicians is essential to ensure appropriate indication for the study, optimal patient preparation, and effective use of the test result in patient clinical management. Currently, there are several appropriate indications for stress myocardial perfusion study; familiarity with appropriate indications is crucial to ensure proper utilization of the test and avoidance of inappropriate use with subsequently excessive cost and radiation.

The indication and finally the choice of the stress modality should be guided by the following principles:

1. Patients who can exercise should exercise in the absence of active angina.
2. In patients with left bundle branch block, vasodilator stress is the preferred mode of stress testing to avoid partially reversible or fixed septal defects on the perfusion image.

Interpretation

The interpretation of myocardial perfusion SPECT images should be a systematic process that routinely includes evaluation of the raw images, perfusion images (location, size, severity, and reversibility of perfusion defects; cardiac chamber sizes; and presence or absence of increased pulmonary uptake), and gated images for global and regional ventricular function.

Perfusion Defects

Myocardial perfusion defects should be identified visual analysis of the reconstructed slices. Perfusion defects should be characterized by their location as they relate to specific myocardial walls, per the 17-segment model. Defect severity is typically expressed qualitatively as mild, moderate, or severe. Defect extent may be qualitatively described as small, medium, or large. Defects whose severity and extent do not change between stress and rest images are categorized as "fixed" or "nonreversible." When perfusion defects are more severe and/or extensive on stress compared with resting images, a qualitative description of the degree of reversibility is required. Reversible defects denote ischemia, whereas fixed defects are the result of scaring or artifacts.

ABNORMAL SINGLE-PHOTON EMISSION COMPUTED TOMOGRAPHIC STUDY

True myocardial perfusion defects are consistent to the following. (1) The defect size or extent (small, medium, and large): small defect represents less than 10%, medium 10%–20%, and large represents greater than or equal 20% of the LV myocardium. (2) Severity of perfusion defect (mild, moderate, and severe): alternatively, the defect size may be estimated as a fraction such as the apical half of the anterior wall. The defect severity is expressed visually as mild, moderate, or severe. A mild defect is identified by a decreased count compared with the adjacent wall with preserved thickness, moderate defect demonstrates wall thinning, and severe defect appears similar to background activity. (3) The extent of reversibility (reversible, irreversible, or reverse redistribution) and (4) the location (based on 17-segment model and coronary artery territory if possible).

Other nonperfusion high-risk markers that should be identified include the following:

- Ventricular dilation: LV enlargement at rest and/or poststress is a high-risk marker. In particular, stress-to-rest LV cavity ratio, also referred to as transient cavity dilatation or transient ischemic dilation, has been described as a marker for high-risk coronary disease. This finding is associated with a higher incidence of multivessel CAD and a worse outcome than in patients not demonstrating this finding.
- Lung uptake: The presence of increased lung uptake after thallium perfusion imaging (lung/heart ratio >0.5) is an indicator of poor prognosis.
- RV uptake and RV dilation: The normal myocardial intensity of the RV is approximately 50% of the peak LV intensity. Increased RV uptake has been associated with the presence of poor long-term outcomes.

Diagnostic Accuracy

The sensitivity for detecting CAD (≥50% stenosis) averages 87% and specificity averages 73%. The normalcy rate is 91% in this pooled analysis. The normalcy rate is a variable that reduces the referral bias inherent in specificity determinations that use patients with angiographically normal coronary arteries at catheterization. The normalcy rate is defined as the percentage of patients with a <5% pretest likelihood of CAD who have normal myocardial perfusion imaging (MPI) studies.

Positron Emission Tomographic Perfusion Imaging

Positron emission tomographic (PET) perfusion imaging is an alternative to SPECT imaging. In most imaging laboratories, PET perfusion imaging is performed with Rb-82, a potassium analog that is transported intracellularly via the sarcolemma membrane. The tracer is eluted from a generator and directly infused into the patient. It has a very short half-life of 75 s, permitting rapid sequential imaging. Because of the short half-life, only vasodilator stress can be used.

The sensitivity and specificity of Rb-82 PET for detection of CAD are 89% and 86%, respectively. Using PET, the detection rate for multivessel disease is 95%. There are several advantages of PET over SPECT imaging. Attenuation correction is more easily performed. PET is characterized by high spatial and contrast resolution and yields better specificity than observed with SPECT. Multivessel disease identification is better achieved.

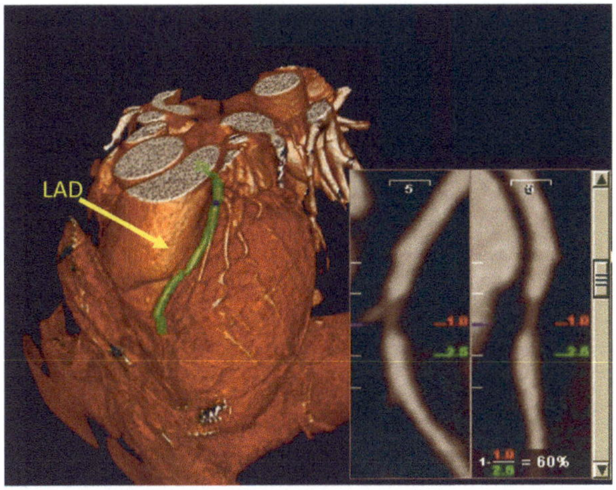

Figure 3.5.1.10 Computed coronary angiography depicting a soft plaque at the left anterior descending (LAD).

THE ROLE OF COMPUTED TOMOGRAPHY IN EVALUATING CORONARY ARTERY DISEASE

Evaluation of the heart and coronary arteries with cardiac CT angiography is revolutionizing cardiac imaging. Nowadays the calcium score plays a crucial role in the evaluation of risk stratification of men or women in addition to the other classical risk scores.

The advent of multidetector CT, particularly with scanners having 64 or more detectors, has continued to improve temporal resolution and allows the acquisition of isotropic voxels. With these scanners, the heart and coronary arteries are routinely imaged as emption-free volume of data. A variety of postprocessing techniques, including multiplanar reformation (MPR), maximum intensity projection, volume rendering, curved reformation, and cine imaging, allow noninvasive assessment of every aspect of the cardiovascular system. This capability requires a thorough understanding of essential coronary arterial and cardiac anatomy (Fig. 3.5.1.10).

MULTIDETECTOR COMPUTED TOMOGRAPHY

Postprocessing Techniques

The interpretation of cardiac CT angiographic studies performed with multidetector scanners requires real-time interaction with the volumetric data set that is generated. The interpretation of the cardiac CT angiographic data is accomplished using a combination of the postprocessing techniques.

Multiplanar Reformation

MPR is the basic tool used to interpret cardiac CT angiographic studies. With use of retrospective electrocardiographic gating, data from specific phases of the cardiac cycle are retrospectively referenced to the electrocardiogram for reconstruction. Because of variations in the orientation of the heart in the thorax, it is often useful to evaluate cardiac structures along the cardiac planes. The multiplanar capabilities of the workstation allow images of the heart and coronary arteries to be manually rotated for optimal evaluation of the cardiac anatomy.

Cine Imaging

Cine images are used to examine the motion and physiologic features of cardiac structures such as the LV and cardiac valves. The capabilities of the multidetector scanner are used to full advantage with this technique, since data from the heart and coronary arteries are typically reconstructed at specific points during the cardiac cycle. This approach is particularly useful for examining LV wall motion and wall thickening and for assessing valve motion in multiple planes. In addition to allowing assessment of myocardial contractility, reconstruction of the cardiac data during both systole and diastole allows determination of quantitative LV functional parameters, including end-diastolic and end-systolic ventricular volumes and stroke volume.

It is worth noting that recently fusion techniques of computed tomographic coronary angiography (CTCA) and 3DE -derived data allows direct visualization of each coronary artery and strain in its territory. Resting strain showed good agreement with stress perfusion, indicating that it may be potentially used to assess hemodynamic impact of coronary stenosis, as an alternative to stress testing that entails additional radiation exposure.

CORONARY ARTERIES

Left Coronary Artery

The left coronary artery (LCA) normally arises from the left sinus of Valsalva near the sinotubular ridge. It courses for a variable distance before giving rise to the LAD and left circumflex (LCx) arteries. The LAD artery courses anterolaterally in the epicardial fat of the anterior interventricular groove and supplies the majority of the LV. The major branches of the LAD artery are the diagonal and septal perforating arteries.

The diagonal branches course laterally and predominantly supply the LV free wall. The septal branches course medially and supply the majority of the interventricular septum, as well as the AV bundle and proximal bundle branch. The LCx artery is the other major branch of the LCA. It courses in the left AV groove, giving rise to obtuse marginal branches, sometimes referred to as lateral branches. The LCx artery and its branches supply the LV free wall and a variable portion of the anterolateral papillary muscle. It usually gives rise to posterolateral artery and PDA branches supplying the diaphragmatic portion of the LV.

In approximately 15% of patients, a third branch, the ramus intermedius (RI) branch, arises at the division of the LCA, resulting in a trifurcation. When present, the RI branch courses laterally toward the LV free wall. Its course is similar to that of a diagonal branch of the LAD artery.

Right Coronary Artery

The RCA normally arises from the right coronary sinus and courses in the right AV groove toward the crux of the heart (the point on the posterior surface of the heart where the AV groove transects the line of the interventricular septum and interatrial septum, forming a cross). In approximately 50%–60% of patients, the first branch of the RCA is a conus artery. The conus artery can also arise directly from the aorta (30%–35% of patients). The conus artery supplies the RV outflow tract (conus arteriosus) and forms the circle of Vieussens, an anastomosis with the LAD arterial circulation. In approximately 58% of patients, the sinoatrial nodal artery arises from the RCA; in the remaining patients (42%), it arises from the LCx artery. Multiple ventricular branches arise from the RCA, the largest of which is called the acute marginal branch.

The origin, course, and segmental coronary artery anatomy are described and used as a "road map" for precise localization of abnormalities. Dominance and the size of the LAD artery, LCx artery, and RCA are described. Typically, the number of diagonal and marginal branches is specified and their disease burden quantified. The overall size of the heart and cardiac chambers should be evaluated. LV function, including ejection fraction and wall motion, should also be assessed. The end-diastolic volume, end-systolic volume, and LV myocardial mass and thickness are easily determined. Other anatomic structures, whether cardiac or extracardiac, are not typically mentioned specifically unless they are abnormal.

ROLE OF CARDIOVASCULAR MAGNETIC RESONANCE IN CORONARY ARTERY DISEASE

Cardiovascular magnetic resonance (CMR) is a noninvasive, nonradiating technique capable of providing highly reproducible information about cardiac function, flow, and tissue characterization. The applications of CMR for the evaluation of chronic CAD include the following.

Evaluation of Myocardial Ischemia

CMR can detect ischemia by two different ways.

1. Observation of wall motion abnormalities (abnormal wall motion and wall thickening) using the stress factor dobutamine (the same way as stress echo, but without the limitation of acoustic window).

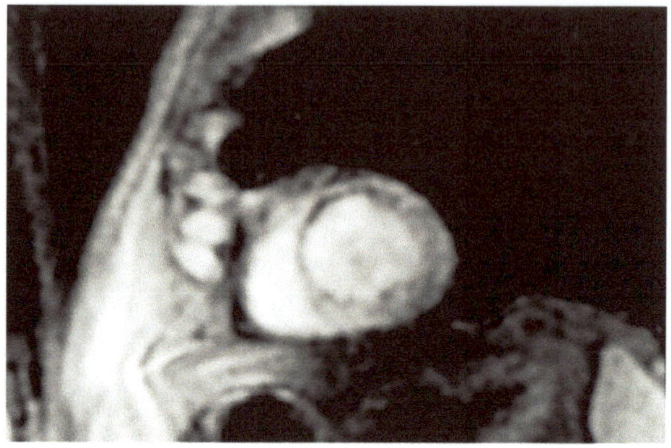
Figure 3.5.1.11 Adenosine stress cardiovascular magnetic resonance identifies hypoperfused area in the anterior wall of the left ventricle.

2. Observation of myocardial perfusion by the first pass of a bolus of a T1-shortening contrast agent (first-pass gadolinium) injected into a peripheral vein. Data acquired during intravenous vasodilator stress (most commonly with adenosine) delineate the underperfused regions associated with myocardial ischemia (Fig. 3.5.1.11). The spatial resolution of CMR MPI of 2–3 mm is greatly superior to those of other imaging modalities, such as nuclear techniques, so that subendocardial ischemia can be more reliably identified. Recent developments led to further improvements in spatial resolution to around 1 mm in the imaging plane. The interpretation of CMR myocardial perfusion studies in clinical practice is most commonly visual, but quantitative approaches are also available and have been validated against X-ray angiography, SPECT, and PET. The MR-IMPACT study in 234 patients reported improved detection of coronary stenosis by CMR compared with SPECT in the first multicenter, multivendor comparison. CE-MARC is the largest, prospective, real-world evaluation of CMR and has established CMR's high diagnostic accuracy in CHD and CMR's superiority over SPECT. It should be adopted more widely than at present for the investigation of CHD.

Evaluation of Fibrosis Using Late Gadolinium-Enhanced Imaging

CMR is the most reliable imaging technique to detect and quantify scar or fibrotic tissue, due to irreversible myocardial damage (viability study). Following acute ischemic injury, the myocardial distribution volume of gadolinium is increased, due to sarcrolemmal rupture and abnormal washout kinetics. The preferred imaging time for scar detection is between 10 and 20 min after contrast agent administration, when the differences between scar, normal myocardium, and blood pool are maximal. This method is referred in the literature as late gadolinium-enhanced (LGE) CMR and is the gold standard for the in vivo assessment of myocardial scar (Fig. 3.5.1.12).

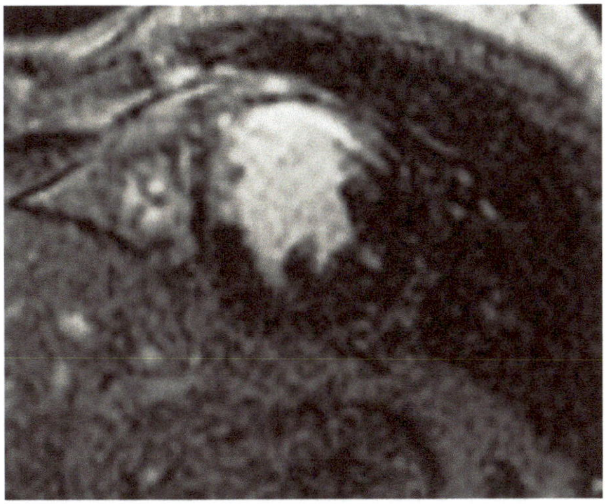

Figure 3.5.1.12 Myocardial scar in the anterior wall of the left ventricle due to anterior myocardial infarction.

Compared with other imaging modalities interpreting the myocardial viability as an all-or-none phenomenon CMR assess the transmural extent of viability across the ventricular wall. CMR not only can detect infarction in as little as 1 cm^3 of tissue, substantially less than other in vivo methods, such as echocardiography and nuclear techniques, but also has shown excellent agreement with histology in animal and human studies. It is of value in the detection of the RV infarction. It is also important to notice that CMR can detect subendocardial myocardial infarction missed by SPECT or PET. The extent of scar on CMR predicts the potential for functional recovery after revascularization.

Left Ventricle–Right Ventricle Function Assessment

CMR measures ventricular volumes and ejection fraction noninvasively and without contrast agent. It is of great value for the assessment of the RV, which is not always adequately assessed by echocardiography. CMR provides 3D images of the heart, which is also feasible with 3DE. Although CMR ejection fraction and volumes are more accurate and reproducible than those of other imaging techniques, there is a good correlation between CMR and these techniques. Echocardiography is still the everyday, bedside tool for ventricular function evaluation, but there is a space for CMR, due to its high reproducibility, to follow individual patients with respect to changes in RV and LV volumes, masses, and functions (dilated cardiomyopathy; valvular, congenital heart disease; RV dysplasia; pulmonary hypertension; etc.).

Recently, T1 mapping at rest and during adenosine stress proved capable of differentiating normal, infarcted, ischemic, and remote myocardium using distinctive T1 mapping profiles and holds promise for ischemia detection without the need for gadolinium contrast.

Specifically, for the evaluation of acute CAD the CMR should also include the following.

Edema Assessment

The interrupted supply of oxygen-rich blood to the myocardium produces an ischemic phenomenon that, if left untreated, may lead to irreversible injury. The amount of inadequately perfused myocardium distal to the coronary artery occlusion is called the area at risk (AAR). With sustained ischemia, the myocardium defined as the AAR may present irreversible injury. Identifying the AAR is the main goal of edema imaging using CMR. According to the study by Aletras et al. the edema area represents the AAR in reperfused myocardial infarction (Fig. 3.5.1.13).

T2 imaging by CMR can detect edema prior to irreversible injury and also identify the AAR beyond the irreversibly injured myocardium. CMR can offer early and more accurate information regarding risk stratification, outcome, and treatment selection for patients with AMI and/or acute coronary syndromes. T2 imaging by delineating myocardium at risk may facilitate translational research and guide the development of novel therapeutic approaches that can potentially restore life to myocardium at risk.

Edema evaluation is very important during AMI, because it is influenced by a low-threshold ischemic injury and is both sensitive and specific to acute injury. Recently, it was documented that the edematous reaction during the first week after ischemia/

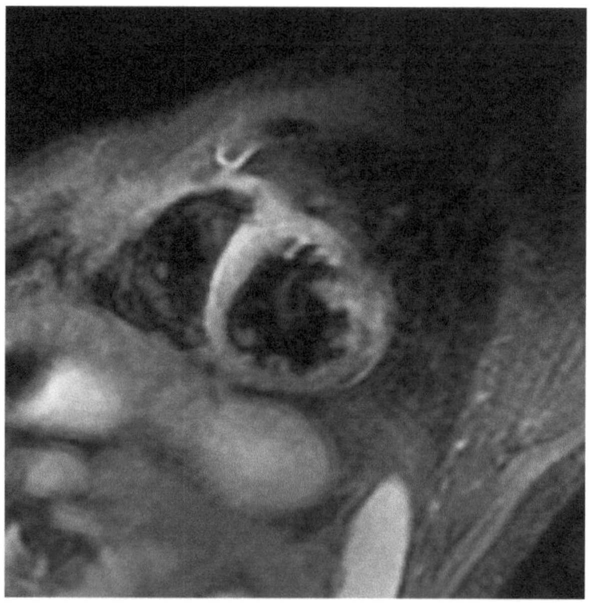

Figure 3.5.1.13 Edema in the anterior wall due to acute myocardial infarction.

reperfusion is not stable, but follows a bimodal pattern. The first wave of edema appears abruptly upon reperfusion and dissipates at 24 h. Conversely, the second wave appears progressively days after ischemia/reperfusion and is maximal at day 7 after reperfusion.

The timing of edema imaging during AMI is important and should be focused within 2–7 days postreperfusion. Some recent investigations addressed questions about the diagnostic validity of edema imaging after acute ST-elevation myocardial infarction (STEMI). Considering the results of various studies we found that the time course of edema post-STEMI is unimodal and not bimodal as post-AMI. Novel CMR precontrast T1 mapping techniques allow voxelwise quantitative assessment of tissue characterization. Although the clinical value of this technique in chronic myocardial scar is yet under evaluation, T1 mapping allows assessment of acute myocardial edema and may provide a better understanding of the pathophysiologic events taking place in acute ischemic injury, with potential to offer additional diagnostic and prognostic. T1 mapping by allowing tissue characterization may allow a comprehensive assessment of ischemic injury and related findings without the need for contrast agents.

Intramyocardial Hemorrhage

Myocardial hemorrhage is the final consequence of severe vascular injury and represents a prognostically significant complication early postmyocardial infarction. It is a therapeutic target to limit reperfusion injury and infarct size post-STEMI, occurs not only after therapeutic but also after spontaneous reperfusion, and can be accurately demonstrated by cardiac magnetic resonance.

According to a recent study of 30 patients with STEMI, the overall incidence of myocardial hemorrhage was 43%, approximately one-quarter of the patients had evidence of myocardial hemorrhage 4–12 h after myocardial infarction (MI), and hemorrhage detection increased progressively, with 20% of new cases at 3 days compared with 4–12 h; hematoma persisted at 10 days and 7 months in 37% and 13% patients, respectively. Myocardial edema, assessed by T2 values, evolved with a bimodal time course in patients with myocardial hemorrhage but with a unimodal time course in patients without myocardial hemorrhage. T2* within the hemorrhagic core followed a similar pattern to T2, with a nadir in both on day 3. During the first 10 days after reperfusion, myocardial edema within the infarct zone and hypointense core had a bimodal distribution in patients with myocardial hemorrhage, whereas myocardial edema had a unimodal progressive increase in patients without myocardial hemorrhage. Finally, changes in edema were inversely related to the extent of hemorrhage and myocardial hemorrhage was associated with sustained reductions in LV ejection fraction and adverse LV remodeling from baseline to 7 months.

Microvascular Obstruction

Microvascular obstruction (MVO) or no-reflow phenomenon is a complication postcoronary reperfusion therapy during AMI. It is considered as a poor prognostic factor and marker of adverse LV remodeling. Although MVO can be assessed by various imaging modalities including electrocardiography, myocardial contrast echocardiography, nuclear scintigraphy, and coronary angiography, evaluation by CMR particularly offers not only diagnosis but also quantification. MVO assessment is part of the routine CMR evaluation of AMI and will play an important role in clinical trials of adjunctive reperfusion agents and strategies.

In humans, clinical no-reflow is characterized by the additional contributing factors of microembolization and inflammatory response. Rupture or erosion of epicardial coronary artery plaques, either spontaneously or by PCI, leads to distal showering of embolic debris, consisting of atherosclerotic plaque and/or thrombotic material. Distal embolization not only contributes to mechanical obstruction of the microvessels but also causes an inflammatory response with the elaboration of vasogenic and thrombogenic factors that further aggravate the existing microvascular dysfunction. This process can finally contribute to no-reflow phenomenon, assessed by MVO.

Imaging within the first few minutes after contrast agent administration is the method of choice to delineate MVO, which prevents contrast delivery to the infarct core and thus results in low signal on T1-weighted imaging. Both acute and old infarctions without MVO retain contrast agent and therefore appear bright (Fig. 3.5.1.14).

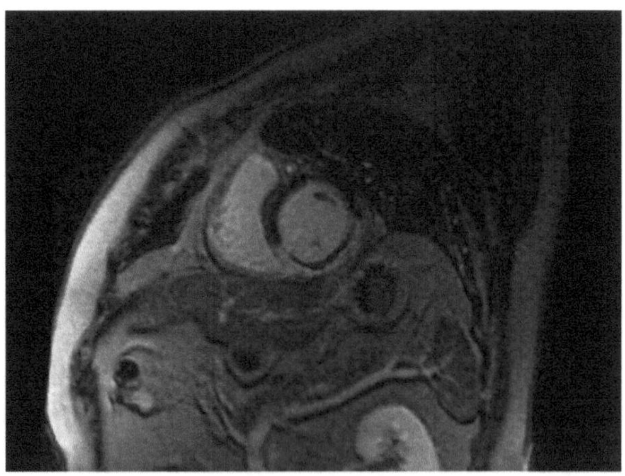

Figure 3.5.1.14 Microvascular obstruction during acute myocardial infarction.

CARDIOVASCULAR MAGNETIC RESONANCE IMAGING APPROACH TO EVALUATE MICROVASCULAR OBSTRUCTION AND INFARCT HEMORRHAGE

Hemorrhage can be evaluated using T2-weighted and T2* imaging. The detection of hemorrhage on magnetic resonance imaging is based on the paramagnetic effects of hemoglobin degradation products. Initially, hemorrhage consists of oxyhemoglobin that lacks paramagnetic properties. However, within the first few days after AMI, oxyhemoglobin denatures into deoxyhemoglobin that has an important paramagnetic effect and therefore can reduce the T2 relaxation time. Deoxyhemoglobin is then gradually converted over the following few days into methemoglobin, which is strongly paramagnetic. After ~2 weeks, methemoglobin is converted into hemosiderin, contained in macrophages, that also leads to low T2 values.

Post-MI hemorrhage can be visualized by CMR as a hypoenhanced area surrounded by elevated signal intensity representing myocardial edema on T2-weighted imaging, and most studies show that hemorrhage is limited to those patients with evidence of MVO. Evaluation of human autopsy specimens in two cases showed a close correlation between pathologic hemorrhage and hypoenhanced signal intensity on T2 sequences. However, MVO without hemorrhage can also cause hypoenhanced areas on T2-weighted sequences and both can appear as rims of persistent hypoenhancement on LGE.

To differentiate MVO with and without hemorrhage, direct T2 quantification could be the method of choice, because it has yielded more robust results for the detection of myocardial edema, compared with conventional T2-weighted short tau inversion recovery. Within the infarct core, T2 mapping demonstrates significant reductions in T2 values corresponding to LGE MVO. Whether or not such a technique can also differentiate between MVO with and without hemorrhage by detecting gradations in reduced T2 values requires further investigation.

Compared to T2 relaxation, which depends primarily on spin—spin interactions, T2* decay is caused by a combination of spin—spin relaxation and magnetic field inhomogeneity. Since iron deposits and blood products cause magnetic field inhomogeneities, T2* imaging has been successfully used to assess myocardial iron deposition in iron overload cardiomyopathies and detect hemorrhage in acute reperfused MI. In a recent experimental canine model, T2* imaging was compared with pathologic analysis of the infarct region on day 3 postreperfusion and significant hemorrhage was limited to animals with the largest infarct sizes and extents of MVO. Hemorrhage was always smaller than the amount of MVO and was contained within it. T2* imaging can facilitate the differentiation between MVO alone and MVO with hemorrhage, because hemorrhage is accompanied by reduced T2* levels, whereas MVO alone will not.

Finally, CMR by using function and edema-fibrosis imaging can reliably assess ventricular and valvular integrity and also myocardial infarction acuity, extent, and coexisting

MVO and hemorrhage. This complete profile can be of great value for risk stratification and treatment strategies.

The role of cardiovascular imaging in CAD extends from pathophysiology of cardiovascular disease to the depiction of coronary arteries, to evaluate the anatomy and function of myocardium and valves, ischemic response, and finally tissue characterization.

FURTHER READING

[1] Focused update of the guideline for the diagnosis and management of patients with stable ischemic heart disease. Circulation 2014;130:1749–67.

[2] O'Gara P, Kushner F, Ascheim D, et al. Guidelines for the management of ST-elevation myocardial infarction. Circulation 2012.

[3] Douglas PS, Garcia MJ, Haines DE, Lai WW, Manning WJ, Patel AR, Picard MH, Polk DM, Ragosta M, Parker Ward R, Weiner RB. Appropriate use criteria for echocardiography. Circulation 2011;57.

[4] Appropriateness criteria for single-photon emission computed tomography myocardial perfusion imaging (SPECT MPI): a report of the American College of Cardiology Foundation Quality Strategic Directions Committee Appropriateness Criteria Working Group and the American Society of Nuclear Cardiology endorsed by the American heart association. J Am Coll Cardiol October 18, 2005;46(8):1587–605.

[5] Appropriatenss criteria for cardiac computed tomography abd cardiac magnetic resonance imaging. JACC 2006;48.

[6] Aggeli C, Lagoudakou S, Felekos I, et al. Two-dimensional speckle tracking for the assessment of coronary artery disease during dobutamine stress echo: clinical tool or merely research method. Cardiovasc Ultrasound October 24, 2015;13:43.

[7] Aggeli C, Felekos I, Tousoulis D, et al. Dobutamine stress contrast echocardiography in patients with coronary artery disease: the prognostic impact of age. Int J Cardiol May 15, 2014;173(3):540–2.

[8] Aggeli C, Pietri P, Vavouranakis M, et al. Sequential thromboembolic events after primary angioplasty in a patient with acute anterior myocardial infarction. Eur Heart J Acute Cardiovasc Care December 2012;1(4):341–3.

[9] Aggeli C, Felekos I, Roussakis G, et al. Value of real-time three-dimensional adenosine stress contrast echocardiography in patients with known or suspected coronary artery disease. Eur J Echocardiogr 2011;12(9):648–55.

[10] Kumar A, Green JD, Sykes JM, et al. Detection and quantification of myocardial reperfusion hemorrhage using T2*-weighted CMR. JACC Cardiovasc Imaging 2011;4:1274–83.

[11] Carrick D, Haig C, Ahmed N, et al. Temporal evolution of myocardial hemorrhage and edema in patients after acute ST-segment elevation myocardial infarction: pathophysiological insights and clinical implications. J Am Heart Assoc February 23, 2016;5(2).

[12] Aletras AH, Tilak GS, Natanzon A, et al. Retrospective determination of the area at risk for reperfused acute myocardial infarction with T2-weighted cardiac magnetic resonance imaging: histopathological and displacement encoding with stimulated echoes (DENSE) functional validations. Circulation 2006;113:1865–70.

[13] Mavrogeni S, Bratis K, Papadopoulos G, et al. The Greek cardiac magnetic resonance experience: a comparison with the EuroCMR Registry. Hell J Cardiol September–October 2013;54(5):355–61.

CHAPTER 3.5.2

Invasive Imaging Techniques

Andreas Synetos, Dimitris Tousoulis
1st Cardiology Clinic, 'Hippokration' General Hospital, National and Kapodistrian University of Athens, School of Medicine, Athens, Greece

Contents

INTRODUCTION

Over the past three decades, significant gains in the understanding of coronary artery disease have guided diagnostic and therapeutic practices. The development of many diagnostic modalities has played a vital role in elucidating the pathophysiology of coronary atherosclerosis. Coronary angiography provides direct visualization of the coronary luminal anatomy and is the gold standard for the diagnosis of coronary artery disease. However, since the development of angiography in the 1960s, the primary method for assessing the lesions that are of physiologic significance has been visual assessment by the operator, which is prone to significant intraobserver and interobserver variability. The significance of a given stenosis is not determined solely by the reduction in luminal diameter because numerous additional factors such as lesion length, shape, and eccentricity affect the flow dynamics of the lesion and thus the physiologic significance. Therefore, coronary angiography cannot solely be relied on to provide the physiologic or clinical significance of a stenosis, particularly when the vessel is narrowed to between 40% and 80% of its normal diameter. Because of the need for a more comprehensive method

Coronary Artery Disease
ISBN 978-0-12-811908-2, http://dx.doi.org/10.1016/B978-0-12-811908-2.00018-0

for determining the anatomic and functional characteristics of a coronary artery lesion, several techniques have been developed to augment standard cineangiography. These novel techniques are centered on the physiologic assessment of lesions and advanced intravascular imaging to provide a more comprehensive anatomic assessment.

The major point of interest of these new modalities is the recognition and treatment of the vulnerable plaque. The angiographic techniques that are routinely used to identify stenotic atherosclerotic lesions are unable to identify high-risk plaques; plaques prone to rupture and cause a cardiovascular event. This is partly due to the fact that the majority of culprit lesions that produce acute cardiovascular syndromes are not severely stenotic, possibly due to significant positive remodeling and reduced protective collateral circulation as well as because the risk of plaque rupture is more closely related to plaque content than plaque size. Recently, the focus of new imaging techniques is to identify the high-risk plaques; the "vulnerable plaques".

In this chapter we will deal with the intravascular imaging modalities that have been used for the evaluation of atheromatosis and they serve as a tool for the optimization of percutaneous coronary interventions (PCIs).

INTRAVASCULAR THERMOGRAPHY

Intravascular thermography is an invasive method that although it is not an imaging technique, it assesses the functional characteristics of the atheromatic plaque and helps to estimate its consistency. Due to the fact that inflammation, one of the most important features of plaque vulnerability, has been closely related to increased heat production, temperature elevation in atherosclerotic plaques may reflect the intensity of local inflammatory process (Fig. 3.5.2.1).

First, Casscells et al. measured ex vivo carotid artery intimal surface temperatures that exhibited thermal heterogeneity, with temperature elevation in several regions that were infiltrated with inflammatory cells predominately macrophages. Stefanadis et al. developed this technique and by the use of dedicated coronary thermographic catheters,

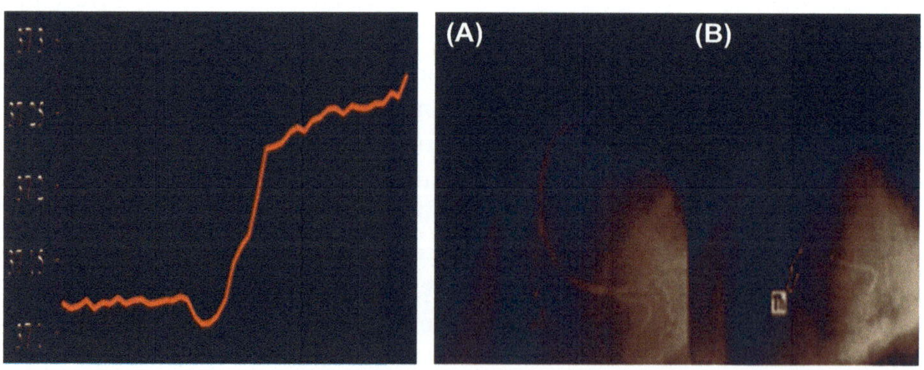

Figure 3.5.2.1 Increased thermal heterogeneity in a culprit, vulnerable plaque lesion of a right coronary artery causing an acute coronary syndrome.

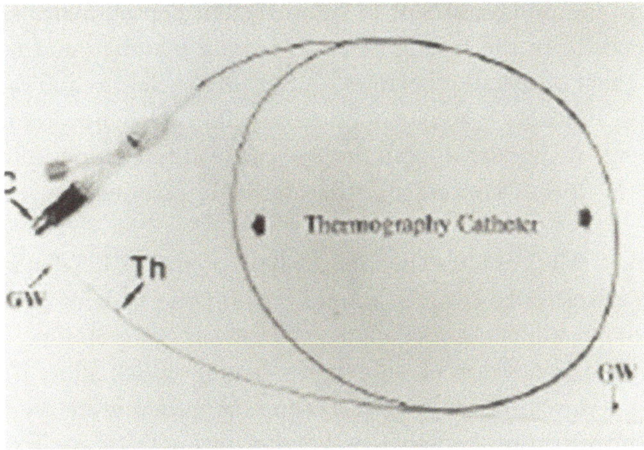

Figure 3.5.2.2 Thermographic catheter with a thermal detector able to detect thermal heterogeneity between two segments of a coronary artery.

thermal heterogeneity between an inflamed plaque and the nearby healthy vascular wall has been detected and correlated with plaque vulnerability (Fig. 3.5.2.2).

In their first human intracoronary thermographic study in 1999, they used a rather large single-channel, thermistor-based catheter, which almost blocked the arterial lumen, therefore, decreasing the heat-dissipating effect of blood flow. In this study, temperature differences between atherosclerotic plaque and healthy vascular wall were $0.10 \pm 0.11°C$ in patients with stable angina, $0.68 \pm 0.34°C$ in those with unstable angina, and $1.47 \pm 0.69°C$ in patients with acute myocardial infarction. Plaque temperature heterogeneity was present in 20%, 40%, and 67% of the patients with stable angina, unstable angina, and acute myocardial infarction, respectively, and did not correlate with the degree of stenosis. Temperature heterogeneity was absent in the control group.

In the presence of acute coronary syndrome (ACS), the positive arterial remodeling and the plaque rupture are correlated with local plaque temperature measurements, whereas thermal heterogeneity is known to be a prognostic factor for long-term outcome after PCIs. Thermography has been used as a tool to evaluate the effect of diet and medications on atherosclerotic plaques. After 3 months of cholesterol lowering diet, plaque thickness remained unchanged, but temperature heterogeneity was significantly decreased and this paralleled plaque.

INTRAVASCULAR ULTRASOUND

Intravascular ultrasound (IVUS) utilizes analysis of radiofrequency backscatter ultrasound signals for the identification of various plaque characteristics that can also be color coded to display the content of various plaques characteristics. It has several clinical implications that include the assessment of intermediate lesions, the assessment of various characteristics of vulnerable plaque, and the guidance of PCIs that include the preassessment of vessel diameter and the postassessment of optimal stent deployment. The experimental use of IVUS extends from the testing of new stenting materials and techniques to the assessment of the effect of specific therapies on the atherosclerosis and on vulnerable plaque. This imaging modality has given impressive boost to new emerging therapies including the proof that aggressive statin therapy can lead to a reduction in the lipid content and increase in fibrous content of atherosclerotic plaques after only 6 months of treatment.

Virtual histology (VH) applies spectral analysis of the IVUS backscatter radiofrequency signals to characterize plaque components on the basis of tissue characteristics such as density, compressibility, concentration of various components, and size. Spectral parameters are applied to develop classification schemes, which allow the differentiation between tissue types in regions of interest. Plaque components are assigned color codes and tissue maps are reconstructed that make the fibrous, fibrous lipid, calcified, and calcified necrotic tissue easy to differentiate. This invasive intravascular imaging has led to the increase of interest of depicting vulnerable plaque in vivo (Fig. 3.5.2.3).

Assessment of Intermediate Lesions

Although coronary arteriography is an invaluable tool for the detection and treatment of coronary artery lesions that result in myocardial ischemia, it has limitations, as discussed previously. IVUS provides real-time cross-sectional images of blood vessels at high resolution and has the ability to detect and localize plaque as well as quantitate plaque burden. IVUS has contributed to our understanding of the biology of atherosclerosis

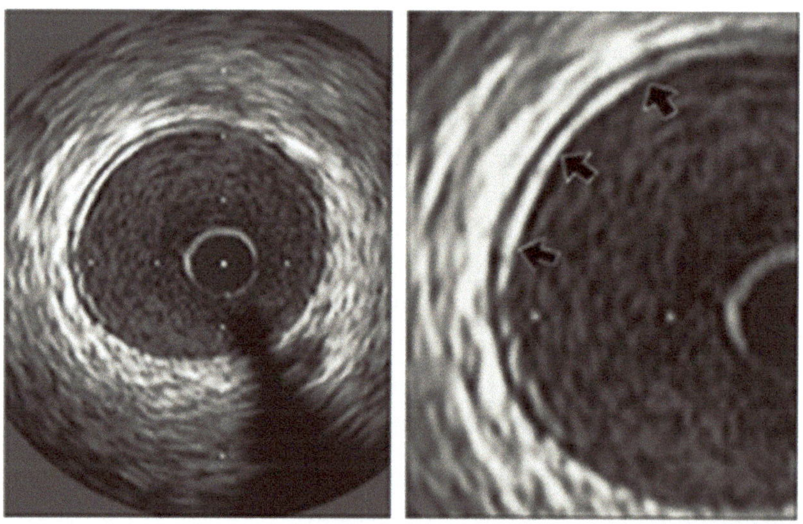

Figure 3.5.2.3 Intravascular imaging with intravascular ultrasound of a normal anterior descending coronary artery. The *arrows* depict the intima and media layer of the vessel.

and restenosis as well as to the improvement of our stenting techniques in order to avoid stent underexpansion and mallaposition. Direct intraluminal visualization of coronary arteries by IVUS is useful during stent deployment and evaluating stent apposition. IVUS has also been useful in determining intermediate lesions. Coronary lesions with minimal lumen area (MLA) of 4.0 mm^2 or less in nonleft main proximal coronary arteries and <6.0 mm^2 in left main coronary artery have been described to have significant stenoses. In comparison to other modality such as fractional flow reserve (FFR), IVUS has the advantage of direct anatomical visualization, whereas FFR can only provide functional information. Additionally, IVUS has been crucial in furthering the fundamental understanding of atherosclerotic disease, especially plaque burden and plaque vulnerability.

Vulnerable Plaque

Culprit plaques causing ACS, studied by IVUS, have more vulnerable characteristics. They appear to be hypoechoic, eccentric, positively remodeled, and relatively free of calcification. In prospective studies of patients with ACS that were followed up by IVUS, plaques related to acute events exhibited an eccentric pattern with the mean percent of plaque area greater than plaque area in patients without ACS. Further studies have verified that large eccentric plaques containing an echolucent zone and exhibiting positive remodeling by IVUS can be unstable. These are important characteristics of the vulnerable plaque, the plaque that is prone to rupture and cause an ACS with thrombus formation. Several studies have examined the utility of IVUS in assessing the prognostic implications of different plaque compositions. Sano and colleagues were the first to use radio frequency (RF)-IVUS to examine patients with stable angina. Patients treated with PCI underwent IVUS of identified non—flow-limiting lesions. Within 30 months, 10 of the studied plaques caused an acute event; these lesions were noted to have increased eccentricity, plaque burden, and lipid tissue. More recently IVUS was used to show that, in patients admitted for stable angina or ACS, a second nontarget event was more likely if the patient had lipid-rich nonobstructive plaques outside the causative lesion. The largest prospective study using three-vessel virtual histology (VH) was the PROSPECT (Providing Regional Observations to Study Predictors of Events in the Coronary Tree) study. The prognostic implications of atheroma burden and plaque composition in 697 patients treated for ACS were studied for 3 years. The primary end point was major adverse cardiovascular events (MACE), which occurred in 149 patients and of which 55 cases involved causative lesions that were previously studied with VH-IVUS. Through multivariate analysis it was determined that the independent predictors of future causative lesions were increased plaque burden, an MLA of less than 4 mm^2, and a thin-cap fibroatheroma (TCFA) phenotype. Several other studies have confirmed the association between plaque composition and future MACE. However, these studies fail to

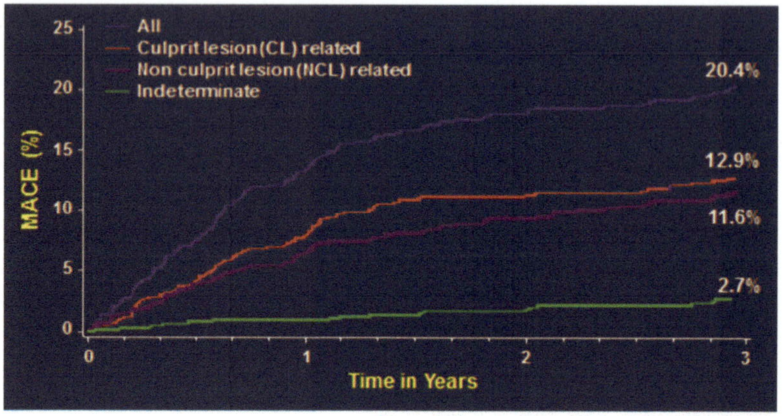

Stone GW et al., N Engl J Med 2011;364:226-35

Figure 3.5.2.4 The prognostic value of identifying the vulnerable plaque.

show a clinically useful predictive accuracy for VH-IVUS; for example, in the PROSPECT study only 18% of the plaques with plaque burden greater than or equal to 70%, MLA less than 4 mm^2, and a TCFA phenotype caused MACE (Fig. 3.5.2.4).

IVUS Guided Percutaneous Coronary Interventions

IVUS has provided valuable information on cross-sectional coronary vascular structure and has played a key role in contemporary stent-based PCIs (by accurately assessing coronary anatomy, assisting in selection of treatment strategy, and defining optimal stenting outcomes. In the bare-metal stent era, randomized trials and meta-analysis demonstrated that IVUS-guided PCI was mainly associated with a lower risk of angiographic restenosis and target vessel revascularization. In the drug-eluting stent era, a recent meta-analysis showed that the risk of death and stent thrombosis was reduced by IVUS guided DES implantation. Up to date some randomized trials, several observational studies and some meta-analysis tried to focus on the possible beneficial effect of IVUS-guided PCI on strong endpoints such as all-cause mortality and cardiovascular death. The majority of the studies showed that IVUS guided PCI is associated with a significantly reduced risk target lesion revascularization, death, myocardial infarction and this is more evident when left main coronary artery disease is involved.

Most recently, the use of IVUS has been implicated in the treatment of chronic total occlusions. It can be used for the identification of the dissection and reentry points and in the optimal stent deployment with minimal underexpansion and malapposition. Moreover it can be used via the retrograde technique especially in patients with renal impairment where contrast injection and contrast induced nephropathy can have tremendous

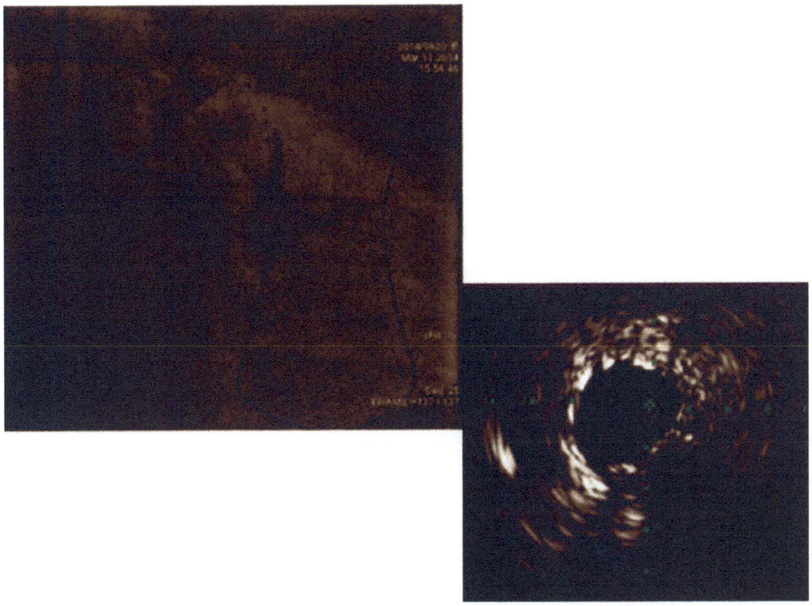

Figure 3.5.2.5 Retrograde approach with the use of intravascular ultrasound for the identification of the dissection-reentry point.

effect of the outcome of the patients even after a successful percutaneous intervention. With the use of this technique, the contrast ejection is minimized resulting in better outcomes (Fig. 3.5.2.5).

Fusion of Angiography and IVUS (ANGUS)

IVUS generates images with a high temporal and spatial resolution and is useful to evaluate both vessel wall morphology and dimensions. However, IVUS imaging is a tomographic technique, which makes it difficult to grasp an overview of an investigated vessel segment. To partly overcome this limitation, multiple images can be acquired during a catheter pullback. Assuming a straight pullback trajectory, 3D stacking of these images is simple and provides a useful overview of the IVUS information. More realistic 3D reconstruction methods have been devised that take into account the curved path of the transducer during pullback. These methods reconstructed the 3D path from multiple biplane X-ray recordings of successive transducer locations or used the vessel centerline as an approximation. Even with these combined techniques some important reconstruction problems related to determination of the true orientation of the IVUS cross-sections and susceptibility to respiratory motion remained.

Using three-dimensional configuration, wall shear stress can be estimated that has a profound influence on vascular biology and localization of atherosclerotic plaque in the presence of systemic risk factors. The often reported inverse relationship between wall thickness and shear stress appears restricted to lumen preservation in vascular remodeling and its disappearance with lumen narrowing suggest a growing importance of a nonshear stress related progression. The application of ANGUS documented the role of shear stress not only in atherosclerosis but also in restenosis after balloon angioplasty. Shear stress can predict neointimal formation in in-stent restenosis and in vascular remodeling after angioplasty. Furthermore, the local increase in shear stress seems to reduce in-stent neointimal formation by 50%. ANGUS was one of the first and most important pioneers in fusion imaging that was evolved to more complicated fusion systems that integrate almost all possible combinations.

IVUS ELASTOGRAPHY AND PALPOGRAPHY

The principles of elastography and palpography are based on the hypothesis that lipid-rich vulnerable plaques might have different mechanical propertied than stable, more fibrotic, plaques. Both are focused on the evaluation of local mechanical tissue properties, based on deformation of tissue. At a defined pressure, soft tissue components deform more than hard components. Consequently, in coronary arteries, regions with soft tissue components within the vessel wall will deform unequally during systole and diastole as the blood pressure (excitation force) varies throughout the heart cycle. The deformation of the plaque is assessed using ultrasound and images obtained at different pressure levels are compared to determine local tissue compression. The radial strain in the tissue is calculated by cross-correlation techniques on the radio-frequency signal and can be displayed as a color-coded image. This information is plotted as an additional image to the IVUS echogram A plaque with a high strain region at the surface with adjacent low strain regions is defined as a vulnerable plaque. In an in vitro study, it was demonstrated that sensitivity and specificity of elastography to detect vulnerable plaques are 88% and 89%, respectively. High correlation between the strain in caps and the amount of macrophages and an inverse relation between the amount of smooth muscle cells and strain was also observed. These results have been also verified in vivo, where it was shown that fatty plaques have an increased mean strain value and high-strain spots are associated with the presence of macrophages.

Based on the similar principle, IVUS palpography is assessing the local mechanical tissue properties of the luminal surface of the vessel wall. One strain value per angle is calculated and plotted as a color-coded contour at the lumen vessel boundary. It has been recently suggested that the distribution of the strain in the three-dimensional (3-D) geometry of an artery is an important tool to identify the presence of vulnerable spots.

In human coronary arteries, the number of highly deformable plaques has been correlated with both clinical presentation and levels of C-reactive protein (CRP). Limitation

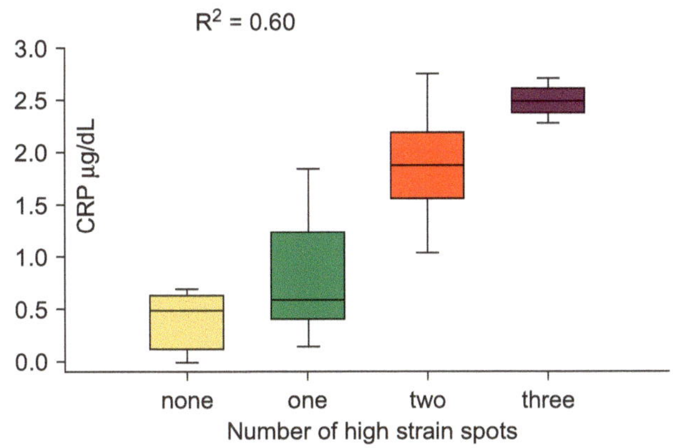

Figure 3.5.2.6 Diagnostic value of palpography. The number of regions with increased strain increases according to the value of C-reactive protein (CRP). *(Based on JACC 9, May 2, 2006;47: 1754–68.)*

of this technique remains rhythm disturbances that may produce poor signal quality and motion artifacts that may also influence the performance of the method of the available techniques (Fig. 3.5.2.6).

RAMAN SPECTROSCOPY

Raman spectroscopy is based on the Raman effect, according to which when incident light (wavelength 750–850 nm) excites molecules in a tissue, the molecules will reflect light in a different wavelength. The reflectant light's wavelength is characteristic for various chemical components and allows detection of the atheromatous plaque chemical synthesis. In particular, Raman spectroscopy can discriminate between the various plaque components, such as elastin, collagen, cholesterol, cholesterol esters, lipids, carotenoids, and calcium apatite deposits. By delivering excitation light and collecting emitted light through flexible optical fibers, fluorescence spectra is collected from a coronary artery and is used to differentiate normal tissue from abnormal tissue. In the experimental setting, Raman spectroscopy has been applied in order to evaluate the effect of atorvastatin and amlodipin on atherosclerotic plaques. Both atorvastatin and amlodipine were found to significantly retard the progression of existing atherosclerotic lesions. Probably, the combination of Raman spectroscopy with other diagnostic techniques may improve our ability for atheromatous plaque assessment and detection of vulnerable plaques. Main limitations of the technique are the absorbance of light by blood and the relative short penetration depth. In addition, Raman spectroscopy has a long acquisition time and yields no information regarding the vessels geometry.

ANGIOSCOPY

Coronary angioscopy was introduced in the 1980s as a means of obtaining a clearer view of the interior of the artery. The angioscope is a miniature fiber optic endoscope that is threaded into the heart through a catheter. After occluding the artery briefly with a balloon and flushing the residual blood from the field of view, the cardiologist can view thrombus, plaque, arterial dissections, and other vascular abnormalities. Although angioscopy gives a rough indication of the composition of a coronary lesion according to its morphology and color, it cannot help in the assessment of below-surface lesions. Coronary angioscopy directly detects intraluminal and coronary arterial surface pathology via optic fibers and it gives information about the lipid content on the plaque (yellow plaques). Coronary angioscopy uses projected white light through thin, flexible glass fibers incorporated into catheters in order to see the color of the arterial surface through a clear saline injection, permitting diagnosis of thrombus, and yellow or white plaques. Luminal irregularities, such as plaque rupture, can be also observed. Atherosclerotic plaques usually appear as white or yellow protrusions into the lumen that may be continuous with the normal vessel wall. Histopathologic analysis of atherectomy specimens of coronary lesions from patients has revealed that white-colored plaques are predominantly fibrous and deep yellow plaques represent atheroma or degenerated fibrous plaque with patchy necrosis. Intimal cholesterol in the arterial wall contains carotenoids, which are yellow-colored and likely give lipid-containing plaques their characteristic yellow color. Platelet-rich thrombus is characterized as white material and fibrin/erythrocyte-rich thrombus as red material protruding into the lumen. Furthermore, yellow plaques are frequently observed at the culprit lesions of ACS. A study followed 552 patients who underwent coronary catheterization and angioscopic examination for the diagnosis of CAD and the number of yellow plaques was counted prospectively. Among the patients, ACS events were detected in 39 patients during a follow-up interval of almost 60 months, and the number of yellow plaques was significantly higher in the patients with an ACS event than in those without Despite these clinical data, the inability to see through the blood because of its opaque nature and the resulting need to remove blood from the visual field remain the primary obstacles to the widespread and routine use of angioscopy to evaluate plaque morphology. In addition, the presence of superficial calcium also correlates with yellowish plaque on coronary angioscopy.

NEAR-INFRARED SPECTROSCOPY

Near-infrared spectroscopy (NIRS) allows the detection of various chemical components of biological specimens based on the absorbance of light by organic molecules. Light is shed on the under study tissue and the reflected spectra from wavelengths between 400 and 2400 nm enables detailed qualitative and quantitative analysis of chemical

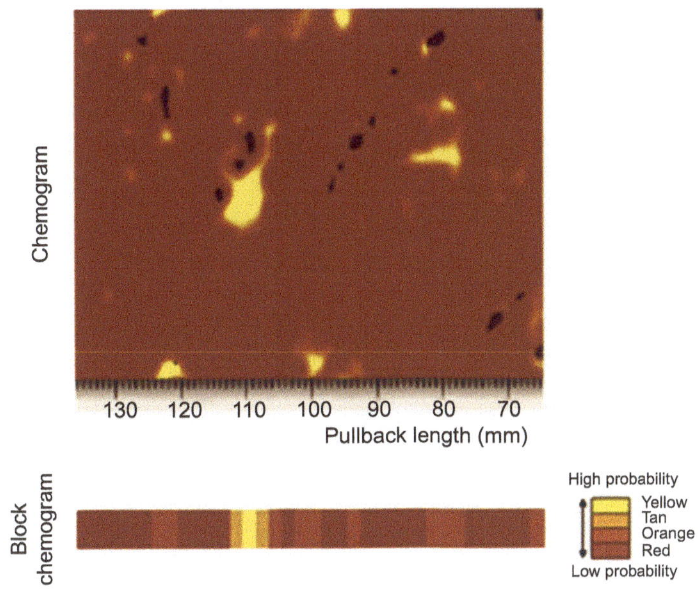

Figure 3.5.2.7 A chemogram of near-infrared spectroscopy depicting a chemical map of an artery wall and indicates the location and intension of coronary lipid.

synthesis. In vitro and ex vivo studies have shown that NIRS can help discriminate between stable and unstable plaques. It has been demonstrated that NIR spectroscopy sensitivity and specificity for histological features of plaque vulnerability were 90% and 93% for lipid pool, 77% and 93% for thin cap, and 84% and 89% for inflammatory cells, respectively. Additionally, an ex vivo study showed that pH measurements are feasible with the use of NIRS. NIRS is a rapidly evolving technique, however, in vivo feasibility has not yet been evaluated. Potential problems that need to be in the future addressed include acquisition time, blood scattering, and influences of pH and temperature (Fig. 3.5.2.7).

OPTICAL COHERENCE TOMOGRAPHY

The potential utility of optical coherence tomography (OCT) as an intravascular imaging technology was recognized soon after the publication of the first OCT images of biological tissue. The results of early studies demonstrated OCT's unique ability to resolve atherosclerotic lesions in microscopic detail, particularly those lesions believed to be prone to sudden rupture. OCT imaging within the beating heart has become practical only recently, owing in large part to technological advances made during the past few years. Features of OCT that make it attractive for intracoronary imaging are high resolution and the small size of fiber-based imaging probes. Healthy coronary arteries have

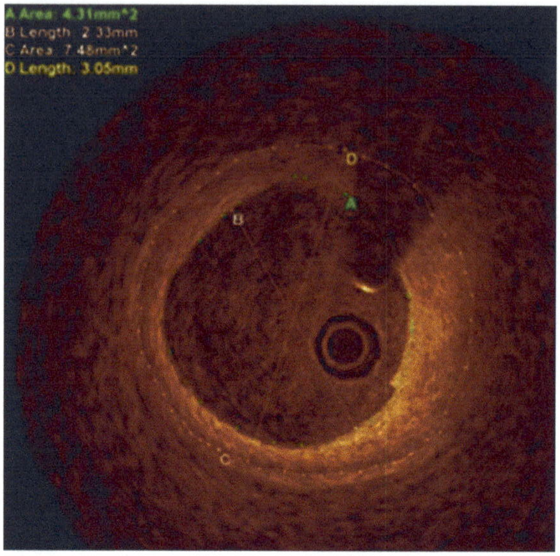

Figure 3.5.2.8 Optical coherence tomography imaging of a proximal part of a coronary artery. Due to the high resolution of the imaging modality, the layers of the artery can be separately visualized and the measurements can be performed with high accuracy.

walls that are only a fraction of a millimeter thick; highly stenosed arteries may have residual lumens with diameters of less than 1 mm. OCT is well suited to imaging structures with these dimensions (Fig. 3.5.2.8).

OCT is a catheter-based technique that subjects the vessel wall to an infrared source, allowing analysis of the returning optical echoes. It has an excellent spatial resolution (5 μm) and therefore is able to distinguish different plaque components such that quantitative measurements of fibrous cap thickness can be made. The intensity of the reflected light is displayed as a false color or gray scale image. The first OCT systems required balloon occlusion of the studied vessel to clear blood, which causes attenuation artifact. This led to an increased risk for arrhythmia. However, current systems utilize a faster frequency-domain analysis instead of time-domain analysis and injection of contrast is sufficient to clear the vessel of blood during the study, essentially eliminating the arrhythmia risk, although increasing the contrast load by 10–30 mL for the procedure.

OCT has been correlated with histology and showed a sensitivity and specificity ranging from 71% to 79% and 97%–98% for fibrous plaques, 95%–96% and 97% for fibrocalcific plaques, and 90%–94% and 90%–92% for lipid-rich plaques, respectively. In a recent study, it was observed that the high contrast and resolution of OCT enables the quantification of macrophages within fibrous caps.

Although OCT has the highest resolution than any currently available imaging technology and the acquisition rates are near video speed, the relatively shallow penetration depth and the interference from intracoronary blood restricts in vivo imaging. Like IVUS, it relies on X-ray angiography for catheter positioning. Dyring the last years OCT balances between a research and a clinical tool for the evaluation of atheromatosis of the coronary arteries.

OCT as a Research Tool

The high-resolution power of OCT has been the focus of interest for the use of OCT as a research tool for the evaluation of the vulnerable plaque. OCT is able to detect in vivo almost all components of the vulnerable plaque, including the lipid pool and the thickness of cap. Cap thickness <65 μm and large lipid core together with increased inflammatory activation and low calcium concentration are characteristic patterns of the vulnerable plaque. OCT has been correlated with intravascular thermography, which is the gold standard for the intravascular assessment of thermal heterogeneity and inflammatory activity of the atheromatic plaque. This correlation is also evident regarding the novel method of the microwave radiometry that measures internal body temperatures noninvasively. Moreover, OCT has a very good correlation with serum inflammatory markers such as CRP and various interleukins. Examination of ex vivo samples has yielded a 71%–96% sensitivity and 90%–98% specificity for plaque type detection, particularly TCFAs, in numerous studies. Nowadays, new algorithms that calculate shear stress from OCT images had been developed, giving new information about the role of local shear stress and the propensity to present with ACS. OCT is the method of choice for the assessment of the efficacy of new technologies including new stents. The effect of bioabsorption of both full bioresorbable and of the stents with bioresorbable polymer can be evaluated with high accuracy with OCT. Moreover OCT had more experimental implications regarding the studies that deal with the duration and the type of dual or triple antiplatelet therapy after an ACS, as this is in close relation with the underexpansion or malapposition of the stent as well as with the stent type.

Optical Coherence Tomography as a Clinical Tool

Besides being an excellent experimental tool, the clinical application of OCT has been questioned during the last years. OCT is a utility that can precisely measure lesion length and vessel diameter in PCI, which is useful in optimizing the size of dilation balloons and stents. Because of the high resolution of OCT, the angle and location of the dissection flap are more accurately identified in OCT systems. The current guidelines exclude OCT from the tools proposed to be used to determine intermediate stenosis severity. D'Ascenzo et al. performed a meta-analysis looking at studies assessing diagnostic accuracy, sensitivity, and specificity of MLD or MLA as derived from IVUS or OCT to detect functionally significant stenosis as determined by FFR. In the OCT studies, they found that neither MLA nor MLD had adequate sensitivity or specificity to guide decisions for revascularization. However, several studies have compared the cut of points of FFR and IVUS with the measurements performed by OCT and have proposed cut off points for OCT in order to assess significant lesions in the proximal coronary vessels.

OCT can be used for the assessment of stent placement and long-term stent outcome because OCT provides a high-resolution image of the stent—vessel interface and has the capacity to detect very thin layers of neointima. Mechanical stent failure (incomplete stent expansion, stent fracture) can be differentiated from impaired healing (absence of strut coverage, absence of homogeneous coverage, or late strut malapposition) by OCT as well. The differentiation of the mechanism of stent failure during an ACS is essential for the proper treatment of the lesion and for the proper pharmaceutical treatment of the patient. Therefore, OCT should be used in such a clinical scenario and this is proposed in the guidelines of revascularization of both the European Society of Cardiology and of the American College of Cardiology and American Heart Association.

OCT can be used with great success for the assessment of the pathophysiology of an ACS, not only in previously stented segments but on native segments as well. This is due to the fact that OCT is the only modality that can depict intracoronary thrombus and can differentiate between plaque rupture and erosion (Fig. 3.5.2.9) Recently at least two studies showed that in case of plaque erosion the conservative treatment of the patients that consists of thrombectomy and dual antiplatelet therapy without stenting are as effective as the traditional stenting of these lesions. However, this approach has to tested in larger populations and has not up to now incorporated into the everyday clinical practice. Similarly OCT can be used to guide PCI by measuring the vessel dimensions and lesion length before the stent deployment. Both pre- and post-PCI OCT have been shown to

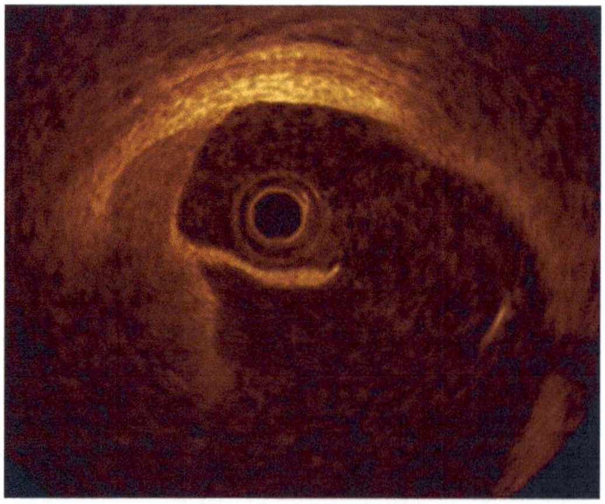

Figure 3.5.2.9 Optical coherence tomography imaging of a coronary vessel with a dissection with the presence of thrombus [1–48].

impact procedural physician decision making. A recent study showed that pre-PCI OCT resulted in altered decision making in 57% of stenoses due to cardiologists selecting different stent lengths and post-PCI OCT prompted further stent optimization with postdilatation or occasionally with additional stents in 27% of stenoses based on findings of malapposition, underexpansion, or edge dissection. Whether OCT-guided PCI can improve the patient's long-term outcome compared to the PCI performed with the guidance of only angiography has been the point of interest of recent studies. It seems that OCT-guided PCI can decrease the need for further target lesion revascularization, but not overall or cardiovascular mortality; therefore, it should be performed in a well-selected patient population.

REFERENCES

[1] Iannaccone M, D'Ascenzo F, Frangieh AH, Niccoli G, Ugo F, Boccuzzi G, Bertaina M, Mancone M, Montefusco A, Amabile N, et al. Impact of an optical coherence tomography guided approach in acute coronary syndromes: a propensity matched analysis from the international FORMIDABLE-CARDIOGROUP IV and USZ registry. Catheter Cardiovasc Interv 2016. http://dx.doi.org/10.1002/ccd.26880.

[2] Chatzizisis YS, Toutouzas K, Giannopoulos AA, Riga M, Antoniadis AP, Fujinom Y, Mitsouras D, Koutkias VG, Cheimariotis G. Doulaverakis C and others. Association of global and local low endothelial shear stress with high-risk plaque using intracoronary 3D optical coherence tomography: introduction of 'shear stress score'. Eur Heart J Cardiovasc Imaging 2016. http://dx.doi.org/10.1093/ehjci/jew134.

[3] Bouki KP, Vlad D, Riga M, Stergiouli I, Toutouzas KP. Improved vascular healing after the successful treatment of very late sirolimus-eluting stent thrombosis with a bare metal stent implantation - a serial optical coherence tomography study. Hell J Cardiol 2016;57(2):129–33.

[4] Synetos A, Papanikolaou A, Toutouzas K, Georgiopoulos G, Karanasos A, Drakopoulou M, Latsios G, Papaioannou S, Tsiamis E, Tousoulis D. Metabolic syndrome predicts plaque rupture in patients with acute myocardial infarction. An optical coherence study. Int J Cardiol 2016;209:139–41.

[5] Iannaccone M, Quadri G, Taha S, D'Ascenzo F, Montefusco A, Omede P, Jang IK, Niccoli G, Souteyrand G, Yundai C, et al. Prevalence and predictors of culprit plaque rupture at OCT in patients with coronary artery disease: a meta-analysis. Eur Heart J Cardiovasc Imaging 2016;17(10):1128–37.

[6] Toutouzas K, Chatzizisis YS, Riga M, Giannopoulos A, Antoniadis AP, Tu S, Fujino Y, Mitsouras D, Doulaverakis C, Tsampoulatidis I, et al. Accurate and reproducible reconstruction of coronary arteries and endothelial shear stress calculation using 3D OCT: comparative study to 3D IVUS and 3D QCA. Atherosclerosis 2015;240(2):510–9.

[7] Bouki KP, Sakkali E, Toutouzas K, Vlad D, Barmperis D, Phychari S, Riga M, Apostolou T, Stefanadis C. Impact of coronary artery stent edge dissections on long-term clinical outcome in patients with acute coronary syndrome: an optical coherence tomography study. Catheter Cardiovasc Interv 2015;86(2):237–46.

[8] Chatzizisis YS, Koutkias VG, Toutouzas K, Giannopoulos A, Chouvarda I, Riga M, Antoniadis AP, Cheimariotis G, Doulaverakis C, Tsampoulatidis I, et al. Clinical validation of an algorithm for rapid and accurate automated segmentation of intracoronary optical coherence tomography images. Int J Cardiol 2014;172(3):568–80.

[9] Hu S, Yonetsu T, Jia H, Karanasos A, Aguirre AD, Tian J, Abtahian F, Vergallo R, Soeda T, Lee H, et al. Residual thrombus pattern in patients with ST-segment elevation myocardial infarction caused by plaque erosion versus plaque rupture after successful fibrinolysis: an optical coherence tomography study. J Am Coll Cardiol 2014;63(13):1336–8.

[10] Toutouzas K, Karanasos A, Riga M, Tsiamis E, Synetos A, Michelongona A, Papanikolaou A, Triantafyllou G, Tsioufis C, Stefanadis C. Optical coherence tomography assessment of the spatial distribution of culprit ruptured plaques and thin-cap fibroatheromas in acute coronary syndrome. EuroIntervention 2012;8(4):477–85.
[11] Toutouzas K, Karanasos A, Stathogiannis K, Synetos A, Tsiamis E, Papadopoulos D, Stefanadis C. A honeycomb-like structure in the left anterior descending coronary artery: demonstration of recanalized thrombus by optical coherence tomography. JACC Cardiovasc Interv 2012;5(6): 688–9.
[12] Prati F, Guagliumi G, Mintz GS, Costa M, Regar E, Akasaka T, Barlis P, Tearney GJ, Jang IK, Arbustini E, et al. Expert review document part 2: methodology, terminology and clinical applications of optical coherence tomography for the assessment of interventional procedures. Eur Heart J 2012; 33(20):2513–20.
[13] Toutouzas K, Karanasos A, Stefanadis C. Pitfalls of angiography in the assessment of atherosclerosis: the role of optical coherence tomography. J Invasive Cardiol 2012;24(5):246–7.
[14] Davlouros PA, Karantalis V, Xanthopoulou I, Mavronasiou E, Tsigkas G, Toutouzas K, Alexopoulos D. Mechanisms of non-fatal stent-related myocardial infarction late following coronary stenting with drug-eluting stents and bare metal stents. Insights from optical coherence tomography. Circ J 2011;75(12):2789–97.
[15] Toutouzas K, Karanasos A, Tsiamis E, Riga M, Drakopoulou M, Synetos A, Papanikolaou A, Tsioufis C, Androulakis A. Stefanadi E and others. New insights by optical coherence tomography into the differences and similarities of culprit ruptured plaque morphology in non-ST-elevation myocardial infarction and ST-elevation myocardial infarction. Am Heart J 2011;161(6):1192–9.
[16] Bouki KP, Katsafados MG, Chatzopoulos DN, Psychari SN, Toutouzas KP, Charalampopoulos AF, Sakkali EN, Koudouri AA, Liakos GK, Apostolou TS. Inflammatory markers and plaque morphology: an optical coherence tomography study. Int J Cardiol 2012;154(3):287–92.
[17] Toutouzas K, Karanasos A, Stefanadis C. Multiple plaque morphologies assessed by optical coherence tomography in a patient with acute coronary syndrome. Heart 2010;96(16):1335–6.
[18] Toutouzas K, Tsiamis E, Karanasos A, Drakopoulou M, Synetos A, Tsioufis C, Tousoulis D, Davlouros P, Alexopoulos D, Bouki K, et al. Morphological characteristics of culprit atheromatic plaque are associated with coronary flow after thrombolytic therapy: new implications of optical coherence tomography from a multicenter study. JACC Cardiovasc Interv 2010;3(5):507–14.
[19] Toutouzas K, Vaina S, Riga MI, Stefanadis C. Evaluation of dissection after coronary stent implantation by intravascular optical coherence tomography. Clin Cardiol 2009;32(7):E47–8.
[20] Stefanadis C, Vavuranakis M, Toutouzas P. Vulnerable plaque: the challenge to identify and treat it. J Interv Cardiol 2003;16(3):273–80.
[21] Stefanadis C, Toutouzas K, Tsiamis E, Vavuranakis M, Tsioufis C, Stefanadi E, Boudoulas H. Relation between local temperature and C-reactive protein levels in patients with coronary artery disease: effects of atorvastatin treatment. Atherosclerosis 2007;192(2):396–400.
[22] Toutouzas K, Stougiannos P, Drakopoulou M, Mitropoulos J, Bosinakou E, Markou V, Latsios G, Karabelas I, Stefanadi E, Stefanadis C. Coronary sinus thermography in idiopathic dilated cardiomyopathy: correlation with systemic inflammation and left ventricular contractility. Eur J Heart Fail 2007; 9(2):168–72.
[23] Toutouzas K, Drakopoulou M, Markou V, Stougianos P, Tsiamis E, Tousoulis D, Stefanadis C. Increased coronary sinus blood temperature: correlation with systemic inflammation. Eur J Clin Invest 2006;36(4):218–23.
[24] Toutouzas K, Drakopoulou M, Mitropoulos J, Tsiamis E, Vaina S, Vavuranakis M, Markou V, Bosinakou E, Stefanadis C. Elevated plaque temperature in non-culprit de novo atheromatous lesions of patients with acute coronary syndromes. J Am Coll Cardiol 2006;47(2):301–6.
[25] Toutouzas K, Drakopoulou M, Stefanadi E, Siasos G, Stefanadis C. Intracoronary thermography: does it help us in clinical decision making? J Interv Cardiol 2005;18(6):485–9.
[26] Toutouzas K, Markou V, Drakopoulou M, Mitropoulos I, Tsiamis E, Stefanadis C. Patients with type two diabetes mellitus: increased local inflammatory activation in culprit atheromatous plaques. Hell J Cardiol 2005;46(4):283–8.

[27] Toutouzas K, Markou V, Drakopoulou M, Mitropoulos I, Tsiamis E, Vavuranakis M, Vaina S, Stefanadis C. Increased heat generation from atherosclerotic plaques in patients with type 2 diabetes: an increased local inflammatory activation. Diabetes Care 2005;28(7):1656–61.

[28] Carlier S, Kakadiaris IA, Dib N, Vavuranakis M, O'Malley SM, Gul K, Hartley CJ, Metcalfe R, Mehran R, Stefanadis C, et al. Vasa vasorum imaging: a new window to the clinical detection of vulnerable atherosclerotic plaques. Curr Atheroscler Rep 2005;7(2):164–9.

[29] Stefanadis C, Toutouzas K, Tsiamis E, Vaina S, Tsioufis C, Stefanadis E, Toutouzas P. The cooling effect of coronary blood flow on heart: a new approach. J Invasive Cardiol 2004;16(9):455–8.

[30] Toutouzas K, Tsiamis E, Vavuranakis M, Stefanadis C. Coronary artery plaque temperature: what do we measure? Eur Heart J 2004;25(11):993–4.

[31] Stefanadis C, Tsiamis E, Vaina S, Toutouzas K, Boudoulas H, Gialafos J, Toutouzas P. Temperature of blood in the coronary sinus and right atrium in patients with and without coronary artery disease. Am J Cardiol 2004;93(2):207–10.

[32] Stefanadis C, Toutouzas K, Vavuranakis M, Tsiamis E, Vaina S, Toutouzas P. New balloon-thermography catheter for in vivo temperature measurements in human coronary atherosclerotic plaques: a novel approach for thermography? Catheter Cardiovasc Interv 2003;58(3):344–50.

[33] Stefanadis C, Toutouzas K, Tsiamis E, Mitropoulos I, Tsioufis C, Kallikazaros I, Pitsavos C, Toutouzas P. Thermal heterogeneity in stable human coronary atherosclerotic plaques is underestimated in vivo: the "cooling effect" of blood flow. J Am Coll Cardiol 2003;41(3):403–8.

[34] Stefanadis C, Toutouzas K, Vaina S, Vavuranakis M, Toutouzas P. Thermography of the cardiovascular system. J Interv Cardiol 2002;15(6):461–6.

[35] Stefanadis C, Toutouzas K, Vavuranakis M, Tsiamis E, Tousoulis D, Panagiotakos DB, Vaina S, Pitsavos C, Toutouzas P. Statin treatment is associated with reduced thermal heterogeneity in human atherosclerotic plaques. Eur Heart J 2002;23(21):1664–9.

[36] Stefanadis C, Toutouzas K, Tsiamis E, Vavuranakis M, Kallikazaros I, Toutouzas P. Thermography of human arterial system by means of new thermography catheters. Catheter Cardiovasc Interv 2001; 54(1):51–8.

[37] Stefanadis C, Toutouzas K, Tsiamis E, Pitsavos C, Papadimitriou L, Toutouzas P. Identification and stabilization of vulnerable atherosclerotic plaques: the role of coronary thermography and external heat delivery. Indian Heart J 2001;53(1):104–9.

[38] Stefanadis C, Toutouzas K, Tsiamis E, Stratos C, Vavuranakis M, Kallikazaros I, Panagiotakos D, Toutouzas P. Increased local temperature in human coronary atherosclerotic plaques: an independent predictor of clinical outcome in patients undergoing a percutaneous coronary intervention. J Am Coll Cardiol 2001;37(5):1277–83.

[39] Stefanadis C, Diamantopoulos L, Dernellis J, Economou E, Tsiamis E, Toutouzas K, Vlachopoulos C, Toutouzas P. Heat production of atherosclerotic plaques and inflammation assessed by the acute phase proteins in acute coronary syndromes. J Mol Cell Cardiol 2000;32(1):43–52.

[40] Stefanadis C, Diamantopoulos L, Vlachopoulos C, Tsiamis E, Dernellis J, Toutouzas K, Stefanadi E, Toutouzas P. Thermal heterogeneity within human atherosclerotic coronary arteries detected in vivo: a new method of detection by application of a special thermography catheter. Circulation 1999; 99(15):1965–71.

[41] Stefanadis C, Toutouzas P. In vivo local thermography of coronary artery atherosclerotic plaques in humans. Ann Intern Med 1998;129(12):1079–80.

[42] Otsuka F, Joner M, Prati F, Virmani R, Narula J. Clinical classification of plaque morphology in coronary disease. Nat Rev Cardiol 2014;11(7):379–89.

[43] Di Vito L, Yoon JH, Kato K, Yonetsu T, Vergallo R, Costa M, Bezerra HG, Arbustini E, Narula J, Crea F, et al. Comprehensive overview of definitions for optical coherence tomography-based plaque and stent analyses. Coron Artery Dis 2014;25(2):172–85.

[44] Prati F, Uemura S, Souteyrand G, Virmani R, Motreff P, Di Vito L, Biondi-Zoccai G, Halperin J, Fuster V, Ozaki Y, et al. OCT-based diagnosis and management of STEMI associated with intact fibrous cap. JACC Cardiovasc Imaging 2013;6(3):283–7.

[45] Karanasos A, Van Mieghem N, van Ditzhuijzen N, Felix C, Daemen J, Autar A, Onuma Y, Kurata M, Diletti R, Valgimigli M, et al. Angiographic and optical coherence tomography insights into bioresorbable scaffold thrombosis: single-center experience. Circ Cardiovasc Interv 2015;(5):8.

[46] Karanasos A, Muramatsu T, Diletti R, Nauta S, Onuma Y, Lenzen M, Nakatani S, Van Mieghem NM, Schultz C, De Jaegere PP, et al. Early and late optical coherence tomography findings following everolimus-eluting bioresorbable vascular scaffold implantation in myocardial infarction: a preliminary report. Hell J Cardiol 2015;56(2):125−35.
[47] Karanasos A, Simsek C, Gnanadesigan M, van Ditzhuijzen NS, Freire R, Dijkstra J, Tu S, Van Mieghem N, van Soest G, De Jaegere P, et al. OCT assessment of the long-term vascular healing response 5 years after everolimus-eluting bioresorbable vascular scaffold. J Am Coll Cardiol 2014; 64(22):2343−56.
[48] Karanasos A, Ligthart J, Witberg K, van Soest G, Bruining N, Regar E. Optical coherence tomography: potential clinical applications. Curr Cardiovasc Imaging Rep 2012;5(4):206−20.

CHAPTER 3.5.3

Functional Assessment of Coronary Lesions in the Cath Lab

Sofia Vaina, Dimitris Tousoulis
1st Cardiology Clinic, 'Hippokration' General Hospital, National and Kapodistrian University of Athens, School of Medicine, Athens, Greece

Contents

INTRODUCTION

Coronary artery bypass grafting (CABG) and percutaneous coronary angioplasty (PCI) have significantly improved quality of life and long-term survival of patients with coronary artery disease (CAD). On the other hand, during the last years, medical treatment with novel agents such as statins, beta-blockers, and newer antiplatelet drugs has also been proven to further prolong life expectancy with sustained good clinical outcome, especially in stable ischemic heart disease. Therefore, it seems imperative to identify those patients that would benefit from revascularization with either CABG or PCI and those patients that would have an excellent prognosis on optimal medical treatment.

Multiple studies have sought to compare optimal medical therapy with revascularization in patients with chronic CAD in order to clarify which treatment approach is appropriate and for whom. COURAGE (Clinical Outcomes Utilizing Revascularization and Aggressive Drug Evaluation) was one of the most significant trials including patients with stable CAD. The study showed that patients randomized to medical therapy had similar long-term clinical outcome to those undergoing PCI [1]. Similar results were demonstrated in BARI 2D (The Bypass Angioplasty Revascularization

Coronary Artery Disease
ISBN 978-0-12-811908-2, http://dx.doi.org/10.1016/B978-0-12-811908-2.00019-2

Investigation 2 Diabetes), which recruited 2368 patients with CAD and diabetes [2]. BARI 2D investigators randomized patients to intensive medical treatment or revascularization. In the particular study, no difference was observed between the two groups in terms of survival (87.8% vs. 88.3%, $P = .97$) or freedom from major cardiovascular events (death, myocardial infarction, or stroke) (77.2% vs. 75.9%, $P = .70$) [2].

As it is concluded from the majority of the studies, even in high-risk populations such as diabetics, long-term prognosis is similar with medical treatment or revascularization. However, it has also been demonstrated that patients with proven extensive ischemia may benefit more from surgical or percutaneous revascularization. A COURAGE subanalysis included 314 patients that underwent stress-gated myocardial perfusion single-photon emission computed tomography before randomization and at follow-up. Those patients who were randomized to PCI were found to have less ischemia after the procedure compared with patients treated with intensive medical therapy [3]. Furthermore, BARI 2D substudy in which 1.505 patients underwent stress myocardial perfusion imaging at follow-up revealed that extended myocardial ischemia is associated with poor prognosis and decreased event-free survival [4]. Hence, the size of ischemia distinguishes high-risk patients and it has been proven that those with >10%–12% left ventricular ischemia have a better outcome with revascularization than medical therapy [3–7].

Even though, there is abundance of evidence that ischemia detection and quantification is important to determine the best treatment strategy for patients with stable CAD and not even half of the patients in whom coronary angiography is performed has proof of ischemia, due to several reasons [8]. Additionally, ad-hoc percutaneous procedures are performed world widely in order to reduce the cost of a second procedure or for patient comfort. Thus, there is a need of a diagnostic test that can be easily implemented in the cath lab and reliably assess the physiological significance of a coronary artery stenosis. Fractional flow reserve (FFR) is a well-established method that has been proven to measure the functional significance of a stenosis. FFR is a pressure-derived index and is the ratio of maximal myocardial blood flow in the presence of an epicardial coronary stenosis to maximal myocardial blood flow if the artery was normal [9]. For example, if FFR is 0.70, it means that maximal myocardial blood flow reaches 70% of its normal value.

FRACTIONAL FLOW RESERVE

Physiological Foundation

FFR is based on a simplified model of the coronary circulation. Coronary blood flow is controlled by regulation of vascular tone that depends on intrinsic myogenic tone, endothelial cell signaling, metabolic effectors, and neurohumoral control [10]. The combination of these underlying mechanisms is characterized as coronary autoregulation. When there is a constant level of myocardial metabolic demands, coronary autoregulation induces dilatation and constriction of coronary arteriole. Consequently, constant coronary blood flow is achieved within a physiological range of coronary perfusion pressures [11]. Accordingly, an increase in myocardial metabolic demands increases blood flow at a constant perfusion pressure. This is called metabolic adaptation. Both coronary autoregulation and metabolic adaptation, which are interrelated, maintain coronary blood flow at a level that fulfills myocardial metabolic demands [12].

In normal coronary arteries without stenosis, blood flow and pressure are constant throughout the coronary arterial bed no matter how distal in the vessel or how small the side-branch. Thus, coronary blood flow depends on the pressure difference across the aorta and the coronary veins and the resistance imposed by the myocardial bed [13]. During exercise or in general when myocardial oxygen demand increases, both coronary blood flow and pressure increase. However, in the presence of a coronary stenosis, blood flow and pressure cannot increase after the lesion. Therefore, coronary blood flow and pressure are lower after the stenosis compared with blood flow and pressure before the stenosis [13]. When maximal hyperemia is achieved, resistance is almost zero and pressure is linearly related to flow [13]. Thus, the ratio of two flows can be derived from the ratio of two pressures as long as these pressures are measured during maximal hyperemia [13]. Pressure measurements are easily obtainable with the use of a specially designed wire that has a pressure sensor located 3 cm proximal to the tip of the wire (Fig. 3.5.3.1). FFR is the ratio of pressure distally to an epicardial coronary stenosis to pressure proximally to the stenosis as measured during maximal hyperemia [13]. In the absence of a stenosis, FFR is 1 and decreases depending on the significance of the stenosis (Fig. 3.5.3.2).

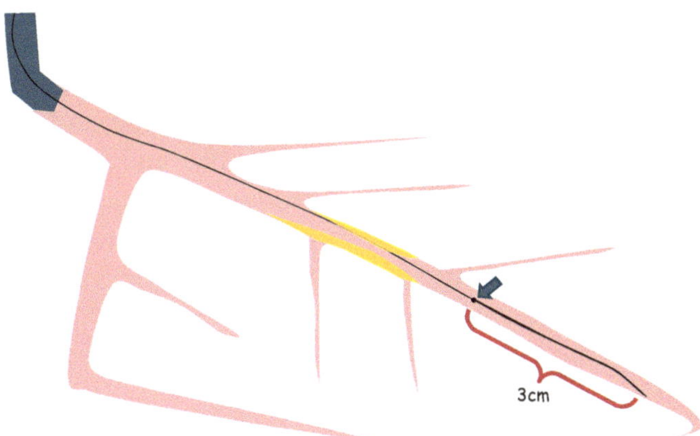

Figure 3.5.3.1 Fractional flow reserve measurements are performed with a specific sensor mounted on a floppy-tipped guide wire. The sensor (*arrow*) is placed at the junction between the 3 cm long radiopaque tip of the wire and the remainder of the wire.

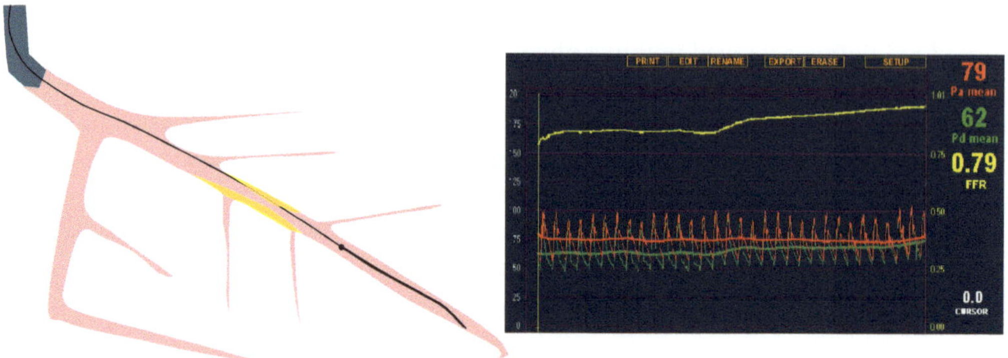

Figure 3.5.3.2 In the absence of a stenosis fractional flow reserve is 1 and decreases depending on the significance of the stenosis.

FFR Measurement Technique

FFR measurements are typically performed after a coronary angiogram to estimate the functional significance of a stenosis or after coronary angioplasty to evaluate the final result. Before the procedure, administration of heparin is mandatory in order to achieve an activated coagulation time-ACT of at least 250 s. Also, abolition of any form of epicardial vasoconstriction is necessary; therefore, intracoronary nitrates should also be administered (a bolus dose of 200 μg isosorbide dinitrate) at least 30 s before measurements [14]. Hyperemia with microvascular vasodilation is induced with the administration of adenosine. Adenosine is administered as a bolus dose intracoronary (40 μg in the right coronary artery and 60 μg in the left coronary artery, increasing the doses incrementally by 30 μg to a maximum of 150 μg) [15], or by intravenous infusion (140 μg/kg/min) [16,17] for 2–3 min or until stable hyperemia. Hyperemia is achieved 10 s after initial intravenous administration of adenosine, whereas maximal hyperemia is achieved after 1 min. Side effects of adenosine include thoracic pain, bradycardia, dyspnea, headache, blood pressure drop, coughing, and patient discomfort. Studies have shown that FFR measurements are similar and independent of the site of adenosine administration. Intracoronary adenosine is associated with less adverse effects and is easily prepared and administered making it an appealing choice. However, for reliable intracoronary adenosine delivery, the catheter must be well seated in the coronary ostium. Thus, intracoronary adenosine should be avoided for the evaluation of ostial lesions. Also, intravenous adenosine results in a prolonged hyperemic state and increases the amount of time available to obtain the FFR measurement. On the other hand, it should be taken under account that intravenous adenosine can influence a patient's hemodynamic state and therefore influence the results of the FFR measurement [18].

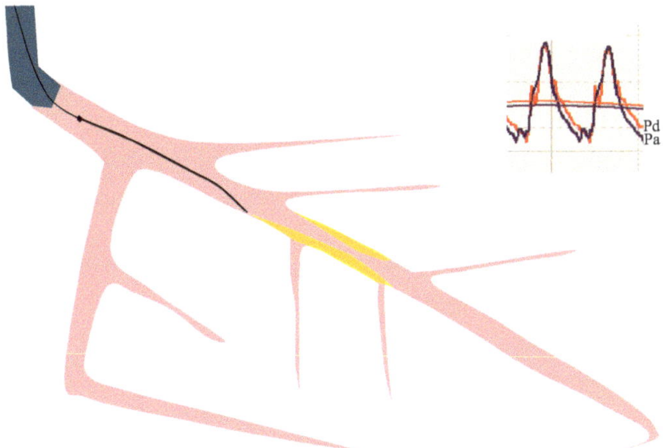

Figure 3.5.3.3 Before advancing the sensor into the vessel beyond the lesion to be studied, the pressures recorded by the sensor (Pd) and by the guiding catheter (Pa) should be equalized.

FFR measurements are performed with a specific sensor mounted on a floppy-tipped guide wire. The sensor is placed at the junction between the 3 cm long radiopaque tip of the wire and the remainder of the wire (Fig. 3.5.3.1). Currently, there are two main producers of the wires and systems, Volcano Corporation (San Diego, CA) and St Jude Medical, Inc. (St. Paul, MN). FFR equipment manufactured by Volcano includes the following: (1) Verrata Pressure Guide Wire, (2) PrimeWire Prestige Plus Guide Wire, and (3) Combowire XT Guide Wire for simultaneous measurement of intravascular pressure and Doppler flow. There are two models, one with Doppler flow and pressure sensors at the tip of the wire and the other with a Doppler flow sensor at the tip and a pressure sensor 1.5 cm from the tip. FFR equipment manufactured by St Jude Medical includes the following: (1) PressureWire X Guidewire.

Before advancing the sensor into the vessel beyond the lesion to be studied, the pressures recorded by the sensor and by the guiding catheter should be equalized (Fig. 3.5.3.3).

Application of FFR

Deferral of Revascularization

Ischemia-driven revascularization in patients with stable CAD improves quality of life and long-term survival. Coronary lesions that are not hemodynamically significant can be treated optimally with medical agents. DEFER (Deferral vs. Performance of PTCA in Patients Without Documented Ischemia) was the first study to show that patients with intermediate stenosis and FFR $\geq$0.75 after deferral of PCI had excellent prognosis [19]. In total, 325 patients enrolled in the study, 144 patients had FFR <0.75 and underwent PCI, 90 patients had FFR ≥0.75 and also underwent PCI and 91 patients had FFR ≥0.75 and were treated medically. At 2-year follow-up, the patients with FFR ≥0.75 that were treated medically had the best clinical outcome compared with the other two groups [19]. As it was expected, patients with FFR <0.75 that underwent PCI had less symptoms compared with those who had FFR ≥0.75 ($P = .001$) [19]. These favorable results were confirmed at 5 and 15 years follow-up, without signs of late "catch-up" phenomenon [20,21].

Appropriateness of Revascularization

DEFER study showed that those coronary lesions that do not induce ischemia can be safely treated with medication. Therefore, in patients with multiple lesions it is important to recognize in which lesions a stent should be implanted for optimal long-term results with the lower possible cost. FAME (Fractional Flow Reserve vs. Angiography for Multivessel Evaluation) a large randomized trial compared treatment based on FFR measurements with treatment guided solely by angiography in patients with multivessel CAD [22]. The study included 1.005 patients. The composite endpoint of major adverse cardiac events, including death, myocardial infarction, and repeat coronary revascularization, at 1 year was 18.3% in the angiography group and 13.2% in the FFR group ($P = .02$). The results of FAME confirmed that routine FFR measurement in patients with multivessel coronary artery improves the prognosis of these patients.

FAME 2 (the Fractional Flow Reserve vs. Angiography for Multivessel Evaluation 2), which was also a randomized, multicenter trial, was undertaken in order to evaluate whether patients with stable CAD and significant stenoses in the coronary arteries had better clinical outcome with FFR-guided PCI compared with medical therapy alone [23]. All patients with significant stenoses and FFR less than 0.8 were randomized either to PCI or optimal medical therapy. The study was terminated prematurely, because of a highly significant difference between PCI and medical therapy in the incidence rates of the primary endpoint, which was the composite of death from any cause, nonfatal myocardial infarction or unplanned hospitalization leading to urgent revascularization [23]. More specifically, the primary endpoint was 4.3% in the PCI group versus 12.7% in the optimal medical group ($P < .001$) [23]. What is interesting is the fact that the primary endpoint excess in the medically treated group was due to increased number of hospitalizations and urgent revascularizations because of acute coronary syndrome.

FAME 2 further established that FFR is a significant tool that helps identify patients in need of coronary revascularization with drug-eluting stent implantation. Other, smaller studies, also demonstrated that FFR guidance is important for coronary bypass surgery [24,25]. Botman et al. showed that at 1 year, the patency of bypass grafts on functionally significant stenoses with FFR <0.75 is significantly higher than the patency of bypass grafts on nonsignificant lesions with FFR >0.75 [24]. Furthermore, in a recent retrospective study, FFR-guided coronary artery bypass graft surgery was associated with a fewer graft anastomoses and a lower rate of on-pump surgery compared with angiography-guided coronary artery bypass graft surgery. Even though, the event rate was similar between the FFR-guided and the angiography-guided surgery during the 36 months follow-up period, FFR-guided surgery was associated with a lower rate of angina [25].

FFR in Different Patient Subsets

Left Main Disease

Significant stenosis in the left main is defined as a diameter stenosis of ≥50%. However, in some patients the severity of left main lesions is not always easily estimated with coronary angiography, as previous studies have shown increased interobserver variability even among very experienced operators [26]. This is due to several reasons such as (1) short left main, (2) deep catheter intubation that may cover the ostium of left main, (3) atherosclerotic lesions of the proximal left anterior descending artery or the proximal circumflex artery that obscure the comparison of the left main lesion with the proximal or distal disease-free vessel. Identification of left main lesion severity is very important in order to decide the optimal therapeutic strategy. Even though left main disease with low SYNTAX (SYNergy between percutaneous coronary intervention with TAXus and cardiac surgery) score can be treated with PCI, it still remains a challenging procedure. Furthermore, coronary artery bypass grafts do not work and tend to occlude if coronary stenosis is not hemodynamically severe. Hamilos et al. confirmed that patients with left main disease and FFR ≥0.80 have excellent prognosis, similar to that of patients with FFR <0.80 who undergo surgical revascularization [26].

Fractional flow reserve measurement is generally easy, but there are some issues that must be taken into account in left main assessment. If there is an ostial lesion the catheter should be not engaged in the ostium and measurements should be performed with intravenous administration of adenosine. A stenosis in left anterior descending artery or in left circumflex or in both vessels can influence the functional severity of a left main stenosis. In such cases, it is preferable to measure left main stenosis by advancing the FFR wire to the disease-free vessel. Nonetheless, one must bear in mind that the true functional severity of the left main lesion is obtained only after alleviated the resistance opposed by the distal lesion. If this is not plausible, an alternative imaging technique such as IVUS or OCT should be considered in order to clarify the severity of the lesion.

Bifurcations

Bifurcations are a subset of lesions that pose a special technical problem when measuring FFR because of their anatomical complexity. Before PCI, FFR measurements are performed in order to evaluate the functional significance of main branch stenosis or isolated side branch stenosis (Table 3.5.3.1) [27]. However, most frequently, the side branch needs evaluation after treatment of the main vessel because in most bifurcations provisional stenting of the main vessel is the preferred technique, even if both the main and

Table 3.5.3.1 FFR during bifurcation intervention

	FFR is useful	FFR is generally not recommended
Preintervention	To assess the functional significance of MB To assess the functional significance of pure SB stenosis	Small SB To determine functional significance of SB when there is a significant MB stenosis SB FFR to predict the functional significance of jailed SB
Post-MB stenting	To assess the functional significance of jailed SB and to predict the outcomes	Small SB Long diffuse, highly angulated or calcified SB SB slow flow
Post-SB angioplasty	To assess SB procedural success and to predict the outcomes after KBI	SB slow flow SB severe dissection
Post-SB stenting	To evaluate residual ischemia	To predict procedural outcomes of complex two stenting

FFR, fractional flow reserve; *KBI*, kissing balloon inflation; *MB*, main branch; *SB*, side branch.

the side branch are diseased [28–31]. Therefore, the FFR wire must be advanced through the struts of the main vessel stent to the jailed side branch and is important to always balance the risk and benefit of FFR measurement for jailed side branches. FFR measurement is considered to be most effective when performed in a large side branch with a short lesion at the ostium [27].

Ostial Stenosis

Stenoses at the ostium of the left main or the right coronary artery are often difficult to evaluate and challenging to treat due to increased restenosis rate. Hence, it is important to recognize which lesion is functionally severe and will benefit the most of a revascularization procedure. Ziaee et al. demonstrated that ostial lesions are overestimated. In 25 patients with ostial lesions angiographically characterized more than >70% only five had FFR <0.75. On the other hand, none of the patients that had less than 70% stenosis had FFR <0.75 [32].

FFR is a useful tool to assess ostial lesions but there are pitfalls and the operator must be careful to keep the catheter away from the ostium during the procedure in order to avoid dumping of the pressure. Additionally, adenosine should be administered intravenously and not intracoronary. If these prerequisites are not fulfilled IVUS or OCT should be considered for lesion evaluation.

Acute Coronary Syndrome

Maximal hyperemia is imperative for linear relationship between pressure and flow, provided that the microcirculation is intact [33]. Myocardial infarction adversely affects coronary microcirculation because of thrombus embolizaton, microvascular stunning, acute ischemic microvascular dysfunction and edema [34]. Hence, although FFR measurement of the culprit lesion in the acute setting has been proven to be safe, the utility of this practice is not established and is not recommended. In non ST elevation myocardial infarction, small studies have shown that FFR is reliable between four and 6 days following the acute event [35,36].

Patients with acute coronary syndrome frequently have multivessel disease. This plays an important role on prognosis. Optimal management of nonculprit vessels during culprit lesion revascularization still remains a controversial issue. It has been suggested that treatment of nonculprit stenoses should be postponed until ischemia is proven by noninvasive stress tests. However, there are several studies that provide substantial evidence to support FFR measurement in nonculprit ambiguous stenoses in the acute coronary syndrome setting [37,38].

INSTANTANEOUS WAVE-FREE RATIO (IFR)

Instantaneous wave-free ratio (iFR) is a recently developed physiological index, which is obtained at rest without the need of hyperemic stimulation. iFR is calculated as the ratio of blood pressure distal and proximal to a coronary artery stenosis during the diastolic wave-free period (Fig. 3.5.3.4) [39]. During mid to late diastole there is a wave-free

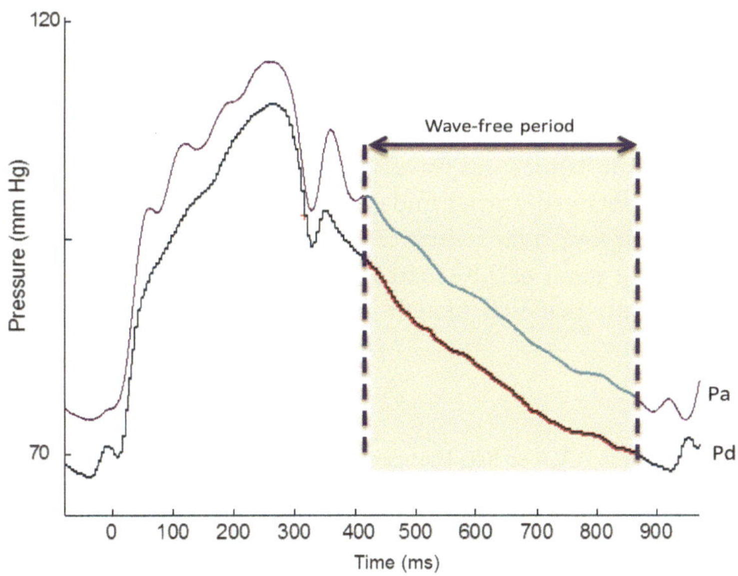

Figure 3.5.3.4 Instantaneous wave-free ratio is calculated as the ratio of blood pressure distal and proximal to a coronary artery stenosis during the diastolic wave-free period. *Pd*, blood pressure distal to the stenosis; *Pa*, blood pressure proximal to the stenosis.

period, which starts 25% of the way into diastole and ends 5msec before the start of diastole [40]. During the wave-free period new pressure wavefronts do not occur and resting microvascular resistance is relatively minimized [40].

The process of iFR measurement is easier than FFR measurement because it only takes about five heartbeats without the need for hyperemic agents to assess lesion significance. Because the introduction of iFR, a hybrid iFR-FFR diagnostic strategy has been proposed [41]. The upper and lower iFR cut-offs of 0.86 and 0.93 can be used to restrict decisions on the basis of iFR to those regions in which its agreement with FFR is very high. FFR measurement is recommended to the intermediate iFR range of values called the "adenosine zone" [41]. Applying this hybrid approach 65.1% of patients may be free from hyperemic agents [42]. However, several trials have demonstrated that an iFR cut point of 0.89 matches an FFR cut point of 0.80 [42,43]. Currently, two randomized trials are undertaken in order to evaluate whether iFR-guided revascularization is noninferior to an FFR-guided therapeutic strategy.

CONCLUSION

Several studies have consistently demonstrated that ischemia-driven revascularization is the best therapeutic approach for patients with stable coronary artery disease or for those with acute coronary syndrome and lesions in the nonculprit artery. However, it is not uncommon patients to undergo coronary catheterization without a noninvasive test documenting ischemia. Hence, it is necessary to have readily available a diagnostic tool in the cath lab that can accurately and reliably assess the physiological significance of a coronary artery stenosis. FFR is a method that has been proven to measure the functional significance of a stenosis after administration of intravenous or intracoronary adenosine. In the absence of a stenosis FFR is 1 and decreases depending on the significance of the stenosis. Instantaneous wave-free ratio is a recently developed physiological index, which is obtained during mid- to late diastole where there is a wave-free period and resting microvascular resistance is relatively minimized. Recent trials have shown that an iFR cut point of 0.89 matches an FFR cut point of 0.80 and accurately identifies those patients that will benefit from revascularization or optimal medical therapy.

REFERENCES

[1] Boden WE, O'Rourke RA, Teo KK, Hartigan PM, Maron DJ, Kostuk WJ, Knudtson M, Dada M, Casperson P, Harris CL, Chaitman BR, Shaw L, Gosselin G, Nawaz S, Title LM, Gau G, Blaustein AS, Booth DC, Bates ER, Spertus JA, Berman DS, Mancini GB, Weintraub WS, Group CTR. Optimal medical therapy with or without PCI for stable coronary disease. N Engl J Med 2007;356:1503–16.

[2] Group BDS, Frye RL, August P, Brooks MM, Hardison RM, Kelsey SF, MacGregor JM, Orchard TJ, Chaitman BR, Genuth SM, Goldberg SH, Hlatky MA, Jones TL, Molitch ME, Nesto RW, Sako EY, Sobel BE. A randomized trial of therapies for type 2 diabetes and coronary artery disease. N Engl J Med 2009;360:2503–15.

[3] Shaw LJ, Berman DS, Maron DJ, Mancini GB, Hayes SW, Hartigan PM, Weintraub WS, O'Rourke RA, Dada M, Spertus JA, Chaitman BR, Friedman J, Slomka P, Heller GV, Germano G, Gosselin G, Berger P, Kostuk WJ, Schwartz RG, Knudtson M, Veledar E, Bates ER, McCallister B, Teo KK, Boden WE, Investigators C. Optimal medical therapy with or without percutaneous coronary intervention to reduce ischemic burden: results from the Clinical Outcomes Utilizing Revascularization and Aggressive Drug Evaluation (COURAGE) trial nuclear substudy. Circulation 2008;117:1283–91.
[4] Shaw LJ, Cerqueira MD, Brooks MM, Althouse AD, Sansing VV, Beller GA, Pop-Busui R, Taillefer R, Chaitman BR, Gibbons RJ, Heo J, Iskandrian AE. Impact of left ventricular function and the extent of ischemia and scar by stress myocardial perfusion imaging on prognosis and therapeutic risk reduction in diabetic patients with coronary artery disease: results from the Bypass Angioplasty Revascularization Investigation 2 Diabetes (BARI 2D) trial. J Nucl Cardiol 2012;19: 658–69.
[5] Task Force M, Montalescot G, Sechtem U, Achenbach S, Andreotti F, Arden C, Budaj A, Bugiardini R, Crea F, Cuisset T, Di Mario C, Ferreira JR, Gersh BJ, Gitt AK, Hulot JS, Marx N, Opie LH, Pfisterer M, Prescott E, Ruschitzka F, Sabate M, Senior R, Taggart DP, van der Wall EE, Vrints CJ, ESCCfP G, Zamorano JL, Achenbach S, Baumgartner H, Bax JJ, Bueno H, Dean V, Deaton C, Erol C, Fagard R, Ferrari R, Hasdai D, Hoes AW, Kirchhof P, Knuuti J, Kolh P, Lancellotti P, Linhart A, Nihoyannopoulos P, Piepoli MF, Ponikowski P, Sirnes PA, Tamargo JL, Tendera M, Torbicki A, Wijns W, Windecker S, Document R, Knuuti J, Valgimigli M, Bueno H, Claeys MJ, Donner-Banzhoff N, Erol C, Frank H, Funck-Brentano C, Gaemperli O, Gonzalez-Juanatey JR, Hamilos M, Hasdai D, Husted S, James SK, Kervinen K, Kolh P, Kristensen SD, Lancellotti P, Maggioni AP, Piepoli MF, Pries AR, Romeo F, Ryden L, Simoons ML, Sirnes PA, Steg PG, Timmis A, Wijns W, Windecker S, Yildirir A, Zamorano JL. 2013 ESC guidelines on the management of stable coronary artery disease: the Task Force on the management of stable coronary artery disease of the European Society of Cardiology. Eur Heart J 2013;34:2949–3003.
[6] Shaw LJ, Iskandrian AE. Prognostic value of gated myocardial perfusion SPECT. J Nucl Cardiol 2004; 11:171–85.
[7] Metz LD, Beattie M, Hom R, Redberg RF, Grady D, Fleischmann KE. The prognostic value of normal exercise myocardial perfusion imaging and exercise echocardiography: a meta-analysis. J Am Coll Cardiol 2007;49:227–37.
[8] Lin GA, Dudley RA, Lucas FL, Malenka DJ, Vittinghoff E, Redberg RF. Frequency of stress testing to document ischemia prior to elective percutaneous coronary intervention. JAMA 2008;300: 1765–73.
[9] Pijls NH, van Son JA, Kirkeeide RL, De Bruyne B, Gould KL. Experimental basis of determining maximum coronary, myocardial, and collateral blood flow by pressure measurements for assessing functional stenosis severity before and after percutaneous transluminal coronary angioplasty. Circulation 1993;87:1354–67.
[10] Tiefenbacher CP, Chilian WM. Heterogeneity of coronary vasomotion. Basic Res Cardiol 1998;93: 446–54.
[11] Mosher P, Ross Jr J, McFate PA, Shaw RF. Control of coronary blood flow by an autoregulatory mechanism. Circ Res 1964;14:250–9.
[12] van de Hoef TP, Meuwissen M, Escaned J, Davies JE, Siebes M, Spaan JA, Piek JJ. Fractional flow reserve as a surrogate for inducible myocardial ischaemia. Nat Rev Cardiol 2013;10:439–52.
[13] De Bruyne B, Sarma J. Fractional flow reserve: a review: invasive imaging. Heart 2008;94:949–59.
[14] Pijls NH, Sels JW. Functional measurement of coronary stenosis. J Am Coll Cardiol 2012;59: 1045–57.
[15] Kern MJ, Lerman A, Bech JW, De Bruyne B, Eeckhout E, Fearon WF, Higano ST, Lim MJ, Meuwissen M, Piek JJ, Pijls NH, Siebes M, Spaan JA, American Heart Association Committee on D and Interventional Cardiac Catheterization CoCC. Physiological assessment of coronary artery disease in the cardiac catheterization laboratory: a scientific statement from the American Heart Association Committee on Diagnostic and Interventional Cardiac Catheterization, Council on Clinical Cardiology. Circulation 2006;114:1321–41.

[16] De Bruyne B, Pijls NH, Barbato E, Bartunek J, Bech JW, Wijns W, Heyndrickx GR. Intracoronary and intravenous adenosine 5'-triphosphate, adenosine, papaverine, and contrast medium to assess fractional flow reserve in humans. Circulation 2003;107:1877–83.
[17] McGeoch RJ, Oldroyd KG. Pharmacological options for inducing maximal hyperaemia during studies of coronary physiology. Catheter Cardiovasc Interventions 2008;71:198–204.
[18] Echavarria-Pinto M, Gonzalo N, Ibanez B, Petraco R, Jimenez-Quevedo P, Sen S, Nijjer S, Tarkin J, Alfonso F, Nunez-Gil IJ, Banuelos C, Quiros A, Fernandez-Ortiz A, Macaya C, Koo BK, Davies J, Escaned J. Low coronary microcirculatory resistance associated with profound hypotension during intravenous adenosine infusion: implications for the functional assessment of coronary stenoses. Circ Cardiovasc Interventions 2014;7:35–42.
[19] Bech GJ, De Bruyne B, Pijls NH, de Muinck ED, Hoorntje JC, Escaned J, Stella PR, Boersma E, Bartunek J, Koolen JJ, Wijns W. Fractional flow reserve to determine the appropriateness of angioplasty in moderate coronary stenosis: a randomized trial. Circulation 2001;103:2928–34.
[20] Pijls NH, van Schaardenburgh P, Manoharan G, Boersma E, Bech JW, van't Veer M, Bar F, Hoorntje J, Koolen J, Wijns W, de Bruyne B. Percutaneous coronary intervention of functionally nonsignificant stenosis: 5-year follow-up of the DEFER Study. J Am Coll Cardiol 2007;49: 2105–11.
[21] Zimmermann FM, Ferrara A, Johnson NP, van Nunen LX, Escaned J, Albertsson P, Erbel R, Legrand V, Gwon HC, Remkes WS, Stella PR, van Schaardenburgh P, Bech GJ, De Bruyne B, Pijls NH. Deferral vs. performance of percutaneous coronary intervention of functionally non-significant coronary stenosis: 15-year follow-up of the DEFER trial. Eur Heart J 2015;36: 3182–8.
[22] Tonino PA, De Bruyne B, Pijls NH, Siebert U, Ikeno F, van' t Veer M, Klauss V, Manoharan G, Engstrom T, Oldroyd KG, Ver Lee PN, MacCarthy PA, Fearon WF, Investigators FS. Fractional flow reserve versus angiography for guiding percutaneous coronary intervention. N Engl J Med 2009;360:213–24.
[23] De Bruyne B, Pijls NH, Kalesan B, Barbato E, Tonino PA, Piroth Z, Jagic N, Mobius-Winkler S, Rioufol G, Witt N, Kala P, MacCarthy P, Engstrom T, Oldroyd KG, Mavromatis K, Manoharan G, Verlee P, Frobert O, Curzen N, Johnson JB, Juni P, Fearon WF, Investigators FT. Fractional flow reserve-guided PCI versus medical therapy in stable coronary disease. N Engl J Med 2012;367:991–1001.
[24] Botman CJ, Schonberger J, Koolen S, Penn O, Botman H, Dib N, Eeckhout E, Pijls N. Does stenosis severity of native vessels influence bypass graft patency? A prospective fractional flow reserve-guided study. Ann Thorac Surg 2007;83:2093–7.
[25] Toth G, De Bruyne B, Casselman F, De Vroey F, Pyxaras S, Di Serafino L, Van Praet F, Van Mieghem C, Stockman B, Wijns W, Degrieck I, Barbato E. Fractional flow reserve-guided versus angiography-guided coronary artery bypass graft surgery. Circulation 2013;128:1405–11.
[26] Hamilos M, Muller O, Cuisset T, Ntalianis A, Chlouverakis G, Sarno G, Nelis O, Bartunek J, Vanderheyden M, Wyffels E, Barbato E, Heyndrickx GR, Wijns W, De Bruyne B. Long-term clinical outcome after fractional flow reserve-guided treatment in patients with angiographically equivocal left main coronary artery stenosis. Circulation 2009;120:1505–12.
[27] Lee JM, Koo BK, Kumsars I, Curzen N, Thondapu V, Barlis P, Escaned J. Coronary fractional flow reserve in bifurcation stenoses: what have we learned? EuroIntervention 2015;11(Suppl. V): V59–63.
[28] Ahn JM, Lee JY, Kang SJ, Kim YH, Song HG, Oh JH, Park JS, Kim WJ, Lee SW, Lee CW, Kim JJ, Park SW, Park SJ. Functional assessment of jailed side branches in coronary bifurcation lesions using fractional flow reserve. JACC Cardiovasc Interventions 2012;5:155–61.
[29] Koo BK. Fractional flow reserve for coronary bifurcation lesions: can fractional flow reserve-guided side branch intervention strategy improve clinical outcomes compared with angiography-guided strategy? JACC Cardiovasc Interventions 2015;8:547–9.
[30] Koo BK, Park KW, Kang HJ, Cho YS, Chung WY, Youn TJ, Chae IH, Choi DJ, Tahk SJ, Oh BH, Park YB, Kim HS. Physiological evaluation of the provisional side-branch intervention strategy for bifurcation lesions using fractional flow reserve. Eur Heart J 2008;29:726–32.

[31] Kumsars I, Narbute I, Thuesen L, Niemela M, Steigen TK, Kervinen K, Sondore D, Holm NR, Lassen JF, Christiansen EH, Maeng M, Jegere S, Juhnevica D, Erglis A, Nordic-Baltic PCIsg. Side branch fractional flow reserve measurements after main vessel stenting: a Nordic-Baltic Bifurcation Study III substudy. EuroIntervention 2012;7:1155−61.

[32] Ziaee A, Parham WA, Herrmann SC, Stewart RE, Lim MJ, Kern MJ. Lack of relation between imaging and physiology in ostial coronary artery narrowings. Am J Cardiol 2004;93:1404−7. A9.

[33] Pijls NH, Tonino PA. The crux of maximum hyperemia: the last remaining barrier for routine use of fractional flow reserve. JACC Cardiovasc Interventions 2011;4:1093−5.

[34] Pijls NH. Fractional flow reserve after previous myocardial infarction. Eur Heart J 2007;28:2301−2.

[35] Samady H, Lepper W, Powers ER, Wei K, Ragosta M, Bishop GG, Sarembock IJ, Gimple L, Watson DD, Beller GA, Barringhaus KG. Fractional flow reserve of infarct-related arteries identifies reversible defects on noninvasive myocardial perfusion imaging early after myocardial infarction. J Am Coll Cardiol 2006;47:2187−93.

[36] De Bruyne B, Pijls NH, Bartunek J, Kulecki K, Bech JW, De Winter H, Van Crombrugge P, Heyndrickx GR, Wijns W. Fractional flow reserve in patients with prior myocardial infarction. Circulation 2001;104:157−62.

[37] Ntalianis A, Sels JW, Davidavicius G, Tanaka N, Muller O, Trana C, Barbato E, Hamilos M, Mangiacapra F, Heyndrickx GR, Wijns W, Pijls NH, De Bruyne B. Fractional flow reserve for the assessment of nonculprit coronary artery stenoses in patients with acute myocardial infarction. JACC Cardiovasc Interventions 2010;3:1274−81.

[38] Banning AS, Gershlick AH. Management of multivessel coronary disease in ST-segment elevation myocardial infarction. Curr Cardiol Rep 2015;17:632.

[39] Escaned J. ADVISE II: a prospective, registry evaluation of iFR vs. FFR. TCT; 2013.

[40] Sen S, Escaned J, Malik IS, Mikhail GW, Foale RA, Mila R, Tarkin J, Petraco R, Broyd C, Jabbour R, Sethi A, Baker CS, Bellamy M, Al-Bustami M, Hackett D, Khan M, Lefroy D, Parker KH, Hughes AD, Francis DP, Di Mario C, Mayet J, Davies JE. Development and validation of a new adenosine-independent index of stenosis severity from coronary wave-intensity analysis: results of the ADVISE (ADenosine Vasodilator Independent Stenosis Evaluation) study. J Am Coll Cardiol 2012;59:1392−402.

[41] Petraco R, Park JJ, Sen S, Nijjer SS, Malik IS, Echavarria-Pinto M, Asrress KN, Nam CW, Macias E, Foale RA, Sethi A, Mikhail GW, Kaprielian R, Baker CS, Lefroy D, Bellamy M, Al-Bustami M, Khan MA, Gonzalo N, Hughes AD, Francis DP, Mayet J, Di Mario C, Redwood S, Escaned J, Koo BK, Davies JE. Hybrid iFR-FFR decision-making strategy: implications for enhancing universal adoption of physiology-guided coronary revascularisation. EuroIntervention 2013;8:1157−65.

[42] Escaned J, Echavarria-Pinto M, Garcia-Garcia HM, van de Hoef TP, de Vries T, Kaul P, Raveendran G, Altman JD, Kurz HI, Brechtken J, Tulli M, Von Birgelen C, Schneider JE, Khashaba AA, Jeremias A, Baucum J, Moreno R, Meuwissen M, Mishkel G, van Geuns RJ, Levite H, Lopez-Palop R, Mayhew M, Serruys PW, Samady H, Piek JJ, Lerman A, Group AIS. Prospective assessment of the diagnostic accuracy of instantaneous wave-free ratio to assess coronary stenosis relevance: results of ADVISE II International, multicenter study (ADenosine vasodilator independent stenosis evaluation II). JACC Cardiovasc Interventions 2015;8:824−33.

[43] Petraco R, Escaned J, Sen S, Nijjer S, Asrress KN, Echavarria-Pinto M, Lockie T, Khawaja MZ, Cuevas C, Foin N, Broyd C, Foale RA, Hadjiloizou N, Malik IS, Mikhail GW, Sethi A, Kaprielian R, Baker CS, Lefroy D, Bellamy M, Al-Bustami M, Khan MA, Hughes AD, Francis DP, Mayet J, Di Mario C, Redwood S, Davies JE. Classification performance of instantaneous wave-free ratio (iFR) and fractional flow reserve in a clinical population of intermediate coronary stenoses: results of the ADVISE registry. EuroIntervention 2013;9:91−101.

CHAPTER 3.6

Novel Antiplatelet Agents

Gerasimos Siasos, Vasiliki Tsigkou, Evangelos Oikonomou, Marina Zaromitidou, Dimitris Tousoulis
1st Cardiology Clinic, 'Hippokration' General Hospital, National and Kapodistrian University of Athens, School of Medicine, Athens, Greece

Contents

LIST OF ABBREVIATIONS

ACS Acute coronary syndrome
ADP Adenosine diphosphate
ASA Aspirin/acetylsalicylic acid
BMS Bare metal stent
CVD Cardiovascular disease
CABG Coronary artery bypass graft surgery
CRP C-reactive protein
DES Drug eluting stent
DAPT Dual antiplatelet treatment
eNOS Endothelial nitric oxide synthase
GI Gastrointestinal
GP Glycoproteins
IL Interleukin
ICAM-1 Intracellular adhesion molecule-1

Coronary Artery Disease
ISBN 978-0-12-811908-2, http://dx.doi.org/10.1016/B978-0-12-811908-2.00020-9

MACE Major adverse cardiovascular events
MMP's Matrix metalloproteases
MI Myocardial infarction
NO Nitric oxide
Non-STEMI Non-ST segment elevation myocardial infarction
PCI Percutaneous coronary intervention
PAR Protease-activated receptor
PPI's Proton pump inhibitors
ROS Reactive oxygen species
STEMI ST-segment myocardial infarction
TXA_2 Thromboxane A_2
TIMP's Tissue inhibitors of metalloproteases
TNF-α Tumor necrosis factor alpha
UA Unstable angina
VCAM-1 Vascular adhesion molecule-1
vWF von Willebrand factor

INTRODUCTION

Coronary artery disease (CAD) is an important cause of death worldwide, especially in the developed societies and affects millions of people. CAD is manifestated either as an acute coronary syndrome (ACS) or as stable CAD. Furthermore, ACS are divided in non-ST elevation myocardial infarction (NSTEMI), ST elevation myocardial infarction (STEMI), and unstable angina. Platelets are key determinants for the control of hemostasis but under excessive stimulation they are associated with pathologic thrombosis [1]. Abnormal platelet function is involved not only in the previous medical conditions but also in the pathophysiology of ischemic stroke or transient ischemic attack and symptomatic peripheral artery disease (PAD) [2].

Many antiplatelet agents have been developed so far, and the most prominent are aspirin, an irreversible inhibitor of cyclooxygenase-1 (COX-1) and the P2Y12 receptor inhibitor clopidogrel. Novel antiplatelet agents are prasugrel, which irreversibly binds to P2Y12 receptor as well as ticagrelor (allosteric modulator of P2Y12 receptor) [3,4]. Also, there is evidence that cangrelor, elinogrel, protease-activated receptor (PAR)-1 inhibitors such as atopaxar and vorapaxar, glycoprotein (GP) IIb/IIIa receptor antagonists such as abciximab, eptifibatide, and tirofiban, and phosphodiesterase (PDE) inhibitors such as cilostazol exert favorable antiplatelet actions. It is of major importance, though, that the differences in the metabolism of these antiplatelet agents among the patients pose a great impact in their efficacy and many genetic assays have been developed to test platelet reactivity under the previous antiplatelet treatment [5].

On the other hand, patients with CAD who undergo coronary angiography with stent placement (PCI) dual antiplatelet treatment (DAPT) for at least 1 year is prescribed depending on the equilibrium between thrombosis and hemorrhage. Moreover, despite the administration of antiplatelet agents a great number of patients still remain at risk of stent thrombosis, myocardial infarction (MI), stroke, and cardiovascular death indicating that these drugs do not control all the pathways of platelet activation in a sufficient manner [6].

In this chapter, we aim to present the most important clinical trials involving the standard antiplatelet regimen as well as novel antiplatelet agents. Furthermore, we will discuss the data relatively to the optimal DAPT duration and the most common genetic polymorphisms affecting antiplatelets' efficacy.

PATHOPHYSIOLOGY OF CAD AND THROMBOSIS: THE ROLE OF PLATELETS

CAD is attributed to many pathologic conditions and especially to atherosclerosis. Atherosclerosis begins since the early adolescence and progresses throughout life especially with the presence of the common atherosclerotic risk factors such as smoking, dyslipidemia, arterial hypertension, diabetes, obesity, family history of CAD, ageing, and hyperomocysteinemia. An important step for the initiation of atherosclerosis is the induction of endothelial dysfunction, which is characterized by deteriorated production of nitric oxide (NO) by endothelial nitric oxide synthase (eNOS) [7]. The dysfunctional endothelium results in an imbalance between pro-oxidant and antioxidant molecules and a consequent overproduction of reactive oxygen species, which are responsible for oxidative modification of low-density lipoprotein (LDL). Moreover, ox-LDL molecules penetrate the vascular wall and are absorbed by macrophages, which are converted to foam cells due to lipid accumulation [8]. Furthermore, a number of proinflammatory cytokines such as C-reactive protein (CRP), tumor necrosis factor alpha (TNF-α), and interleukins (IL) such as IL-1, IL-6, and adhesion molecules such as vascular adhesion molecule-1 and intracellular adhesion molecule-1 are expressed by the damaged endothelium and result in increased attachment of white blood cells on the vascular wall [9]. Also, with the progression of the disease the atheromatous lesion is formed inside the vessel wall, which consists of a lipid core rich in cholesterol and triglycerides, extracellular accumulation of cholesterol, and cell debris covered by a fibrous cap made of collagen. Stable atherosclerotic plaques consist of a small lipid core and a thick fibrous cap in comparison to unstable atherosclerotic plaques, which consist of a thick lipid core and a thin fibrous cap that is more prone to rupture. Moreover, unstable atherosclerotic plaques exhibit increased expression of matrix metalloproteases (MMPs) especially, MMP-2, MMP-9, and decreased expression of their tissue inhibitors, which result in the degradation of the fibrous cap, the formation of the thrombus, and the occlusion of the vessel [10]. ACS occurs due to an abrupt reduction in blood flow in a myocardial vessel which results in a total or partial occlusion. The normal hemostatic procedure and the formation of the arterial thrombus consist of three steps, initially, the increased vasomotor response of the damaged endothelium, followed by platelet aggregation, finally the induction of the coagulation cascade.

Platelets are a nucleated cells which are produced by megakaryocytes in bone barrow and circulate in the blood for approximately 7—10 days [11]. Platelets express molecules, which are responsible for platelet aggregation such as glycoproteins GPIa/IIa and GP IIb/IIIa. More specifically, von Willebrand factor binds to GPIIb/IIIa platelet receptors and mediates platelet aggregation. Furthermore, activated platelets release a series of active molecules such as thrombin, adenosine diphosphate (ADP) (which binds to the platelet P2Y12 receptor), and thromboxane A_2 (TXA_2) initiating a vicious cycle of platelet stimulation [12,13]. Additionally, endothelial damage results in the expression of tissue factor and the adhesion of platelets in the subenthothelial collagen [14]. The final step is the conversion of fibrinogen to fibrin through the actions of thrombin, which stabilizes the blood clot [15]. The previous actions of thrombin are mediated through PAR-1 and PAR-4, which are expressed in platelet surface, with most prominent actions of PAR-1 [16].

On the other hand, platelets are also implicated in the chronic progression of atherosclerosis due to platelet—endothelium interactions, which result in increased inflammation in the vascular wall. Activated platelets secrete proinflammatory markers such as CRP, MMPs, and interleukins, which predispose the rupture of the atherosclerotic plaque [17]. Therefore, the simultaneous inhibition of multiple targets on platelet surface is an important goal for the regulation of both acute and stable CAD and the most widely used pharmacologic agents aiming at the before-mentioned pivotal pathways of platelet activation are aspirin, P2Y12 ADP receptor antagonists, GPIIb/IIIa antagonists, and PAR-1 inhibitors [18,19].

ASPIRIN

Aspirin (acetylsalicylic acid) is an irreversible inhibitor of COX-1 resulting in decreased production of TXA_2. There is a nonlinear relation between aspirin dose and COX-1 inhibition, which results in rapid recovery of platelet function within 3—4 days upon drug withdrawal [20]. Apart from the antiplatelet actions, aspirin also exerts anti-inflammatory actions and improves endothelial dysfunction through COX-1 independent actions [21].

Approximately 20% of the US citizens are prescribed aspirin daily for the primary prevention of CAD [22]. Moreover, it is a cost-effective choice for the secondary prevention of CAD too [23]. Aspirin is a standard treatment in ACS and especially in patients with STEMI undergoing percutaneous coronary intervention (PCI). The current guidelines indicate the administration of 150—300 mg orally as soon as the symptoms arise before coronary angiography continued by a daily maintenance dose of 75 mg for at least 9—12 months [24]. Moreover, aspirin reduces the overall risk for serious cardiovascular events by 15% in patients with ACS [25].

The administration of aspirin is associated with decreased rates of ischemic events not only in patients with ACS as well as in patients undergoing elective PCI [4]. Furthermore, it is used in combination to P2Y12 ADP receptor antagonists (clopidogrel, prasugrel, and ticagrelor) and is referred to as dual antiplatelet therapy (DAPT) [26].

Aspirin has a few contraindications such as a history of intolerance, active bleeding, active peptic ulcer, or gastrointestinal (GI) or genito-urinary bleeding, active retinal bleeding and severe untreated hypertension [27]. Moreover, it is usually prescribed along with gastro-protection with proton pump inhibitors (PPIs) due to the risk of upper GI bleeding especially in patients uptaking other antiplatelets or anticoagulants [28]. Relatively to the risk of bleeding, a meta-analysis indicated that aspirin decreases by 23% the risk of ischemia and increases the risk of major extracranial bleeding by 40% in doses 160–325 mg/day contrary to 50% in those treated with aspirin 75–150 mg/day [25]. Moreover, the CURRENT-OASIS7 trial indicated that 75–100 mg of aspirin was as effective as 300–325 mg in the reduction of the ischemic events with an increase in minor bleedings [29].

In conclusion, many studies indicate the phenomenon of aspirin resistance or aspirin nonresponsiveness means that the administration of aspirin fails to reduce the risk of cardiovascular disease (CVD) complications [30,31]. There are many functional tests that assess aspirin responsiveness and predict the risk of future cardiovascular events [32].

ADP-RECEPTOR INHIBITORS (P2Y12 RECEPTOR INHIBITORS)

The P2Y12 receptor is a G-inhibitory-protein receptor in the platelet membrane, which is activated by ADP and results in the inhibition of adenylyl cyclase, thereby inducing platelet aggregation [33,34]. Moreover, the activation of P2Y12 receptor hinders the antiplatelet effects of prostacyclin. There is no doubt that P2Y12 receptor is an important mediator for increased platelet reactivity and many pharmacological agents have been developed so far aiming at this target such as clopidogrel as well as the novel antiplatelet agents such as prasugrel and ticagrelor [33]. Also, the first generation of P2Y12 receptor inhibitors was the theinopyridineticlopidine, which was removed due to its life-threatening effects of neutropenia and thrombotic thrombocytopenic purpura.

CLOPIDOGREL

Clopidogrel is a prodrug that binds irreversibly to P2Y12 receptors and has the disadvantage of delayed onset of action [33]. There are many generics of clopidogrel; however, there seems to be no significant differences relatively to their safety profile, pharmacokinetics, and resistance among the patients [35,36]. Moreover, loading doses of clopidogrel usually do not reach the ultimate platelet inhibition in patients with STEMI who receive thrombolytic therapy in comparison to the other antiplatelet agents [37].

Although DAPT with aspirin and clopidogrel was the most popular combination for many years there exist many limitations relatively to clopidogrel such as delayed onset, irreversible effect on P2Y12 receptor, as well as drug-to-drug interactions especially in patients who uptake a combination of drugs [38–41]. Many polymorphisms affect the responsiveness to clopidogrel such as the polymorphisms of ATP-binding cassette subfamily B member 1 (ABCB1) and paraxonase-1 (PON-1) gene as well as of CYP450 liver enzymes [42]; however, the CYP450 polymorphisms are the most studied (especially CYP2C19) and this is further discussed at a separate part of this chapter [43,44]. Moreover, clopidogrel efficacy is also affected by the gender, the ethnicity, and comorbidities such as the liver disease [36].

Many clinical trials have examined the effects of clopidogrel in patients with ACS as well as stable CAD with conflicting results. To begin with, the GRAVITAS trial revealed that higher doses of clopidogrel in patients undergoing PCI with drug eluting stents (DES) induced not-withstanding improvements in platelet inhibition and no statistically significant differences in the risk of death from cardiovascular causes, nonfatal MI or stent thrombosis [45]. Moreover, the CAPRIE trial indicated that monotherapy with clopidogrel resulted in greater efficacy than monotherapy with aspirin with similar risk of bleeding [46].

Also, in the COMMIT study, it was revealed that clopidogrel along with aspirin decreased the risk of stroke, reinfarction and cardiovascular death in patients with STEMI in comparison to those treated with aspirin with no significant differences in the risk of bleeding among the two study groups [47]. The CHARISMA trial examined the effects of combined therapy with aspirin and clopidogrel versus aspirin monotherapy in the secondary prevention of CVD as well as in the primary prevention. The results indicated no significant differences among the study groups relatively to the primary end point although a post hoc analysis indicated a beneficial effect of the combination especially in the high-risk patients with previous MI, ischemic stroke, and symptomatic PAD with a nonsignificant increase in the risk of severe bleeding [48]. Additionally to these, in the CLARITY study, it was revealed that the administration of clopidogrel before fibrinolytic therapy in patients with STEMI who underwent PCI resulted in 46% decrease in ischemic events at 30 days, without an increase in bleeding [49].

On the other hand, results from the CURE trial indicated that the combination of aspirin plus clopidogrel in patients with NSTEMI resulted in significantly decreased risk of CVD death at the expense of increased bleeding rates [3]. Similarly, the CURRENT-OASIS-7 trial revealed that the administration of double dose of clopidogrel reduced cardiovascular death, MI, or stroke at 30 days irrespectively of the aspirin dose (high dose or low dose) in patients with ACS (STEMI or NSTEMI) who underwent PCI. Moreover, increased rates of major bleeding in the group of double-dose clopidogrel were found, and these rates were unaffected by the dose of aspirin administrated [50].

PRASUGREL

Prasugrel is a third generation P2Y12 receptor inhibitor, which belongs to the family of thienopyridines and binds irreversibly to the P2Y12 receptor [33]. Moreover, it is more efficient than clopidogrel and has a more predictable clinical outcome because of the different metabolism of the two drugs as prasugrel is not associated with CYP450 [36]. In a study of Japanese patients with stable CAD it was found that DAPT with 3.75 mg of prasugrel for 2 weeks reduced platelet reactivity in comparison to DAPT with clopidogrel. Also, the assessment of platelet reactivity with the measurement of P2Y12 reactivation units indicated that prasugrel is not affected by CYP2C19 polymorphisms in comparison to clopidogrel rendering prasugrel an important option especially for patients at risk of stent thrombosis and ischemic events [51].

Many studies have examined the antiplatelet efficacy of prasugrel. To begin with, the TRITON-TIMI 38 study indicated the beneficial effects of prasugrel plus aspirin in patients with ACS treated with PCI in comparison to the combination of aspirin plus clopidogrel. Prasugrel plus aspirin significantly decreased the rates of death from cardiovascular causes, nonfatal stroke, and nonfatal MI depicting increased risk of major bleeding, life-threatening bleeding, need of blood transfusions, and coronary artery bypass graft (CABG) related bleeding. However, a beneficial effect in patients >75 years, weighting <60 kg with a prior stroke or transient ischemic attack, which are the main contraindications of prasugrel, was not found [4].

On the other hand, the PRINCIPLE-TIMI 44 trial, examined the effects of loading loses of 60 mg prasugrel versus 600 mg clopidogrel in patients undergoing elective PCI as well as the effects of the administration of a maintenance dose of 10 mg prasugrel versus 150 mg clopidogrel. The results indicated that a loading dose of 60 mg prasugrel significantly inhibits platelet reactivity as well as a maintenance dose of 10 mg in comparison to clopidogrel with nonsignificant increase in bleeding events [52]. Additionally to this, the Excelsior LOAD study examined the effects of the administration of a loading dose of prasugrel 30 mg or 60 mg over a loading dose of clopidogrel 600 mg in patients undergoing elective PCI. The results indicated that 55% of the patients who received clopidogrel had increased platelet reactivity compared to 37% and 33% in the groups of prasugrel with no important differences in the risk of bleeding [53].

TICAGRELOR

Ticagrelor, an acyclo-pentyl-triazolo-pyrimidine, is an oral, nonthienopyridine, second generation, reversible P2Y12 inhibitor [37]. It is an active drug, which does not undergo hepatic metabolism and acts through an allosteric modulation of the receptor in a dose-dependent manner [54]. Ticagrelor has also non-P2Y12 receptor-mediated antiplatelet effects, such as the increase in plasma adenosine, which results in cardioprotective actions despite the induction of bronchoconstriction and dyspnea [55]. Moreover, it is contraindicated in patients with previous hemorrhagic stroke and hepatic dysfunction. Additionally, it has a more predictable and potent inhibition of platelet aggregation compared to the other antiplatelet agents and is associated with decreased risk of bleeding due to the rapid reversal of platelet inhibition after the short discontinuation of the therapy, which is beneficial especially in patients undergoing surgery [56].

In general, results from many studies indicate that long-term DAPT with aspirin and ticagrelor have a beneficial effect in the reduction of the risk of major adverse cardiovascular events (MACE) relative to the administration of aspirin monotherapy, although ticagrelor is associated with increased bleeding, especially in nonfatal bleeding [57]. For example, the PLATO trial indicated that DAPT with ticagrelor plus aspirin versus clopidogrel plus aspirin for 1 year led to a 20% reduction in all-cause mortality especially in patients who undergo PCI and was associated with a significant decrease in the death from CVD, stroke, and MI with an increase in the risk of non-CABG-related major bleeding [58]. Moreover, the PEGASUS-TIMI 54 trial proved that the administration of aspirin and ticagrelor (60 mg or 90 mg twice daily) in patients with recent history of MI in comparison to aspirin plus placebo significantly decreased the rates of MI, stroke, and cardiovascular death with a concomitant increase in bleeding events [59]. Also, the ONSET/OFFSET study compared the efficacy of ticagrelor (loading dose 180 mg, maintenance dose 90 mg twice daily) against clopidogrel (loading dose 600 mg, maintenance dose 75 mg) in patients treated with aspirin with stable CAD. The results revealed that ticagrelor resulted in faster platelet inactivation 5.8 h earlier than clopidogrel, which was maintained at the first 24 h within the 6-week period of treatment [55]. Besides, in Hispanic patients with stable CAD it was indicated that the administration of ticagrelor (loading dose of 180 mg followed by a maintenance dose of 90 mg twice daily) in comparison to clopidogrel (loading dose of 600 mg followed by a maintenance dose of 75 mg once daily) for 7—9 days significantly reduced platelet reactivity 2 h post the administration. The beneficial effects for the diabetic patients were attributed to the anti-ischemic actions of ticagrelor as well as the improvement of insulin resistance [60].

On the other hand, the ATLANTIC study indicated that prehospital administration of ticagrelor in patients with STEMI did not improve ST-segment elevation before PCI and coronary perfusion prior to angiography in comparison to the in-hospital administration possibly due to the delayed onset of action of ticagrelor and prasugrel when administrated orally. However, prehospital treatment was associated with decreased rates of definite stent thrombosis within the first 24 h as well as at 30 days [61].

Finally, the SWAP-2 study examined the effects of the administration of ticagrelor (loading dose 180 mg, maintenance dose 90 mg twice daily) and the switch to prasugrel (loading dose 60 mg, maintenance dose 10 mg daily). The results indicated significantly higher platelet reactivity in the group of prasugrel [62].

ELINOGREL

Elinogrel is a third generation, reversible and competitive inhibitor of P2Y12 receptor administrated both orally and intravenously, which depicts a more stable and safe profile among the patients [63]. Moreover, the antiplatelet effect is reversed within 24 h if given orally and within 8 h if given intravenously. Studies in mice indicate the beneficial antiplatelet actions of elinogrel; more specifically, it reduces the production of thrombus and exerts optimal platelet inhibition compared to clopidogrel [64]. Also, the intravenous administration of elinogrel results in 87% decrease of thrombus formation in a dose-dependent manner, with a rapid and reversible effect and could, therefore, be used as a bridge in patients undergoing surgery who uptake theienopyridines.

To begin with, the ERASE-MI trial revealed that the administration of several doses of elinogrel intravenously in patients with STEMI resulted in no significant rates of bleeding [63]. Similarly to this, the INNOVATE-PCI study examined the effects of clopidogrel versus elinogrel given intravenously and then continued orally in patients undergoing elective PCI. The results showed no significant differences among the study groups relatively to the ischemic events although elinogrel was associated with higher rates of bleeding requiring medical intervention especially at vascular access sites, as well as >3-fold increase in liver enzymes, dyspnea, and serious adverse events [65].

CANGRELOR

Cangrelor is a competitive and reversible inhibitor of the P2Y12 receptor usually administrated intravenously which inactivates ADP-receptor pathway and is useful especially in patients with nausea and vomiting [66]. Moreover, it has a short half-life of 3–6 min and should only be given in hospital [67]. Additionally, switching to another oral antiplatelet agent such as clopidogrel or prasugrel might cause drug-to-drug interactions [33]. The rapid onset of its action as well as the offset of the antiplatelet effect has indicated a potent efficiency of cangrelor as a bridge to the patients who require surgery and are treated with other antiplatelets. For example, the BRIDGE trial indicated that the administration of cangrelor in patients with ACS undergoing CABG who were treated with thienopyridines (stopped for at least 48 h) exhibited an important inhibition of platelet reactivity without any effects in bleeding risk [68].

In vitro studies in canine carotid arteries indicate that cangrelor decrease artery occlusion from 100% to 17% with a concomitant decrease in thrombus weight by 83% [69]. Moreover, the CHAMPION PCI and CHAMPION PLATFORM studies examined the effects of the administration of cangrelor versus a higher loading dose of clopidogrel in patients undergoing PCI. In the first study, the drugs were administrated within 30 min of the procedure, whereas in the second study cangrelor was administrated at the beginning and clopidogrel in the end of the procedure. The results, after further statistical analysis of pooled data, indicated a beneficial effect of cangrelor especially in patients with NSTEMI [70]. Finally, in the CHAMPION PHOENIX trial it was revealed that cangrelor reduces MI and stent thrombosis in both men and women undergoing urgent or elective PCI more powerfully than clopidogrel without increased risk of bleeding [71].

GP IIB/IIIA RECEPTOR ANTAGONISTS

The inhibition of GP IIb/IIIa receptor is an important step in the control of platelet activation [22]. According to American College of Cardiology (ACC)/American Heart Association (AHA) guidelines GPIIb/IIIa receptor antagonists such as abciximab, tirofiban, and eptifibatide are administrated to patients with concomitant ischemia, or in patients at high risk who undergo PCI [72]. Abciximab has a greater affinity for GP IIb/IIIa receptor in comparison to the other drugs as well as a longer half-life [73].

In ISAR REACT 2 trial, it was found that the addition of abciximab in patients with ACS treated with clopidogrel prior to PCI resulted in beneficial outcomes [74]. Also, the EPIC trial revealed that abciximab decreased the risk for MACE with a concomitant increase in bleeding [75]. Moreover, in the EPISTENT trial it was found that the administration of abciximab in patients undergoing elective or urgent PCI resulted in decreased rates of MI and death at 6 months especially in those who underwent coronary artery stenting and not balloon angioplasty [76].

In a retrospective study, abciximab depicted a successful efficacy despite the increased rates of bleeding in CAD patients at high risk for thrombosis undergoing PCI [77]. Besides, in the ISAR-SWEET study it was revealed that abciximab decreased the risk of stent restenosis at 1-year follow-up in patients with diabetes undergoing elective PCI with bare metal stent (BMS) [78]. On the other hand, in a prospective study abciximab did not alter the incidence of MACE in comparison to the standard therapeutical options in patients at low thrombotic risk undergoing elective or urgent PCI [79].

As for eptifibatide, in the PURSUIT study it was revealed that the administration of eptifibatide in patients with ACS decreased MACE and increased bleeding related to invasive procedures [80]. Also, the administration of high-dose, intravenous eptifibatide in patients prior to elective PCI resulted in decreased rates of MACE at 30-days follow up without important increase in the risk of bleeding [81].

Results from the RESTORE trial revealed that the administration of tirofiban decreased by 16% the 30-day mortality without statistically significant differences in the risk of bleeding and thrombocytopenia [82]. Similarly, the administration of tirofiban in poor responders to aspirin and clopidogrel undergoing elective PCI indicated a diminution in MACE at 30-days as well as in postprocedural MI [83].

PAR-1 ANTAGONISTS

Another important mechanism of platelet activation is the stimulation of platelets by thrombin through the activation of protease activated receptors (PAR) which are G-proteins associated with the cellular responses involved in thrombosis and hemostasis [84]. There are four types of PAR expressed in the surface of platelets, especially PAR-1 along with PAR-4 and PAR-1exerts the most powerful actions [85]. Therefore, the administration of PAR-1 antagonists is a reasonable strategy to antithrombotic treatment [86].

Vorapaxar is a reversible PAR-1 antagonist, given orally with a long half-time [87]. Results from the TRA 2P-TIMI 50 trial indicated favorable effects of vorapaxar at 3-year follow-up on reducing MACE in comparison to the standard therapy with aspirin and clopidogrel, with an outstanding increase in bleeding especially in patients with prior stroke due to increased rates of intracranial hemorrhage [88]. Therefore, it is contraindicated to patients at high risk of bleeding such as in those with a history of stroke, intracranial hemorrhage and a transient ischemic attack as well as in those with renal or hepatic impairment [89]. Also, a subanalysis from this study indicated that long-term administration of vorapaxar decreased stent thrombosis as well as very late stent thrombosis in patients with stable CAD [90]. Similarly to these, a meta-analysis indicated that vorapaxar decreased the rates of MI at the expense of increased risk of hemorrhagic stroke in patients undergoing PCI or CABG [91]. On the other hand, TRA-PCI trial proved that vorapaxar along with aspirin and clopidogrel did not increase TIMI major or minor bleeding risk in patients undergoing elective PCI [92].

Atopaxar, is an oral, selective inhibitor of PAR-1 receptor. In vitro studies indicate that atopaxar also regulates the expression of GP IIb/IIIa, platelet endothelial adhesion molecule (PECAM)-1, vitronectin and thrombospondin [93]. Results from the J-LANCELOT trial in Japanese patients with ACS or high-risk CAD indicated nonsignificant differences in the rates of bleeding requiring medical intervention among the patients treated with atopaxar versus standard treatment as well as similar rates of MACE [94].

PDE INHIBITORS (PDE INHIBITORS)

Cilostazol is an orally given selective inhibitor of cyclic adenosine monophosphate (cAMP)-specific type 3 PDE, which is expressed in human erythrocytes [95]. The increased levels of cAMP result in the activation of protein kinase A (PKA), which hinders platelet aggregation; moreover, it stimulates NO production by eNOS and decreases the vasoconstrictive actions mediated by angiotensin-II [95]. Also, cilostazol exhibits lipid-lowering actions, as well as positive inotropic and chronotropic effects [96].

The results of a meta-analysis of 34 trials indicated that the administration of cilostazol was more beneficial than the intensification of DAPT and decreased the rates of MACE, target vessel revascularization and stent thrombosis in patients undergoing PCI with no high rates of bleeding [97]. Also, low-dose cilostazol in addition to the standard DAPT in patients undergoing PCI for complex lesions resulted in diminished rates of stent thrombosis, MI and MACE at 1 year follow-up without an increase in bleeding [98]. On the other hand, cilostazol along with DAPT did not have a powerful effect in patients with ACS [99]. Therefore, there is a need of further investigation regarding this field.

OPTIMAL DAPT DURATION

There are great differences relatively to the optimal duration of DAPT and European Society of Cardiology (ESC) as well as AHA/ACC have published relative guidelines. The final choice should be made after estimation of the type of the implanted stent (BMS or DES), the medical history of ACS or stable CAD, the estimation of the patients' bleeding risk and the need of anticoagulants [100]. Also, the duration of triple therapy with oral anticoagulant, aspirin 75–100 mg and clopidogrel 75 mg should be determined after the calculation of the thrombotic risk (CHA_2DS_2-VASc score) and the risk of bleeding (HAS-BLED score) [100,101]. The administration of oral antiplatelet agents should reach a balance between the risk of thrombosis and hemorrhage due to the fact that bleeding is associated with increased mortality especially in patients with ACS [102,103]. Moreover, the clinicians should also take into consideration the existence of comorbidities such as diabetes, extremely low body weight as well as impaired kidney and liver disease [104]. Current ESC guidelines indicate that there is a beneficial effect of ticagrelor and prasugrel over clopidogrel in patients with STEMI undergoing PCI and therefore clopidogrel should be the only choice when the previous drugs are contraindicated or not available [105].

Aspirin is highly recommended after coronary artery stenting due to its powerful actions against cardiovascular death, nonfatal MI and hospitalization for ACS indicated by COURAGE and FAME-2 trials [106,107]. Moreover, aspirin discontinuation is only suggested in cases of enormously increased risk of bleeding in patients undergoing a surgery, although most studies indicate that the risk of bleeding does not exceed the thrombotic risk [108].

On the one hand, the EXCELLENT trial examined the effects of 3-month DAPT with aspirin and clopidogrel in patients with CAD in comparison to 12 months DAPT. The results indicated that short-term DAPT did not affect the risk of thrombosis although it significantly decreased the risk of bleeding indicating that in patients at low thrombotic risk there is no need for prolonged DAPT [109]. On the other hand, meta-analyses of PRODIGY, OPTIMIZE, RECENT, and EXCELLENT trials revealed that long-term DAPT is related to increased risk of major bleeding with similar effects on cardiovascular death or MI compared to short-term DAPT [110]. Contrary to these findings, the CURE trial revealed that 12-month treatment with DAPT in patients with NSTEMI resulted in decreased rates of the primary end point in comparison to 3-month DAPT [3]. Finally, the CHARISMA trial depicted that the administration of DAPT beyond the first year might be beneficial in patients with prior MI, ischemic stroke or symptomatic PAD [111].

Relatively to the novel antiplatelets, the ARCTIC-Interruption trial indicated that prolonged DAPT with aspirin and prasugrel for 6–18 months in patients with stable CAD resulted in increased risk of major and minor bleeding than DAPT with aspirin and clopidogrel [112]. Also, the PEGASUS-TIMI 54 trial showed that prolongation of DAPT with aspirin and ticagrelor for >12 months in patients with history of MI is associated with increased major bleeding and has not a beneficial effect in mortality although it hinders the ischemic events [113]. Moreover, the DAPT trial revealed that the prolongation of DAPT with aspirin and clopidogrel or prasugrel for 12–30 months significantly decreased ischemic events irrespectively of the type of stent placed or the type of P2Y12 inhibitor prescribed [104].

PHARMACOGENETICS AND EFFICACY OF ORAL ANTIPLATELET AGENTS

Recent studies indicate that there are many polymorphisms affecting the individual's responsiveness to antiplatelet treatment [39,114]. To begin with, aspirin metabolism is mainly affected by PlA1/A2 polymorphism of GPIIIa gene and the polymorphisms of COX-1 gene. A meta-analysis of Goodman et al. indicated that relatively to the PlA1/A2 polymorphism, there were no significant differences in aspirin resistance in patients with CAD, although a poorer response to aspirin was found in healthy subjects [115]. Moreover, in another meta-analysis it was found that the A-842G/C50T polymorphism of COX-1 is not significantly related to aspirin resistance [116]. Similarly, polymorphisms of COX-2, GPIa, GPIba, GPVI, P2Y1, P2Y12, PEAR1, CYP2C19*2, and *17 also proved to be clinically unimportant [117–119].

Clopidogrel is metabolized in the liver into the inactive carboxylic acid by carboxylesterases 1 and 2 and the remaining clopidogrel is further metabolized in the liver in a two-step process which involves CYP2C19, CYP2B6, and CYP1A2 enzymes. Firstly, clopidogrel is converted into 2-oxo-clopidogrel and then into the active metabolite through the actions of CYP3A4/5, CYP2B6, CYP2C9, and CYP2C19 [120]. As mentioned above, CYP2C19 is the most important determinant of clopidogrel metabolism and many studies indicate that the carriers of CYP2C19*2 dysfunctional allele exhibit low responsiveness to both low (300 mg) and high (900 mg) doses of clopidogrel [108,121,122]. At least one CYP2C19 loss-of-function alleles is associated with poor response to clopidogrel and this is a finding shown in 30%–70% of Asians, 11%–35% of Caucasians and 14%–25% of African-Americans and the poorest efficacy is depicted in those who have two pathological genes [41,123]. On the one hand, CYP2C19*2 loss-of-function genotype is associated with increased ischemic events in patients with prior MI as well as with stent thrombosis [124–126]. On the other hand, there are studies which indicate that CYP2C19 does not affect the incidence of MACE, MI, or stroke in patients with ACS [127]. Also, PPIs interfere with CYP2C19; however, data from the COGENT study revealed no significant increase in ischemic events in patients who received omeprazole [128]. Finally, CYP2C19*17, which is a gain-of-function allele, exists in 15%–30% of Caucasian and African populations and <5% in Asian populations and is associated with increased response to the drug [129]. Many studies indicate that carriers of CYP2C19*17 alleles have diminished risk of MACE despite an increased risk of bleeding [130].

Moreover, as mentioned in another part of this article, polymorphisms of ABCB1 in patients treated with clopidogrel, which encodes P-glycoprotein, especially ABCB1 3435C→T, is related to increased risk of cardiovascular death, MI, and stroke [131]. Besides, the Q192R genotype of the PON1 gene, which is involved in the last step of clopidogrel metabolism was not associated with high platelet reactivity and the incidence of MACE [132]. Furthermore, Siasos et al. indicated that C34 T polymorphism of P2Y12 receptor is related to the incidence of adverse cardiovascular outcomes in patients with CAD undergoing PCI especially in the carriers of at least one C34T decreased-function allele [5,133]. Last but not least, recent studies indicate that micro-RNA's are involved in the metabolism and responsiveness of clopidogrel such as miR-223, which is associated with poorer response to clopidogrel in a Chinese population [134].

On the other hand, CYP2C19 as well as ABCB1 polymorphisms in patients with ACS or in the healthy are not associated with the metabolism of prasugrel or ticagrelor [108,131,135,136]. Additionally, the efficacy of prasugrel is no related to the polymorphisms of CYP2C9, CYP2B6, CYP3A5, CYP1A2, and MDR1 [131]. Finally, the RAPID GENE trial examined the effects of CYP2C19*2 polymorphisms in patients undergoing PCI who were treated with clopidogrel and prasugrel. The results indicated that only clopidogrel depicted 30% increased platelet reactivity [137].

In conclusion, there is still little evidence about the utility of genetic testing relatively to patient's responsiveness to antiplatelets and we should also take into consideration that not only genetic parameters but also epigenetics, transcriptomics, metabolomics, as well as metagenomics are involved in platelet responsiveness as well as environmental factors creating a complicated impact on individual's platelet reactivity.

Table 3.6.1 summarizes the major studies relatively to P2Y12 receptor inhibitors.

Table 3.6.2 summarizes the main characteristics of P2Y12 receptor inhibitors.

Fig. 3.6.1 summarizes the mechanisms of action of the most common antiplatelet agents.

Table 3.6.1 Summary of the major studies

Trial	Trial design	Intervention	Primary outcomes	Secondary outcomes
GRAVITAS [45]	Randomized, double-blind, active control	600 mg initial dose clopidogrel then 150 mg daily versus standard dose clopidogrel 75 mg daily-no additional loading dose	Death from cardiovascular causes, nonfatal myocardial infarction, or stent thrombosis 2.3% versus 2.3% (HR 1.01, 95% CI 0.58–1.76, $P = 0.97$	Severe or moderate bleeding was not increased 1.4% versus. 2.3%, HR 0.59, 95% CI 0.31 –1.11, $P = 0.10$
CAPRIE [46]	Randomized, double blind	75 mg once daily versus aspirin 325 mg once daily	Risk of ischemic stroke, myocardial infarction, or vascular death 5.32% versus 5.83%	Severe rash (0.26% vs. 0.10%), diarrhea (0.23% vs. 0.11%), upper gastrointestinal discomfort (0.97% vs. 1.22%), intracranial hemorrhage (0.33% vs. 0.47%), and gastrointestinal hemorrhage (0.52% vs. 0.72%)
CURE [3]	Randomized, double blind	Clopidogrel 300 mg followed by 75 mg daily plus aspirin versus aspirin plus placebo	Death from cardiovascular causes, nonfatal myocardial infarction, or stroke 9.3% versus 11.4% (HR 0.80, 95 CI 0.72–0.90, $P < .001$) refractory ischemia 16.5% versus 18.8% (HR 0.86, 95 CI 0.79–0.94, $P < .001$)	Major bleeding 3.7% versus 2.7% HR 1.38, $P = .001$

Continued

Table 3.6.1 Summary of the major studies—cont'd

Trial	Trial design	Intervention	Primary outcomes	Secondary outcomes
TRITON-TIMI 38 [4]	Randomized, double blind, active-control	Prasugrel 60 mg load then 10 mg daily versus clopidogrel 300 mg load then 75 mg daily	Death from cardiovascular causes, nonfatal myocardial infarction, or nonfatal Stroke in 9.9% versus 12.1% (HR 0.81, 95% CI 0.73–0.90, $P < .001$	Nonfatal myocardial infarction 7.4 versus 9.5% ($P < .001$), urgent target-vessel revascularization (2.5% vs. 3.7%; $P < .001$), and stent thrombosis (1.1% vs. 2.4%, $P < .001$) Major bleeding 2.4% versus 1.8% (HR 1.32, 95% CI 1.03–1.68, $P = .03$) life-threatening bleeding 1.4% versus 0.9%, $P = .01$including nonfatal bleeding (1.1% vs. 0.9%, HR 1.25, $P = .23$ and fatal bleeding (0.4% vs. 0.1%, $P = .002$)
PLATO [58]	Randomized, double-blind	Ticagrelor (180-mg loading dose, 90 mg twice daily versus clopidogrel (300-to-600-mg loading dose, 75 mg daily	Death from vascular causes, myocardial infarction, or stroke 9.8% versus 11.7% (HR 0.84, 95% CI 0.77–0.92, $P < .001$	Myocardial infarction (5.8% vs. 6.9% $P = .005$) Death from vascular causes (4.0% vs. 5.1%, $P = .001$) Death from any cause (4.5%, vs. 5.9%, $P < .001$) Major bleeding 11.6% versus 11.2%, $P = .43$) Major bleeding not related to coronary-artery bypass grafting (4.5% vs. 3.8%, $P = .03$)

Table 3.6.2 Key characteristics of P2Y12 receptor inhibitors

Key Characteristics	Clopidogrel	Prasugrel	Ticagrelor
Prodrug	Yes	No	No
Drug generation	First	First	Second
P2Y12 inhibition	Irreversible	Irreversible	Reversible
Onset of action (hours)	2–8 h	30 min–4 h	30 min–2 h
Maximum P2Y12 inhibition (hours)	6–8 h	3 h	3 h
Inter-individual variability	Large	Small	Small
Offset of action	7–10 days	7–10 days	3–5 days
Dose	300–600 mg loading, 75 mg daily	60 mg loading, 10 mg daily	180 mg loading, 90 g twice daily
Side effects	Bleeding, neutropenia, thrombotic thrombopenic purpura	Bleeding	Bleeding, bronchospasm, bradycardia

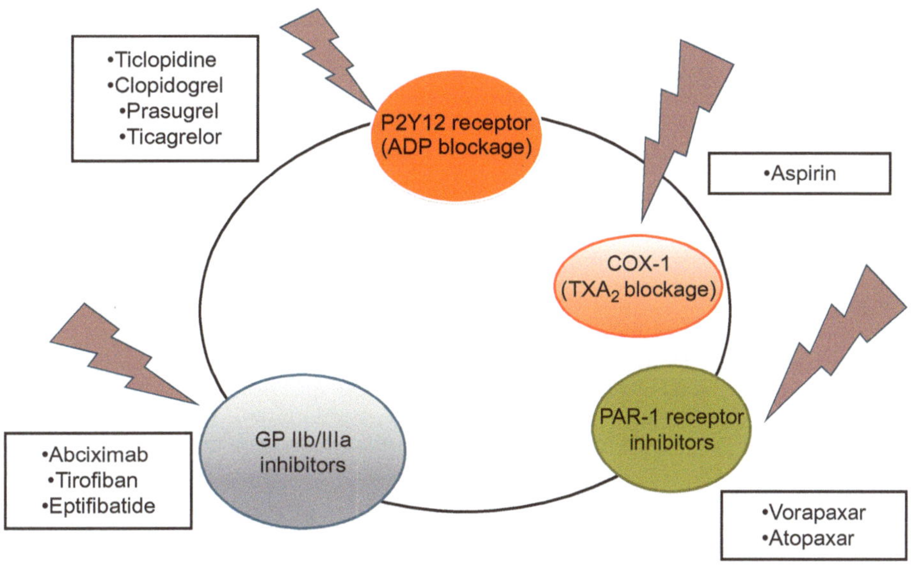

Figure 3.6.1 Mechanism of action of the most common antiplatelet agents.

CONCLUSION

CAD is the leading cause of mortality and morbidity worldwide and is presented either as ACS or as stable CAD. Optimal antiplatelet treatment is of major importance for both these entities. Aspirin and clopidogrel are the most popular administrated antiplatelet agents worldwide; however, recent data favor the use of novel antiplatelets such as prasugrel and ticagrelor. Moreover, there has been a lot of discussion about the therapeutic advantages of elinogrel, cangrelor as well as GP IIb/IIIa receptor antagonists, PAR-1 antagonists and PDE inhibitors. Besides, the duration of DAPT is a major subject of debate among several studies as well as the variability of antiplatelet efficacy among the patients and therefore further investigation should be done to this direction.

REFERENCES

[1] Jennings LK. Mechanisms of platelet activation: need for new strategies to protect against platelet-mediated atherothrombosis. Thromb Haemostasis 2009;102(2):248–57.
[2] Davi G, Patrono C. Platelet activation and atherothrombosis. N Engl J Med 2007;357(24):2482–94.
[3] Yusuf S, Zhao F, Mehta SR, Chrolavicius S, Tognoni G, Fox KK, et al. Effects of clopidogrel in addition to aspirin in patients with acute coronary syndromes without ST-segment elevation. N Engl J Med 2001;345(7):494–502.
[4] Wiviott SD, Braunwald E, McCabe CH, Montalescot G, Ruzyllo W, Gottlieb S, et al. Prasugrel versus clopidogrel in patients with acute coronary syndromes. N Engl J Med 2007;357(20):2001–15.
[5] Siasos G, Kioufis S, Oikonomou E, Zaromitidou M, Maniatis K, Vavuranakis M, et al. Impact of C34T P2Y12 ADP receptor polymorphism and smoking status on cardiovascular outcome in coronary artery disease patients receiving clopidogrel. Int J Cardiol 2016;210:161–3.
[6] Wallentin L, Becker RC, Budaj A, Cannon CP, Emanuelsson H, Held C, et al. Ticagrelor versus clopidogrel in patients with acute coronary syndromes. N Engl J Med 2009;361(11):1045–57.
[7] Oikonomou E, Siasos G, Zaromitidou M, Hatzis G, Mourouzis K, Chrysohoou C, et al. Atorvastatin treatment improves endothelial function through endothelial progenitor cells mobilization in ischemic heart failure patients. Atherosclerosis 2015;238(2):159–64.
[8] Davignon J. Beneficial cardiovascular pleiotropic effects of statins. Circulation 2004;109(23 Suppl 1): III39–43.
[9] Bellosta S, Ferri N, Bernini F, Paoletti R, Corsini A. Non-lipid-related effects of statins. Ann Med 2000;32(3):164–76.
[10] Tousoulis D, Oikonomou E, Economou EK, Crea F, Kaski JC. Inflammatory cytokines in atherosclerosis: current therapeutic approaches. Eur Heart J 2016;37(22):1723–32.
[11] Kaushansky K. The molecular mechanisms that control thrombopoiesis. J Clin Invest 2005;115(12): 3339–47.
[12] Sorrentino S, Studt JD, Medalia O, Tanuj Sapra K. Roll, adhere, spread and contract: structural mechanics of platelet function. Eur J Cell Biol 2015;94(3–4):129–38.
[13] Penz S, Reininger AJ, Brandl R, Goyal P, Rabie T, Bernlochner I, et al. Human atheromatous plaques stimulate thrombus formation by activating platelet glycoprotein VI. FASEB J 2005;19(8): 898–909.
[14] Furie B, Furie BC. Mechanisms of thrombus formation. N Engl J Med 2008;359(9):938–49.
[15] Dubois C, Steiner B, Kieffer N, Reigner SC. Thrombin binding to GPIbalpha induces platelet aggregation and fibrin clot retraction supported by resting alphaIIbbeta3 interaction with polymerized fibrin. Thromb Haemostasis 2003;89(5):853–65.
[16] Schror K, Bretschneider E, Fischer K, Fischer JW, Pape R, Rauch BH, et al. Thrombin receptors in vascular smooth muscle cells - function and regulation by vasodilatory prostaglandins. Thromb Haemostasis 2010;103(5):884–90.
[17] Larsen SB, Grove EL, Wurtz M, Neergaard-Petersen S, Hvas AM, Kristensen SD. The influence of low-grade inflammation on platelets in patients with stable coronary artery disease. Thromb Haemostasis 2015;114(3):519–29.
[18] Androulakis E, Norrington K, Bakogiannis C, Lioudaki E, Siasos G, Tousoulis D. The impact of antiplatelet treatment on endothelial function. Curr Pharm Des 2016;22(29):4512–8.

[19] Oikonomou E, Vogiatzi G, Papamikroulis GA, Naar L, Siasos G, Mourouzis K, et al. Antiplatelet therapy in acute coronary syndromes. Evidence based medicine. Curr Pharm Des 2016;22(29):4519–36.

[20] Santilli F, Rocca B, De Cristofaro R, Lattanzio S, Pietrangelo L, Habib A, et al. Platelet cyclooxygenase inhibition by low-dose aspirin is not reflected consistently by platelet function assays: implications for aspirin "resistance". J Am Coll Cardiol 2009;53(8):667–77.

[21] Doroszko A, Szahidewicz-Krupska E, Janus A, Jakubowski M, Turek A, Ilnicka P, et al. Endothelial dysfunction in young healthy men is associated with aspirin resistance. Vasc Pharmacol 2015;67–69: 30–7.

[22] Varon D, Spectre G. Antiplatelet agents. Hematol Am Soc Hematol Educ Program 2009:267–72.

[23] Galper BZ, Wang YC, Einstein AJ. Strategies for Primary Prevention of Coronary Heart Disease Based on Risk Stratification by the ACC/AHA Lipid Guidelines, ATP III Guidelines, Coronary Calcium Scoring, and C-Reactive Protein, and a Global Treat-All Strategy: A Comparative–Effectiveness Modeling Study. PLoS One 2015;(9):10. e0138092.

[24] Serebruany VL, Pershukov IV. The differences between the guidelines of the European Society of Cardiology and the American College of Cardiology/American Heart Association for oral P2Y12 inhibitor therapy in the management of patients with acute coronary syndromes. Kardiologiia 2013;53(7):70–7.

[25] Antithrombotic Trialists C. Collaborative meta-analysis of randomised trials of antiplatelet therapy for prevention of death, myocardial infarction, and stroke in high risk patients. BMJ 2002;324(7329): 71–86.

[26] Keach JW, Yeh RW, Maddox TM. Dual antiplatelet therapy in patients with stable ischemic heart disease. Curr Atheroscler Rep 2016;18(1):5.

[27] Braunwald E, Antman EM, Beasley JW, Califf RM, Cheitlin MD, Hochman JS, et al. ACC/AHA guidelines for the management of patients with unstable angina and non-ST-segment elevation myocardial infarction: executive summary and recommendations. A report of the American College of Cardiology/American Heart Association task force on practice guidelines (committee on the management of patients with unstable angina). Circulation 2000;102(10):1193–209.

[28] Venerito M, Kandulski A, Malfertheiner P. Dilemma between gastroprotection and cardiovascular prevention. Dtsch Medizinische Wochenschr 2010;135(44):2193–8.

[29] Investigators C-O, Mehta SR, Bassand JP, Chrolavicius S, Diaz R, Eikelboom JW, et al. Dose comparisons of clopidogrel and aspirin in acute coronary syndromes. N Engl J Med 2010;363(10): 930–42.

[30] Navaratnam K, Alfirevic A, Alfirevic Z. Low dose aspirin and pregnancy: how important is aspirin resistance? BJOG 2016;123(9):1481–7.

[31] Tousoulis D, Siasos G, Stefanadis C. Aspirin resistance: what the cardiologist needs to know? Int J Cardiol 2009;132(2):153–6.

[32] Dretzke J, Riley RD, Lordkipanidze M, Jowett S, O'Donnell J, Ensor J, et al. The prognostic utility of tests of platelet function for the detection of 'aspirin resistance' in patients with established cardiovascular or cerebrovascular disease: a systematic review and economic evaluation. Health Technol Assess 2015;19(37):1–366.

[33] Wijeyeratne YD, Heptinstall S. Anti-platelet therapy: ADP receptor antagonists. Br J Clin Pharmacol 2011;72(4):647–57.

[34] Briasoulis A, Telila T, Palla M, Siasos G, Tousoulis D. P2Y12 receptor antagonists: which one to choose? A systematic review and meta-analysis. Curr Pharm Des 2016;22(29):4568–76.

[35] Neubauer H, Kruger JC, Lask S, Endres HG, Pepinghege F, Engelhardt A, et al. Comparing the antiplatelet effect of clopidogrel hydrogensulfate and clopidogrel besylate: a crossover study. Clin Res Cardiol 2009;98(9):533–40.

[36] Siasos G, Tousoulis D, Stefanadis C. Multiple mechanisms affect the clopidogrel response. J Am Coll Cardiol 2009;53(10):900–1. author reply 1.

[37] Alexopoulos D, Barampoutis N, Gkizas V, Vogiatzi C, Tsigkas G, Koutsogiannis N, et al. Crushed versus integral tablets of ticagrelor in ST-segment elevation myocardial infarction patients: a randomized pharmacokinetic/pharmacodynamic study. Clin Pharmacokinet 2016;55(3):359–67.

[38] Ueno M, Ferreiro JL, Angiolillo DJ. Update on the clinical development of cangrelor. Exp Rev Cardiovasc Ther 2010;8(8):1069–77.
[39] Oikonomou E, Papageorgiou N, Papaioannou S, Toutouzas K, Latsios G, Siasos G, et al. Genetic testing and antiplatelet treatment: still way to go? Int J Cardiol 2015;187:63–5.
[40] Siasos G, Oikonomou E, Zaromitidou M, Kioufis S, Kokkou E, Mourouzis K, et al. Clopidogrel response variability is associated with endothelial dysfunction in coronary artery disease patients receiving dual antiplatelet therapy. Atherosclerosis 2015;242(1):102–8.
[41] Siasos G, Zaromitidou M, Oikonomou E, Mourouzis K, Tsalamandris S, Kioufis S, et al. Effects of CYP2C19 polymorphism on endothelial function, arterial stiffness and inflammation in coronary artery disease patients under clopidogrel treatment. Curr Pharm Des 2015;21(34):5041–6.
[42] Bonello L, Camoin-Jau L, Mancini J, Bessereau J, Grosdidier C, Alessi MC, et al. Factors associated with the failure of clopidogrel dose-adjustment according to platelet reactivity monitoring to optimize P2Y12-ADP receptor blockade. Thromb Res 2012;130(1):70–4.
[43] Siasos G, Tousoulis D, Stefanadis C. CYP2C19 genotype and outcomes of clopidogrel treatment. N Engl J Med 2011;364(5):481–2.
[44] Tousoulis D, Siasos G, Zaromytidou M, Papageorgiou N, Stefanadi E, Oikonomou E, et al. The role of the cytochrome P450 polymorphisms in clopidogrel efficacy and clinical utility. Curr Med Chem 2011;18(3):427–38.
[45] Price MJ, Berger PB, Teirstein PS, Tanguay JF, Angiolillo DJ, Spriggs D, et al. Standard- vs high-dose clopidogrel based on platelet function testing after percutaneous coronary intervention: the GRAVITAS randomized trial. JAMA 2011;305(11):1097–105.
[46] Committee CS. A randomised, blinded, trial of clopidogrel versus aspirin in patients at risk of ischaemic events (CAPRIE). CAPRIE Steering Committee. Lancet 1996;348(9038):1329–39.
[47] Chen ZM, Jiang LX, Chen YP, Xie JX, Pan HC, Peto R, et al. Addition of clopidogrel to aspirin in 45,852 patients with acute myocardial infarction: randomised placebo-controlled trial. Lancet 2005; 366(9497):1607–21.
[48] Bhatt DL, Fox KA, Hacke W, Berger PB, Black HR, Boden WE, et al. Clopidogrel and aspirin versus aspirin alone for the prevention of atherothrombotic events. N Engl J Med 2006;354(16): 1706–17.
[49] Sabatine MS, Cannon CP, Gibson CM, Lopez-Sendon JL, Montalescot G, Theroux P, et al. Effect of clopidogrel pretreatment before percutaneous coronary intervention in patients with ST-elevation myocardial infarction treated with fibrinolytics: the PCI-CLARITY study. JAMA 2005;294(10): 1224–32.
[50] Mehta SR, Tanguay JF, Eikelboom JW, Jolly SS, Joyner CD, Granger CB, et al. Double-dose versus standard-dose clopidogrel and high-dose versus low-dose aspirin in individuals undergoing percutaneous coronary intervention for acute coronary syndromes (CURRENT-OASIS 7): a randomised factorial trial. Lancet 2010;376(9748):1233–43.
[51] Teng R, Oliver S, Hayes MA, Butler K. Absorption, distribution, metabolism, and excretion of ticagrelor in healthy subjects. Drug Metab Dispos 2010;38(9):1514–21.
[52] Wiviott SD, Trenk D, Frelinger AL, O'Donoghue M, Neumann FJ, Michelson AD, et al. Prasugrel compared with high loading- and maintenance-dose clopidogrel in patients with planned percutaneous coronary intervention: the prasugrel in comparison to clopidogrel for inhibition of platelet activation and aggregation-thrombolysis in myocardial infarction 44 trial. Circulation 2007; 116(25):2923–32.
[53] Hochholzer W, Amann M, Titov A, Younas I, Loffelhardt N, Riede F, et al. Randomized comparison of different thienopyridine loading strategies in patients undergoing elective coronary intervention: the ExcelsiorLOAD trial. JACC Cardiovasc Interventions 2016;9(3):219–27.
[54] Burki NK, Lee LY. Blockade of airway sensory nerves and dyspnea in humans. Pulm Pharmacol Ther 2010;23(4):279–82.
[55] Gurbel PA, Bliden KP, Butler K, Tantry US, Gesheff T, Wei C, et al. Randomized double-blind assessment of the ONSET and OFFSET of the antiplatelet effects of ticagrelor versus clopidogrel in patients with stable coronary artery disease: the ONSET/OFFSET study. Circulation 2009; 120(25):2577–85.

[56] Joshi RR, Hossain R, Morton AC, Ecob R, Judge HM, Wales C, et al. Evolving pattern of platelet P2Y12 inhibition in patients with acute coronary syndromes. Platelets 2014;25(6):416–22.
[57] Parker WA, Storey RF. Long-term antiplatelet therapy following myocardial infarction: implications of PEGASUS-TIMI 54. Heart 2016;102(10):783–9.
[58] Held C, Asenblad N, Bassand JP, Becker RC, Cannon CP, Claeys MJ, et al. Ticagrelor versus clopidogrel in patients with acute coronary syndromes undergoing coronary artery bypass surgery: results from the PLATO (Platelet Inhibition and Patient Outcomes) trial. J Am Coll Cardiol 2011; 57(6):672–84.
[59] Bonaca MP, Bhatt DL, Cohen M, Steg PG, Storey RF, Jensen EC, et al. Long-term use of ticagrelor in patients with prior myocardial infarction. N Engl J Med 2015;372(19):1791–800.
[60] Price MJ, Clavijo L, Angiolillo DJ, Carlson G, Caplan R, Teng R, et al. A randomised trial of the pharmacodynamic and pharmacokinetic effects of ticagrelor compared with clopidogrel in Hispanic patients with stable coronary artery disease. J Thromb Thrombolysis 2015;39(1):8–14.
[61] Alexopoulos D, Moulias A, Koutsogiannis N, Xanthopoulou I, Kakkavas A, Mavronasiou E, et al. Differential effect of ticagrelor versus prasugrel on coronary blood flow velocity in patients with non-ST-elevation acute coronary syndrome undergoing percutaneous coronary intervention: an exploratory study. Circ Cardiovasc Interventions 2013;6(3):277–83.
[62] Angiolillo DJ, Curzen N, Gurbel P, Vaitkus P, Lipkin F, Li W, et al. Pharmacodynamic evaluation of switching from ticagrelor to prasugrel in patients with stable coronary artery disease: results of the SWAP-2 Study (Switching Anti Platelet-2). J Am Coll Cardiol 2014;63(15):1500–9.
[63] Berger JS, Roe MT, Gibson CM, Kilaru R, Green CL, Melton L, et al. Safety and feasibility of adjunctive antiplatelet therapy with intravenous elinogrel, a direct-acting and reversible P2Y12 ADP-receptor antagonist, before primary percutaneous intervention in patients with ST-elevation myocardial infarction: the Early Rapid ReversAl of platelet thromboSis with intravenous Elinogrel before PCI to optimize reperfusion in acute Myocardial Infarction (ERASE MI) pilot trial. Am Heart J 2009;158(6):998–1004.e1.
[64] Andre P, DeGuzman F, Haberstock-Debic H, Mills S, Pak Y, Inagaki M, et al. Thienopyridines, but not elinogrel, result in off-target effects at the vessel wall that contribute to bleeding. J Pharmacol Exp Ther 2011;338(1):22–30.
[65] Leonardi S, Rao SV, Harrington RA, Bhatt DL, Gibson CM, Roe MT, et al. Rationale and design of the randomized, double-blind trial testing INtraveNous and Oral administration of elinogrel, a selective and reversible P2Y(12)-receptor inhibitor, versus clopidogrel to eVAluate Tolerability and Efficacy in nonurgent Percutaneous Coronary Interventions patients (INNOVATE-PCI). Am Heart J 2010;160(1):65–72.
[66] Norgard NB. Cangrelor: a novel P2Y12 receptor antagonist. Exp Opin Invest Drugs 2009;18(8): 1219–30.
[67] Ferreiro JL, Angiolillo DJ. New directions in antiplatelet therapy. Circ Cardiovasc Interventions 2012;5(3):433–45.
[68] Angiolillo DJ, Firstenberg MS, Price MJ, Tummala PE, Hutyra M, Welsby IJ, et al. Bridging antiplatelet therapy with cangrelor in patients undergoing cardiac surgery: a randomized controlled trial. JAMA 2012;307(3):265–74.
[69] Huang J, Driscoll EM, Gonzales ML, Park AM, Lucchesi BR. Prevention of arterial thrombosis by intravenously administered platelet P2T receptor antagonist AR-C69931MX in a canine model. J Pharmacol Exp Ther 2000;295(2):492–9.
[70] White HD, Chew DP, Dauerman HL, Mahaffey KW, Gibson CM, Stone GW, et al. Reduced immediate ischemic events with cangrelor in PCI: a pooled analysis of the CHAMPION trials using the universal definition of myocardial infarction. Am Heart J 2012;163(2). 182.e4–190.e4.
[71] O'Donoghue ML, Bhatt DL, Stone GW, Steg PG, Gibson CM, Hamm CW, et al. Efficacy and safety of cangrelor in women versus men during percutaneous coronary intervention: insights from the cangrelor versus standard therapy to achieve optimal management of platelet inhibition (CHAMPION PHOENIX) trial. Circulation 2016;133(3):248–55.
[72] Braunwald E, Antman EM, Beasley JW, Califf RM, Cheitlin MD, Hochman JS, et al. ACC/AHA guidelines for the management of patients with unstable angina and non-ST-segment elevation

myocardial infarction. A report of the American College of cardiology/American heart association task force on practice guidelines (committee on the management of patients with unstable angina). J Am Coll Cardiol 2000;36(3):970–1062.
[73] Muniz-Lozano A, Rollini F, Franchi F, Angiolillo DJ. Update on platelet glycoprotein IIb/IIIa inhibitors: recommendations for clinical practice. Ther Adv Cardiovasc Dis 2013;7(4):197–213.
[74] Kastrati A, Mehilli J, Neumann FJ, Dotzer F, ten Berg J, Bollwein H, et al. Abciximab in patients with acute coronary syndromes undergoing percutaneous coronary intervention after clopidogrel pretreatment: the ISAR-REACT 2 randomized trial. JAMA 2006;295(13):1531–8.
[75] Bledzka K, Smyth SS, Plow EF. Integrin alphaIIbbeta3: from discovery to efficacious therapeutic target. Circ Res 2013;112(8):1189–200.
[76] Lincoff AM, Califf RM, Moliterno DJ, Ellis SG, Ducas J, Kramer JH, et al. Complementary clinical benefits of coronary-artery stenting and blockade of platelet glycoprotein IIb/IIIa receptors. Evaluation of Platelet IIb/IIIa Inhibition in Stenting Investigators. N Engl J Med 1999;341(5):319–27.
[77] Matos V, Marques AM, Oliveira H, Ramos D, Lopes P, Camacho M, et al. Adjuvant therapy with a glycoprotein IIb-IIa inhibitor (abciximab) in coronary angioplasties with a high thrombotic risk. Port J Cardiol 1998;17(12):1001–5. Revista portuguesa de cardiologia: orgao oficial da Sociedade Portuguesa de Cardiologia.
[78] Kastrati A, Mehilli J, Schuhlen H, Dirschinger J, Dotzer F, ten Berg JM, et al. A clinical trial of abciximab in elective percutaneous coronary intervention after pretreatment with clopidogrel. N Engl J Med 2004;350(3):232–8.
[79] Price DJ, Campbell PG, Sutton AG, Grech ED, Davies A, Hall JA, et al. Selective use of abciximab in coronary stenting: overall outcomes can still be equivalent to those in the EPISTENT treatment group. Int J Cardiovasc Interventions 2001;4(1):15–20.
[80] Lincoff AM, Harrington RA, Califf RM, Hochman JS, Guerci AD, Ohman EM, et al. Management of patients with acute coronary syndromes in the United States by platelet glycoprotein IIb/IIIa inhibition. Insights from the platelet glycoprotein IIb/IIIa in unstable angina: receptor suppression using integrilin therapy (PURSUIT) trial. Circulation 2000;102(10):1093–100.
[81] Fischell TA, Attia T, Rane S, Salman W. High-dose, single-bolus eptifibatide: a safe and cost-effective alternative to conventional glycoprotein IIb/IIIa inhibitor use for elective coronary interventions. J Invasive Cardiol 2006;18(10):487–91.
[82] Effects of platelet glycoprotein IIb/IIIa blockade with tirofiban on adverse cardiac events in patients with unstable angina or acute myocardial infarction undergoing coronary angioplasty. The RESTORE Investigators. Randomized Efficacy Study of Tirofiban for Outcomes and REstenosis. Circulation 1997;96(5):1445–53.
[83] Valgimigli M, Campo G, de Cesare N, Meliga E, Vranckx P, Furgieri A, et al. Intensifying platelet inhibition with tirofiban in poor responders to aspirin, clopidogrel, or both agents undergoing elective coronary intervention: results from the double-blind, prospective, randomized Tailoring Treatment with Tirofiban in Patients Showing Resistance to Aspirin and/or Resistance to Clopidogrel study. Circulation 2009;119(25):3215–22.
[84] Coughlin SR. Protease-activated receptors in hemostasis, thrombosis and vascular biology. J Thromb Haemostasis 2005;3(8):1800–14.
[85] Vu TK, Hung DT, Wheaton VI, Coughlin SR. Molecular cloning of a functional thrombin receptor reveals a novel proteolytic mechanism of receptor activation. Cell 1991;64(6):1057–68.
[86] Kahn ML, Nakanishi-Matsui M, Shapiro MJ, Ishihara H, Coughlin SR. Protease-activated receptors 1 and 4 mediate activation of human platelets by thrombin. J Clin Invest 1999;103(6):879–87.
[87] Fala L. Zontivity (Vorapaxar), first-in-class PAR-1 antagonist, receives FDA approval for risk reduction of heart attack, stroke, and cardiovascular death. Am Health Drug Benefits 2015;8(Spec Feature): 148–51.
[88] Scirica BM, Bonaca MP, Braunwald E, De Ferrari GM, Isaza D, Lewis BS, et al. Vorapaxar for secondary prevention of thrombotic events for patients with previous myocardial infarction: a prespecified subgroup analysis of the TRA 2 degrees P-TIMI 50 trial. Lancet 2012;380(9850):1317–24.

[89] French SL, Arthur JF, Lee H, Nesbitt WS, Andrews RK, Gardiner EE, et al. Inhibition of protease-activated receptor 4 impairs platelet procoagulant activity during thrombus formation in human blood. J Thromb Haemostasis 2016;14(8):1642–54.
[90] Bonaca MP, Scirica BM, Braunwald E, Wiviott SD, O'Donoghue ML, Murphy SA, et al. Coronary stent thrombosis with vorapaxar versus placebo: results from the TRA 2 degrees P-TIMI 50 trial. J Am Coll Cardiol 2014;64(22):2309–17.
[91] Sharma A, Lavie CJ, Helft G, Chatterjee S, Goel S, Garg A, et al. Role of vorapaxar after coronary revascularization. Am J Cardiol 2016;117(7):1059–64.
[92] Becker RC, Moliterno DJ, Jennings LK, Pieper KS, Pei J, Niederman A, et al. Safety and tolerability of SCH 530348 in patients undergoing non-urgent percutaneous coronary intervention: a randomised, double-blind, placebo-controlled phase II study. Lancet 2009;373(9667):919–28.
[93] Serebruany VL, Kogushi M, Dastros-Pitei D, Flather M, Bhatt DL. The in-vitro effects of E5555, a protease-activated receptor (PAR)-1 antagonist, on platelet biomarkers in healthy volunteers and patients with coronary artery disease. Thromb Haemostasis 2009;102(1):111–9.
[94] Goto S, Ogawa H, Takeuchi M, Flather MD, Bhatt DL, Investigators JL. Double-blind, placebo-controlled Phase II studies of the protease-activated receptor 1 antagonist E5555 (atopaxar) in Japanese patients with acute coronary syndrome or high-risk coronary artery disease. Eur Heart J 2010;31(21):2601–13.
[95] Biscetti F, Ferraccioli G, Flex A. New therapeutic effects of cilostazol in patients with ischemic disorders. Curr Vasc Pharmacol 2015;13(3):399–404.
[96] Schror K. The pharmacology of cilostazol. Diabetes Obes Metab 2002;4(Suppl. 2):S14–9.
[97] Bangalore S, Singh A, Toklu B, DiNicolantonio JJ, Croce K, Feit F, et al. Efficacy of cilostazol on platelet reactivity and cardiovascular outcomes in patients undergoing percutaneous coronary intervention: insights from a meta-analysis of randomised trials. Open Heart 2014;1(1):e000068.
[98] Chen J, Meng H, Xu L, Liu J, Kong D, Chen P, et al. Efficacy and safety of cilostazol based triple antiplatelet treatment versus dual antiplatelet treatment in patients undergoing coronary stent implantation: an updated meta-analysis of the randomized controlled trials. J Thromb Thrombolysis 2015;39(1):23–34.
[99] Geeganage C, Wilcox R, Bath PM. Triple antiplatelet therapy for preventing vascular events: a systematic review and meta-analysis. BMC Med 2010;8:36.
[100] Vavuranakis M, Kalogeras K, Vrachatis D, Kariori M, Moldovan C, Mpei E, et al. Antithrombotic therapy in patients undergoing TAVI with concurrent atrial fibrillation. One center experience. J Thromb Thrombolysis 2015;40(2):193–7.
[101] Windecker S, Kolh P, Alfonso F, Collet JP, Cremer J, Falk V, et al. 2014 ESC/EACTS Guidelines on myocardial revascularization. Kardiologia Polska 2014;72(12):1253–379.
[102] Rao SV, O'Grady K, Pieper KS, Granger CB, Newby LK, Van de Werf F, et al. Impact of bleeding severity on clinical outcomes among patients with acute coronary syndromes. Am J Cardiol 2005;96(9):1200–6.
[103] Siasos G, Mourouzis K, Oikonomou E, Vavuranakis M, Vogiatzi G, Briasoulis A, et al. Duration of dual antiplatelet therapy after coronary stenting. Curr Pharm Des 2016;22(29):4583–95.
[104] Kozuma K, Otsuka M, Ikari Y, Uehara Y, Yokoi H, Sano K, et al. Clinical and angiographic outcomes of paclitaxel-eluting coronary stent implantation in hemodialysis patients: a prospective multicenter registry: the OUCH-TL study (outcome in hemodialysis of TAXUS Liberte). J Cardiol 2015;66(6):502–8.
[105] Task Force on the management of STseamiotESoC, Steg PG, James SK, Atar D, Badano LP, Blomstrom-Lundqvist C, et al. ESC Guidelines for the management of acute myocardial infarction in patients presenting with ST-segment elevation. Eur Heart J 2012;33(20):2569–619.
[106] Acharjee S, Teo KK, Jacobs AK, Hartigan PM, Barn K, Gosselin G, et al. Optimal medical therapy with or without percutaneous coronary intervention in women with stable coronary disease: a pre-specified subset analysis of the Clinical Outcomes Utilizing Revascularization and Aggressive druG Evaluation (COURAGE) trial. Am Heart J 2016;173:108–17.

[107] De Bruyne B, Pijls NH, Kalesan B, Barbato E, Tonino PA, Piroth Z, et al. Fractional flow reserve-guided PCI versus medical therapy in stable coronary disease. N Engl J Med 2012;367(11):991–1001.
[108] Tantry US, Bliden KP, Wei C, Storey RF, Armstrong M, Butler K, et al. First analysis of the relation between CYP2C19 genotype and pharmacodynamics in patients treated with ticagrelor versus clopidogrel: the ONSET/OFFSET and RESPOND genotype studies. Circ Cardiovasc Genet 2010;3(6):556–66.
[109] Gwon HC, Hahn JY, Park KW, Song YB, Chae IH, Lim DS, et al. Six-month versus 12-month dual antiplatelet therapy after implantation of drug-eluting stents: the Efficacy of Xience/Promus Versus Cypher to Reduce Late Loss After Stenting (EXCELLENT) randomized, multicenter study. Circulation 2012;125(3):505–13.
[110] Palla M, Briasoulis A, Siddiqui F, Alesh I, Afonso L. Long (>12 months) and short (<6 months) versus standard duration of dual antiplatelet therapy after coronary stenting: a systematic review and meta-analysis. Am J Ther July/August 2017;24(4):e468–76.
[111] Bhatt DL, Flather MD, Hacke W, Berger PB, Black HR, Boden WE, et al. Patients with prior myocardial infarction, stroke, or symptomatic peripheral arterial disease in the CHARISMA trial. J Am Coll Cardiol 2007;49(19):1982–8.
[112] Collet JP, Silvain J, Barthelemy O, Range G, Cayla G, Van Belle E, et al. Dual-antiplatelet treatment beyond 1 year after drug-eluting stent implantation (ARCTIC-Interruption): a randomised trial. Lancet 2014;384(9954):1577–85.
[113] Mauri L, Kereiakes DJ, Yeh RW, Driscoll-Shempp P, Cutlip DE, Steg PG, et al. Twelve or 30 months of dual antiplatelet therapy after drug-eluting stents. N Engl J Med 2014;371(23):2155–66.
[114] Siasos G, Zaromitidou M, Oikonomou E, Vavuranakis M, Tsigkou V, Papageorgiou N, et al. Genetics in the clinical decision of antiplatelet treatment. Curr Pharm Des 2017;23(9):1307–14.
[115] Goodman T, Ferro A, Sharma P. Pharmacogenetics of aspirin resistance: a comprehensive systematic review. Br J Clin Pharmacol 2008;66(2):222–32.
[116] O'Gara PT, Kushner FG, Ascheim DD, Casey Jr DE, Chung MK, de Lemos JA, et al. 2013 ACCF/AHA guideline for the management of ST-elevation myocardial infarction: a report of the American College of Cardiology Foundation/American Heart Association Task Force on practice guidelines. Circulation 2013;127(4):e362–425.
[117] Yasmina A, de Boer A, Klungel OH, Deneer VH. Pharmacogenomics of oral antiplatelet drugs. Pharmacogenomics 2014;15(4):509–28.
[118] Kunicki TJ, Williams SA, Nugent DJ, Harrison P, Segal HC, Syed A, et al. Lack of association between aspirin responsiveness and seven candidate gene haplotypes in patients with symptomatic vascular disease. Thromb Haemostasis 2009;101(1):123–33.
[119] Al-Azzam SI, Alzoubi KH, Khabour OF, Tawalbeh D, Al-Azzeh O. The contribution of platelet glycoproteins (GPIa C807T and GPIba C-5T) and cyclooxygenase 2 (COX-2G-765C) polymorphisms to platelet response in patients treated with aspirin. Gene 2013;526(2):118–21.
[120] Cavallari LH, Jeong H, Bress A. Role of cytochrome P450 genotype in the steps toward personalized drug therapy. Pharmacogenomics Pers Med 2011;4:123–36.
[121] Collet JP, Hulot JS, Anzaha G, Pena A, Chastre T, Caron C, et al. High doses of clopidogrel to overcome genetic resistance: the randomized crossover CLOVIS-2 (Clopidogrel and Response Variability Investigation Study 2). JACC Cardiovasc Interventions 2011;4(4):392–402.
[122] Siasos G, Oikonomou E, Vavuranakis M, Kokkou E, Mourouzis K, Tsalamandris S, et al. Genotyping, platelet activation, and cardiovascular outcome in patients after percutaneous coronary intervention: two pieces of the puzzle of clopidogrel resistance. Cardiology 2017;137(2):104–13.
[123] Tousoulis D, Siasos G, Zaromitidou M, Oikonomou E, Maniatis K, Kioufis S, et al. The impact of CYP2C19 genotype on cardiovascular events and platelet reactivity in patients with coronary artery disease receiving clopidogrel. Int J Cardiol 2013;168(2):1594–6.
[124] Collet JP, Hulot JS, Pena A, Villard E, Esteve JB, Silvain J, et al. Cytochrome P450 2C19 polymorphism in young patients treated with clopidogrel after myocardial infarction: a cohort study. Lancet 2009;373(9660):309–17.

[125] Vavuranakis M, Kalogeras K, Dagres N, Kariori M, Vrachatis D, Moldovan C, et al. Residual platelet reactivity after clopidogrel loading in ST-elevation myocardial infarction patients undergoing a delayed catheterization. Impact on long term clinical events. Int J Cardiol 2014;176(3):1292–3.
[126] Siasos G, Oikonomou E, Zaromitidou M, Kioufis S, Vavuranakis M, Maniatis K, et al. High platelet reactivity is associated with vascular function in patients after percutaneous coronary intervention receiving clopidogrel. Int J Cardiol 2014;177(1):192–6.
[127] Pare G, Mehta SR, Yusuf S, Anand SS, Connolly SJ, Hirsh J, et al. Effects of CYP2C19 genotype on outcomes of clopidogrel treatment. N Engl J Med 2010;363(18):1704–14.
[128] Bhatt DL, Cryer BL, Contant CF, Cohen M, Lanas A, Schnitzer TJ, et al. Clopidogrel with or without omeprazole in coronary artery disease. N Engl J Med 2010;363(20):1909–17.
[129] Li-Wan-Po A, Girard T, Farndon P, Cooley C, Lithgow J. Pharmacogenetics of CYP2C19: functional and clinical implications of a new variant CYP2C19*17. Br J Clin Pharmacol 2010;69(3): 222–30.
[130] Li Y, Tang HL, Hu YF, Xie HG. The gain-of-function variant allele CYP2C19*17: a double-edged sword between thrombosis and bleeding in clopidogrel-treated patients. J Thromb Haemostasis 2012;10(2):199–206.
[131] Mega JL, Close SL, Wiviott SD, Shen L, Walker JR, Simon T, et al. Genetic variants in ABCB1 and CYP2C19 and cardiovascular outcomes after treatment with clopidogrel and prasugrel in the TRITON-TIMI 38 trial: a pharmacogenetic analysis. Lancet 2010;376(9749):1312–9.
[132] Reny JL, Combescure C, Daali Y, Fontana P, Group PONM-A. Influence of the paraoxonase-1 Q192R genetic variant on clopidogrel responsiveness and recurrent cardiovascular events: a systematic review and meta-analysis. J Thromb Haemostasis 2012;10(7):1242–51.
[133] Siasos G, Kioufis S, Oikonomou E, Zaromitidou M, Vavuranakis M, Maniatis K, et al. Impact of C34T P2Y12 genotype on endothelial function and arterial stiffness in patients after percutaneous coronary intervention receiving clopidogrel. Int J Cardiol 2014;177(3):1073–5.
[134] Shi R, Ge L, Zhou X, Ji WJ, Lu RY, Zhang YY, et al. Decreased platelet miR-223 expression is associated with high on-clopidogrel platelet reactivity. Thromb Res 2013;131(6):508–13.
[135] Brandt JT, Close SL, Iturria SJ, Payne CD, Farid NA, Ernest 2nd CS, et al. Common polymorphisms of CYP2C19 and CYP2C9 affect the pharmacokinetic and pharmacodynamic response to clopidogrel but not prasugrel. J Thromb Haemostasis 2007;5(12):2429–36.
[136] Wallentin L, James S, Storey RF, Armstrong M, Barratt BJ, Horrow J, et al. Effect of CYP2C19 and ABCB1 single nucleotide polymorphisms on outcomes of treatment with ticagrelor versus clopidogrel for acute coronary syndromes: a genetic substudy of the PLATO trial. Lancet 2010;376(9749): 1320–8.
[137] Donohue MM, Tirschwell DL. Implications of pharmacogenetic testing for patients taking warfarin or clopidogrel. Curr Neurol Neurosci Rep 2011;11(1):52–60.

CHAPTER 3.7

Primary Percutaneous Coronary Intervention

Konstantinos Toutouzas, Odysseas Kaitozis, Dimitris Tousoulis
1st Cardiology Clinic, 'Hippokration' General Hospital, National and Kapodistrian University of Athens, School of Medicine, Athens, Greece

Contents

INTRODUCTION

Despite major advances in therapy, cardiovascular diseases and especially acute coronary syndromes remain the leading cause of death worldwide [1]. Currently, ST elevation myocardial infarction (STEMI) comprises 25%–40% of myocardial

Coronary Artery Disease
ISBN 978-0-12-811908-2, http://dx.doi.org/10.1016/B978-0-12-811908-2.00021-0

infarction (MI) presentations [2]. During the last decades both in-hospital and 1-year mortality from STEMI have decreased significantly, but cardiovascular outcomes could still be improved.

HISTORICAL OVERVIEW

Clinical features of acute MI (AMI) were described in 1912 in a landmark presentation by James Herrick [3]. He proposed that AMI was caused by a sudden obstruction of the coronary arteries and advised bed rest as treatment. For decades, treatment consisted of 6 weeks of absolute bed rest and in-hospital mortality averaged about 50%. Over the past decades, mortality rate was reduced to 5% by the introduction of coronary care units, defibrillation, aspirin as well as reperfusion with thrombolysis and mainly primary percutaneous coronary intervention (pPCI). The concept of reperfusion was introduced in 1959 by Boucek et al. [4] when he used a nonspecific fibrolytic agent infused in the aortic root of patients with electrocardiogram (ECG) evidence of STEMI demonstrating ECG evidence of reperfusion. A major breakthrough came when Sones demonstrated that selective coronary arteriography was feasible in conscious sedated patients [5]. But it was years later in 1977 that Gruntzig performed the first percutaneous transluminal balloon angioplasty in humans. At the same time (1979) De Wood et al. [6] proved in a landmark observational study that total occlusion of the coronary arteries by a thrombus in the first hours of AMI is the cause of STEMI rebutting the mainstream hypothesis which was supported by many researchers at the time that coronary thrombosis may well play a minor role or none at all, in precipitating a fatal coronary event and that myocardial necrosis comes first and that coronary thrombosis is secondary [7]. As a result the "open artery hypothesis" was established and the modern era of reperfusion therapy was about to begin. Rentrop [8] demonstrated that intracoronary occlusions could be successfully opened with intracoronary streptokinase (SK) therapy and Khaja [9] that SK was superior to placebo in opening occluded coronary arteries. The later was the first randomized trial in the field of reperfusion. Soon after intracoronary SK therapy was linked with increased 1-year survival compared to placebo by Kennedy et al. [10]. Soon it became clear that unfortunately thrombolysis had a serious drawback which was bleeding occurring at the vascular puncture site as well as remote areas like the brain. Major intracranial bleeding occurred in 1%—2% of patients receiving fibrolytic therapy and in most of the cases it proved fatal for the patients. In addition it was shown that complete reperfusion was achieved in almost 70% of the cases [11], as in 30% of the cases the therapy was unsuccessful with incomplete or no reperfusion. Therefore researchers turned to mechanical reperfusion as an alternative. Many years after Gruntzig's first attempt it became clear that early opening of the infarct-related coronary artery improved survival [12]. Mechanical reperfusion with coronary angioplasty was first tested in combination with thrombolysis by several studies [13,14] and has proven to be harmful. In the years that followed there was a significant number of publications that compared coronary angioplasty with thrombolytic therapy and showed that angioplasty was associated

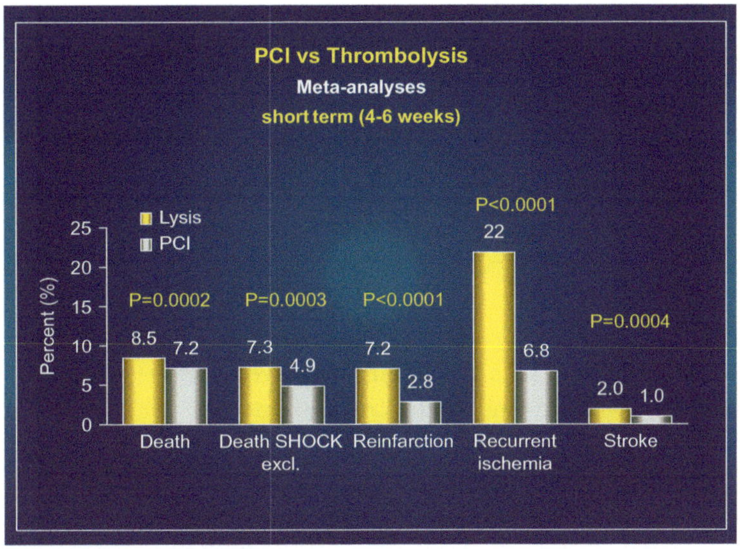

Figure 3.7.1 Percutaneous coronary intervention versus thrombolysis. *(Lancet May 20, 2006; 367(9523):1656.)*

with increased survival and decreased rate of reinfarction. Keeley et al. summarized these studies in a meta-analysis in 2006 demonstrating a reduction in mortality (7.4% vs. 5.5%,$P = .003$), reduction in reinfarction (6.8% vs. 2.5%, $P < .0001$) and a reduction in risk of stroke (2% vs. 1%, $P = .0004$) for angioplasty therapy [15] (Fig. 3.7.1).

The meta-analysis proved that "facilitated PCI" being associated in a large number of patients with increased mortality, bleeding, reinfarction, and stroke should not be routinely applied. Nowadays pPCI has become the treatment of choice in AMI [2]. Outcomes after pPCI have improved by using technologically improved stents, better adjunctive pharmacotherapy and improved procedural techniques such as thrombus aspiration. Furthermore, improved logistics with prehospital triage in the ambulance and skilled personnel have made pPCI accessible for almost all patients with STEMI. Because then pPCI is considered to be superior and safer to thrombolytic therapy in patients with STEMI.

Pathophysiology

In order to improve our treatment options for patients with STEMI, we must first try to understand the pathophysiology of acute coronary syndromes. Acute coronary syndrome consists of a spectrum of clinical conditions ranging from unstable angina, non-STEMI (NSTEMI), and STEMI. All these conditions are characterized by the common pathophysiology of a disrupted atherosclerotic plaque.

Atherosclerotic plaques develop gradually over years. They begin with the accumulation of low-density lipoprotein cholesterol and saturated fat in the intima of blood vessels. This is followed by the adhesion of leukocytes to endothelium, then diapedesis and entry into the intima, where they accumulate lipids and become foam cells. Foam cells are a rich source of proinflammatory mediators and there is strong evidence that support a central role for inflammation in all phases of the atherosclerotic process which includes the activation of numerous different cell types and chemoattractant agents such as proinflammatory cytokines, macrophages, and T lymphocytes as well as mast cells, neutrophils and dendritic cells [16]. The lesion up to this point is referred to as a fatty streak, and may be reversible to a certain extent. Subsequent evolution involves migration of smooth muscle cells from the media, and their proliferation and deposition of extracellular matrix, including proteoglycans, interstitial collagen, and elastin fibers. Some of the smooth muscle cells in advanced plaques exhibit apoptosis. Plaques often develop areas of calcification as they evolve. The plaque initially evolves with the artery remodeling outwards, followed by encroachment on the arterial lumen.

In the majority of cases the syndrome occurs when atherosclerotic plaque rupture, fissures, or ulcerates and precipitates thrombus formation, resulting in an acute total or near-total arterial occlusion. Alternatively, a piece of thrombus may break off leading to downstream vessel occlusion. Platelets play a central role in the development of thrombi and subsequent ischemic events, and this process of platelet-mediated thrombus formation involves adhesion, activation, and aggregation. Plaque rupture exposes subendothelial collagen, a highly thrombogenic material, which serves as a site of platelet adhesion, activation, and aggregation. Activated platelets undergo a series of steps including: shape change, adhesion to endothelial cells of vessels, aggregation, and the secretion of granules that perpetuate the cycle [17]. Fibrinogen and thrombospondin are secreted from a-granules. Within 1 min of activation, the presence of fibrinogen and thrombospondin results in platelet aggregation through the linking of glycoprotein (GP) IIb/IIIa complexes [18]. An adhesive GP, von Willebrand factor (vWF), allows platelets to stay attached to the subendothelial vessel wall (via GP Ib) despite high shear forces. Following adhesion, platelets are activated to secrete a variety of agonists which are proaggregatory molecules, such as thrombin, serotonin, adenosine diphosphate (ADP), and thromboxane A2 (TXA2) (Fig. 3.7.2).

These agonists, which further augment the platelet activation process, bind to specific receptor sites on the platelets to activate the GP IIb/IIIa receptor complex, the final common pathway to platelet aggregation [19]. Once activated, the GP IIb/IIIa receptor undergoes a conformational change that enables it to bind with fibrinogen (Fig. 3.7.3).

Vascular injury also exposes subendothelial tissue factor, which activates the coagulation cascade leading to thrombin generation. Thrombin converts fibrinogen to fibrin, an essential component of arterial thrombus, and is one of the most potent platelet activators by binding to protease-activated receptors (PARs) on the platelet membrane. Notably, only a modest amount of thrombin is produced as a result of the coagulation cascade, whereas the surface of activated platelets is the main source of circulating thrombin [20].

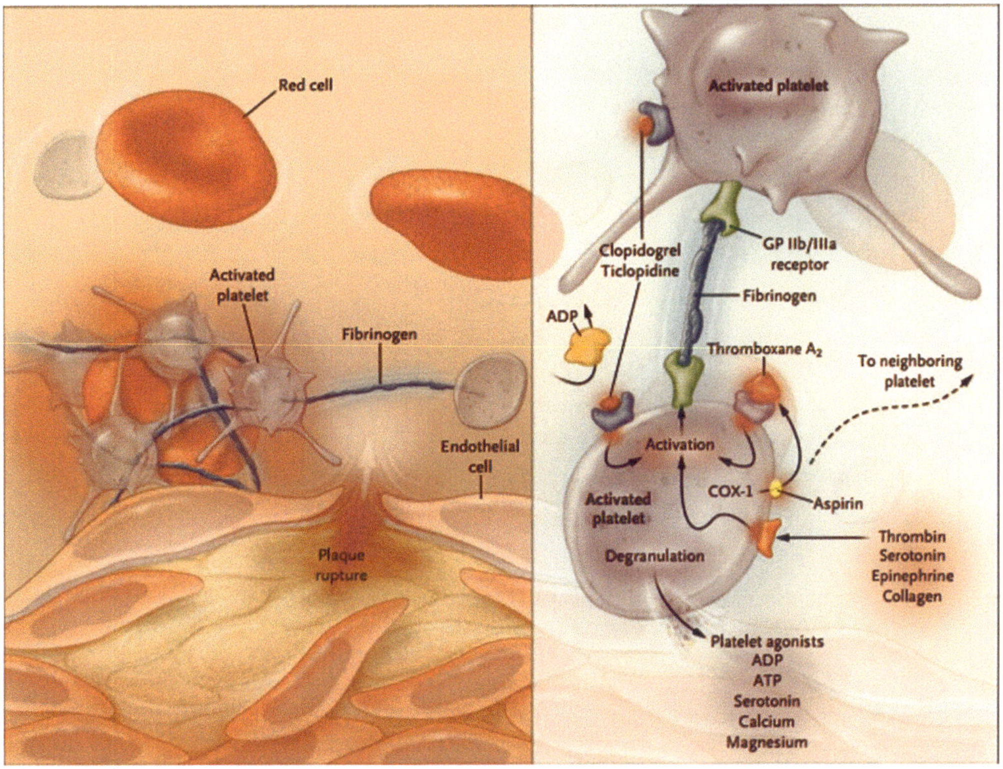

Figure 3.7.2 The disruption of an atherosclerotic plaque results in exposure of highly thrombogenic material. In patients with atherothrombosis, the activation of platelets and coagulation are inseparable, reciprocally self-amplifying processes. The inhibition of platelets alone does not block the coagulation activators. *ADP*, adenosine diphosphate; *COX-1*, cyclooxygenase-1; *GP*, glycoprotein. *(Adapted after Mohler ER. N Engl J Med 2007;357:293–6.)*

Progression of postinfarct myocardial pathology can lead to the occurrence of possible characteristic complications at predictable times after the initial event. Although there may be no apparent visible alterations in the gross morphological appearance of infarcted tissue for at least 6 h after the onset of cell death, changes in cell biochemistry and ultrastructure begin to show abnormalities within 20 min of ischemia. Myocardial ischemia can cause an immediate loss of contractility in the affected myocardium, leading to hypokinesis. Contractility in the remaining viable myocardium increases, a process termed hyperkinesis. Cell damage is progressive, becomingly increasingly irreversible over about 12 h. Therefore, this period can provide a window of opportunity during which thrombolysis and reperfusion may salvage some of the infarct. Between four and 12 h after cell death starts, the infarcted myocardium begins to undergo coagulation necrosis, a process characterized by cell swelling, organelle breakdown, and protein denaturation. Between four and 7 days following a STEMI the infarcted myocardium is especially soft and prone

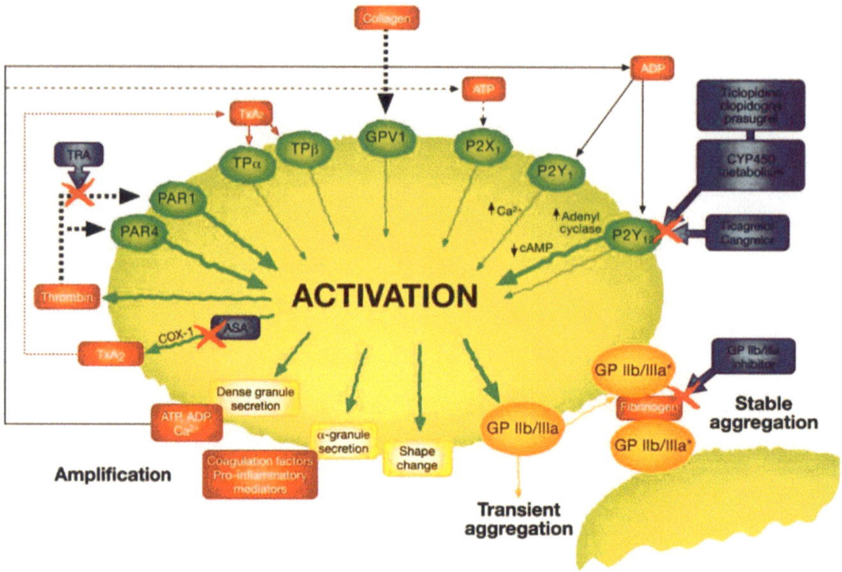

Figure 3.7.3 Mechanisms of platelet activation and potential receptor site blockage by various antiplatelet agents. *(Wallentin L. Eur Heart J 2009;30:1964–77)*

to rupturing, an event usually fatal, and can occur at any time during the initial first 2 weeks and is responsible for about 10% of STEMI mortality. By about 2–3 months following the infarction, the area has healed, leaving a thinned, firm and pale grey non-contracting region of the ventricular wall. Over the course of several months, there is progressive dilatation, not only of the infarct zone, but also of healthy myocardium. This process of ventricular remodeling is caused by an increase in end-diastolic wall stress. Infarct expansion puts patients at a substantial risk for the development of complications.

Overall the environment of an occluded coronary artery in STEMI is characterized by high clot burden and excessive platelet activation. In 20% of cases even it has been described that this hostile environment can lead to reocclusion of the artery and subsequently reinfarction. Mechanical reperfusion with angioplasty has been associated with significantly fewer reperfusion rates compared to thombolysis.

PRE-PROCEDURAL CONSIDERATIONS

Time to Reperfusion

The importance of time in the evolution of an acute coronary syndrome has been well established. In the 1970s, Reimer and Jennings published a series of experimental studies in dogs after acute coronary occlusion, in which they examined the relation between duration of ischemia, area at risk, collateral blood flow, and final infarct size. Their results introduced the concept of "wave front phenomenon" of myocardial death. This concept suggests that infarct size increases in a transmural wave front extending from the endocardium to the epicardium with increasing duration of coronary occlusions and with increasing severity of ischemia. During the initial 15–30 min of sustained coronary occlusion, necrosis starts to develop in the subendocardium, with the necrotic region marching outward toward the epicardium within the next 3–6 h, eventually spanning the entire ventricular wall. In some areas (generally at the edges of the infarct) the myocardium is stunned (reversibly damaged) and can eventually recover if blood flow is restored. Coronary occlusions lasting < 6 h result in subendocardial infarcts but when coronary occlusion exceeds 6 h, infarcts become transmural with an infarct size encompassing the entire area at risk [21]. Decades later, further investigation has demonstrated that the maximum value for reperfusion therapy is within 60–180 min of onset of occlusion and only modest benefit can be expected thereafter [22]. This concept has defined modern therapeutic strategies for STEMI treatment. Guidelines that suggest to minimize delays between the onset of the artery occlusion, the diagnosis of STEMI

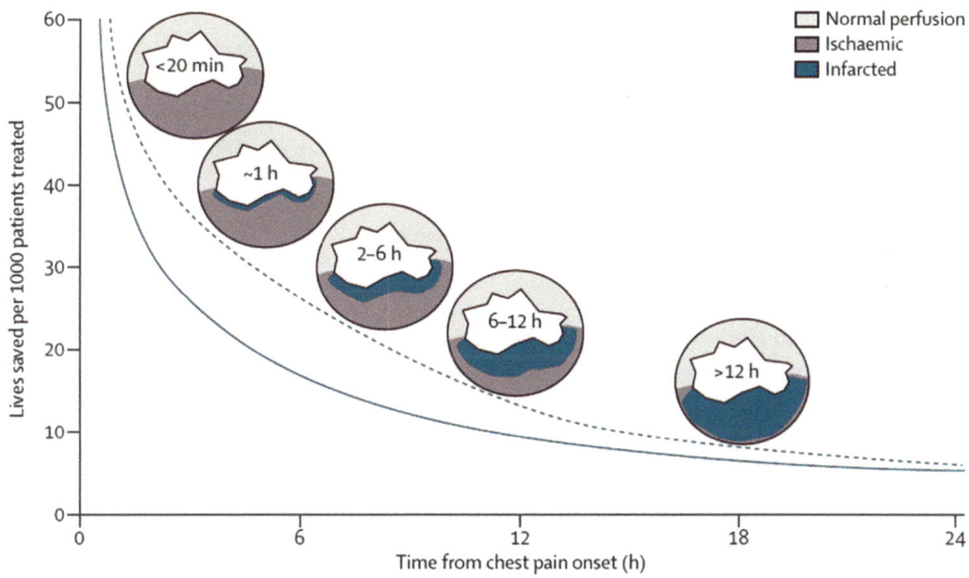

Figure 3.7.4 Relation between time to reperfusion and benefit accrued [23]. Heart cross-sections show the left ventricle at increasing durations of epicardial coronary artery occlusion. This idea of progressive necrosis is modeled on experimental data. The benefit of reperfusion in terms of lives saved at 35 days (*solid line*) is time dependent and modeled on benefit seen in patients randomized to thrombolysis versus placebo [45]. In the first hour after symptom onset, reperfusion significantly reduces mortality. However, benefit slows rapidly, possibly because of progressive infarction shown in the heart cross-sections. The relation between an effective cardioprotective intervention and reperfusion is depicted by the *dotted line*.

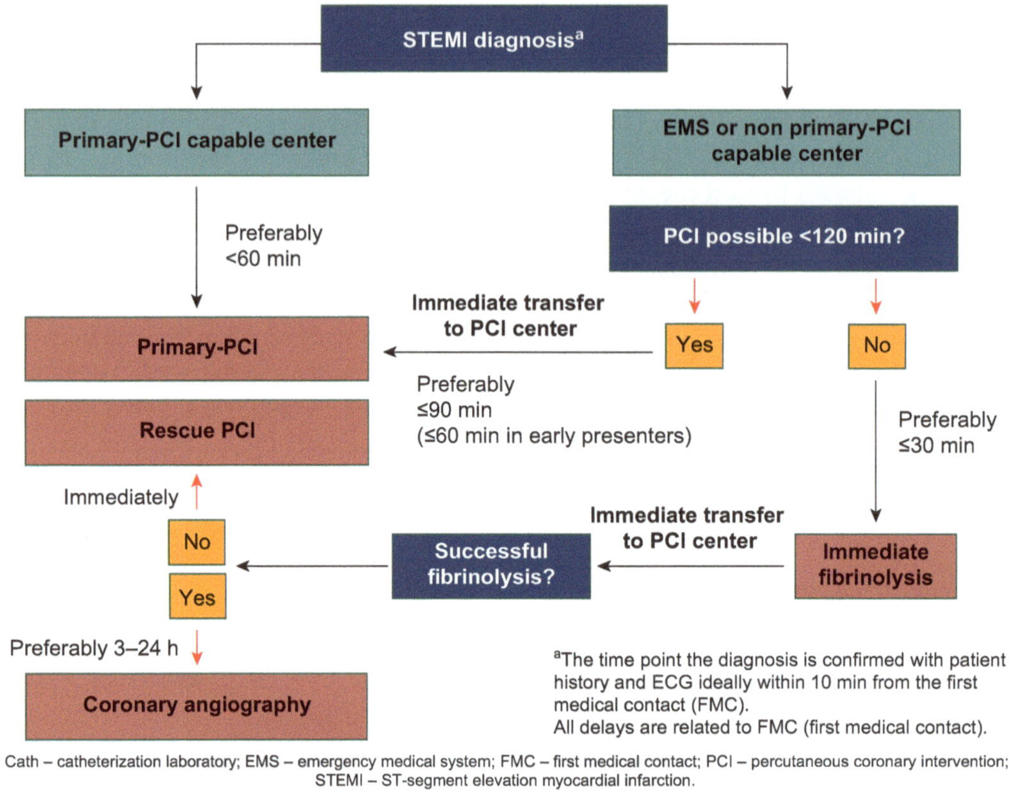

Figure 3.7.5 European Society of Cardiology guidelines for the management of acute myocardial infarction in patients presenting with ST-segment elevation [2].

and the reperfusion therapy, ideally by angioplasty (pPCI or by thrombolysis) in order to reduce the duration of ischemia, save viable myocardium in the risk area and improve clinical outcomes [23] (Figs. 3.7.4 and 3.7.5).

Aside from time, there are other factors that determine the final infarction size and should be taken into account in optimizing treatment strategies. One of them is the development of collateral flow. It has been well recognized that well-developed collateral flow prevents extensive central ischemic zone necrosis and is found mostly in patients with non-ST segment elevation MI or in patients with right coronary occlusions compared with occlusions in LAD or circumflex.

Another factor that can determine the infarct size in STEMI is remote preconditioning of the myocardium. When a cuff is inflated around a limb to cause a series of 5-min periods of ischemia, it is supposed that this kind of remote ischemia causes humoral and neural changes that alter mitochondrial function systemically, which makes the myocardium more resistant to ischemia-reperfusion injury [24]. In 2010, it was demonstrated that remote preconditioning reduced the mean salvage index (reduction of the final infarct area as a proportion of the at risk myocardium on single-photon emission computed tomography [SPECT]) in 333 randomized STEMI patients (0.69 vs. 0.57, 95% CI 0.01 to 0.21, $P = .033$) [25]. The long-term follow-up showed that remote preconditioning reduced the primary outcome of major adverse cardiac events (MACE) (18.1% vs. 27.5%, hazards ratio [HR] 0.62, 95% confidence interval [CI] 0.39 to 0.99, $P = .045$). There were other clinical trials like LIPSIA-CONDITIONING trial that support postconditioning alone does not provide benefit [26]. A large trial powered for clinical outcomes is required to investigate any benefits of remote preconditioning in STEMI before it could be recommended in routine practice.

PRE-PROCEDURAL PHARMACOLOGICAL THERAPY

Oral antithrombotic therapy is the cornerstone of treatment of patients with STEMI. Several classes of antithrombotic, both antiplatelet and anticoagulant, agents available for oral administration are currently approved for clinical use for the management of patients with STEMI (Table 3.7.1). These agents include cyclooxygenase 1 (COX1) inhibitors (aspirin); P2Y12-receptor antagonists (clopidogrel, prasugrel, ticagrelor, and ticlopidine); PAR1 inhibitors (vorapaxar); and factor Xa inhibitors (rivaroxaban). The use of aspirin in combination with a P2Y12-receptor antagonist is commonly known as standard dual antiplatelet therapy (DAPT).

Antiplatelet Therapy

As demonstrated earlier in the pathophysiology of STEMI, occlusive coronary thrombus is the result of adhesion of a small collection of activated platelets at the site of intimal disruption in an unstable atherosclerotic plaque. After an atherosclerotic plaque rupture, thrombosis occurs through platelet adhesion, activation, and aggregation. Each of these three phases represents a potential target for the development of pharmacologic antiplatelet agents (Fig. 3.7.6). Inhibitors of platelet adhesion are still under investigation and not yet approved for clinical use. Inhibitors of platelet aggregation (i.e., intravenous GP IIb/IIIa inhibitors) are reserved only for patients undergoing pPCI. On the other hand inhibitors of platelet activation processes represent the mainstay treatment for the acute and long-term prevention of recurrent ischemic events in STEMI patients.

ADP stimulates platelet activation through two G-protein coupled receptors, P2Y1 and P2Y12 [27]. Although binding of ADP to both receptors is required for complete platelet aggregation, P2Y12 is the predominant receptor involved in ADP-stimulated platelet activation of the GP IIb/IIIa receptor [28]. Binding of ADP to P2Y1 stimulates activation of the GP IIb/IIIa receptor resulting in calcium mobilization, platelet shape change, and transient platelet aggregation [29]. Binding of ADP to P2Y12 stimulates

Table 3.7.1 Pharmacological properties of parenteral anticoagulants and indication in pPCI

Feature	Aspirin	Clopidogrel	Prasugrel	Ticagrelor	Vorapaxar	Rivaroxaban[a]
Target	COX_1	$P2Y_{12}$ receptor	$P2Y_{12}$ receptor	$P2Y_{12}$ receptor	PAR_1	Factor Xa
Type of blockade	Irreversible	Irreversible	Irreversible	Reversible	Reversible	Reversible
Dose	150–325 mg LD; 81–100 mg once-daily MD	600 mg LD: 75 mg once-daily MD	60 mg LD: 10 mg once-daily MD	180 mg LD: 90 mg twice-daily MD	2.08 mg once-daily MD	2.5 mg twice-daily MD
Prodrug	No	Yes	Yes	No[b]	No	No
Onset of action	60 min	2–8 h	30 min–4 h[c]	30 min–4 h[c]	>12 h[d]	2–4 h
Offset of action	7–10 days	7–10 days	7–10 days	3–5 days	4–8 weeks	12 h
Drug interactions	NSAIDs	CYP2C19 inhibitors	No	CYP3A inhibitors or inducers, drugs using P-glycoprotein transporter	CYP3A inhibitors or inducers, drugs using P-glycoprotein transporter	CYP3A4 inhibitors or inducers. P-glycoprotein transporter inhibitors
Timing of administration	Immediately after presentation	At presentation or at time of pPCI	At presentation or at time of pPCI	At presentation or at time of pPCI	After stabilization[e]	After stabilization (>24 h after admission)
Contraindications	Hypersensitivity	Hypersensitivity, active pathological bleeding	Prior CVA. high risk of bleeding, hypersensitivity	Prior ICH, high risk of bleeding, severe hepatic dysfunction, hypersensitivity	Prior ICH or CVA. high risk of bleeding, hypersensitivity	Prior CVA, CrCl <15 mL/min high risk of bleeding, severe hepatic dysfunction, treatment with other anticoagulant, hypersensitivity

COX, cyclooxygenase; *CrCL*, creatinine clearance; *CVA*, cerebrovascular accident; *CYP*, cytochrome P4S0; *ICH*, intracranial hemorrhage; *LD*, loading dose; *MD*, maintenance dose; *NSAID*, nonsteroidal anti-inflammatory drug; *PAR*, protease-activated receptor; *pPCI*, primary percutaneous coronary intervention; *STEMI*, ST-segment elevation myocardial infarction.

[a]Approved only by the European Medicines Agency; not approved by the Food and Drug Administration (FDA) for this use.

[b]Although most ticagrelor-mediated antiplatelet effects are direct, approximately 30%–40% are attributed to an active metabolite (AR-C124910XX).

[c]Depending on clinical setting.

[d]If administered without a loading dose.

[e]Two weeks after the acute event according to European Medicines Agency: no timing prespecified by the FDA.

Antithrombotic therapy for patients with STEMI undergoing pPCI. Nat Rev Cardiol: Macmillan Publishers Limited, part of Springer Nature; 2017.

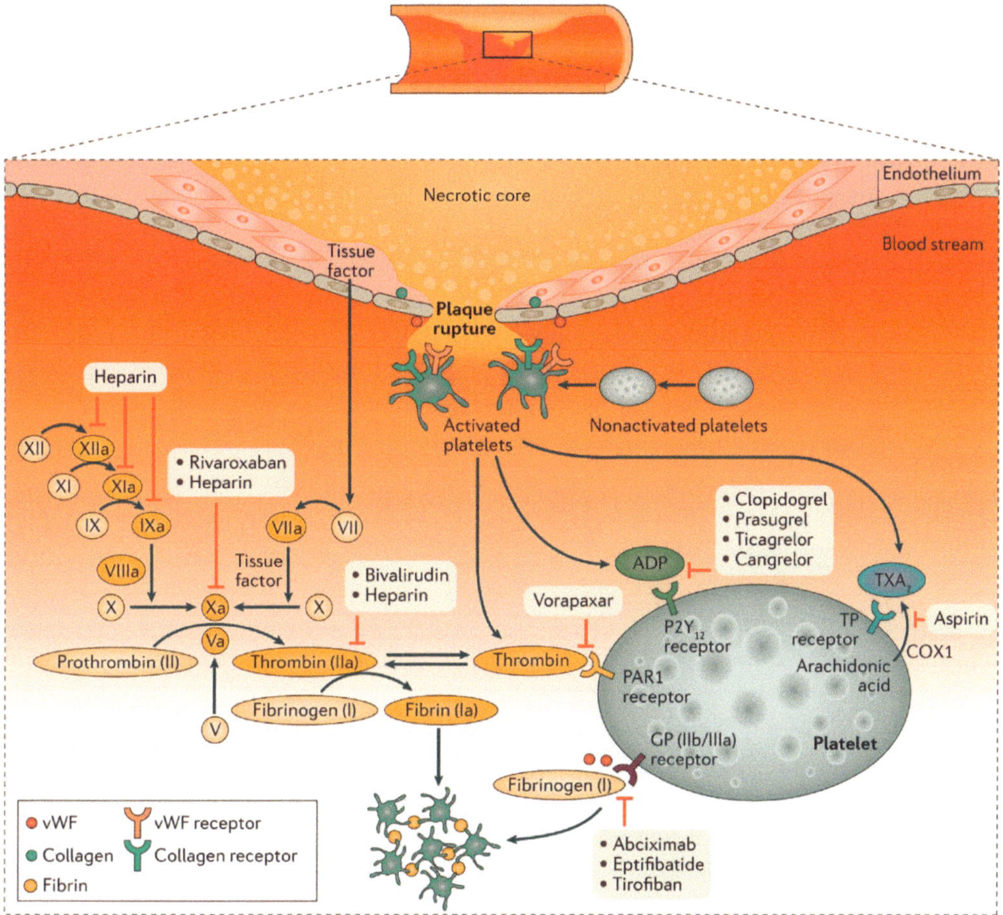

Figure 3.7.6 Mechanism of thrombus formation during ST-segment elevation myocardial infarction, and targets of currently available antithrombotic agents. *(Antithrombotic therapy for patients with STEMI undergoing primary PCI. Nat Rev Cardiol: Macmillan Publishers Limited, part of Springer Nature; 2017.)*

activation of the GP IIb/IIIa receptor resulting in enhanced platelet degranulation and thromboxane production and prolonged platelet aggregation (Fig. 3.7.3). Moreover, activated platelets can synthesize prostanoids, primarily TXA2 from arachidonic acid released from membrane phospholipids through rapid coordinated activation of phospholipase(s), COX-1 and TX-synthase. At least four distinct platelet proteins represent the target of reversible and irreversible inhibitors with variable antiplatelet effects, i.e., COX-1, GPIIb/IIIa, the PGH2/TXA2(TP) receptor, and the ADP receptor P2Y12 [30].

Aspirin

TXA2 is synthesized from arachidonic acid and released by platelets in response to a variety of stimuli (for example, thrombin, collagen, and ADP) and in turn induces irreversible platelet aggregation [31]. Aspirin (acetylsalicylic acid) selectively acetylates the hydroxyl group of a single serine residue at position 529 within the polypeptide chain of platelet prostaglandin G/H synthase 1 [32], causing the irreversible loss of its COX activity. By the irreversible inhibition of this enzyme (COX1) aspirin blocks the production of TXA2 and as a result platelet activation.

The ISIS-2 [33] (Second International Study of Infarct Survival) Collaborative Group was a multicenter, multinational, randomized, double-blind, placebo-controlled randomized trial of 17,187 cases of suspected AMI. One hundred and twenty-seven patients were randomized to one of four groups involving SK or aspirin. In the trial, aspirin therapy was associated with a significant reduction in vascular mortality in patients with suspected acute MI, either as a stand-alone therapy or in combination with SK (9.4% vs. placebo 11.8%, $P < .00001$). Today aspirin is considered to be the first-line therapy in patients with STEMI, and nonenteric-coated aspirin (150—325 mg) should be administered promptly after presentation (class I recommendation). The aspirin loading dose should preferably be given orally including chewing, to ensure complete inhibition of TXA2-dependent platelet aggregation. An intravenous formulation can be used in patients who are unable to swallow.

There are no large, prospective, randomized trials randomizing patients undergoing pPCI to either low (160 mg) versus high doses (320 mg) of aspirin. Data from the GUSTO I and Global Utilization of Streptokinase and Tissue Plasminogen Activator for Occluded Coronary Arteries III trials as well as results from nonrandomized retrospective studies comparing 30-day mortality and bleeding risks show that low dose of aspirin may be as effective as, and perhaps safer than high dose for the acute treatment of STEMI [34]. In conclusion the antithrombotic effect of aspirin is saturable at low to high and that despite a half-life of approximately 20 min in the human circulation, the antithrombotic effect of aspirin is observed with dosing intervals of 24—48 h, reflecting the permanent nature of platelet COX-1 inactivation and the duration of TXA2 suppression.

P2Y12-receptor Inhibitors

As described before ADP exerts its effects on platelets through the G-protein-coupled P2Y12 and P2Y1 receptors, with the former having the major role in platelet activation and aggregation. The combination of a P2Y12 receptor antagonist and aspirin can produce a degree of platelet inhibition higher than that achieved by each single agent, and this synergistic effect has been shown to be beneficial in patients undergoing pPCI [35].

Thienopyridines are a subcategory of antiplatelet medications that prevent platelet aggregation through the binding of select, extracellular cysteine residues on the P2Y12 receptor located on the platelet membrane. Thienopyridine antiplatelet agents interfere with platelet activation and aggregation induced by ADP. The first available P2Y12 receptor inhibitor was the first-generation thienopyridineticlopidine, which in combination with aspirin was superior to both aspirin alone and aspirin plus warfarin for prevention of thrombotic events in patients undergoing PCI. However, its use has been largely abandoned owing to its frequent adverse effects, including life-threatening hematological disorders, and because of the better clinical outcomes achieved with clopidogrel [36].

Other members of the thienopyridine class of antiplatelet agents are clopidogrel and prasugrel available. Both agents are prodrugs and require conversion to an active metabolite to exhibit an antiplatelet effect. The active metabolite binds irreversibly to the P2Y12 receptor, blocking the binding of ADP and thereby inhibiting platelet activation and aggregation.

Clopidogrel

The Clopidogrel as Adjunctive Reperfusion Therapy—Thrombolysis in Myocardial Infarction 28 trial [37] was a double-blind, randomized, placebo-controlled trial that randomized 3491 STEMI patients treated with standard thrombolytic therapy, aspirin, and heparin to either clopidogrel 300 mg loading dose followed by 75 mg/day for 30 days or to placebo. This study showed that there was a 36% reduction in the clopidogrel group compared to placebo for the primary endpoint of infarct-related occlusion of arteries on angiography or death or MI recurrence before angiography which was performed two to 8 days after lysis. In addition there was also a significant reduction of 20% in the major cardiovascular events (cardiovascular death, recurrent MI or recurrent ischemia requiring emergent revascularization) within 30 days of presentation.

In the COMMIT trial [38], patients ($n = 45,852$) with acute MI, mainly STEMI, were randomly assigned to receive either clopidogrel (75 mg per day) or matching placebo, plus aspirin. Clopidogrel therapy significantly reduced thrombotic events (9.2% vs. 10.1%; odds ratio [OR] 0.91, 95% CI 0.86−0.97, $P = .002$) and death from any cause (7.5% vs. 8.1%; OR 0.93, 95% CI 0.87−0.99, $P = .03$). On the basis of these data, clopidogrel has been for many years the standard of care, in addition to aspirin, in patients with STEMI. Today clopidogrel (300−600 mg loading dose and 75 mg daily maintenance dose) is recommended (class I level of recommendation) for the treatment of patients with STEMI. Available data also suggest that a 600 mg loading dose should be preferred in patients undergoing pPCI.

Clopidogrel (300−600 mg loading dose and 75 mg daily maintenance dose) is currently recommended (class I) for the treatment of patients with STEMI. Available data also suggest that a 600 mg loading dose should be preferred in patients undergoing pPCI1−4. In patients with STEMI treated with fibrinolytics, clopidogrel is also the only P2Y12-receptor inhibitor currently recommended and should be administered as a 300 mg loading dose followed by 75 mg per day in patients aged ≤75 years, and 75 mg per day [42].

However several factors have raised concerns about the response variability of clopidogrel. These actors include genetic (CYP polymorphisms), clinical (poor absorption, drug–drug interactions, acute coronary syndrome [ACS], diabetes mellitus, obesity, chronic kidney disease), and cellular (upregulation of P2Y12 pathway) factors [39]. Importantly, several pharmacodynamic studies have shown that approximately 30%–40% of patients have high platelet reactivity while on treatment with clopidogrel, which translates into higher rates of cardiovascular events, including stent thrombosis [38]. These findings underscore the need for more potent antiplatelet agents, especially for with STEMI undergoing pPCI and having a high thrombotic burden, which has led to the clinical development of prasugrel and ticagrelor.

Prasugrel

Prasugrel is a third-generation thienopyridine. As mentioned above prasugrel is a prodrug and, after intestinal absorption, requires a single-step oxidation process through hepatic CYP to generate its active metabolite, which irreversibly binds to the P2Y12 receptor. The metabolic conversion of prasugrel is more efficient compared to clopidogrel which results higher bioavailability [40]. These pharmacological properties theoretically translate into faster onset of action, enhanced platelet inhibition, and lower interindividual variability compared with clopidogrel.

In a subgroup analysis of the TRITON-TIMI 38 trial with patients ($n = 3534$) with STEMI undergoing PCI [41], prasugrel reduced the primary end point at 30 days compared with clopidogrel (6.5% vs. 9.5%; HR 0.68, 95% CI 0.54–0.87, $P = .002$). The main secondary end point of cardiovascular death, MI, or urgent target-vessel revascularization at 30 days, as well as stent thrombosis, were also significantly reduced. It should be noted that STEMI patients were divided in those enrolled within 12 h of onset of symptoms (pPCI), and those enrolled between 12 h and 14 days after symptom onset (secondary PCI). No significant benefit with prasugrel over clopidogrel was shown for patients receiving pPCI, and the effects seen in patients with STEMI as a whole were mainly driven by patients who received secondary PCI. However, the effects on the primary end point were consistent between pPCI and secondary PCI at 15 months ($P = .15$ for interaction), with a tendency toward a difference in treatment effect at 30 days ($P = .06$ for interaction). This trend was attenuated after periprocedural MI events were excluded from the analysis. Prasugrel (60 mg loading dose and 10 mg daily maintenance dose) is currently a class I recommendation in patients with STEMI undergoing reperfusion with pPCI, and can be administered in this setting even before coronary anatomy has been established [42].

Direct-Acting $P2Y_{12}$ Inhibitors

As discussed above, thienopyridines are indirectly acting platelet inhibitors where the active metabolites irreversibly bind to the P2Y12 receptor. Thus, the delayed onset of action of these drugs is a disadvantage especially during PCI. Newer, direct-acting P2Y12 inhibitors like cangrelor and ticagrelor change the conformation of the P2Y12 receptor resulting in reversible inhibition of the receptor.

Ticagrelor

Ticagrelor is the first in a new chemical class, the cyclo-pentyl-triazolo-pyrimidines and is chemically distinct from the thienopyridines, such as clopidogrel and prasugrel. It is administered orally, it has a reversible P2Y12 receptor inhibitory effect and it has a more rapid onset and with a more pronounced platelet inhibition that is nearly double that of clopidogrel.

In the PLATelet Inhibition and Patient Outcomes study, a subset of 8430 patients who were in the midst of STEMI and were scheduled for pPCI were randomized to ticagrelor or clopidogrel in addition to aspirin [43]. The subanalysis revealed that the clopidogrel treatment arm compared to ticagrelor resulted in a reduction of cardiovascular events (composite of CV death, heart attack and stroke) for up to a year (ticagrelor vs. clopidogrel, 9.3% vs. 11.0%, $P = .02$). There was a statistically significant reduction in MI (4.7% vs. 6.1%, $P = .01$). In addition no increase in major bleeding (9.0% vs. 9.3%, $P = .63$) was observed.

However, it should be taken into account that ticagrelor also has non-P2Y12-mediated effects; in particular, the drug increases plasma levels of adenosine owing to inhibition of cellular uptake. These effects might contribute to the overall benefits of ticagrelor, but have also been suggested to contribute to its nonbleeding adverse effects, such as higher incidence of dyspnea and ventricular pauses [44].

Ticagrelor (180 mg loading dose and 90 mg twice-daily maintenance dose) is currently recommended (class I) for the treatment and prevention of secondary atherothrombotic events in patients with STEMI undergoing PPC.

Glycoprotein IIb/IIIa Inhibitors

GPIs compete with vWF and fibrinogen for GPIIb/IIIa receptor binding and provide fast and potent antiplatelet effects inhibiting platelet aggregation [45]. They have been shown to improve ischemic outcomes in patients with STEMI undergoing pPCI. Nevertheless, the clinical benefit of GPIs was mainly shown in the era before routine use of DAPT. In these trials, GPI therapy was associated with a reduction in MACE at the expense of increased bleeding and thrombocytopenia. To date, the administration of GPIs is an accepted treatment option in patients undergoing pPCI but with a reduced role as available data support selective (that is, bail-out in the event of large residual thrombus) rather than routine use. Intracoronary administration of GPIs has been tested in several small studies and shown some benefits, as compared with intravenous administration, but these results have not been confirmed in large-scale clinical trials.

Currently, the American College of Cardiology/American Heart Association (ACC/AHA) guidelines give a class IIa recommendation for GPIs at the time of pPCI (abciximab, double-bolus eptifibatide, or high-bolus-dose tirofiban) in selected patients with STEMI receiving unfractionated heparin (UFH) (with or without stenting or clopidogrel pretreatment), and a class IIb indication for intracoronary abciximab [46]. The ESC guidelines instead give a class IIa recommendation for GPIs for bailout use only (e.g., in the event of evidence of no-reflow or thrombotic complications) [42].

Several studies have also investigated whether early, upstream use of GPIs as a facilitating strategy to early reperfusion can improve clinical outcomes in STEMI. However, results were not consistent, and a clear benefit of this approach has not been demonstrated. Both the ACC/AHA and ESC guidelines give a class IIb (level of evidence B) recommendation for upstream use of GPIs.

Cangrelor

Cangrelor is an adenosine triphosphate analog which reversibly and directly, without any biotransformation, inhibits the P2Y12 receptor. It is characterized by rapid onset of action, great degree of platelet inhibition, and rapid onset of action. Cangrelor has an extremely short half-life due to rapid deactivation by plasma ectonucleotidases. It is the first such drug to be administered intravenously. There were two major phase-3 trials (CHAMPION PCI t and the CHAMPION PLATFORM) [47] were stopped early due to negative results and insufficient evidence for clinical effectiveness. It is still being studied as a potential bridge for patients on clopidogrel who need to go off the drug to undergo surgery or for patients undergoing pPCI who are not pretreated with oral P2Y12-receptor antagonists, especially for those with clinical or angiographic factors of high thrombotic risk. This strategy would also allow bridging of the gap associated with the delayed onset of action of oral P2Y12-receptor inhibitors, including prasugrel and ticagrelor, which require several hours before they reach their full platelet inhibitory effects.

Anticoagulant Therapy

Anticoagulant therapy is mandatory in PPC as STEMI is characterized by a high prothrombotic milieu and an increased risk of thrombotic complications. Anticoagulant drugs available for parenteral use are indirect thrombin (factor IIa) and factor Xa inhibitors, which include UFH and low molecular weight heparin (LMWH); the direct thrombin inhibitor bivalirudin; and the direct factor Xa inhibitor fondaparinux. All of these agents have different mechanisms of action, binding specificity, pharmacokinetic and pharmacodynamic consistency. Because of the high rate of catheter thrombosis recorded in the OASIS-6 trial [48], fondaparinux is not recommended as background anticoagulant for pPCI.

Unfractionated Heparin

UFH is a sulfated polysaccharide with a molecular weight range of 3000–30,000 Da. Its major anticoagulant effect is based on inactivation of thrombin and activated factor X (factor Xa) through an antithrombin-dependent mechanism. Its activity is based on a unique pentasaccharide with a high affinity binding sequence to antithrombin III (ATIII). The heparin–antithrombin complex inactivates not only thrombin but also factors Xa, IXa, XIa, and XIIa. By inactivating thrombin, heparin not only prevents fibrin formation but also inhibits thrombin-induced activation of platelets and of factors V and VIII.

Historically, UFH has been widely used as an anticoagulant in the treatment of STEMI for decades. However disadvantages to the use of UFH include its sometimes difficult-to-manage effects on coagulation because of its narrow therapeutic window, necessitating the need for continuous monitoring of coagulation, the potential for inducing platelet activation. Other major limitations of UFH use include its variable pharmacokinetic profile with high intrapatient and interpatient variability, and the increased risk of heparin-induced thrombocytopenia (HIT).

Although no placebo-controlled trials have evaluated the use of UFH in pPCI, intravenous UFH titrated to target-recommended activated clotting times is a familiar and extensively practiced strategy for anticoagulant therapy at the time of pPCI. UFH currently has a class I indication for anticoagulation during pPCI in both the ACC/AHA and ESC guidelines [42,46] on the basis of expert consensus.

Low-Molecular-Weight Heparin

LMWHs are derived from UFH by depolymerization and consists of only short chains of polysaccharides having an average molecular weight of less than one-third the molecular weight of UFH. The lower molecular weight results to pharmacological and pharmacokinetic advantages over unfractionated heparin such as a reduced capacity to bind to protein and cells and a more predictable pharmacokinetic profile, a longer plasma half-life, high bioavailability and a lower incidence of HIT [49]. Like UFH, LMWHs produce their anticoagulant effect by enhancing antithrombin activity. Laboratory coagulation monitoring is generally not needed, and dosing is weight-adjusted. LMWHs have predominantly renal clearance, and their use is generally not recommended in patients with creatinine clearance of <30 mL/min. **However, LWMHs are considered an attractive potential replacement for UFH because of the convenient subcutaneous route of administration and reliable anticoagulation effects**. Enoxaparin led to a significant 41% relative reduction in the composite main secondary end point of death, recurrent ACS, or urgent revascularization (7% vs. 11%; $P = .015$), as well as other secondary end points, without increasing the risk of major bleeding (5% vs. 5%; $P = .79$). It has been demonstrated that crossover between anticoagulants or simultaneous administration of both was associated with worst outcomes [50]. In total the ATOLL trial provided neutral results for the use of enoxaparine [47]. So the ACC/AHA guidelines do not give any recommendation for the use of enoxaparin in pPCI [46]. On the other hand the ESC guidelines give a class IIa indication for enoxaparin on the basis of a potential benefit in the secondary end point [42].

Thrombin Inhibitors

Bivalirudin

Bivalirudin is a specific and reversible direct thrombin inhibitor. Bivalirudin is a short, synthetic peptide that is potent, highly specific, and a reversible inhibitor of thrombin and it does not have the many limitations seen with indirect thrombin inhibitors, such as heparin. It inhibits both circulating and clot-bound thrombin as well as inhibiting thrombin-mediated platelet activation and aggregation [51]. Several clinical trials support the efficacy and safety of bivalirudin in patients with ACS undergoing PCI [52]. However, the benefit of bivalirudin use in patients with STEMI undergoing pPCI has been debatable.

In the HORIZONS-AMI trial [53] bivalirudin was compared to heparin in STEMI patients undergoing pPCI. Bivalirudin significantly reduced the risk of net adverse clinical events at 30 days (9.2% vs. 12.1%; relative risk [RR] 0.76, 95% CI 0.63–0.92, $P = .005$), which was mainly due to a 40% reduction in major bleeding (4.9% vs. 8.3%; RR 0.60, 95% CI 0.46–0.77, $P < .001$). It should be noted that MACE were similar between the two treatments. On the other hand there was the HEAT-pPCI trial [54] where patients with STEMI undergoing pPCI were randomly assigned to receive either bivalirudin or UFH. Bivalirudin use compared with UFH use significantly increased the incidence of MACE (8.7% vs. 5.7%; RR 1.52, 95% CI 1.09–2.13, $P = .02$), mainly driven by a significantly higher rate of reinfarction, with no significant difference in major bleeding (3.5% vs. 3.1%; $P = .59$). It should be noted that Bivalirudin use has associated in some but not all trials with an increased rate of acute stent thrombosis which has been related to the duration of infusion after the end of pPCI in the different studies.

Overall, results of clinical trials show that bivalirudin compared with heparin yields similar rates of MACE and net adverse clinical events at 30 days, and reduces major bleeding, with a signal toward a reduction in all-cause and cardiac mortality, at the price of an increased risk of acute stent thrombosis. Currently, bivalirudin, with or without previous heparin therapy, has a class I (level of evidence B) recommendation in the ACC/AHA guidelines [42,46]. The ESC guidelines give a class IIa recommendation. The different levels of recommendations in these guidelines might be owing to the fact that the ESC guidelines incorporated the results of later trials like HEAR-pPCI, which led to a downgrading of bivalirudin from a class I recommendation to the current class IIa.

Procedural Considerations

Radial Access

Overall, the femoral artery is the most common route of access for Coronary angiography and pPCI in the United States; however, the use of radial access is increasing. In selected labs in the United States and in some parts of Europe, radial artery access exceeds 90% [55].

The proposed benefits of radial approach are the reduction of vascular complications, reduction of major bleeding and possibly mortality. In addition, without the need for lengthy postprocedural bed rest, earlier ambulation and therefore discharge are possible, potentially reducing hospital costs and improving patient satisfaction [56]. However it should be noted that there are concerns about increases in procedure time, radiation exposure, and access failure in patients undergoing radial artery catheterization that could outbalance the benefits of decreased vascular complications.

Two randomized, controlled trials reported equal or better outcomes with transradial access than with femoral access in STEMI patients undergoing pPCI. In the Radial vs. Femoral Randomized Investigation in ST-Elevation Acute Coronary Syndrome study, a multicenter randomized trial radial access was associated with significantly lower rates of cardiac mortality (5.2% vs. 9.2%) and bleeding (7.8% vs. 12.2%) than femoral access in STEMI patients [57].

The Radial vs. Femoral Access for Coronary Intervention trial compared the efficacy and bleeding outcomes of radial and femoral access separately in patients with STEMI and non-STEMI (NSTEMI) [58]. Radial access was associated with reduced all-cause mortality (1.3% vs. 3.2%) and reduced death/MI/stroke (2.7% vs. 4.6%) in STEMI patients but not in NSTEMI patients. In both STEMI and NSTEMI groups, radial access was associated with significantly reduced ACUITY (Acute Catheterization and Urgent Intervention Triage strategY) major bleeding and major vascular access site complications.

The MATRIX trial compared outcomes of radial access with those of femoral access in 8404 patients with acute coronary syndrome [59]. The radial group had lower all-cause mortality as well as Bleeding Academic Research Consortium major bleeding unrelated to coronary artery bypass grafting.

The ESC guidelines on STEMI patients recommend preference of radial over femoral access, if it is performed by an experienced radial operator (class IIa, level B).

Multivessel Disease

During angiography, 50% of STEMI patients have multivessel disease. Previous evidence had suggested that PCI of nonculprit arteries should not be performed at the time of pPCI due to catecholamine influx and local vasoactive substances which cause prominent coronary vasoconstriction [60]. The PRAMI trial illustrated that multivessel pPCI reduced cardiac death, MI or refractory angina from 22% to 9% (HR 0.35, 95% CI 0.21 to 0.58, $P < .0001$) [61]. These findings were supported by the CvLPRIT trial, which found that patients randomized to complete revascularization (either at pPCI or before discharge) had significantly lower rates of all-cause mortality, MI, heart failure or ischemia-driven revascularization within 12 months (10.0% vs. 21.2%, HR 0.45, 95% CI 0.24–0.84, $P = .009$) [62]. Most recently, the DANAMI-3 PRIMULTI trial showed that fractional flow reserve (FFR) guided multivessel PCI reduced mortality, MI or revascularization (13% vs. 22%, HR 0.56, CI 0.38–0.83, $P = .004$), driven mainly by less revascularization in the multivessel PCI arm [63]. The available data is inconclusive in determining if nonculprit PCI during pPCI is beneficial for hard outcomes.

Thrombus Aspiration

In the setting of pPCI the detachment of thrombus for the ruptured atherosclerotic plaque can lead to distal coronary embolization with subsequent development of microvascular obstruction("slow or no re-flow" phenomena with a low Myocardial Blush Grade). This in turn has been associated with significant loss of myocardial viability, adverse ventricular remodeling and death due to heart failure [64]. Within this background, several strategies to manage intracoronary thrombosis and distal embolization have been developed. Thrombus aspiration (TA) emerged as an attractive adjunctive strategy in pPCI for STEMI. The proposed benefits of TA in STEMI may include reduction of the risk of thrombus distal embolization with development of microvascular obstruction, reduction of local thrombus burden improving stent deployment, reduction of the subsequent risk of late incomplete stent apposition and stent thrombosis, more reliable vessel sizing and stent diameter/length selection. Randomized trials evaluating safety and efficacy of TA in STEMI yielded controversial results over time. Although it seems evident that the use of TA is associated with a significant improvement in myocardial salvage measured with different angiographic, electrocardiographic, or functional parameters, its clinical impact is to date still unclear. The TA during pPCI in acute myocardial infarction (TAPAS) Trial [65] and some meta-analyses found that aspiration thrombectomy during STEMI improves myocardial reperfusion and procedural outcomes, reducing no-reflow, mortality, and distal embolization. On the other hand the larger TASTE trial demonstrated no differences in mortality at STEMI patients. However, results of the TASTE trial do not entirely exclude a potential clinical benefit of this procedure. In fact, stent thrombosis and reinfarction were numerically reduced by approximately 50% in patients randomized to TA. Moreover, there was no evidence of safety concerns with such strategy. Finally, in Routine Aspiration Thrombectomy with PCI versus PCI Alone in Patients with STEMI trial (TOTAL trial), routine use of TA did not reduce the primary endpoint defined as the composite of cardiovascular death, recurrent MI, cardiogenic shock, or class IV heart failure within 180 days [66]. Moreover, routine TA was associated with a significant increase in stroke. The TOTAL optical coherence tomography substudy also found there was no difference in the residual thrombus volume despite aspiration.

Conclusively routine adjunctive thrombus aspiration compared with conventional pPCI may reduce the risk of late MACE in patients with STEMI but increase risk in stroke. Due to the ambivalent results from conflicting trials, use of TA as a routine strategy in patients with STEMI cannot be supported. However, thrombus aspiration might be an attractive strategy in patients with low stroke risk and high lesion thrombus burden. Further studies are needed in order to clarify which patient or lesions subsets might benefit the most from this procedural technique.

Device Therapy

Intra-aortic balloon pump (IABP) counterpulsation has been thought to enhance both myocardial and peripheral perfusion by providing mechanical hemodynamic support. Although IABP was thought to be a useful adjunct in cardiogenic shock after STEMI when patients do not quickly stabilize with pharmacological therapy, its routine use is no longer recommended [67] mainly because of the results of the IABPSHOCK II trial. The study found no difference for routine use of IABP in pPCI in 30-day mortality in patients with refractory cardiogenic shock.

Mechanical LV assist devices (LVADs) can often be used when cardiogenic shock is refractory to reperfusion, inotropes, fluids and IABP [56]. The Impella is the most minimally invasive LVAD, which is a catheter-based microaxial pump that can be inserted percutaneously or after surgical cutdown. The distal end of the Impella sits in the left ventricle and it produces flows into the aorta of up to 4–5 L/min [68]. A meta-analysis compared Impella and Tandem Heart percutaneous LVADs with IABP in AMI or decompensated heart failure and found improved cardiac index, mean arterial pressure and pulmonary arterial wedge pressure, but no difference in 30-day mortality [69]. Mechanical LV assist devices are not recommended as first-line for patients with an acutely ischemic myocardium but may be considered on an individual basis, especially in patients with preexisting or established infarction and as a bridge to more definitive therapy [42,46]. Furthermore, extracorporeal membrane oxygenation (ECMO) has been used in IABP-refractory cardiogenic shock. An observational study found ECMO reduced mortality (72.0% vs. 39.1%, $P = .008$) [70]. IABP, LVADs, and ECMO remain options for shocked patients, but have limited data evidence for routine use.

REFERENCES

[1] WHO Fact sheet N8310, updated June 2011, http://www.who.int/mediacentre/factsheets/fs310/en/index.html.

[2] Wijns W, Kolh P, Danchin N, et al. Guidelines on myocardial revascularization: the Task Force on Myocardial Revascularization of the European Society of Cardiology (ESC) and the European Association for Cardio-Thoracic Surgery (EACTS). Eur Heart J 2010;31.

[3] Herrick JB. Clinical features of sudden obstruction of the coronary arteries. JAMA 1983;250:1757–65.

[4] Boucek RJ, Murphy Jr WP. Segmental perfusion of the coronary arteries with fibrinolysis in man following a myocardial infarction. Am J Cardiol 1960;6:525—33.
[5] Sones Jr FM. Nihon Igaku Hoshasen Gakkai Zasshi 1968;287(6):714—9.
[6] DeWood MA, Spores J, Notske R, et al. Prevalence of total coronary occlusion during the early hours of transmural myocardial infarction. N Engl J Med 1980;303:897—902.
[7] Roberts WC, Ferrans VJ. The role of thrombosis in the etiology of atherosclerosis (a positive one) and in precipitating fatal ischemic heart disease (a negative one). Semin Thromb Hemost 1976;2:123—35.
[8] Rentrop KP. Development and pathophysiological basis of thrombolytic therapy in acute myocardial infarction:part II 1977—1980/the pathogenetic role of thrombus is established by the Gottingen pilot studies of mechanical interventions and intracoronary thombolysis in acute myocardial infarction. J Intev Cardiol 1998;11(4):265—85.
[9] Khaja E, Walton Jr JA, Brymer JE, et al. Intracoronary fibrolytic therapy in acute myocardial infarction: report of a prospective randomized trial. N Eng J Med 1983;308:1305—11.
[10] Kennedy JW, Ritchie JL, Davis KB, et al. Western Washington randomized trial of intracoronary streptokinase in acute myocardial infarction. N Eng J Med 1983;390:1477—82.
[11] de Lemos JA, Antman EM, Giugliano RP, McCabe CH, Murphy SA, Van de Werf F, Gibson CM, Braunwald E. ST-segment resolution and infarct-related artery patency and flow after thrombolytic therapy. Thrombolysis in myocardial infarction (TIMI) 14 investigators. Am J Cardiol 2000;85(3): 299—304.
[12] Dalen JE, Gore JM, Braunwald E, et al. Six- and twelve-month follow-up of the phase I Thrombolysis in myocardial infarction (TIMI) trial. Am J Cardiol 1988;62:179—85.
[13] Topol EJ, Califf RM, George BS, et al. A randomized trial of immediate versus delayed angioplasty after intravenous tissue plasminogen activator in acute myocardial infarctio. N Eng J Med 1987; 317:581—8.
[14] Timi Research Group. Immediate vs delayed catheterization and angioplasty following thrombolytic therapy for acute myocardial infarction. JAMA 1988;260:2849—58.
[15] Keeley EC, Boura JA, Grines CL. Comparison of primary and facilitated percutaneous coronary interventions for ST-elevation myocardial infarction: quantitative review of randomised trials. Lancet 2006;367(9510):579—88.
[16] Tousoulis D, Kampoli AM, Papageorgiou N, et al. Pathophysiology of atherosclerosis: the role of inflammation. Curr Pharm Des 2011;17(37):4089—110.
[17] Clappers N, Brouwer MA, Verheugt FW. Antiplatelet treatment for coronary heart disease. Heart 2007;93:258—65.
[18] Savi P, Herbert JM. ADP receptors on platelets and ADP-selective antiaggregating agents. Med Res Rev 1996;16:159—79.
[19] Schafer AI. Antiplatelet therapy. Am J Med 1996;101:199—209.
[20] Brummel KE, Paradis SG, Butenas S, Mann KG. Thrombin functions during tissue factor-induced blood coagulation. Blood 2002;100:148—52.
[21] Reimer KA, Jennings RB. The 'wavefront phenomenon' of myocardial ischemic cell death. II. Transmural progression of necrosis within the framework of ischemic bed size (myocardium at risk) and collateral flow. Lab Invest 1979;40:633—44.
[22] Gersh BJ, Stone GW, White HD, et al. Pharmacological facilitation of Primary Percutaneous intervention for acute myocardial infarction. Is the slope of the curve the shape of the future? JAMA 2005;293:979—86.
[23] Windecker S, Bax JJ, Myat A, Stone GW, Marber MS. Future treatment strategies in ST-segment elevationmyocardial infarction. Lancet 2013;382(9892):644—57.
[24] Sloth A, Schmidt M, Munk K, Kharbanda R, Redington A, Schmidt M, et al. For the CONDI investigators. Improved long-term clinical out- comes in patients with ST-elevation myocardial infarction undergoing remote ischaemic conditioning as an adjunct to primary percutaneous coronary intervention. Eur Heart J 2014;35(3):168—75.
[25] Bøtker HE, Kharbanda R, Schmidt MR, Bøttcher M, Kaltoft AK, Terkelsen CJ, et al. Remote ischaemic conditioning before hospital admission, as a complement to angioplasty, and effect on myocardial salvage in patients with acute myocardial infarction: a randomised trial. Lancet 2010;375:727—34.

[26] Eitel I, Stiermaier T, Rommel KP, Fuernau G, Sandri M, Mangner N, et al. Cardioprotection by combined intrahospital remote ischaemic perconditioning and postconditioning in ST-elevation myocardial infarction: the randomized LIPSIA-CONDITIONING trial. Eur Heart J 2015;36(44): 3049–57.
[27] Intravenous NPA for the treatment of infarcting myocardium early; InTIME-II, a double-blind comparison of single-bolus lanoteplasevs accelerated alteplase for the treatment of patients with acute myocardial infarction. Eur Heart J 2000;21:2005–13.
[28] Davi G, Patrono C. Platelet activation and atherothrombosis. N Engl J Med 2007;357:2482–94.
[29] Dorsam RT, Kunapuli SP. Central role of the P2Y12 receptor in platelet activation. J Clin Invest 2004;113:340–5.
[30] Paul BZ, Jin J, Kunapuli SP. Molecular mechanism of thromboxane A(2)- induced platelet aggregation. Essential role for p2t(ac) and alpha(2a) receptors. J Biol Chem 1999;274:29108–14.
[31] Hamberg M, Svensson J, Samuelsson B. Thromboxanes: a new group of biologically active compounds derived from prostaglandin endoperoxides. Proc Natl Acad Sci USA 1975;72:2294–8.
[32] Patrono C. Aspirin and human platelets: from clinical trials to acetylation of cyclooxygenase and back. Trends Pharmacol Sci 1989;10:453–8.
[33] ISIS-2 (Second International Study of Infarct Survival) Collaborative Group. Randomised trial of intravenous streptokinase, oral aspirin, both, or neither among 17,187 cases of suspected acute myocardial infarction: ISIS-2. Lancet 1988;2:349–60.
[34] Sagar KA, Smyth MR. A comparative bioavailability study of different aspirin formulations using online multidimensional chromatography. J Pharm Biomed Anal 1999;21:383–92.
[35] Schömig A, et al. A randomized comparison of antiplatelet and anticoagulant therapy after the placement of coronary-artery stents. N Engl J Med 1996;334:1084–9.
[36] Bhatt DL, et al. Meta-analysis of randomized and registry comparisons of ticlopidine with clopidogrel after stenting. J Am Coll Cardiol 2002;39:9–14.
[37] Sabatine MS, Cannon CP, Gibson CM, et al. Addition of clopidogrel to aspirin and fibrinolytic therapy for myocardial infarction with ST-segment elevation. N Engl J Med 2005;352:1179–89.
[38] Chen ZM, et al. Addition of clopidogrel to aspirin in 45 852 patients with acute myocardial infarction: randomised placebo-controlled trial. Lancet 2005;366:1607–21.
[39] Angiolillo DJ, et al. Variability in individual responsiveness to clopidogrel: clinical implications, management, and future perspectives. J Am Coll Cardiol 2007;49:1505–16.
[40] Sugidachi A, et al. The greater in vivo antiplatelet effects of prasugrel as compared to clopidogrel reflect more efficient generation of its active metabolite with similar antiplatelet activity to that of clopidogrel's active metabolite. J Thromb Haemost 2007;5:1545–51.
[41] Udell JA, et al. Prasugrel versus clopidogrel in patients with ST-segment elevation myocardial infarction according to timing of percutaneous coronary intervention: a TRITON-TIMI 38 subgroup analysis (Trial to Assess Improvement in Therapeutic Outcomes by Optimizing Platelet Inhibition with Prasugrel-Thrombolysis in Myocardial Infarction 38). JACC Cardiovasc Interv 2014;7:604–12.
[42] Steg PG, James SK, Atar D, Badano LP, Blömstrom-Lundqvist C, Borger MA, et al. ESC Guidelines for the management of acute myocardial infarction in patients presenting with ST-segment elevation. Eur Heart J 2012;33(20):2569–619.
[43] Wallentin L, Becker RC, Budaj A, et al. Ticagrelor versus clopidogrel in patients with acute coronary syndromes. N Engl J Med 2009;361:1045–57.
[44] Bonaca MP, et al. Long-term use of ticagrelor in patients with prior myocardial infarction. N Engl J Med 2015;372:1791–800.
[45] Muñiz-Lozano A, Rollini F, Franchi F, Angiolillo DJ. Update on platelet glycoprotein IIb/IIIa inhibitors: recommendations for clinical practice. Ther Adv Cardiovasc Dis 2013;7:197–213.
[46] O'Gara, et al. 2013 ACCF/AHA guideline for the management of ST-elevation myocardial infarction. Circulation January 29, 2013;127(4):e362–425.
[47] Harrington RA, Stone GW, McNulty S, et al. Platelet inhibition with cangrelor in patients undergoing PCI. N Engl J Med 2009;361:2318–29.
[48] Yusuf S, et al. Effects of fondaparinux on mortality and reinfarction in patients with acute ST-segment elevation myocardial infarction: the OASIS-6 randomized trial. JAMA 2006;295:1519–30.

[49] Garcia DA, Baglin TP, Weitz JI, Samama MM. Parenteral anticoagulants: antithrombotic therapy and prevention of thrombosis, 9th ed.: American college of chest physicians evidence-based clinical practice guidelines. Chest 2012;141:e24S−43S.
[50] Montalescot G, et al. Intravenous enoxaparin or unfractionated heparin in primary percutaneous coronary intervention for ST-elevation myocardial infarction: the international randomised open-label ATOLL trial. Lancet 2011;378:693−703.
[51] Xiao Z, Theroux P. Platelet activation with unfractionated heparin at therapeutic concentrations and comparisons with a low-molecular weight heparin and with a direct thrombin inhibitor. Circulation 1998;97:251−6.
[52] Stone GW, et al. Bivalirudin for patients with acute coronary syndromes. N Engl J Med 2006;355: 2203−16.
[53] Stone GW, et al. Bivalirudin during primary PCI in acute myocardial infarction. N Engl J Med 2008; 358:2218−30.
[54] Shahzad A, et al. Unfractionated heparin versus bivalirudin in primary percutaneous coronary intervention (HEAT-PPCI): an open-label, single centre, randomised controlled trial. Lancet 2014;384: 1849−58.
[55] Bertrand OF, Rao SV, Pancholy S, Jolly SS, Rodes-Cabau J, Larose E, Costerousse O, Harmon M, Mann T. Transradial approach for coronary angiography and interventions: results of the first international transradial practice survey. JACC Cardiovasc 2010;3:1022−31.
[56] Mann JT, Cubeddu MG, Schneider JE, Arrowood M. Right radial access for PTCA: a prospective study demonstrates reduced complications and hospital charges. J Invasive Cardiol 1996;8:40D−4D.
[57] Romagnoli E, Biondi-Zoccai G, Sciahbasi A, Politi L, Rigattieri S, Pendenza G, et al. Radial versus femoral randomized investigation in ST-segment elevation acute coronary syndrome: the RIFLE-STEACS (Radial versus Femoral Randomized Investigation in ST-Elevation Acute Coronary Syndrome) study. J Am Coll Cardiol December 18, 2012;60(24):2481−9.
[58] Jolly SS, Yusuf S, Cairns J, Niemela K, Xavier D, Widimsky P, et al. Radial versus femoral access for coronary angiography and intervention in patients with acute coronary syndromes (RIVAL): a randomised, parallel group, multicentre trial. Lancet 2011;377(9775):1409−20.
[59] Valgimigli M, Gagnor A, Calabró P, et al. MATRIX Investigators. Radial versus femoral access in patients with acute coronary syndromes undergoing invasive management: a randomised multicentre trial. Lancet June 20, 2015;385(9986):2465−76.
[60] Hanratty C, Koyama Y, Rasmussen H, Nelson G, Hansen P, Ward M. Exaggeration of nonculprit stenosis severity during acute myocardial infarction: implications for immediate multivessel revascularization. J Am Coll Cardiol 2002;40(5):911−6.
[61] Wald DS, Morris JK, Wald NJ, Chase AJ, Edwards RJ, Hughes LO, et al. Randomized trial of preventive angioplasty in myocardial infarction. N Engl J Med 2013;369(12):1115−23.
[62] Gershlick AH, Khan JN, Kelly DJ, Greenwood JP, Sasikaran T, Curzen N, et al. Randomized trial of complete versus lesion-only revascularization in patients undergoing primary percutaneous coronary intervention for STEMI and multivessel disease: the CvLPRIT trial. J Am Coll Cardiol 2015; 65(10):963−72.
[63] Engstrom T, Kelbæk H, Helqvist S, Høfsten D, Kløvgaard L, Holmvang L, et al. Complete revascularisation versus treatment of the culprit lesion only in patients with ST-segment elevation myocardial infarction and multivessel disease (DANAMI-3—PRIMULTI): an open-label, randomised controlled trial. Lancet 2015;386(9994):665−71.
[64] Henriques JP, et al. Incidence and clinical significance of distal embolization during primary angioplasty for acute myocardial infarction. Eur Heart J 2002;23(14):1112−7.
[65] Svilaas T, Vlaar PJ, van der Horst IC, Diercks GF, de Smet BJ, van den Heuvel AF, Anthonio RL, Jessurun GA, Tan ES, Suurmeijer AJ, Zijlstra F. Thrombus aspiration during primary percutaneous coronary intervention. N Engl J Med 2008;358:557−67.
[66] Jolly SS, Cairns JA, Yusuf S, Meeks B, Pogue J, Rokoss MJ, Kedev S, Thabane L, Stankovic G, Moreno R, Gershlick A, Chowdhary S, Lavi S, Niemel€a K, Steg PG, Bernat I, Xu Y, Cantor WJ, Overgaard CB, Naber CK, Cheema AN, Welsh RC, Bertrand OF, Avezum A, Bhindi R, Pancholy S, Rao SV, Natarajan MK, ten Berg JM, Shestakovska O, Gao P,

Widimsky P, Dzavık V. Randomized trial of primary PCI with or without routine manual thrombectomy. N Engl J Med 2015;372:1389–98.
[67] Windecker S, Kolh P, Alfonso F, Collet J, Cremer J, Falk V, et al. 2014 ESC/EACTS guidelines on myocardial revascularization: the Task Force on Myocardial Revascularization of the European Society of Cardiology (ESC) and the European association for Cardio-Thoracic surgery (EACTS). Eur Heart J 2014;46(4):517–92.
[68] Engstrom AE, Cocchieri R, Driessen AH, Sjauw KD, Vis MM, Baan J, et al. The Impella 2.5 and 5.0 devices for ST-elevation myocardial infarction patients presenting with severe and profound cardiogenic shock: the Academic Medical Center intensive care unit experience. Crit Care Med 2011;39: 2072–9.
[69] Cheng J, den Uil C, Hoeks S, van der Ent M, Jewbali L, van Domburg R, et al. Percutaneous left ventricular assist devices vs. intra-aortic balloon pump counterpulsation for treatment of cardiogenic shock: a meta-analysis of controlled trials. Eur Heart J 2009;30(17):2102–8.
[70] Sheu JJ, Tsai TH, Lee FY, Fang HY, Sun CK, Leu S, et al. Early extracorporeal membrane oxygenator-assisted primary percutaneous coronary intervention improved 30-day clinical outcomes in patients with ST-segment elevation myocardial infarction complicated with profound cardiogenic shock. Crit Care Med 2010;38(9):1810–7.

CHAPTER 3.8

Novel Developments in Interventional Strategies

Emmanouil Vavouranakis, Konstantinos Kalogeras, Dimitris Tousoulis
1st Cardiology Clinic, 'Hippokration' General Hospital, National and Kapodistrian University of Athens, School of Medicine, Athens, Greece

Contents

INTRODUCTION

Cardiovascular disease is the leading cause of death worldwide. According to the World Health Organization, 30% of all global death is attributed to cardiovascular disease, and despite the considerable progress in the therapeutic strategies, the number will continue to increase in the next decades. Medical therapy remains one of the principal therapeutic strategies for treatment of cardiovascular diseases in primary and secondary prevention. However, percutaneous transluminal interventions have emerged as a therapeutic approach in most cardiovascular diseases. Specifically in coronary artery disease, coronary percutaneous interventions have rapidly become a prevalent treatment with impressive results with regard to efficacy and safety.

Since Stephen Hales placed the first catheter in the left or right ventricle of a living horse in 1711, it took more than 200 years until the procedure was applied to humans. Later in 1929, Werner Forssmann underwent the first human cardiac catheterization,

Coronary Artery Disease
ISBN 978-0-12-811908-2, http://dx.doi.org/10.1016/B978-0-12-811908-2.00022-2

while the first human coronary intervention was conducted in 1977 with a balloon dilation, performed by Andreas Grüntzig. By that time, interventional cardiology was consolidated and rapid development followed. Few years later, the first coronary bare metal stent (BMS) implantation was performed by Jacques Puel. Stent implantation has successfully overcome negative remodeling and has decreased the rate of restenosis to approximately 15%–30% compared with 30%–60% after balloon angioplasty. Finally, coronary drug-eluting stents (DESs) were introduced, and these diminished the rate of restenosis to even lower levels with simultaneous achievement of optimal angiographic and hemodynamic results.

In this struggle against coronary atheromatosis, interventional cardiology has produced novel methods in the field of materials and devices, as well as in the field of coronary imaging. Thus, following second-generation DESs, bioresorbable coronary scaffolds as well as self-expanding and conical stents have been developed and applied in a variety of situations. Furthermore, demanding techniques and advanced materials have been developed for the treatment of chronic total occlusions (CTOs). Finally, intravascular imaging has been remarkably enhanced by the development of intravascular ultrasonography (IVUS) and optical coherence tomography.

BIORESORBABLE STENTS

The evolution of percutaneous coronary intervention (PCI) from balloon angioplasty to BMS and finally to DES has resulted in a progressive decline in the rates of occlusion and restenosis. These advancements have enabled the treatment of demanding lesions and yielded excellent procedural outcomes. Nevertheless, DESs have significant limitations including the risk of neoatherosclerosis, late stent thrombosis, and a lower but still remaining rate of restenosis. All of them affect patients' long-term prognosis. The common pathophysiological mechanism behind these events may be attributed to the metallic implant that mechanically distorts and constrains the arterial wall, impairs coronary vasomotion, and prohibits adaptive coronary remodeling. Permanent metal and polymer implants may also result in chronic inflammation, promote neoatherosclerosis, and manifest strut fractures. Thus once the implanted stent has completed its initial task of preventing occlusion, sealing vascular dissection, or inhibiting restenosis, it would be optimal to get rid of a metallic frame within the coronary artery. To overcome these drawbacks, bioresorbable scaffolds (BRSs) were introduced in PCI. These devices have the ability to resolve after their implantation allowing restoration of vessel wall structural integrity and physiological function. The concept of using BRSs is attractive as the scaffold will provide the transient support and after the healing period, it will degrade and be resorbed completely, leaving the vessel with a healthy endothelium and normal vasomotion. Furthermore, it is supposed that the risk of late stent thrombosis will be significantly reduced, as there will not be any residual material left inside the vessel after degradation. BRS implantation has also the advantage of unfailing covered side branches,

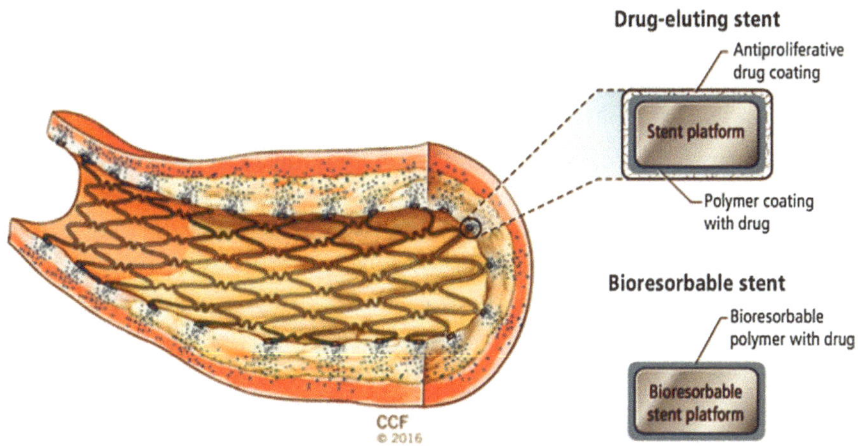

Figure 3.8.1 Structure of drug-eluting and bioresorbable stents.

restoring access in the future. Moreover, its use avoids "full metal jacket" stenting, restoring late bypass surgery options (Fig. 3.8.1).

BRS design includes three overlapping functionality phases that occur over time after implantation. The first phase, which is *revascularization*, involves the treatment of an ischemic coronary stenosis in a similar way to metallic DES. During the second phase, *restoration*, the molecular weight is progressively reduced followed by mass loss as well. With polymeric scaffolds, molecular weight loss occurs by hydrolysis and depolymerization, followed by metabolism of lactate into carbon dioxide and water via the Krebs cycle. In contrast, magnesium scaffolds are absorbed by surface erosion through a series of oxidation–reduction reactions, resulting in magnesium hydroxide and magnesium cations that are renally excreted. Finally, *resorption*, is the third phase, which includes the full recovery of the vascular architecture and functionality. With current available bioresorbable stents, this phase may require several years (Fig. 3.8.2).

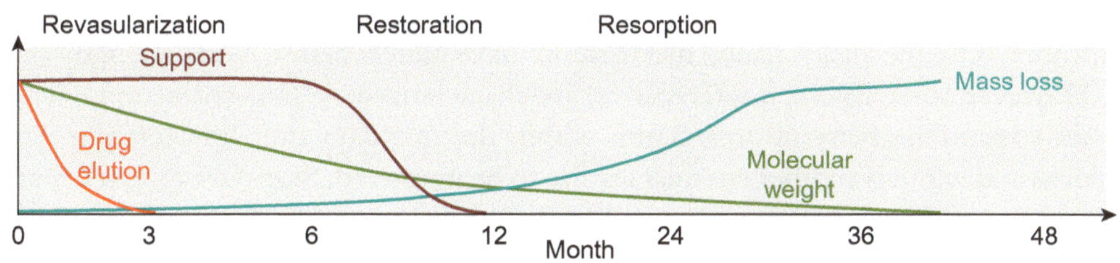

Figure 3.8.2 Phases of BRS functionality. Revascularization, restoration, and resorption are the three phases of BRS functionality. Revascularization includes mechanical support and drug delivery. During eestoration, radial rigidity and mechanical restraint are lost, whereas cyclic pulsatility and vasomotor return. Finally, resorption is characterized by mass loss with return of adaptive vascular remodeling response.

The majority of the materials used in BRSs are polymers such as poly-L-lactic acid (PLLA), although corrodible metals such as magnesium and iron have also been applied. The three most studied devices are the PLLA-based Absorb Bioresorbable Vascular Scaffold (BVS; Abbott Vascular), DESolve (Elixir Medical), and the magnesium-based DREAMS/Magmaris (Biotronik) scaffold. PLLA polymers exhibit approximately 100-fold lower tensile modulus than steel material. To overcome the diminished strength, BRSs require up to 240% thicker struts, leading to limited expansion and deliverability. At the same time, these polymeric devices are prone to fractures when overdilated. Due to these characteristics, BRS implantation requires extensive vessel preparation particularly in more calcified lesions. To elaborate, they are bulky enough, which makes delivery and deployment a demanding process especially in tortuous or calcified vessels. Based on this, BRS implantation requires appropriate scaffold selection and optimal implantation, which usually is facilitated with the use of intravascular imaging. Particularly in small vessels, higher early rates of scaffold thrombosis or target vessel—related myocardial infarction may be observed with BRS.

Problems With First-Generation Bioresorbable Scaffolds

Regarding the main problems characterizing the first-generation BRSs, the time required for strut absorption is a main feature to deal with. Specifically, the supportive role for BVS (PLLA-based scaffold) is short-lived, diminishing rapidly after the first 6 months and expected to be completely disappearing within the first year. On the other hand, BVS struts persist long after their functionality period, retaining their molecular weight up to 3 years from implantation. Similar to BVS, biodegradable magnesium alloy and polymer scaffolds maintain mechanical integrity for 6 months, although they totally degrade by 12 months. The issue of various absorption speed can be confronted by adding different alloying elements to magnesium. This way, the absorption speed of various magnesium alloys can be altered, targeting to an optimal degradation period.

Furthermore, radial strength is an adding issue of first-generation BRS that has to be dealt with. PLLA scaffolds have poor radial (hoop) strength, making lesion preparation mandatory. On the other hand, metal stents have much better radial strength (Mg, Fe). However, some BRS stents (DESolve) are characterized by "self-correction," which is a self-expansion ability up to 0.3 mm within the first days after implantation. Strut thickness and width is another essential feature to be improved. Scaffold strength depends on the strut width and thickness, on the shape of the hoops, and on the number of hoops per unit length. Stent deliverability and crossing profile depends mainly on strut width more than thickness, whereas width limits crimping. First-generation BRSs have a variety of strut thickness dimensions up to 157 μm in PLLA Absorb BVS and strut width ranging from 191 μm in BVS to 20 μm in magnesium-based Magmaris. However, strut thickness is steadily reduced in latest-generation BRS, up to 100 μm in Absorb next-generation scaffold.

Additionally, strut fracture resistance is a problem of BRS that is being gradually resolved. Newer scaffolds are more expandable, which means they are more fracture resistant. To elaborate, the overexpansion limit extends to 0.75 mm for Absorb next-generation scaffold and to 1.5 mm for DESolve scaffold. Finally, limited visibility is a problem of first-generation BRSs that has to be confronted, to achieve optimal implantation. Based on the aforementioned issues, it is obvious that next-generation BRSs ought to be faster in absorption, better in radial strength, thinner in struts, more expandable with fracture resistance, and better in angiographic visibility.

The race to create the optimal bioabsorbable vascular scaffold includes more than 20 companies, developing 23 stents. Strut material development includes apart from PLLA, alternative materials like magnesium alloys (DREAMS/Magmaris), tyrosine polycarbonate (Fantom/ReZolve), and nitride iron (Lifetech nitride iron stent). To this direction, next-generation Absorb BVS is characterized by thinner struts (<100 μm), expanded range of diameters and lengths, larger expansion limit, broader pressure working range, and shorter resorption time.

Outcomes

The relative safety and efficacy of Absorb BVS compared with contemporary DES has been demonstrated through published randomized controlled trials with 1 and 2 years of follow up in almost 3800 patients. Bioabsorbable stent implantation in very small vessels defined as <2.5 mm by visual estimate or <2.25 mm by QCA (quantitative coronary analysis) is associated with increased rates of device thrombosis and target vessel myocardial infarction, which may be confronted with improved patient and lesion selection and lesion preparation.

Most clinical data for BRS are based on studies with the Absorb BVS (Abbott Vascular, CA), which has been compared in randomized trials against best-in-class cobalt-chromium everolimus-eluting stent (Xience CoCr-EES). A meta-analysis that included six trials and compared the 1-year follow-up angiographic results favored CoCr-EES with lower in-device late lumen loss over BVS. The 1-year clinical outcome of both devices showed similar overall patient- and device-orientated composite endpoints. The 3-year follow-up results from the ABSORB II trial revealed that vasomotor reactivity was not statistically different between the two groups and that late lumen loss was significantly larger in the BVS arm, thereby not meeting the criteria for noninferiority. The long-term comparative efficacy and safety data for BVS are awaited from the ABSORB IV trial, the primary endpoint of which is a landmark analysis of target lesion failure, tested first for noninferiority of Absorb BVS to CoCr-EES with reflex testing to superiority. However, recent clinical data underline the importance of appropriate implantation on clinical outcomes. Specifically, postdilatation at high pressure seems to be safe and currently recommended in PCI with BVS. The optimal outcome is ensured by preparing and sizing the lesion appropriately.

Regarding antithrombotic treatment after bioresorbable stent implantation, a more aggressive antiplatelet therapy might be required, although there are currently no data evaluating the efficacy of newer agents, like prasugrel or ticagrelor compared with clopidogrel. The current recommendation is dual antiplatelet therapy for at least 12 months in all patients. However, more data are required, although it seems reasonable to use more potent antiplatelet agents in patients treated with bioabsorbable stents, especially in procedures that are complex enough and in patients characterized by low bleeding risk.

In conclusion, the use of BRSs can potentially deal with disadvantages of DESs by temporally opening up the diseased vessel before completely being resorbed after the vessel heals. However, the increased strut thickness and the consequent marginally increased rate of scaffold thrombosis remains a concern and makes uncertain the efficacy of first-generation BRSs. However, second-generation scaffolds are addressing the problems of the first-generation BRSs. Fabricating a thin strut polymeric BRS with sufficient radial strength still remains a technical challenge that should be achieved through progress in processing of polymers. Regarding clinical utility, it remains uncertain whether BRSs can rival the short-term outcomes of current DESs. However, long-term benefits of BRSs remain attractive, but unproven.

POLYMER-FREE DRUG-COATED CORONARY STENTS

Although DESs have significantly reduced rates of restenosis when compared with BMSs, issues regarding late stent thrombosis and dual antiplatelet therapy remain controversial. Late stent thrombosis is mainly attributed to delayed arterial healing and local hypersensitivity reactions potentially related to the drug or the polymer. Thus dual antiplatelet treatment is prolonged with concomitant ramifications on safety and bleeding complications. Therefore an ideal stent should provide the safety profile of BMS and the antirestenotic effectiveness of DES. With this in mind, a polymer-free stent has been developed and tested in randomized clinical trials.

The BioFreedom stent (Biosensors, Europe) is a polymer-free and carrier-free drug-coated stent that transfers umirolimus (also known as biolimus A9), a highly lipophilic sirolimus analog, into the vessel wall during a period of 1 month. BA9 is a 31-membered triene macrolide lactone derivative of sirolimus, which is a potent antiproliferative agent, developed for vascular applications. The BioFreedom stent uses a stainless-steel BioFlex II stent platform with a textured abluminal surface, onto which the BA9 in solvent is applied (Fig. 3.8.3). Contrary to the BioFreedom stent, currently available polymer-coated DESs release the drug over a period of several months. The BioFreedom stent has three main technical and functional characteristics. It releases the antiproliferating drug directly from the surface without a polymer ingredient in the coating. In this way, tissue responses associated with polymer are reduced. Since BA9 is the only coating component that

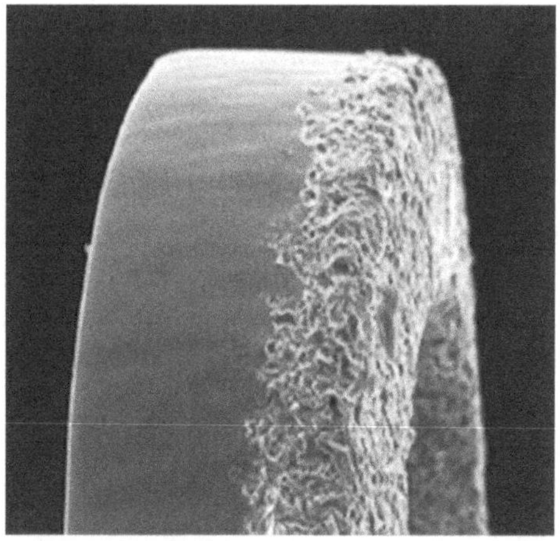

Figure 3.8.3 Illustration of a polymer-free stent. In the abluminal (outer) side, there is a microstructured porous surface, whereas in the luminal (inner) side, there is polished surface.

delivers the drug, when it is completely absorbed by the vessel wall, the stent turns into a BMS, with all the long-term advantages this means. Second, BA9 is a highly lipophilic drug, which easily crosses the cell membrane to achieve therapeutic effects in its target tissue. Third, the drug coating is restricted only to the abluminal surface of the stent. Thus the drug release is directed almost entirely into the vessel wall. Subsequently, there is little drug release on the luminal surface of the stent, where endothelial cell growth and healing are expected to occur.

Polymer-free BA9-coated stents have been tested in a prospective, multicenter randomized clinical trial. The BioFreedom trial randomized patients undergoing PCI into BA9-coated stents versus first-generation paclitaxel-eluting stents. The BioFreedom stent demonstrated noninferiority compared with DESs in terms of in-stent late lumen loss. At 5 years' clinical follow-up, the clinical event rates of MACE (major cardiovascular adverse events) were similar among the groups without occurrence of stent thrombosis.

The same stent has been used in high-bleeding-risk patients who are often treated with BMS followed by a limited 1 month period of dual antiplatelet therapy. In a double-blind control trial 2466 patients at high bleeding risk (Crusade score = 34) were randomized to either the polymer-free drug-coated stent or a very similar BMS. All patients were treated with dual antiplatelet therapy for 1 month. At almost 1-year follow-up, the polymer-free umirolimus-coated stent was superior to the BMS with respect to the primary safety endpoint of composite cardiac death, myocardial infarction, and stent thrombosis. Moreover, it was superior regarding the primary efficacy endpoint of target lesion revascularization. This result was mainly driven by the lower rate of myocardial infarction among the patients who received a Biofreedom stent. At 2 years' follow-up of these high-bleeding-risk patients undergoing PCI, the aforementioned safety and efficacy benefits of polymer-free stent over BMSs were maintained. However, the major bleeding rates and the coronary thrombotic events were similar among the two groups.

In a Leeders Free trial subsidy including patients undergoing PCI due to an acute coronary syndrome (ACS), it was shown that at 12-month follow-up, treatment with the polymerless BA9 stent was more effective than that with BMS. To elaborate, the target lesion revascularization events were significantly less in the BA9 group when compared with the BMS group. Furthermore, the cumulative incidence of cardiac death, myocardial infarction, or definite or probable stent thrombosis was diminished, driven mainly by significantly lower rates of cardiac mortality.

Despite the fact that initially only BMSs were recommended for treatment of high-bleeding-risk patients, the European guidelines currently suggest a short-period use of new-generation DESs. However, the use of polymer-free drug-coated BA9 stents seems to be safe and effective enough in this high-risk group of patients, since they simultaneously combine the valuable features of vessel wall drug apposition and early endothelialization. Undoubtedly, polymer-free drug-coated stents could be an excellent alternative to the conventional DESs.

SELF-EXPANDING CORONARY STENTS

Coronary stenting utilizing second-generation DESs represents the standard of care for treatment of coronary artery disease in ACSs, as well as in stable coronary artery disease. Nevertheless, conventional stent implantation may lead very often to suboptimal results, especially in complex lesions. Stent malapposition and a possible subsequent stent thrombosis may occur in cases of vessel bifurcation, ectasia, or excessive calcification. Furthermore, in cases of primary PCI, vasoconstriction or the presence of thrombus may lead to stent undersizing and finally implantation of a wrong size stent with adverse outcome.

The limitations of balloon-expandable stents include conformability, in terms of inadequate support of the vascular wall. A fundamental requirement is the ability to conform to the curved configuration of arteries, providing simultaneously a sustained and uniform support to the wall. Furthermore, the large diameter of mainly the left main (LM) and its significant tapering across the length of the vessel is an imperative limitation for conventional balloon-expandable DES. Postdilatation is required very often to achieve optimal expansion and apposition. Applying this technique may cause severe stent deformation and coverage of important side branches. The strategy of implanting two different duplicating stents does not seem to provide any additional benefit. All the aforementioned mechanical properties are of particular importance when dealing with bifurcation lesions. In addition, vascular remodeling and vasospasm can make the appropriate choice of stent size quite challenging, especially in ACS

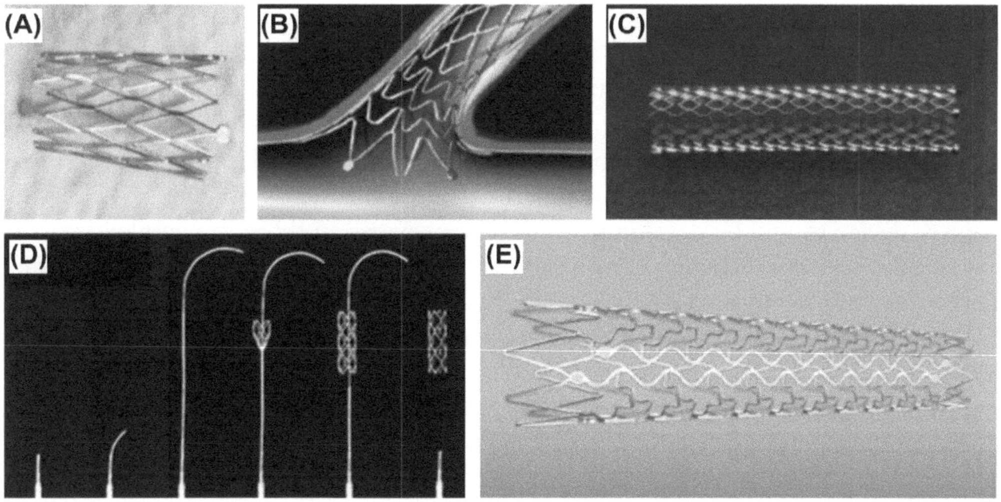

Figure 3.8.4 Overview of contemporary self-expanding stent platforms. (A) Axxess stent (B) Cappella stent (C) STENTYS stent (D) CardioMind Sparrow stent (E) vProtect stent.

conditions or CTO treatment. Finally, stent over- or underexpansion may influence adequate drug delivery from the stent scaffold.

Based on these kinds of limitations, coronary self-expanding stents have been developed and introduced (Fig. 3.8.4). Its clinical use is initially restricted to challenging lesions and particularly for LM coronary artery disease. The STENTYS S.A., (Paris, France) is a self-apposing, nitinol, paclitaxel-eluting stent, with a nominal strut width of 68 μm, which allows adequate expansion in the main vessel, due to its property of adaptation to the vessel size. In addition, it facilitates side branch access, thanks to its Z-shaped mesh that is linked by small interconnections, which can be disconnected by balloon inflation between the struts. The STENTYS is deployed by withdrawal of a retractable sheath and is available in three lengths (17, 22, and 27 mm) with diameters suitable for vessels ranging from 2.5 to 3.0 mm (small), 3.0–3.5 mm (medium), and 3.5–4.5 mm (large). Apart from a bare metal version, the stent is also available in a paclitaxel-eluting version and in a sirolimus-eluting version (1.4 $\mu g/mm^2$ of stent), all of which are incorporated in a coating (ProTeqtor), a durable polymer matrix of polysulfone and soluble polyvinylpyrrolidone that acts as an excipient.

Regarding clinical data available about self-expanding coronary stents, most clinical trials conducted with patients who received this kind of stent have been carried out as feasibility and safety studies that were single-arm studies and included only a limited number of patients. However, for the STENTYS, larger series of randomized controlled studies are now available, like the APPOSITION trials. Despite the fact that all these trials have been underpowered to document clinical outcomes, they have shown to a great extent that self-expandable stents are at least as safe and effective as conventional platforms. Furthermore, it has been shown that their implantation is feasible in most of the situations. Thus initial clinical trials comparing the performance of self-expanding versus conventional stents have demonstrated encouraging results and proof of concept, despite the fact of not primarily studying clinical outcomes.

A lot of challenges have to be confronted in the future to establish wide range use of self-expandable scaffolds. Regarding mechanical improvements, a smaller profile of the delivery device should be developed, to facilitate navigation through complex anatomy and enable 5 Fr compatibility. Furthermore, visibility has to be enhanced by adding markers or improving radiopacity. Meanwhile, longer devices are needed for diffuse disease and reduced strut profile and density is required to enable treatment of small vessels.

INTRAVASCULAR ULTRASONOGRAPHY

Coronary angiography has been considered for many years as the "gold standard" method for the assessment of coronary artery disease. However, coronary angiography provides a two-dimensional depiction of the coronary artery lumen, causing problems in the accurate measurement of the luminal stenosis. Furthermore, due to the process of positive remodeling (outward expansion of the vessel wall), which is often observed in coronary artery disease, angiography can underestimate the burden of the atherosclerotic plaque and fail to identify characteristics of the plaques, which are predisposing to the development of ACSs and sudden death. Thin-cap fibroatheromas (TCFAs) consist of a lipid-laden necrotic core beneath a fibrous cap thinner than 65 μm, which contains scant smooth muscle cells and numerous macrophages. These TCFAs are susceptible to plaque rupture, but generally cause mild to moderate coronary obstruction on conventional arteriography. The aforementioned drawbacks were confronted with the introduction of intravascular imaging methods. Indeed, IVUS provides direct visualization of the arterial wall.

IVUS is the most widely used intravascular imaging technique. Catheter-based IVUS has been used over the last two decades to image coronary arteries for atherosclerotic lesions, to evaluate the lumen, plaque dimensions, and features of plaque vulnerability and to guide intervention and optimize stent deployment. IVUS is based on the emission, attenuation, and backscattering of ultrasonic, waves which are converted to electrical signals and then processed and depicted as an image. According to the amplitude of the backscattered signal, a gray scale IVUS image is formed.

At present there are two types of IVUS catheters: mechanical (rotational) and phased array. Mechanical catheters emit ultrasound in the range of 20—40 MHz and phased array transducers emit in the range of 10—20 MHz. The receiving signal provides delineation of the cross-sectional anatomy of the coronary wall in real time with 70—200 μm axial resolution, 200—400 μm lateral resolution, and 5—10 mm imaging depth. Phased array catheters are easier to use and additionally can be pulled back manually without mechanical pullback, but have the disadvantage of reduced resolution. When visualized, the appearance of the IVUS image depends on the composition of the plaque. To elaborate, the highly calcified plaques have a bright appearance with acoustic shadowing, whereas a lipid-rich plaque is less echodense.

In recent years, information coming from the spectral analysis of IVUS data has been added to gray scale reconstructions to get a more detailed characterization of plaque morphology as a color-coded map. Based on this principal, three main postprocessing virtual histology methods have been developed: Virtual histology IVUS (VH-IVUS, Volcano Therapeutics, Rancho Cordova, CA, United States), iMAP-IVUS (Boston Scientific Corp, Fremont, CA, United States), and integrated backscatter IVUS.

The IVUS-based elastography technique, intravascular palpography, is a technique that measures the mechanical strain of the arterial wall and has the potential to differentiate between fibrous and fatty plaque components and detect high-stress regions. Applying palpography, it is possible to assess the local mechanical properties during arterial deformation caused by intraluminal pressure, which can be used to perform high-risk plaque assessment. However, when used in clinical studies in patients with ACS, the reliability of using intravascular palpography and ultrasonic spectrum analysis to detect vulnerable plaque was diminished. This was mainly due to the insufficient resolution of IVUS to reliably characterize different tissue types and to precisely detect TCFAs at such small scales.

Regarding clinical applications of the IVUS, at this time its most common uses are to determine optimal stent apposition and measurement of the minimal lumen area (MLA) of a lesion. IVUS can provide the operator with the ability to visualize reference vessel diameter, identify landing zones, quantitate the residual luminal diameter, and confirm that stent struts are properly expanded. Especially during PCIs, IVUS utility is imperative when stent underexpansion is suspected, in long lesions with uncertain reference vessel size, or when stent expansion is threatened due to extreme calcification. Finally, during LM PCI, use of IVUS is of great importance. The strength of IVUS is its ability to facilitate real-time decision making and immediate decisions have long-term prognostic outcomes. It has been shown from a meta-analysis that IVUS-guided revascularization results in more stents, longer stents, and larger stents. However, these more aggressive features were associated with decreased risk of composite endpoint of target lesion revascularization, death, myocardial infarction, and stent thrombosis.

Several studies have examined the utility of IVUS in assessing the prognostic implications of different plaque compositions. The largest prospective trial using VH-IVUS was the Providing Regional Observations to Study Predictors of Events in the Coronary Tree (PROSPECT) trial, which enrolled 697 patients with ACS (secondary prevention). The PROSPECT trial studied the prognostic implications of atheroma burden and plaque composition in patients with ACS over 3 years. Three characteristics were identified as predictive of subsequent cumulative MACE rate: plaque burden ≥70%, a minimal lumen diameter ≤4.0 mm, and a TCFA phenotype. In the absence of any of these three IVUS parameters, the risk of MACE was 0.3%, whereas if two or more of these features are seen, the MACE risk is 10%–18% over the next 3 years. Similar to this were the findings of the VIVA (Vascular endothelial growth factor in Ischemia for Vascular Angiogenesis) trail, conducted in patients with ACS and stable coronary artery disease.

Based on the aforementioned data, use of IVUS has been included as a class IIa indication in the European Society of Cardiology guidelines for revascularization (2014) with recommendation in selected patients to optimize stent implantation or assess severity and optimize treatment of unprotected LM lesions. Finally, it is suggested as a class IIb recommendation to assess the mechanisms of stent failure.

IVUS is useful for PCI guidance under certain circumstances. However, its prognostic significance remains to be determined. Its inherent limitations, such as moderate resolution, artifacts, and noise, along with the limited reliability of VH-IVUS needs to be improved before the expansion of its indications.

Multifrequency Intravascular Ultrasonography Imaging System

Recently, a multifrequency IVUS imaging system, which integrates a conventional IVUS transducer (35 MHz) with an ultra-high-frequency IVUS transducer (80–150 MHz) into one single catheter, was reported to obtain in vitro coronary artery images. The multifrequency IVUS system was able to achieve a more comprehensive visualization of the coronary artery by imaging with deep penetration the whole plaque volume using the low-frequency transducer and by measuring the thin fibrous cap using the ultra-high-frequency transducer. The multifrequency IVUS system is a promising alternative to the existing intravascular imaging modalities because of its low cost and its relative simplicity. Such multifrequency imaging would not only offer to operators a multiscale morphological information of the vessel wall but also additionally enhance molecular imaging and acoustic angiography to image the proliferation of vasa vasorum in the atherosclerotic plaques.

Near-Infrared Spectroscopy Imaging System

Near-infrared spectroscopy (NIRS) was the first intravascular imaging technique developed to achieve lipid content characterization within plaques. The development of this modality was based on the effect that different tissue compositions have different optical absorption and scattering effect on near-infrared (NIR) light (400–2400 nm). The lipid content characterization can be achieved by analyzing the absorbance of emitted NIR light at different wavelengths. A major drawback of this technique is the fact that it lacks the quantitative information about the size and location of the lipid core. Thus the clinical utility of the method is still limited. However, hybrid imaging systems have been developed like the TVC Imaging System (Infraredx, Burlington, Massachusetts), which is the first commercially available multimodal imaging system that fully assembles an IVUS transducer and an optical NIRS fiber into a single catheter, providing simultaneously IVUS–NIRS images without the need for flushing. Despite the advantages of the system, several limitations have to be overcome.

Intravascular Photoacoustics

Intravascular photoacoustics (IVPA) is an imaging method that detects the acoustic waves generated by thermal expansion induced by pulsed light to provide unique optical absorption contrast at ultrasound resolution. Because lipid has a distinct absorption spectrum in the NIT wavelength range, IVPA imaging is a promising technique for detecting and quantifying the amount of lipid in atherosclerotic plaques. IVPA quantifies only the size of lipid components, a key indicator of vulnerable plaque, and also automatically incorporates the IVUS image to clarify ambiguity such as photoacoustic signals generated by calcification. Trying to combine IVUS and IVPA systems in a single modality, led to low imaging speeds (>25 s per frame), a limitation that prevents the clinical utility of IVPA.

CHRONIC TOTAL OCCLUSIONS

A CTO is defined as a complete interruption of antegrade coronary flow [thrombolysis in myocardial infarction (TIMI) 0], which is angiographically documented, and persists for more than 3 months duration. The detection of the occlusion needs a careful examination of coronary angiography, since bridging collaterals may provide antegrade flow and mislead the diagnosis. Moreover, the presence of intraluminal microchannels may complicate the interpretation of the angiography. Thus multiple angiographic views should be taken, to avoid these kinds of false interpretations. Regarding the prevalence of CTO, it ranges between 10% and 30%. Depending on the prior history of myocardial infarction, CTO can be met in 30%–50% of these patients. Patients with a history of myocardial infarction develop the occlusion as a result of the acute phase, whereas in the remaining patients, CTO is likely to occur. Studies utilizing magnetic resonance imaging with late gadolinium enhancement in patients with CTO revealed a high proportion (86%) of patients with prior myocardial infarction. However, in recent studies the CTO prevalence was diminished, possibly due to the extended application of revascularization methods. Furthermore, CTOs are very common (more than 50%) among patients treated with coronary artery bypass grafting (CABG). Among patients with acute ST segment elevation myocardial infarction (STEMI), the incidence is 13%, while the presence of a CTO in a non–infarct-related artery was found to strongly and independently predict early and late mortality.

The basic pathological features of a CTO include the proximal cap, the body, and the distal cap. The proximal cap is mainly fibrotic or calcified and is either of tapering type or not. The body of the occlusion has often some kind of neovascularization and consists of fibrous tissue, atheroma, calcified tissue, and lymphocyte infiltrates. Finally, the distal cap, which is opacified via contrast injections in the collateral circulation, has more frequently a tapered morphology. Due to this morphology, the promotion of the PCI wire with retrograde techniques is facilitated.

Collateral circulation is an imperative feature of CTO. This circulation consists of interarterial connections that provide blood flow to a vascular bed whose original supply is compromised. They are mainly developed due to shear forces, which develop along the pressure gradient through the recruitment of preexisting interarterial connections, and they are valuable for preservation of the integrity of the territory supplied by the obstructed vessel. The value of collateral circulation is important during vessel occlusion, when they are often immediately recruitable, even preventing this way a myocardial infarction when an occlusion gradually develops. The period generally required for the development of the collateral's functional capacity is 2–12 weeks, and they mostly remain below angiographic resolution (200 μm). However, the functional supply of even angiographically well-developed collaterals is considered by definition suboptimal, with less than 10% of the collateral channels achieving a normal coronary flow reserve during pharmacological stress. Some grading systems for assessing collateral circulation have been proposed. According to Werner et al.'s classification, CC grade 0 = no continuous connection between collateral supplying and receiving vessel, CC1 = threadlike continuous connection, CC2 = side branch-like connection, and CC3 = $\geq$ 1 mm diameter of direct connection. In case revascularization of CTO is conducted, collateral circulation is diminished, starting immediately after reestablishment of antegrade flow. However, acute subsequent reocclusion of the CTO vessel would lead to an ACS, as the recruitment of collaterals is not instantaneous in most patients.

Like any revascularization method, the ultimate goal of CTO is to achieve symptomatic relief and prognosis improvement. However, symptoms related to a CTO are often nontypical, as patients adopt their functionality to their symptoms, leading to limited exercise capacity. There are data from nonrandomized studies that relate successful CTO recanalization with angina improvement. Additionally, there are limited randomized data supporting the improvement of left ventricle function after CTO treatment. However, some studies have shown a regional function effect, which starts 1–4 weeks after revascularization. Regarding the prognostic impact of successful CTO recanalization, a number of nonrandomized studies report favorable long-term outcome and decreased need for CABG. In a meta-analysis published in 2015, including 28.486 patients with 29.315 CTO PCI procedures, successful CTO revascularization was compared with unsuccessful procedures. According to this meta-analysis, during a mean follow-up period of 3.1 years, successful CTO PCI was associated with lower mortality [odds ratio (OR) 0.52], less residual angina (OR, 0.38), lower risk for stroke (OR, 0.72), less need for subsequent CABG (OR, 0.18), and lower risk for major adverse cardiac events (OR, 0.59). On the other hand, there was no difference in the incidence of target vessel revascularization [OR, 0.66; 95% confidence interval (CI), 0.36−1.23] or myocardial infarction (OR, 0.73; 95% CI, 0.52−1.03). To elaborate, it seems that the target vessel affects the prognostic outcome with a recent series reporting improved long-term survival with successful PCI of a CTO in the left anterior descending and the left circumflex, but not the right coronary artery. Multicenter randomized trials are expected to further investigate the prognostic impact of CTO treatment. Until then, the indications for CTO revascularization are not supposed to differ from those for subocclusive lesions. Patients with poorly controlled angina symptoms with optimal medical treatment may have indications for revascularization. Furthermore, technical features, like the dimension of the occluded artery and the presence of other critically narrowed arteries, may influence the decision about recanalization.

When treating a CTO, lesion characteristics play an imperative role in the likelihood of successful revascularization. The main lesion characteristics, which have been blamed for failure, are the increasing age of the occlusion, the presence of calcium, the excessive tortuosity of the vessel, the presence of a nontapered stump, the long length of the occlusion, the existence of side branches at the occlusion entry, the existence of bridging collaterals, and finally, the lack of visibility of path regarding the distal vessel. However, following the typical learning curve pattern, the operators' experience has increased and simultaneously, integrated techniques have evolved, increasing the success rates from 60% to 80%. Thus the main factors attributing to failure are considered the occlusion length and the severity of the calcification. Based on the aforementioned data, a grading system for the related difficulty has been proposed. The J-CTO score includes the following five parameters: length greater than 20 mm, presence of a greater than 45 degrees bend within the occlusion, presence of intralesional calcification, delineation of a stump at the proximal end, and previous failed attempt. Lesions with a score of 0−1 have a success rate of greater than 90% and a shorter time for wire crossings is required in most cases. The success progressively falls when the J-CTO score increases, and lesions with score greater than 3 have a 73% success and prolonged wire crossing times. Contrary to antegrade conventional techniques, the aforementioned predictive factors do not apply for retrograde techniques. The predictive factors of failure are considered to be related to collateral circulation characteristics. These characteristics are the presence and quality of the collaterals, their location in the septum or in the epicardium, their continuity and tortuosity, and the angle of the collateral anastomosis with the CTO vessel. Multislice computed tomography has been utilized to access lesion characteristics prior to the operation. Nevertheless, its application is not recommended for routine use as a preprocedural imaging screening.

The CTO recanalization is based on three basic techniques: antegrade, retrograde, and dissection reentry (Fig. 3.8.5). These techniques are complementary to each other, and they can be alternatively applied and combined as necessary. The transprocedural

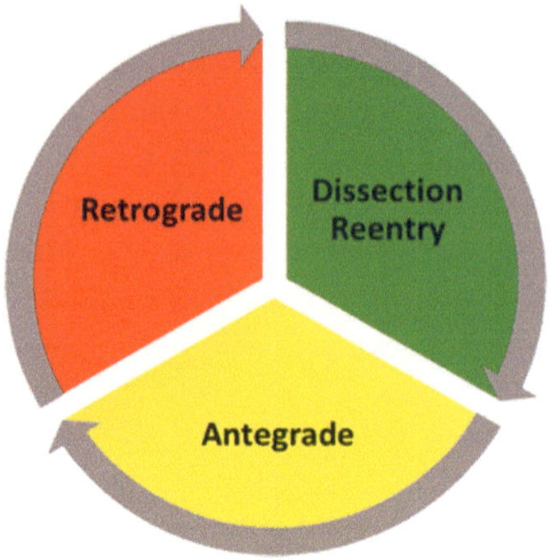

Figure 3.8.5 The continuum of chronic total occlusion percutaneous coronary intervention.

rapid switching from antegrade to retrograde approach is considered to be a hybrid approach. Antegrade techniques are almost always applied during a CTO procedure. It is widely accepted that they should be used first during the operation. Usually a single wire is advanced over a microcatheter. There are two variations: the wire escalation approach (stiff—stiffer—stiffest) and the "step up—step down" approach (stiff—soft—stiff). When the single wire fails, the parallel wire technique is applied, keeping the first wire in the subintimal space and advancing a second, stiffer wire toward the distal true lumen. Retrograde techniques were significantly developed during the last 10 years. According to the controlled antegrade retrograde subintimal tracking, an intentional collateral channel crossing is conducted. Most often, the septal collaterals are used and a channel dilatation with 1.5-mm balloon is performed. The advent of dedicated microcatheters for channel crossing, such as the Corsair from Asahi and the Finecross from Terumo, outweigh the disadvantage of the need for septal dilatation. In both techniques the main principle is the connection of the antegrade and retrograde subintimal spaces. This is achieved with the facilitation of recently introduced guiding catheter extensions (Guideliner, Vascular Solutions, Inc., Minneapolis, Minnesota; and Guidezilla, Boston Scientific, USA) with or without IVUS use. At the final step of the procedure a long 330-cm, 0.010-in. (0.26-mm)-diameter wire is externalized making a loop and providing excellent support to complete the PCI in an antegrade way. Finally, the subintimal tracking and reentry technique includes dissection and reentry of the vessel. Either antegrade or retrograde, with rapid steering of a stiff polymeric wire in the subintimal space, knuckled within the body of the occlusion, the wire dissects and reenters the lumen. Dedicated for this dissect-and-reentry technique devices have been developed (CrossBoss coronary catheter, Boston Scientific, USA). Simultaneously with the aforementioned techniques, intravascular imaging (IVUS and OCT) is used for their facilitation. By applying these techniques, the success rates have soared to 90%. However, the data about long-term outcomes of CTO PCIs with these techniques are limited and there is a need for further investigation.

Regarding the complications rates, these are comparable to those of nonocclusive coronary artery disease. However, major concerns arise about the radiation exposure and the contrast usage, which should be kept as low as possible. Periprocedural complications include pericardial tamponade and myocardial injury, identified with measurement of myocardial biomarkers.

In conclusion, CTO revascularization has made great strides during the last few years, mainly due to the development of new materials and devices and due to the experience gained by the operators. However, there is still an underutilization of CTO PCIs, limited to 5%–6% of all the revascularization procedures. Further evidence from randomized controlled studies and beneficial outcomes are required to encourage operators to get more involved in such demanding operations.

PLAQUE MODIFICATION

Rotational Atherectomy

The use of PCI has increased dramatically since the first case was described by Andreas Grunzig back in the early 1980s. Evolution of various techniques and newer materials has allowed the percutaneous treatment of an increasingly number of patients suffering from coronary artery disease. Thus the majority of coronary lesions can be treated nowadays, despite their complexity and severity. Coronary calcification is an important characteristic of the atherosclerotic plaque that impairs the successful percutaneous intervention. Severely calcified coronary lesions are extremely challenging for operators and are one of the main reasons of a failed or a complicated procedure. If compared with noncalcified coronary lesions, calcified lesions are associated with a higher procedural risk and worse clinical outcomes. Calcification can lead to unsuccessful advance of materials and devices (stent, balloon) through the stenosis or impede the appropriate dilatation of the balloon. Subsequently, the stent apposition is inappropriate and can be associated with increased risk of adverse events after PCI, like stent restenosis and thrombosis. These characteristics of calcified lesions demand the development and application of specific methods and materials.

The modification of the atherosclerotic plaque has been proposed and applied mainly by the use of rotational atherectomy (RA). Percutaneous transluminal coronary RA is a technique that uses a diamond-coated burr to debulk atherosclerotic plaques and vessel calcification by causing fragmentation of the plaque. The diamond-coated burr disperses the plaque into microfragments, causing an increase of the luminal diameter. These fragments, which average 5 μm in diameter, predominantly pass into the capillary circulation and are then taken up by the reticuloendothelial system. Subsequently, this plaque modification by RA permits a safe and effective stent deployment. After treatment with RA, the luminal morphology of the lesions is smooth and nonendothelialized. Contrary to RA treatment, when balloon predilatation is utilized for plaque modification, increase in lumen diameter is achieved through plaque squeezing and vessel wall stretching. However, in the modern era of DESs, RA application is relatively infrequent and only 3%–5% of coronary interventions include it.

Regarding pathophysiology, coronary calcification occurs mainly in the intimal or medial layer of the artery and increases the vessel stiffness. These calcifications are definitely associated with atherosclerosis, characterized by inflammation, fibrosis, lipid accumulation, and a high proportion of necrotic tissue. Due to these characteristics, major adverse clinical events, like cardiovascular death and myocardial infarction, increase along with higher coronary artery calcium scores. In practice, apart from difficulties in stent implantation, severe calcification often requires expansion in high pressure, which significantly increases the unlikely possibility of dissection, perforation, or balloon rupture. Furthermore, the advancement of a DES across a calcified stenosis may cause damage to the stent polymer coating drug and an inadequate diffusion to the vessel wall.

Despite the promising results of RA during the balloon angioplasty era, the restenosis rates with RA prior to BMS implantation remained as high at 20%–30%. In the DES era, neointimal formation has been significantly reduced and the rates of restenosis and target lesion revascularization have declined. Data from noncontrolled observational studies applying RA and DES have shown a rate of procedural and angiographic success of more than 90%. Regarding the follow-up period, the rates of MACE varied from 4.4% to 17.7%. However, it should be kept in mind that these studies particularly included patients of high-agenda complex lesions, as well as with more comorbidities. Complications of RA in the DES era are infrequent and include vessel dissection, no flow or slow flow, and perforation. Finally, the occurrence of definite and probable stent thrombosis differed significantly among the studies, from 1% to 4.8%, although these rates are higher compared with those usually observed in noncalcified lesions.

Only two randomized controlled studies have compared the outcomes in patients receiving DES with or without RA use. Clavijo et al. have shown that there are no differences in rates of clinical success regarding in-hospital and 6-month follow-up outcomes between the two groups. The Rotational Atherectomy Prior to TAXUS Stent Treatment for Complex Native Coronary Artery Disease study showed that in patients characterized by complex calcified coronary lesions and similar baseline characteristics, late lumen loss was greater in patients treated with PES and RA at 9 months follow-up, whereas clinical outcomes were similar among the two groups. However, the clinical significance of the study included the fact that RA cannot improve the antiproliferative effect of DES, but that it remains an important tool for complex calcified lesions, which significantly improves the procedural success. Additionally, RA may be associated with greater extent of neointimal hyperplasia and thus should be facilitated as a default strategy regarding complex calcified lesions. Nevertheless, the second-generation DES, characterized by thinner stent strut platforms and more biocompatible polymers, fit perfectly for use in the treatment of calcified lesions with RA and may lead to improved clinical outcomes.

Generally, RA should be used in bulky calcified plaques, which are unfavorable for stent deployment. On the other hand, it should be avoided in mild or moderate calcified lesions since it may result in greater tissue injury and increased neointimal formation.

A new atherectomy device, the orbital system (Cardiovascular Systems, Inc. St. Paul, MN, USA), has been developed. It is a novel rotational device that uses a diamond-coated crown on a coil consisting of three spiral wires to orbit eccentrically within the vessel lumen. With this device, the operator is capable of controlling the depth and diameter of ablation by adjusting the velocity (80,000–200,000 rpm). Orbital atherectomy has been successfully used to treat calcified peripheral lesions. In ORBIT 1 trial a 98% device success rate and a 94% procedural success rate was reported with this newer device. At 3-year follow-up, a total of 18.2% of MACE, including 3% of TLR, 6.1% of non-Q wave, and 9.1% of cardiac, death was observed.

Cutting Balloon

Apart from RA, severely coronary calcified lesions have been treated with cutting balloon (CB) angioplasty. By applying these kinds of balloons, except for typical balloon angioplasty benefits, longitudinal incisions are created within the target plaque, interrupting the elastic and fibrous components of the neointima. Simultaneously, the risk of device slippage across the calcified lesion is minimized. However, there is a slightly increased risk of vessel injury. In a randomized controlled trial that compared CB angioplasty to conventional balloon angioplasty, there was no difference in MACE or in stent restenosis rates, but CB reduced balloon slippage, the need for additional stenting, and the number of balloons used.

Regarding plaque modification in general, applying both RA and CB angioplasty before DES implantation, a randomized controlled trial compared RA plus CB with RA alone in patients with severe calcified lesions. The mean minimum stent cross-sectional areas after PCI of the RA + CB group was significantly larger than that of the RA group, whereas patients in the first group achieved significantly larger acute cross-sectional area gain after PCI relative to the RA group. The periprocedural complications were similar among the two arms, but at the 1-year follow-up the RA plus CB group had a lower rate of revascularization for restenosis of the target vessel and MACE than did the RA group.

In conclusion, calcified coronary lesions pose a great technical challenge for operators and are associated with poor clinical outcomes. Plaque modification prior to PCI is often necessary to achieve the optimal result. RA application, in combination with DES, has improved significantly the procedural success. Simultaneously, intravascular imaging techniques are required for guidance and result evaluation in such demanding procedures. In addition, various lesion preparation strategies, including CB and orbital system, offer operators more options to treat severely calcified lesions.

CONCLUSIONS

According to the aforementioned data, it seems clear that interventional cardiology has made great strides in the field of coronary atheromatosis. Novel materials and devices like bioabsorbable scaffolds, polymer-free stents, and self-expanding stents have been developed during the last few years, offering additional options in the treatment of coronary artery disease. Moreover, novel techniques like RA and CTO treatment techniques ensure the success in treating demanding lesions, which were considered untreatable until recently. Intravascular imaging methods, like IVUS, optimize the procedure and enhance further the rates of success. Undoubtedly, interventional cardiology has a valuable contribution in the struggle against atheromatosis and promises even greater achievements for the future.

FURTHER READING

[1] Jackson D, Tong D, Layland J. A review of the coronary applications of the drug coated balloon. Int J Cardiol January 1, 2017;226:77–86.

[2] Kawamoto H, Ruparelia N, Tanaka A, Chieffo A, Latib A, Colombo A. Bioresorbable scaffolds for the management of coronary bifurcation lesions. JACC Cardiovasc Interv May 23, 2016;9(10):989–1000.

[3] Tenekecioglu E, Farooq V, Bourantas CV, Silva RC, Onuma Y, Yılmaz M, Serruys PW. Bioresorbable scaffolds: a new paradigm in percutaneous coronary intervention. BMC Cardiovasc Disord February 12, 2016;16:38.

[4] Ang HY, Bulluck H, Wong P, Venkatraman SS, Huang Y, Foin N. Bioresorbable stents: current and upcoming bioresorbable technologies. Int J Cardiol February 1, 2017;228:931–9.

[5] Ellis SG, Riaz H. Bioresorbable stents: the future of interventional cardiology? Cleve Clin J Med November 2016;83.

[6] Kereiakes DJ, Onuma Y, Serruys PW, Stone GW. Bioresorbable vascular scaffolds for coronary revascularization. Circulation July 12, 2016;134(2):168–82.

[7] Rizik DG, Hermiller JB, Kereiakes DJ. Bioresorbable vascular scaffolds for the treatment of coronary artery disease: clinical outcomes from randomised controlled trials. Catheter Cardiovasc Interv November 2016;88(S1):21–30.

[8] Campos CM, Lemos PA. Bioresorbable vascular scaffolds: novel devices, novel interpretations, and novel interventions strategies. Catheter Cardiovasc Interv July 1, 2014;84(1):46–7.

[9] Tenekecioglu E, Bourantas CV, Abdelghani M, Sotomi Y, Suwannasom P, Tateishi H, Onuma Y, Yılmaz M, Serruys PW. Optimisation of percutaneous coronary intervention: indispensables for bioresorbable scaffolds. Expert Rev Cardiovasc Ther September 2016;14(9):1053–70.

[10] Lopez JJ, Arain SA, Madder R, Parekh N, Shroff AR, Westerhausen D. Techniques and best practices for optical coherence tomography: a practical manual for interventional cardiologists. Catheter Cardiovasc Interv November 1, 2014;84(5):687–99.

[11] Yu Y, Zhou Y, Ma Q, Jia S, Wu S, Sun Y, Liu X, Zhao Y, Liu Y, Shi D. The conical stent in coronary artery improves hemodynamics compared with the traditional cylindrical stent. Int J Cardiol January 15, 2017;227:166–71.

[12] Buccheri D, Piraino D, Andolina G, Cortese B. Understanding and managing in-stent restenosis: a review of clinical data, from pathogenesis to treatment. J Thorac Dis October 2016;8.

[13] Costa RA, Abizaid A, Mehran R, Schofer J, Schuler GC, Hauptmann KE, Magalhães MA, Parise H, Grube E, BioFreedom FIM clinical trial Investigators. Polymer-free biolimus a9-coated stents in the treatment of de novo coronary lesions: 4- and 12-month angiographic follow-up and final 5-year clinical outcomes of the prospective, multicenter BioFreedom FIM clinical trial. JACC Cardiovasc Interv January 11, 2016;9(1):51–64.

[14] Garot P, Morice MC, Tresukosol D, Pocock SJ, Meredith IT, Abizaid A, Carrié D, Naber C, Iñiguez A, Talwar S, Menown IB, Christiansen EH, Gregson J, Copt S, Hovasse T, Lurz P, Maillard L, Krackhardt F, Ong P, Byrne J, Redwood S, Windhövel U, Greene S, Stoll HP, Urban P, LEADERS FREE Investigators. 2-Year outcomes of high bleeding risk patients after polymer-free drug-coated stents. J Am Coll Cardiol January 17, 2017;69(2):162–71.

[15] Tada N, Virmani R, Grant G, Bartlett L, Black A, Clavijo C, Christians U, Betts R, Savage D, Su SH, Shulze J, Kar S. Polymer-free biolimus a9-coated stent demonstrates more sustained intimal inhibition, improved healing, and reduced inflammation compared with a polymer-coated sirolimus-eluting cypher stent in a porcine model. Circ Cardiovasc Interv April 2010;3(2): 174–83.

[16] Naber CK, Urban P, Ong PJ, Valdes-Chavarri M, Abizaid AA, Pocock SJ, Fabbiocchi F, Dubois C, Copt S, Greene S, Morice MC, LEADERS FREE Investigators. Biolimus-A9 polymer-free coated stent in high bleeding risk patients with acute coronary syndrome: a Leaders Free ACS sub-study. Eur Heart J May 17, 2016;38.

[17] Urban P, Meredith IT, Abizaid A, Pocock SJ, Carrié D, Naber C, Lipiecki J, Richardt G, Iñiguez A, Brunel P, Valdes-Chavarri M, Garot P, Talwar S, Berland J, Abdellaoui M, Eberli F, Oldroyd K, Zambahari R, Gregson J, Greene S, Stoll HP, Morice MC, LEADERS FREE Investigators. Polymer-free drug-coated coronary stents in patients at high bleeding risk. N Engl J Med November 19, 2015;373(21):2038–47.

[18] Urban P, Abizaid A, Chevalier B, Greene S, Meredith I, Morice MC, Pocock S. Rationale and design of the LEADERS FREE trial: a randomized double-blind comparison of the BioFreedom drug-coated stent vs the Gazelle bare metal stent in patients at high bleeding risk using a short (1 month) course of dual antiplatelet therapy. Am Heart J May 2013;165(5):704–9.

[19] Wykrzykowska JJ, Garg S, Onuma Y, de Vries T, Goedhart D, Morel MA, van Es GA, Buszman P, Linke A, Ischinger T, Klauss V, Corti R, Eberli F, Wijns W, Morice MC, di Mario C, van Geuns RJ, Juni P, Windecker S, Serruys PW. Value of age, creatinine, and ejection fraction (ACEF score) in assessing risk in patients undergoing percutaneous coronary interventions in the 'All-Comers' LEADERS trial. Circ Cardiovasc Interv February 1, 2011;4(1):47–56.

[20] Silenzi S, Grossi P, Mariani L, Aimi A, Marchese P, Moretti L. A real world single centre experience using the STENTYS self-expanding coronary stent. Int J Cardiol April 15, 2016;209:57–9.

[21] Briguori C, Visconti G, Donahue M, Focaccio A, Mitomo S, Kawamoto H, Nakamura S. The STENTYS® paclitaxel-eluting stent in the treatment of unprotected distal left main. Catheter Cardiovasc Interv September 2015;86(3):E131–9.

[22] Ma T, Zhou B, Hsiai TK, Shung KK. A review of intravascular ultrasound-based multimodal intravascular imaging: the synergistic approach to characterising vulnerable plaques. Ultrason Imaging September 2016;38(5):314–31.

[23] Munnur RK, Nerlekar N, Wong DT. Imaging of coronary atherosclerosis in various susceptible groups. Cardiovasc Diagn Ther August 2016;6(4):382–95.

[24] Groves EM, Seto AH, Kern MJ. Invasive testing for coronary artery disease: FFR, IVUS, OCT, NIRS. Heart Fail Clin January 2016;12(1):83–95.

[25] Celeng C, Takx RA, Ferencik M, Maurovich-Horvat P. Non-invasive and invasive imaging of vulnerable coronary plaque. Trends Cardiovasc Med August 2016;26(6):538–47.

[26] Pathan F, Negishi K. Prediction of cardiovascular outcomes by imaging coronary atherosclerosis. Cardiovasc Diagn Ther August 2016;6(4):322–39.

[27] Hoang V, Grounds J, Pham D, Virani S, Hamzeh I, Qureshi AM, Lakkis N, Alam M. The role of intracoronary plaque imaging with intravascular ultrasound, optical coherence tomography, and near-infrared spectroscopy in patients with coronary artery disease. Curr Atheroscler Rep September 2016;18(9):57.

[28] Jegere S, Narbute I, Erglis A. Use of intravascular imaging in managing coronary artery disease. World J Cardiol June 26, 2014;6(6):393–404.

[29] Christakopoulos GE, Christopoulos G, Carlino M, Jeroudi OM, Roesle M, Rangan BV, Abdullah S, Grodin J, Kumbhani DJ, Vo M, Luna M, Alaswad K, Karmpaliotis D, Rinfret S, Garcia S, Banerjee S, Brilakis ES. Meta-analysis of clinical outcomes of patients who underwent percutaneous coronary interventions for chronic total occlusions. Am J Cardiol May 15, 2015;115(10):1367–75.

[30] Christopoulos G, Menon RV, Karmpaliotis D, Alaswad K, Lombardi W, Grantham A, Patel VG, Rangan BV, Kotsia AP, Lembo N, Kandzari D, Carlson H, Garcia S, Banerjee S, Thompson CA, Brilakis ES. The efficacy and safety of the "hybrid" approach to coronary chronic total occlusions: insights from a contemporary multicenter US registry and comparison with prior studies. J Invasive Cardiol September 2014;26(9):427–32.

[31] Sianos G, Konstantinidis NV, Di Mario C, Karvounis H. Theory and practical based approach to chronic total occlusions. BMC Cardiovasc Disord February 9, 2016;16:33. http://dx.doi.org/10.1186/s12872-016-0209-3.

[32] Vaquerizo B, Serra A, Miranda F, Triano JL, Sierra G, Delgado G, Puentes A, Mojal S, Brugera J. Aggressive plaque modification with rotational atherectomy and/or cutting balloon before drug-eluting stent implantation for the treatment of calcified coronary lesions. J Interv Cardiol June 2010;23(3):240–8.

[33] Li Q, He Y, Chen L, Chen M. Intensive plaque modification with rotational atherectomy and cutting balloon before drug-eluting stent implantation for patients with severely calcified coronary lesions: a pilot clinical study. BMC Cardiovasc Disord May 26, 2016;16:112.

[34] Tian W, Lhermusier T, Minha S, Waksman R. Rational use of rotational atherectomy in calcified lesions in the drug-eluting stent era: review of the evidence and current practice. Cardiovasc Revasc Med March 2015;16(2):78–83.

[35] Stankovic G, Milasinovic D. Rotational atherectomy in clinical practice: the art of tightrope walking. Circ Cardiovasc Interv November 2016;9(11).

[36] Kokkinidis DG, Waldo SW, Armstrong EJ. Treatment of coronary artery in-stent restenosis. Expert Rev Cardiovasc Ther March 2017;15(3):191–202.

[37] Couper LT, Loane P, Andrianopoulos N, Brennan A, Nanayakkara S, Nerlekar N, Scott P, Walton AS, Clark DJ, Duffy SJ, Ajani AE, Reid C, Shaw JA, Melbourne Interventional Group (MIG) Investigators. Utility of rotational atherectomy and outcomes over an eight-year period. Catheter Cardiovasc Interv October 2015;86(4):626–31.

INDEX

'*Note*: Page numbers followed by "f" indicate figures and "t" indicate tables.'

A

O

S